A guide to identifying and classifying yeasts

A guide to identifying and classifying yeasts

J. A. BARNETT
School of Biological Sciences, University of East Anglia, Norwich, England

R. W. PAYNE
Statistics Department, Rothamsted Experimental Station, Harpenden, Hertfordshire, England

D. YARROW
Yeast Division, Centraalbureau voor Schimmelcultures, Julianalaan 67A, Netherlands

CAMBRIDGE UNIVERSITY PRESS
CAMBRIDGE
LONDON · NEW YORK · MELBOURNE

Published by the Syndics of the Cambridge University Press
The Pitt Building, Trumpington Street, Cambridge CB2 1RP
Bentley House, 200 Euston Road, London NW1 2DB
32 East 57th Street, New York, NY 10022, USA
296 Beaconsfield Parade, Middle Park, Melbourne 3206, Australia

© Cambridge University Press 1979

First published 1979

Printed in Great Britain by
Galliard (Printers) Ltd.
Great Yarmouth, Norfolk

Library of Congress Cataloguing in Publication Data
Barnett, James Arthur, 1923–
A guide to identifying and classifying yeasts
Includes bibliographical references and indexes
1. Yeast fungi–Identification. 2. Yeast fungi–Classification. 3. Fungi–Identification.
4. Fungi–Classification. I. Payne, R. W., joint author. II. Yarrow, D., joint author. III. Title.
QK617.5.B37   589'.23   79-11136
ISBN 0 521 22762 3

The Naming of Cats is a difficult matter,
   It isn't just one of your holiday games;
You may think at first I'm as mad as a hatter
When I tell you, a cat must have THREE DIFFERENT NAMES.
T. S. Eliot

For some unknown reason it seems to be impossible
to discuss nomenclatural problems calmly and with
the quiet detachment expected of scientists...
S. T. Cowan

Corrigenda
Throughout, for *Debaryomyces melissophila*, read *Debaryomyces melissophilus*; for *Debaryomyces polymorpha*, rear *Debaryomyces polymorphus*; for *Sporidibolus*, read *Sporidiobolus*; and for *Torulopsis cantarelii*, read *Torulopsis cantarellii*.

In Chapters 7 and 8, *Zendera ovetensis* should be followed by two asterisks, thus Zendera ovetensis**, indicating that this species is included in *The Yeasts* (Lodder, 1970) under another name.

On p. 67, *Torulopsis maris* is described incorrectly as being unable to utilize methanol. However, this mistake does not affect correct identification by any of the keys (1, 2, 10 and 12). A note about *Torulopsis maris* has been added to key no. 12.

Contents

	Acknowledgements	ix
	The way the book has been printed: publisher's note	x
1	**Introduction**	1
2	**How yeasts are classified**	3
	Biological classification and the yeasts	3
	Characteristics used in classifying yeasts	4
	Naming yeasts	6
	Renaming yeasts	6
	'Objective' classifications	7
3	**Characteristics of the genera**	9
4	**Newly accepted and renamed genera and species**	15
5	**A simple guide to laboratory methods for identifying yeasts**	39
	Microscopical appearance of non-filamentous vegetative cells	39
	Microscopical examination for filamentous growth	40
	Microscopical examination for ballistoconidia	40
	Microscopical examination for ascospores	40
	Assessing the ability to use organic compounds as sole source of carbon for aerobic growth	41
	Assessing the ability to use nitrogen compounds for aerobic growth	43
	Assessing the ability to use certain sugars anaerobically	43
	Assessing vitamin requirements	44
	Detecting production of extracellular starch-like compounds	44
	Assessing growth at high concentrations of D-glucose	44
	The media	45
6	**Characteristics of the species**	47
	Numbered characteristics used in Chapters 6, 7 and 8	47
7	**The keys**	72
	How to use the keys	72
	List of keys	73
	1. All yeasts that do not ferment D-glucose: physiological and microscopical tests	77
	2. All yeasts that do not ferment D-glucose: physiological tests only	89
	3. All yeasts that ferment D-glucose: physiological and microscopical tests	102
	4. All yeasts that ferment D-glucose: physiological tests only	120
	5. Ascosporogenous yeasts with spherical, oval or reniform ascospores: physiological and microscopical tests	138
	6. Ascosporogenous yeasts with hat- or Saturn-shaped ascospores: physiological and microscopical tests	143
	7. Ascosporogenous yeasts with ascospores of other shapes: physiological and microscopical tests	145
	8. Basidiomycetous yeasts: physiological and microscopical tests	146
	9. Yeasts with pink colonies: physiological and microscopical tests	147
	10. Yeasts that utilize hydrocarbons: physiological and microscopical tests	148
	11. Yeasts that utilize hydrocarbons: physiological tests only	151
	12. Yeasts that utilize methanol: physiological tests only	154
	13. Yeasts most commonly isolated clinically: physiological tests only	155
	14. Extended list of yeasts isolated clinically: physiological tests only	157
	15. Yeasts associated with food: physiological and microscopical tests	160
	16. Yeasts associated with wine and wine-making: physiological and microscopical tests	167
	17. Yeasts found in brewing: physiological and microscopical tests	172
	18. Yeasts found in brewing: physiological tests only	174
8	**The identification of particular species**	177

9	**How the computing was done**	277
	Irredundant test sets	277
	How the keys were produced	278
	Sets of tests to identify individual yeasts	280
10	**Alternatives to keys**	281
	Polyclaves	281
	Computer-based identification systems	281
	Identification by matching	282
	Keys based on groups of tests	282
	Appendix: Recently altered names of yeasts species	283
	Glossary	285
	References	288
	Addendum: List of new taxa	301
	Author index	303
	Index of taxa	305
	Subject index	307

Acknowledgements

We are most grateful for invaluable help, suggestions and criticisms from friends and colleagues, in particular from S. A. Barnett, J. C. Gower, Barbara Kirsop, A. G. Kitchell, A. P. Sims and R. P. White. In addition, the staff of the Library of the University of East Anglia, especially those concerned with interlibrary loans, have given constant generous assistance with bibliographical problems.

Finally, special acknowledgement is due to the members of the staff of the Cambridge University Press for their patient work in producing this book.

The way the book has been printed: publisher's note

The table of test results in chapter 6, the keys in chapter 7 and the identification tables in chapter 8 were all produced by computer at Rothamsted Experimental Station, using the Genkey program. In order to minimize errors, the computer output in the form of magnetic tape was used to control the typesetting of these parts of the book on a VIP photosetter, which was also used to set the text conventionally. The typeface is Times New Roman.

In Chapters 6 and 8, the abbreviations of the names of the tests are not always self-consistent. This is because these abbreviations were made for the initial computer print-out in which, unlike the final typeset version, each letter occupies the same space.

1
Introduction

This book is principally to help in overcoming the difficulties in identifying yeasts. To this end, we present 18 identification keys and various tables, all derived from the information in the records of the Yeast Division of the Centraalbureau voor Schimmelcultures at Delft. The information is tabulated in Chapter 6. The keys and tables are not available elsewhere.

The *Guide* is intended for use with the current major work on yeast taxonomy, the second edition of *The Yeasts. A Taxonomic Study*, edited by J. Lodder (1970). At the time of writing, the third edition, edited by N. J. W. Kreger-van Rij, is in preparation.

A New Key to The Yeasts (Barnett & Pankhurst, 1974) included the first computer-made key to all the yeasts. This key used only physiological tests. The favourable reception it was given has encouraged us to diversify, and to produce several kinds of key, of which some use both physiological tests and the microscopical appearance of the yeasts. Moreover, some keys are for particular groups of yeasts and some are designed for the needs of different kinds of user. In the time between compiling the two works, *A New Key* and *A Guide*, about 70 new species were accepted by the Centraalbureau voor Schimmelcultures; and in all 151 new species have been accepted since *The Yeasts* was written. There have also been changes in names, and amalgamations and divisions of species and genera. The changes are included in the tables and keys. They are also each described briefly, with references, in Chapter 4, which details 189 specific names that were not accepted when *The Yeasts* was compiled.

We have taken advantage of new, improved methods of computing and arranging identification keys. In particular, it is now easier to select the smallest sets of tests that can provide as complete identification as can be achieved with the available results. Our compact form of key is easy to read and is economical of space. Further economy is achieved by reticulation: sometimes the reader is referred back to an earlier part of the key, instead of forward; this avoids printing parts of the key more than once.

In Chapter 7 we give keys to all the yeasts except three genera, *Cyniclomyces*, *Oosporidium* and *Pityrosporum*, which require very special conditions for cultivation. There are also keys to particular groups, some of special interest to workers in medicine or in different parts of industry. These groups include yeasts that utilize hydrocarbons or methanol, yeasts isolated clinically, and yeasts associated with food and food manufacture, with wine-making or with brewing.

Chapter 8 provides a means of confirming or rejecting the supposed identity of a yeast; for this purpose there are minimal lists of physiological tests, with their results for each of the 439 species detailed in this book. The lists give maximum differentiation of each species from the other 438. The effectiveness of this discrimination is limited only by the available results of the physiological tests.

Chapter 5 gives the bare essentials of laboratory methods for identifying yeasts. It should help those who lack experience of such work and who have difficulty in choosing which methods to use. Where we give alternative procedures, we have tried to explain their different merits.

The table of generic characteristics in Chapter 3 is intended to aid people who wish to keep abreast of the constantly changing face of yeast classification. Many of the taxonomic changes are a nuisance to people who are concerned with yeasts in industry, medicine and various kinds of research. Some colleagues who are experimentalists hold that the frequent changes of names indicate yeast taxonomists' incompetence, irresponsibility and obsession with trivialities. So in Chapter 2 we give an account of how yeasts are classified. We hope that this exposition will reduce misunderstanding.

Nevertheless, yeast taxonomists often seem indifferent to the inconvenience caused by frequent changes in nomenclature. Consider what has happened to *Zygosaccharomyces fermentati* (Naganishi, 1928): this species was altered to *Saccharomyces cerevisiae* by Lodder & Kreger-van Rij in 1952, back to *Zygosaccharomyces fermentati* by Kudriavzev in 1954, to *Saccharomyces montanus* by Phaff, Miller & Shifrine in 1956, to *Torulaspora manchurica* (van der Walt & Johannsen, 1975*d*), and finally (?) back again to *Zygosaccharomyces fermentati* (von Arx, Rodrigues de Miranda, Smith & Yarrow, 1977). Each name could have been used correctly for the same organism in different publications. Similarly, it may be exceedingly difficult for a biochemist who reads a paper on *Saccharomyces carlsbergensis*, published in 1965, to appreciate that this is the same kind of yeast as that called *Saccharomyces uvarum* in 1975 and *Saccharomyces cerevisiae* in 1978. Each name was formally correct at the time. Such confusion of names can only retard our knowledge of yeasts. Some of the confusion comes from the practice of publishing 'new combinations' of names. Lloyd (1905) rebuked the mycologist who treated a new combination 'as a foot ball to kick his own name forward.' What would he have said of the authors of some recently accepted names of this kind that have been published without even a formal justification?

J. A. Barnett
November 1978

2
How yeasts are classified

Biological classification and the yeasts
Biological systematics includes (i) taxonomy, the study of classification, (ii) classification, the grouping of organisms into taxa [singular, taxon], (iii) identification, the comparison of unnamed with similar, named individuals, (iv) nomenclature, the naming of accepted taxa.

Charles Darwin (1859) wrote: '... I look at the term species as one arbitrarily given for the sake of convenience to a set of individuals closely resembling each other.' Similarly, some taxonomists regard species as arbitrarily defined, artificial categories, without existence in nature; but for other taxonomists species have an objective reality. For example, Kudriavzev (1954, p. 80) wrote about yeast species:

> However, these are not abstract species, which are usually defined according to various combinations of arbitrary heritable characteristics not associated with the conditions of life of their possessors. Instead, with the help of the same morphological method used earlier, there have been demonstrated completely real groups of organisms, differing between themselves within the limits of each genus by the conditions of life (the habitat occupied by them) and also according to the totality of their specific adaptations to these conditions [Translated from Russian.]

When a group of organisms reproduces sexually, those classified as belonging to the same species can usually interbreed and produce fully fertile offspring; but those assigned to different species either cannot be crossed at all or, if they can, produce infertile offspring such as mules. These observations have led to the notion that interfertility can be a test of whether two individuals belong to the same species. But some think differently. On the one hand, according to Wickerham & Burton (1956), where practicable, a test of the ability of yeasts to hybridize is important taxonomically 'since it may demonstrate relationship or lack of relationship between ... species'. Accordingly, van der Walt (1970c) accepted that separate species of *Saccharomyces* are interfertile. On the other hand, for the genus *Metschnikowia*, Pitt & Miller (1970) embraced the famous dictum of Mayr (1942): 'Species are groups of actually or potentially interbreeding natural populations, which are reproductively isolated from other such groups.' For classifying yeasts, this principle has limited application because (i) in many yeasts only asexual reproduction is known and (ii) many strains are self-fertile. The recent use of techniques of hybridizing nucleic acids of different strains (see Price, Fuson & Phaff, 1978) should make possible some crude assessment of closeness of evolutionary descent, but will not solve the problem of defining species satisfactorily. Some taxonomists think that such comparisons between genotypic rather than phenotypic characters will provide a more stable classification.

Present-day classification of the yeasts is based on strains. A strain is made up of the descendants of a single isolation in pure culture, properly from a single colony. It is often assumed, without evidence, that such a colony derives from a single cell. Furthermore, it is impracticable to cultivate some strains without their forming ascospores, especially when first isolated. So, in practice, strains may deviate considerably from the ideal of being clones, which would be derived vegetatively from a single cell and would be genetically homogenous (Cowan, 1968; *International Code of Nomenclature of Bacteria*, 1975). Most yeast strains are specified by a number; for example, two strains of *Saccharomyces cerevisiae* are CBS (Centraalbureau voor Schimmelcultures) 1172 and CBS 1234. However often taxonomists alter the names of the yeasts, these numbers do not change;

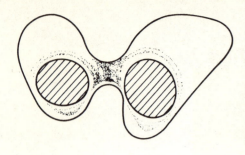

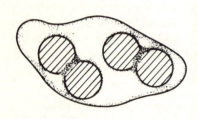

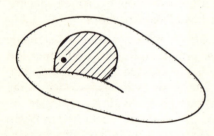

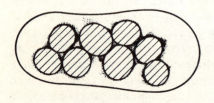

and the characteristics of most strains remain remarkably constant during many years of cultivation.

A species may have only one strain, but many include more than one. For example, at present, only one strain of *Brettanomyces abstinens* has been reported, whereas the Centraalbureau voor Schimmelcultures maintains over 200 strains of *Saccharomyces cerevisiae*. Just as strains are grouped into species, so species with certain features in common are arranged in genera, while groups of genera form families. Families are assembled in orders, orders in classes, and classes in divisions. This system conforms with the *International Code of Botanical Nomenclature* (1978). The higher taxa are designated by standard suffixes, such as *-aceae* for families and *-ales* for orders. The genus *Saccharomyces* belongs to the family Saccharomycetaceae of the order Endomycetales.

The names of the genera are singular nouns with a capital letter. When the name of a species is given, that of its genus is always given too. Thus a specific name is binomial, one part peculiar to the species, the other designating the genus, as *Candida albicans*. Here the generic name, *Candida*, is followed by the specific epithet, *albicans*. The latter has a lower case initial letter and must have a Latin suffix, which is (i) adjectival (e.g. *albicans*), or (ii) of a noun in apposition (as in *Brettanomyces lambicus*) or (iii) of a genetive noun (*Saccharomyces cerevisiae*). In a scientific publication, the first mention of any species should give its name in full, and should specify the accepted authority for that name. For instance, writing '*Zygosaccharomyces fermentati* Saito' makes it clear that this is the yeast described by Saito (1923) and not the one of the same name described by Naganishi (1928). '*Saccharomyces fermentati* (Saito) Lodder et Kreger-van Rij' means that this species was first described by Saito (1923) in another genus and was later transferred to *Saccharomyces* by Lodder & Kreger-van Rij (1952). '*Zygosaccharomyces florentinus* Castelli ex Kudrjawzew' is written thus because although the name of this species was coined by Castelli (1938), it was not validly published until Kudrjawzew [Kudriavzev] (1960) gave a Latin description.

Characteristics used in classifying yeasts
The chief characteristics used to classify yeasts are (i) the microscopical appearance of the cells, (ii) the mode of sexual reproduction, (iii) certain physiological (especially nutritional) activities and (iv) certain biochemical features.
Microscopical appearance. Taxonomists examine yeast cells microscopically and consider their size and shape, how they reproduce vegetatively (by multilateral, bipolar or unipolar budding, by fission, by forming filaments), and the form, structure and mode of formation of asci, ascospores and teliospores, if any of these are to be found (Fig. 2.1). The shape of the cells often indicates the mode of vegetative reproduction. For example, cells budding multilaterally are round to cylindrical; the unipolar cells of *Malassezia* (*Pityrosporum*) may be bottle-shaped; bipolar *Hanseniaspora* and *Saccharomycodes* species often have lemon-shaped cells. *Trigonopsis variabilis* forms both triangular cells which bud from their corners and elliptical cells which bud multilaterally. Triangular, clover-leaf-shaped and polymorphic blastoconidia also occur with the filaments of some species of *Pichia* and *Candida* (Fig. 2.2). Yeasts of the asexual (that is, 'imperfect') genus *Selenozyma* are characterized by having crescentic cells. *Schizosaccharomyces* species reproduce vegetatively by fission only, whereas both budding and fission are found in the genus *Trichosporon*, the hyphal filaments of which break up into arthroconidia (Fig. 2.3). New cells of *Sterigmatomyces* are produced on narrow stalks, sterigmata (Fig. 2.4), each of which breaks in the middle when the daughter-cell separates, so that older cells that have produced a number of daughter-cells have these projections scattered over the surface. The genera *Bullera* and *Sporobolomyces* are characterized by ballistospores formed on sterigmata (Fig. 2.5). In various genera which form filaments, certain of their features are used in classification, that is, whether the filaments are pseudohyphae or true septate hyphae, whether the septa have

2
How yeasts are classified

Biological classification and the yeasts
Biological systematics includes (i) taxonomy, the study of classification, (ii) classification, the grouping of organisms into taxa [singular, taxon], (iii) identification, the comparison of unnamed with similar, named individuals, (iv) nomenclature, the naming of accepted taxa.

Charles Darwin (1859) wrote: '... I look at the term species as one arbitrarily given for the sake of convenience to a set of individuals closely resembling each other.' Similarly, some taxonomists regard species as arbitrarily defined, artificial categories, without existence in nature; but for other taxonomists species have an objective reality. For example, Kudriavzev (1954, p. 80) wrote about yeast species:

> However, these are not abstract species, which are usually defined according to various combinations of arbitrary heritable characteristics not associated with the conditions of life of their possessors. Instead, with the help of the same morphological method used earlier, there have been demonstrated completely real groups of organisms, differing between themselves within the limits of each genus by the conditions of life (the habitat occupied by them) and also according to the totality of their specific adaptations to these conditions [Translated from Russian.]

When a group of organisms reproduces sexually, those classified as belonging to the same species can usually interbreed and produce fully fertile offspring; but those assigned to different species either cannot be crossed at all or, if they can, produce infertile offspring such as mules. These observations have led to the notion that interfertility can be a test of whether two individuals belong to the same species. But some think differently. On the one hand, according to Wickerham & Burton (1956), where practicable, a test of the ability of yeasts to hybridize is important taxonomically 'since it may demonstrate relationship or lack of relationship between ... species'. Accordingly, van der Walt (1970c) accepted that separate species of *Saccharomyces* are interfertile. On the other hand, for the genus *Metschnikowia*, Pitt & Miller (1970) embraced the famous dictum of Mayr (1942): 'Species are groups of actually or potentially interbreeding natural populations, which are reproductively isolated from other such groups.' For classifying yeasts, this principle has limited application because (i) in many yeasts only asexual reproduction is known and (ii) many strains are self-fertile. The recent use of techniques of hybridizing nucleic acids of different strains (see Price, Fuson & Phaff, 1978) should make possible some crude assessment of closeness of evolutionary descent, but will not solve the problem of defining species satisfactorily. Some taxonomists think that such comparisons between genotypic rather than phenotypic characters will provide a more stable classification.

Present-day classification of the yeasts is based on strains. A strain is made up of the descendants of a single isolation in pure culture, properly from a single colony. It is often assumed, without evidence, that such a colony derives from a single cell. Furthermore, it is impracticable to cultivate some strains without their forming ascospores, especially when first isolated. So, in practice, strains may deviate considerably from the ideal of being clones, which would be derived vegetatively from a single cell and would be genetically homogenous (Cowan, 1968; *International Code of Nomenclature of Bacteria*, 1975). Most yeast strains are specified by a number; for example, two strains of *Saccharomyces cerevisiae* are CBS (Centraalbureau voor Schimmelcultures) 1172 and CBS 1234. However often taxonomists alter the names of the yeasts, these numbers do not change;

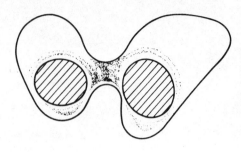

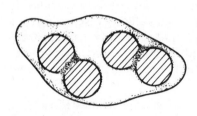

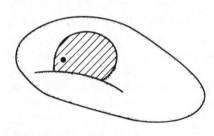

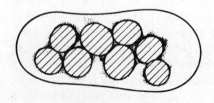

and the characteristics of most strains remain remarkably constant during many years of cultivation.

A species may have only one strain, but many include more than one. For example, at present, only one strain of *Brettanomyces abstinens* has been reported, whereas the Centraalbureau voor Schimmelcultures maintains over 200 strains of *Saccharomyces cerevisiae*. Just as strains are grouped into species, so species with certain features in common are arranged in genera, while groups of genera form families. Families are assembled in orders, orders in classes, and classes in divisions. This system conforms with the *International Code of Botanical Nomenclature* (1978). The higher taxa are designated by standard suffixes, such as *-aceae* for families and *-ales* for orders. The genus *Saccharomyces* belongs to the family Saccharomycetaceae of the order Endomycetales.

The names of the genera are singular nouns with a capital letter. When the name of a species is given, that of its genus is always given too. Thus a specific name is binomial, one part peculiar to the species, the other designating the genus, as *Candida albicans*. Here the generic name, *Candida*, is followed by the specific epithet, *albicans*. The latter has a lower case initial letter and must have a Latin suffix, which is (i) adjectival (e.g. *albicans*), or (ii) of a noun in apposition (as in *Brettanomyces lambicus*) or (iii) of a genetive noun (*Saccharomyces cerevisiae*). In a scientific publication, the first mention of any species should give its name in full, and should specify the accepted authority for that name. For instance, writing '*Zygosaccharomyces fermentati* Saito' makes it clear that this is the yeast described by Saito (1923) and not the one of the same name described by Naganishi (1928). '*Saccharomyces fermentati* (Saito) Lodder et Kreger-van Rij' means that this species was first described by Saito (1923) in another genus and was later transferred to *Saccharomyces* by Lodder & Kreger-van Rij (1952). '*Zygosaccharomyces florentinus* Castelli ex Kudrjawzew' is written thus because although the name of this species was coined by Castelli (1938), it was not validly published until Kudrjawzew [Kudriavzev] (1960) gave a Latin description.

Characteristics used in classifying yeasts

The chief characteristics used to classify yeasts are (i) the microscopical appearance of the cells, (ii) the mode of sexual reproduction, (iii) certain physiological (especially nutritional) activities and (iv) certain biochemical features.

Microscopical appearance. Taxonomists examine yeast cells microscopically and consider their size and shape, how they reproduce vegetatively (by multilateral, bipolar or unipolar budding, by fission, by forming filaments), and the form, structure and mode of formation of asci, ascospores and teliospores, if any of these are to be found (Fig. 2.1). The shape of the cells often indicates the mode of vegetative reproduction. For example, cells budding multilaterally are round to cylindrical; the unipolar cells of *Malassezia* (*Pityrosporum*) may be bottle-shaped; bipolar *Hanseniaspora* and *Saccharomycodes* species often have lemon-shaped cells. *Trigonopsis variabilis* forms both triangular cells which bud from their corners and elliptical cells which bud multilaterally. Triangular, clover-leaf-shaped and polymorphic blastoconidia also occur with the filaments of some species of *Pichia* and *Candida* (Fig. 2.2). Yeasts of the asexual (that is, 'imperfect') genus *Selenozyma* are characterized by having crescentic cells. *Schizosaccharomyces* species reproduce vegetatively by fission only, whereas both budding and fission are found in the genus *Trichosporon*, the hyphal filaments of which break up into arthroconidia (Fig. 2.3). New cells of *Sterigmatomyces* are produced on narrow stalks, sterigmata (Fig. 2.4), each of which breaks in the middle when the daughter-cell separates, so that older cells that have produced a number of daughter-cells have these projections scattered over the surface. The genera *Bullera* and *Sporobolomyces* are characterized by ballistospores formed on sterigmata (Fig. 2.5). In various genera which form filaments, certain of their features are used in classification, that is, whether the filaments are pseudohyphae or true septate hyphae, whether the septa have

List of figures

P.4
Fig. 2.1. (*a*) *Zygosaccharomyces bailii*, ascus with two ascospores. (*b*) *Saccharomycodes ludwigii*, ascus with four ascospores. (*c*) *Hansenula anomala*, ascus with one hat-shaped ascospore with brim. (*d*) *Schizosaccharomyces octosporus*, ascus with eight ascopores. (*e*) *Rhodosporidium malvinellum*, teliospore with hypha.

P.5
Fig. 2.2. (*a*) *Pichia kudriavzevii*, pseudohyphae with blastoconidia. (*b*) *Candida albicans*, true hypha with blastoconidia.

Fig. 2.3. *Geotrichum fermentans*, true hypha breaking up into arthroconidia.

Fig. 2.4. *Sterigmatomyces halophilus*, conidium formed on sterigma.

Fig. 2.5. *Sporobolomyces roseus*, formation of a ballistoconidium (after Buller, 1958).

Characteristics used in classifying yeasts

pores, and the occurrence and arrangement of the blastoconidia. Kreger-van Rij and Veenhuis (1971) used electron microscopy to examine the structure of the cell walls and the way buds are formed; in both features, ascosporogenous yeasts differed from basidiomycetous yeasts because, unlike the walls of ascosporogenous yeasts, those of the basidiomycetous yeasts were lamellate.

Sexual reproduction. Yeasts may reproduce sexually (i) by ascospores or (ii) by teliospores (Fig. 2.1). Yeasts that reproduce by teliospores are classified partly by structural features, such as clamp connections; the forms taken by sporophores and teliospores are also used. Such yeasts are held (e.g. by Kreger-van Rij, 1973) to be related to basidiomycetous fungi because of resemblances in the sexual cycles. For ascosporogenous yeasts, taxonomic importance is given to whether asci are formed (i) from a vegetative (probably diploid) cell, (ii) from two conjugating (perhaps haploid) cells or (iii) from a mother-cell that has conjugated with its bud. For yeasts with asci borne on filaments, the arrangement of the asci, whether in chains (catenulate) or bunches (botryose), may be used to distinguish between genera, such as *Hormoascus* or *Ambrosiozyma*. Some asci, such as those of *Lodderomyces* and *Saccharomyces*, are burst by the germinating ascospores; others, such as those of *Kluyveromyces*, rupture early and release the intact ascospores. The number of ascospores in each ascus, their shape (round, oval, hat-, kidney-, Saturn- and needle-shaped) and whether the ascospore walls are smooth or rough, are features which are also used to separate genera, as well as to separate species within a genus. Kreger-van Rij (1964) studied the structure of the ascospores by electron microscopy. As a result the genus *Wingea* was created by van der Walt (1967) from the former species *Pichia robertsii*. Similarly, an electron microscopic study of the ascospores of *Saccharomyces transvaalensis* led van der Walt (1978) to create the genus *Pachytichospora*.

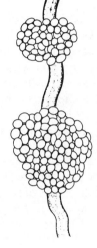

Physiological features. Physiological characteristics used for classifying yeasts are chiefly the following: the ability (i) to 'ferment' certain sugars semi-anaerobically; (ii) to grow aerobically with various compounds, each as sole major source of either carbon or nitrogen ('assimilation' tests); (iii) to grow without an exogenous supply of certain vitamins; (iv) to grow in the presence of 50% or 60% (usually w/v) D-glucose or 10% (w/v) NaCl + 5% (w/v) D-glucose; (v) to grow at 37°C; (vi) to grow in the presence of cycloheximide (actidione); (vii) to split fat; (viii) to produce polysaccharide which reacts like starch with iodine; (ix) to hydrolyse urea; (x) to form acid. Such characteristics are used chiefly for differentiating between species. But sometimes genera are distinguished by these features: by definition, yeasts of the genus *Cryptococcus* utilize inositol as a carbon source for growth, yeasts of the genus *Dekkera* produce acid, yeasts of the genus *Hansenula* use nitrate, *Saccharomyces* 'ferment' D-glucose, and so on.

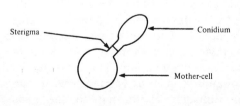

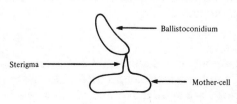

Biochemical characteristics. Studies of certain biochemical characters may also influence taxonomic decisions, for example the chemical structure of the cell walls (for review, see Phaff, 1971; Bonaly, 1974), particularly of the wall mannans (Gorin & Spencer, 1970; Ballou, 1974). Another example comes from observations on the kind of ubiquinone (coenzyme Q) present in different yeasts. This coenzyme participates in the transport of electrons from organic substrates to oxygen in the respiratory chain of mitochondria. It is a reversibly reducible quinone with a long isoprenoid side chain. The chain may have 6 to 10 isoprene units and, accordingly, the coenzymes are designated Q_6 to Q_{10}. According to Yamada, Nojiri, Matsuyama & Kondo (1976), some *Debaryomyces* species have Q_6 and others Q_9. As a result of this evidence of heterogeneity within the genus *Debaryomyces*, general approval has not been forthcoming for the proposal of van der Walt & Johannsen (1975d) to unite the species of *Debaryomyces* with some species of *Saccharomyces* and transfer them to the genus *Torulaspora*.

Naming yeasts

The naming of yeasts is governed by the *International Code of Botanical Nomenclature* (1978) which lays down the requirements for publishing the name of a new taxon and for determining the correct name when a change is made. Yeast taxonomists profess to follow this code in all aspects save one, namely Article 9 note 3, which rules that the type of the name must be a dried herbarium specimen. ('Type' is here used as the Greek word τύπος in the sense of 'the original pattern'.) Species of yeast are distinguished from one another mainly by physiological characteristics, and type material which does not permit such characteristics to be examined is useless. Some important provisions of the Code are discussed below.

A Latin description must be given for any taxon published on or after 1 January 1935 (Article 36). If this description is omitted, the name is a *nomen nudum* [naked name] but may be 'clothed' and hence validly published at a later date by the effective publication of a Latin description. Publication of names from 1 January 1958 must indicate the type.

The type of a species is a strain and the type of a genus is a species. There are four important kinds of type. (i) The holotype of a yeast species is the strain which the author of the name of that species used or designated as the type. (ii) When the author designates more than one type, but no holotype, these are syntypes. (iii) The strain chosen subsequently as the nomenclatural type from these syntypes is the lectotype. (iv) A neotype is one chosen as the nomenclatural type, when holotype, lectotype and syntypes are lost. A valid type may not be changed, except a neotype, which may be changed when the lost holotype is found.

The purpose of a type is to determine the correct name when, for example, two or more species are united, or one is divided. The name that must be used is then the first valid name that had been published. The effective date of publication is that when all the requirements of valid publication are met. Two examples follow. First, *Saccharomyces beticus* of Marcilla, Alas & Feduchy (1939) remained *nomen nudum* and hence invalid until Santa María (1970) published a Latin description; so the recognized date for this species is 1970 and not 1939. Secondly, *Candida ravautii*, *Candida brumptii* and *Candida catenulata* are shortly to be combined in a single species (S. A. Meyer, personal communication). Although Langeron & Guerra (1935) described the first two of these species, neither seems ever to have had a Latin description and, therefore, *Candida catenulata*, published later but with a valid name by Diddens & Lodder (1942), has priority and will be the correct name for the species. Article 37 of the *International Code* recommends particularization of the type after the words *typus* or *holotypus* at the end of the Latin description. The source of the strain is usually given and, also, the strain number in each collection where it is deposited.

There are several kinds of name which, although published and valid, do not accord with the Rules of Nomenclature and so are illegitimate. Such illegitimate names include homonyms, which are identical to previously existing legitimate names based on different types, and tautonyms in which the specific epithet repeats the generic name. An example of a tautonym could be given by *Torulopsis candida* when it is transferred to the genus *Candida*; since *Candida candida* is illegitimate, the species will be called *Candida famata*, the second name in the line of priority. *Candida famata* will be a new combination. This is written in a late Latin form, *nova combinatio*. So when Rodrigues de Miranda (1975) transferred *Sterigmatomyces acheniorum* Buhagiar et Barnett to the genus *Rhodotorula*, he wrote the specific name as *Rhodotorula acheniorum* (Buhagiar et Barnett) Rodrigues de Miranda *nov. comb*.

Renaming yeasts

A name in current use may be changed for one or more of the following reasons. (i) The name was introduced instead of another which had priority; although the Code is relatively modern, its provisions are retroactive and, further, some recent workers have ignored the rules and others have misinterpreted them. (ii) Some names have been misapplied because the types have not been properly established. (iii) Authors have been unaware that the taxon they describe already possessed a description and name which, when found, is given preference.

A good illustration of changes made for nomenclatural reasons is given by the name *Torula*, which was once used as the generic name of the yeast currently known as *Candida utilis*. Turpin (1838) introduced the name *Torula* for yeasts, although Persoon had already used it for black moulds in 1796 (Ciferri, 1925). So *Torula* Turpin, being a homonym of *Torula* Persoon, was illegitimate. Noticing this error, Berlese (1895) created the genus *Torulopsis* to replace *Torula* Turpin. Unfortunately, Berlese's publication was generally overlooked until Ciferri (1925) and Ciferri & Redaelli (1929) drew attention to it and amended the diagnosis to include anascosporogenous, fermenting or non-fermenting, non-filamentous yeasts with round or oval cells. Lodder (1934) again amended the diagnosis, excluding all red pigmented yeasts. Berlese (1895) had designated *Torulopsis rosea* as the type of the genus; but as his strain of this species was not maintained, there was no longer an existing type strain to give the unequivocal information about the identity of the species. From Berlese's description, Lodder (1938) suggested that the species was *Torulopsis pulcherrima* (now *Metschnikowia pulcherrima*), but later authors have doubted this identification. Lodder & Kreger-van Rij (1952) suggested conserving the name *Torulopsis*, and designated *Torulopsis colliculosa* Hartmann as the new type of the genus. But as they did not take the necessary formal steps to have *Torulopsis* included in the *nomina conservanda*, this was not done. Hence, *Torulopsis* Lodder et Kreger-van Rij is an illegitimate homonym of *Torulopsis* Berlese; furthermore it was published without a Latin description and so is invalid. To sum up, *Torula* Turpin is an illegitimate homonym of *Torula* Persoon; *Torulopsis* Berlese is *nomen dubium*, because it lacks an identifiable type species; and *Torulopsis* Lodder et Kreger-van Rij is an illegitimate homonym of *Torulopsis* Berlese.

Nonetheless, van Uden & Vidal-Leiria (1970) continued to use the name *Torulopsis*, whilst admitting (van Uden & Buckley, 1970) that *Torulopsis* is distinguished from *Candida* only by a totally inadequate intergeneric criterion, namely the capacity of *Candida* species to form pseudohyphae. Van Uden & Buckley (1970) discussed the prospect of merging all the species of *Candida* and *Torulopsis* into a single genus. They wrote:

> This measure, logical though it may be, appeared to us rather inadvisable since it would make necessary the provisory [?provisional] renaming of a great number of species. This would inevitably lead to confusion and justified irritation among the increasing number of workers in various fields who use or encounter yeasts of this group. We have therefore maintained *Candida* and *Torulopsis* as separate genera.

Nonetheless, because *Torulopsis* species are without legality, Yarrow & Meyer (1978) have now proposed transferring them to the genus *Candida* and have amended the diagnosis of *Candida* to include non-filamentous species.

'Objective' classifications

For many years, intuitive and subjective decisions have been predominant in the classification of yeasts. But since the recent advances in the numerical taxonomy of bacteria (Sneath & Sokal, 1973) some yeast taxonomists have hoped to introduce greater objectivity into their practice. Measurements have been made of similarities and differences between numbers of strains, by analysing quantitatively the results of many tests (reviewed by Campbell, 1974). Other taxonomists have wished to base their classification on the capacity for interbreeding (e.g. Johannsen & van der Walt, 1978) or DNA sequence complementarity (e.g. Meyer, Smith & Simione, 1978). But, as can be seen in the following examples, even these two 'objective' approaches sometimes give conflicting results. This should not evoke surprise since the two kinds of classification are based on different kinds of characteristic, phenotypic and genotypic.

Blagodatskaja & Kocková-Kratochvílová (1973) based two taxa, *Candida pseudolipolytica* and *Candida lipolytica* var. *thermotolerans*, on numerical analysis of the yeasts' characteristics; but Yarrow (1975) crossed the type strains with each other and with strains of *Saccharomycopsis lipolytica*, and considered these new taxa to be merely strains of *Saccharomycopsis lipolytica* of opposite sex. Similarly, numerical analysis of *Saccharomyces* species led Campbell (1972) to group *Saccharomyces fermentati* with *Saccharomyces rosei* and *Saccharomyces pretoriensis*, but to separate them from *Saccharomyces inconspicuus*, *Saccharomyces delbrueckii*, *Saccharomyces vafer* and *Saccharomyces kloeckerianus*. By contrast, after measuring DNA hybridization, Price, Fuson & Phaff (1978) held *Saccharomyces fermentati*, *Saccharomyces rosei*, *Saccharomyces inconspicuus* and *Saccharomyces vafer* to be synonymous with *Saccharomyces delbrueckii*, whilst *Saccharomyces kloeckerianus* and *Saccharomyces pretoriensis* were assigned to separate species.

3
Characteristics of the genera

This chapter summarizes the characteristics of the currently accepted genera. The information is given in Table 3.1, with the genera arranged alphabetically. References to *The Yeasts* are to the second edition (Lodder, 1970).

Table 3.1. *Summary of the characteristics of the currently accepted genera*

Genus	Vegetative cells	Sexual reproduction	Physiology	No. of species	Comments and references
Aciculoconidium	Budding cells; septate hyphae with chains of conidia. Terminal conidia are needle-shaped	None	Slight fermentation. Nitrate −	1	(King & Jong, 1976.) The sole species was classified as *Trichosporon aculeatum* in *The Yeasts*, p. 1312
Aessosporon	Budding cells; ballistoconidia; sometimes pseudo- and true hyphae	Teliospores form promycelia with sporidia	Fermentation − Nitrate +/−	2	Sexual states of *Bullera* and *Sporobolomyces* species. (Van der Walt, 1970a, 1973)
Ambrosiozyma	Budding cells; true hyphae with plugged septal pores	Asci, in lateral clusters, each with up to four hat- or Saturnoid-shaped ascospores	Fermentation weak. Nitrate −	3	(Van der Walt, 1972; von Arx, 1972.) One species was classified in *Endomycopsis* in *The Yeasts*, p. 190
Arthroascus	Septate hyphae, no blastoconidia	Asci develop from swollen hyphal cells, each with one to four hat-shaped ascospores	Fermentation − Nitrate −	1	(Von Arx, 1972.) The sole species was classified as *Endomycopsis javanensis* in *The Yeasts*, p. 186
Botryoascus	Septate hyphae, clavate conidia, no blastoconidia	Asci carried laterally on hyphae in small bunches, each with up to four hat-shaped ascospores	Fermentation − Nitrate −	1	(Von Arx, 1972)
Brettanomyces	Budding cells, often ogival in shape; sometimes pseudohyphae	None	Fermentation − Nitrate +/− Acetic acid produced aerobically from glucose. Slow growth	6	(*The Yeasts*, p. 863.) Asexual states of *Dekkera*
Bullera	Budding cells; ballistoconidia; sometimes pseudohyphae	None	Fermentation − Nitrate +/−	3	(*The Yeasts*, p. 815; Stadelmann, 1975)
Candida	Budding cells; pseudo- and sometimes septate hyphae	None	Fermentation +/− Nitrate +/−	104	(*The Yeasts*, p. 893)
Citeromyces	Budding cells; no hyphae	Asci, each with one spherical warty ascospore. Heterothallic	Fermentation +	1	(*The Yeasts*, p. 121)
Cryptococcus	Budding cells, usually with capsule of polysaccharide	None	Fermentation − Inositol +	7	(*The Yeasts*, p. 1088.) *Cryptococcus laurentii* may form hyphae and chlamydospores (Kurtzman, 1973)

Table 3.1 Continued

Genus	Vegetative cells	Sexual reproduction	Physiology	No. of species	Comments and references
Cyniclomyces	Budding from or near poles of the cells. Pseudohyphae	Asci, each with one to four oval to cylindrical ascospores	Fermentation weak. Complex growth requirements. Cultures short lived	1	The sole species was classified as *Saccharomycopsis guttulata* in *The Yeasts*, p. 728. See van der Walt & Scott (1971b). Restricted to guts of rabbits. Omitted from keys; no growth in standard media
Debaryomyces	Budding cells; sometimes pseudohyphae	Asci, each with one to four spherical or oval warty ascospores. Conjugation between cell and bud	Fermentation +/− Nitrate −	11	(*The Yeasts*, p. 129)
Dekkera	Budding cells, often ogival in shape; sometimes pseudohyphae	Asci, each with one to four hat-shaped ascospores	Fermentation + Nitrate +/− Acetic acid produced aerobically from glucose. Growth slow	2	(*The Yeasts*, p. 157)
Filobasidium	Budding cells, pseudo- and septate hyphae with dolipores	Sporophores bearing whorls of basidiospores at the apex, which is slightly swollen	Fermentation +/− Nitrate +/− Inositol +	3	(Olive, 1968; Rodrigues de Miranda, 1972; Kwon-Chung, 1977)
Geotrichum	True septate hyphae and arthroconidia. No budding cells	None	Fermentation +/−	9	(Von Arx, 1977.) Asexual states of the filamentous fungus, *Dipodascus*. The three species included in this guide were classified in *The Yeasts* in the genus *Trichosporon*
Guilliermondella	Budding cells; septate hyphae	Asci terminal, intercalary or lateral in chains on hyphae, each with two to four lunate to fusiform ascospores	Fermentation weak. Nitrate −	1	(Von Arx, 1972.) The sole species was classified as *Endomycopsis selenospora* in *The Yeasts*, p. 200
Hanseniaspora	Bipolar budding, lemon-shaped or oval cells; rarely pseudohyphae	Asci, each with one to four spherical or hat-shaped ascospores	Fermentation + Nitrate −	6	(Meyer, Smith & Simione, 1978)
Hansenula	Budding cells; sometimes pseudo- or septate hyphae	Asci, each with one to four hat- or Saturn-shaped ascospores	Fermentation +/− Nitrate +	30	(*The Yeasts*, p. 226)
Hormoascus	Budding cells; true septate hyphae	Asci in chains in a brush-like arrangement on ascophores, each with two to four hat-shaped ascospores	Fermentation weak. Nitrate +	1	(Von Arx, 1972.) The sole species was classified as *Endomycopsis platypodis* in *The Yeasts*, p. 197
Hyphopichia	Septate hyphae with blastoconidia on denticles	Asci, each with one to four hat-shaped ascospores. Heterothallic	Fermentation weak. Nitrate −	1	(Von Arx & van der Walt, 1976.) The sole species was classified as *Endomycopsis burtonii* in *The Yeasts*, p. 174
Kluyveromyces	Budding cells; pseudohyphae	Asci evanescent, each with 1–60 spherical, oval or reniform ascospores	Fermentation + Nitrate −	16	(*The Yeasts*, p. 316; Martini, 1973; Johannsen & van der Walt, 1978)

Table 3.1 Continued

Genus	Vegetative cells	Sexual reproduction	Physiology	No. of species	Comments and references
Leucosporidium	Apical budding; pseudohyphae and septate hyphae	Teliospores, each producing promycelium with lateral or terminal sporidia. Clamp connections	Fermentation +/− Nitrate + No pink pigmentation. Starch + Psychrophilic	6	(*The Yeasts*, p. 776; Fell, 1974)
Lipomyces	Budding cells, usually with polysaccharide capsule. No hyphae	Asci, each with 1–16 oval, dark ascospores	Fermentation − Nitrate − Starch + Inositol +/−	5	(*The Yeasts*, p. 379; Nieuwdorp, Bos & Slooff, 1974)
Lodderomyces	Budding cells; pseudohyphae	Asci persistent, each with one to two smooth, oval ascospores	Fermentation + Nitrate − *n*-Decane + *n*-Hexadecane +	1	(*The Yeasts*, p. 403)
Metschnikowia	Budding cells, rarely lunate or falcate; usually pseudohyphae	Asci club-shaped, each with one or two needle-shaped ascospores	Fermentation + Nitrate −	6	(*The Yeasts*, p. 408)
Nadsonia	Bipolar budding on a broad base; large cells, lemon-shaped or oval	Asci, each with one to two rough, spherical ascospores	Fermentation + Nitrate −	2	(*The Yeasts*, p. 430)
Nematospora	Budding cells; pseudo- and septate hyphae	Asci, each with up to 16 fusiform ascospores each with a whip-like appendage	Fermentation + Nitrate −	1	(*The Yeasts*, p. 440)
Oosporidium	Budding on a broad base combined with septation. Pseudohyphae. No arthroconidia	None	Fermentation − Nitrate + Growth slow. Colonies pink	1	(*The Yeasts*, p. 1161.) Omitted from keys; does not grow in standard media
Pachysolen	Budding cells; sometimes pseudohyphae	Asci, each formed at distal end of tubular extension of a cell, contain two to four hat-shaped ascospores	Fermentation + Nitrate +	1	(*The Yeasts*, p. 448)
Pachytichospora	Budding cells; sometimes pseudohyphae	Asci persistent, each with one to two spherical or oval ascospores, formed from diploid cell	Fermentation + Nitrate −	1	(Van der Walt, 1978.) The sole species was classified as *Saccharomyces transvaalensis* in *The Yeasts*, p. 696
Phaffia	Budding cells; chlamydospores; sometimes pseudohyphae	None	Fermentation slow Nitrate − Starch produced. Carotenoid pigments synthesized	1	(Miller, Yoneyama & Soneda, 1976)
Pichia	Budding cells; usually pseudohyphae, occasionally septate hyphae	Asci, each with one to four, rarely up to eight, smooth spherical, hat- or Saturn-shaped ascospores	Fermentation +/− Nitrate −	51	(*The Yeasts*, p. 455)
Pityrosporum	Monopolar budding on a broad base (phialoconidia); sometimes hyphae	None	Fermentation − Many strains require olive oil in medium	2	(*The Yeasts*, p. 1167.) Omitted from keys; no growth on standard media. Found on skin of man and other mammals

Table 3.1 Continued

Genus	Vegetative cells	Sexual reproduction	Physiology	No. of species	Comments and references
Rhodosporidium	Budding cells; sometimes pseudo- and septate hyphae	Teliospores, each forming promycelium with lateral or terminal sporidia. Clamp connections may occur. Most species are heterothallic	Fermentation − Nitrate +/− Inositol +/− Starch production +/− Carotenoid pigments synthesized	8	(*The Yeasts*, p. 803; Fell, 1974.) Sexual states of *Rhodotorula* and *Cryptococcus* species
Rhodotorula	Budding cells	None	Fermentation − Nitrate +/− Inositol − Starch production +/− Carotenoid pigments synthesized	11	(*The Yeasts*, p. 1187)
Saccharomyces	Budding cells; sometimes pseudohyphae	Asci persistent, each formed direct from diploid cell, with one to four, rarely more, smooth spherical or oval ascospores	Fermentation + Nitrate −	7	(*The Yeasts*, p. 555)
Saccharomycodes	Bipolar budding, large lemon-shaped or elongate cells. Sometimes pseudohyphae	Asci, each with usually four smooth spherical ascospores, each with a narrow ledge	Fermentation + Nitrate −	1	(*The Yeasts*, p. 719)
Saccharomycopsis	Budding cells; septate hyphae	Asci, each with two to four oblate or hat-shaped ascospores	Fermentation +/− Nitrate −	6	(Van der Walt & Scott, 1971b.) Some species were classified in *Endomycopsis* in *The Yeasts*
Sarcinosporon	Budding cells; septate hyphae. Sarcina-like septation of cells in different planes	None	Fermentation − Nitrate −	1	(King & Jong, 1975.) The sole species was classified as *Trichosporon inkin* in *The Yeasts*, p. 1337
Schizoblastosporion	Bipolar budding on oval or cylindrical cells	None	Fermentation − Nitrate −	1	(*The Yeasts*, p. 1224)
Schizosaccharomyces	Cells divide, no budding; sometimes septate hyphae	Asci, each with up to eight spherical oval or falcate ascospores	Fermentation + Nitrate −	5	(*The Yeasts*, p. 733.) Grow poorly in standard medium for growth tests
Schwanniomyces	Budding cells; sometimes simple pseudohyphae	Asci, each with one to two warty spherical ascospores, each with an equatorial ledge	Fermentation + Nitrate −	1	(*The Yeasts*, p. 756)
Selenozyma	Budding cells, lunate or hemispherical, also some oval cells	None	Fermentation +/− Nitrate −	2	(Yarrow, 1969b; Von Arx, Rodrigues de Miranda, Smith & Yarrow, 1977)
Sporidiobolus	Budding cells; ballistoconidia; septate hyphae	Dark chlamydospores, hyphae with clamp connections	Fermentation − Nitrate + Carotenoid pigments synthesized	2	(*The Yeasts*, p. 822)

Table 3.1 Continued

Genus	Vegetative cells	Sexual reproduction	Physiology	No. of species	Comments and references
Sporobolomyces	Budding cells; ballistoconidia; sometimes pseudo- and septate hyphae	None (but see note in last column)	Fermentation – Nitrate +/– Usually carotenoid pigments are synthesized	11	(*The Yeasts*, p. 831.) Mating has been reported between *Sporobolomyces odorus*, *Sporobolomyces hispanica* and *Sporobolomyces salmonicolor*, and in *Sporobolomyces pararoseus* (Bandoni, Lobo & Brezden, 1971; Bandoni, Johri & Reid, 1975)
Stephanoascus	Budding cells; pseudo- and septate hyphae with conidia on denticles	Asci, each formed after conjugation between cells of two hyphae, contain one to four usually hat-shaped ascospores. Heterothallic	Fermentation – Nitrate –	1	(Smith, van der Walt & Johannsen, 1976)
Sterigmatomyces	Conidia on sterigmata; sometimes hyphae	None	Fermentation – Nitrate +/–	7	(*The Yeasts*, p. 1229)
Sympodiomyces	Spherical or oval, one or two celled conidia formed on conidiophores in sympodulae; septate hyphae	None	Fermentation – Nitrate –	1	(Fell & Statzell, 1971)
Torulaspora	Budding cells; no hyphae	Persistent asci, each with one to four usually rough ascospores. Conjugation usually between cell and its bud	Fermentation + Nitrate –	3	(Van der Walt & Johannsen, 1975*d*.) Classified as *Saccharomyces* in *The Yeasts*
Torulopsis	Budding cells; occasionally pseudohyphae	None	Fermentation +/– Nitrate +/– Inositol –	58	(*The Yeasts*, p. 1235.) The name is invalid and the species will be moved to *Candida* (Yarrow & Meyer, 1978)
Trichosporon	Budding cells; pseudo- and septate hyphae and dolipores, arthroconidia	None	Fermentation – Nitrate +/–	8	(*The Yeasts*, p. 1309)
Trigonopsis	Mainly triangular cells with budding from corners, also oval cells with multilateral budding	None	Fermentation – Nitrate –	1	(*The Yeasts*, p. 1353)
Wickerhamia	Bipolar budding, cells apiculate with blunt ends, oval or elongate	Asci, each with 1–16 ascospores, which are shaped like jockeys' caps	Fermentation + Nitrate –	1	(*The Yeasts*, p. 767)
Wickerhamiella	Budding cells; no hyphae	Asci, each with one oblong rough very small ascospore	Fermentation – Nitrate +	1	(Van der Walt & Liebenberg, 1973.) The sole species was classified as *Torulopsis domercqii* in *The Yeasts*, p. 1258
Wingea	Budding cells; no hyphae	Asci, each with one to four oblong smooth ascospores	Fermentation + Nitrate –	1	(*The Yeasts*, p. 772)
Zendera	Septate hyphae with arthroconidia	Asci, each with two to eight oval smooth ascospores	Fermentation – Nitrate –	2	(Redhead & Malloch, 1977.) The sole species was classified as *Endomycopsis ovetensis* in *The Yeasts*, p. 192

Table 3.1 Continued

Genus	Vegetative cells	Sexual reproduction	Physiology	No. of species	Comments and references
Zygosaccharomyces	Budding cells; sometimes pseudohyphae	Asci, formed by conjugating cells, each contain one to four spherical or oval ascospores	Fermentation + Nitrate −	8	Classified as *Saccharomyces* species in *The Yeasts*. *Zygosaccharomyces* (Barker, 1901a,b), rejected by Lodder & Kreger-van Rij (1952), now restored (von Arx, Rodrigues de Miranda, Smith & Yarrow, 1977)

4
Newly accepted and renamed genera and species

This chapter lists and briefly describes genera and species accepted in this volume but not in the second edition of *The Yeasts* (Lodder, 1970). Both new and renamed taxa are included. The names, to which full references are given, were accepted by the Yeast Division of the Centraalbureau voor Schimmelcultures at the beginning of 1978. No description is given of species already described under another name in *The Yeasts* (Lodder, 1970).

ACICULOCONIDIUM King et Jong

King & Jong (1976) created this genus of the *Fungi Imperfecti* to accommodate the single species, *Trichosporon aculeatum*. There are budding cells, mycelia of branched, septate hyphae and chains of round to oval blastoconidia. The terminal conidia are needle-shaped, rounded proximally and pointed distally.

ACICULOCONIDIUM ACULEATUM (Phaff, Miller et Shifrine) King et Jong

Do Carmo-Sousa (1970) described this species as *Trichosporon aculeatum*. The yeast is no longer retained in *Trichosporon* because it forms needle-shaped holoblastic conidia and does not form arthroconidia of the kind characteristic of *Geotrichum* and *Trichosporon*.

AESSOSPORON van der Walt

Van der Walt (1970a) created this genus to accommodate a sexually reproducing strain of *Sporobolomyces salmonicolor*. Later, he changed the description of the genus (van der Walt, 1973). Vegetative cells reproduce by budding and by ballistoconidia. Pseudohyphae and true hyphae may be formed, but not binucleate, clamp-bearing hyphae. There are no basidiocarps. Teliospores, smooth, hyaline, round to oval, thick-walled, containing lipid globules, form non- or one-septate promycelia on which one to four budding, sessile sporidia are formed. There is no anaerobic fermentation of sugars; carotenoid pigments may occur and starch may be produced.

AESSOSPORON DENDROPHILUM van der Walt

This species is the sexually reproducing form of *Bullera dendrophila* van der Walt et Scott (1970).

Aessosporon dendrophilum has spherical or oval budding cells, 4.0 to 8.0 by 5.0 to 16 μm, round to oval, symmetrical ballistoconidia, and may form pseudohyphae and true hyphae. Teliospores are smooth, round, oval, apiculate or pear-shaped, 5.0 to 9.5 by 8.0 to 19 μm. Two strains of *Bullera dendrophila* were isolated from frass in subcortical galleries of Bupristidae larvae infesting a dead specimen of *Dichrostachys cinerea* in the Transvaal. Van der Walt (1973) derived two diploid strains of *Aessosporon dendrophilum* from the type strain of *Bullera dendrophila*.

AESSOSPORON SALMONICOLOR (Fischer et Brebeck) van der Walt

Phaff (1970b) described this species as *Sporobolomyces salmonicolor*. *Aessosporon salmonicolor* is a form that reproduces sexually by teliospores (van der Walt, 1970a). (Other forms may be designated *Sporidiobolus* species.)

AMBROSIOZYMA van der Walt

This ascosporogenous genus was described by van der Walt (1972) and discussed by von Arx (1972). Its species reproduce vegetatively both by budding and by fission; they form pseudohyphae, true septate hyphae and blastoconidia. The hyphal septa have single, central, dolipore-like bodies. Asci are usually terminal, intercalary or lateral. Each ascus has up to four ascospores, which are hat-shaped, spherical or hemispherical with equatorial or sub-equatorial ledges. Yeasts of this genus have been found chiefly associated with Ambrosia beetles.

AMBROSIOZYMA CICATRICOSA (Scott et van der Walt) van der Walt

Synonym
Pichia cicatricosa Scott et van der Walt (1971*b*)

Description
Ambrosiozyma cicatricosa forms septate hyphae with large septal pores. Budding is on a broad base. Budding cells are oval or elongate, 3.0 to 5.0 by 6.0 to 15.0 μm. Pseudohyphae and true hyphae are formed. Asci occur singly in nodal or terminal clusters at the distal ends of hyphae. Each ascus usually has two ascospores, which are large, hat-shaped with a prominent brim, and produced on V-8 agar and potato–glucose agar. Strains were isolated from tunnels of wood-eating Ambrosia beetles *Xyleborus torquatus* and *Platypus externedentatus* infesting *Cussonia umbellifera* and *Macaranga capensis* in northern Natal. Scott & van der Walt (1971*b*) examined five strains; also see van der Walt (1972).

AMBROSIOZYMA MONOSPORA (Saito) van der Walt

Kreger-van Rij (1970*b*) described this species as *Endomycopsis monospora*. With the rejection of the generic name, *Endomycopsis* (von Arx, 1972), van der Walt (1972) transferred the species to *Ambrosiozyma*.

AMBROSIOZYMA PHILENTOMA van der Walt, Scott et van der Klift

Ambrosiozyma philentoma has spherical, oval or elongate cells, 3.0 to 8.0 by 3.0 to 16.0 μm, and branching hyphae with single pore bodies in the septa. Abundant hyphae and pseudohyphae bear clusters of blastoconidia. Asci, found on V-8 agar, maize meal agar and malt agar, occur singly or in short chains at hyphal tips. Each ascus has one to four ascospores, which are hat-shaped with a brim. Van der Walt (1972) isolated five strains from tunnels of Ambrosia beetles *Xyleborus aemulus* infesting *Rapanea melanophoeos* in the Cape Province.

ARTHROASCUS von Arx

Von Arx (1972) created this ascosporogenous genus to accommodate the single species, *Endomycopsis javanensis*. Colonies are composed of a network of branched, septate hyphae, without blastoconidia. The asci, which develop from swollen hyphal cells and are in chains, disintegrate readily, each liberating one to four hat-shaped, hyaline ascospores.

ARTHROASCUS JAVANENSIS (Klöcker) von Arx

Kreger-van Rij (1970*b*) described this species as *Endomycopsis javanensis*. Rejecting the generic name *Endomycopsis*, von Arx (1972) transferred the species to *Arthroascus*, emphasizing the taxonomic importance of the arrangement of asci in chains and the hat-shaped ascospores.

BOTRYOASCUS von Arx

Von Arx (1972) created this ascosporogenous genus in order to accommodate the single species, *Saccharomycopsis synnaedendra*. Colonies are composed of branched, septate hyphae. Asci are borne laterally on the hyphae, often stalked, single or in short chains, mostly spherical or pear-shaped. Each ascus has two to eight hat-shaped ascospores. Pear-shaped conidia, with truncated bases, are borne at the tips of erect conidiogenous cells in short sympodulae.

BOTRYOASCUS SYNNAEDENDRUS (Scott et van der Walt) von Arx

Synonyms
Saccharomycopsis synnaedendra Scott et van der Walt (van der Walt & Scott, 1971*c*)

Pichia microspora Batra (1971)

Description

Botryoascus synnaedendrus forms pseudohyphae and branched, septate hyphae, 1.5 to 3.5 μm in diameter, with sparse, small blastoconidia. Asci, singly, in pairs or in short chains occur laterally on the hyphae, often on small projections, or terminally on small side branches. Also there may be intercalary asci. Most asci are round to oval, 5 to 7 μm in diameter, with four hat-shaped ascospores, 1.5 to 2.0 by 2.5 to 3.0 μm, including the prominent 'brim'. The strains were homothallic. Van der Walt & Scott (1971c) isolated them from fungi associated with wood-eating Scolytoidea infesting trees in Natal and Cape Province, and Batra (1971) from fungi associated with *Crossotarsus wollastoni* infesting *Dipterocarpus indicus* in Mysore.

BRETTANOMYCES ABSTINENS Yarrow et Ahearn

The cells of *Brettanomyces abstinens* are oval or elongate, 2.0 to 4.5 by 4.0 to 8.3 μm. Slide cultures give branched chains of elongated cells. One strain has been isolated from American ginger ale. Yarrow & Ahearn (1971) examined one strain.

BRETTANOMYCES NAARDENENSIS Kolfschoten et Yarrow

The cells of *Brettanomyces naardenensis* are oval or cylindrical, 1.5 to 3.0 by 3.0 to 22.0 μm. Slide cultures give branched chains of elongated cells. Kolfschoten & Yarrow (1970) isolated 12 strains from commercial lemonade or mineral water (pH 2.6 to 3.2).

BULLERA Derx

Stadelmann (1975) changed the description of this genus to include species forming asymmetrical ballistoconidia and hyphae. Consequently, the only character separating this genus from *Sporobolomyces* is the formation of visible pink carotenoid pigment.

BULLERA PIRICOLA Stadelmann

The cells are oval to oblong, 3.7 to 7.3 by 8.0 to 14.5 μm (mean 5.2 by 11.0 μm). Short pseudohyphae and some septate hyphae are formed. Ballistoconidia are either (i) ovoid to oblong with obtuse and tapering ends or (ii) kidney-shaped with a short protuberance, 3.7 to 3.8 by 8.7 to 17.0 μm, (mean 5.5 by 12.6 μm). Stadelmann (1975) isolated three strains from living pear leaves.

CANDIDA AMYLOLENTA van der Walt, Scott et van der Klift

Candida amylolenta has spherical or oval cells, 3.0 to 6.5 by 3.0 to 9.0 μm, which are often encapsulated. Slide cultures give pseudohyphae. Van der Walt, Scott & van der Klift (1972) isolated two strains in the Transvaal from the frass of beetles *Sinoxylon ruficorne* in *Dichrostachys cinerea* and *Enneadesmus forficulus* in *Dombeya rotundifolia*.

CANDIDA BOLETICOLA Nakase

Candida boleticola has spherical, oval or elongate cells, 2.5 to 5.0 by 2.5 to 14.0 μm, and forms pseudohyphae. Nakase (1971c) isolated eight strains, six from mushrooms.

CANDIDA BOMBI Montrocher

Candida bombi has spherical, oval or cylindrical cells, 1.5 to 6.0 μm by 3.5 to 11.0 μm, and some pseudohyphae. Montrocher (1967) isolated two strains from a bee (*Bombus* sp.) in the Savoie, France.

CANDIDA BRASSICAE Amano, Goto et Kagami

The cells are oval to cylindrical, 2.5 to 5 by 3 to 17.5 μm, and form pseudohyphae. Amano, Goto & Kagami (1975) isolated one strain from cabbage trash.

CANDIDA BUINENSIS Soneda et Uchida

Oval cells measure 2.0 to 3.6 by 2.5 to 4.0 μm; pseudohyphae bear spindle-shaped blastoconidia. Soneda & Uchida (1971) isolated this yeast from gelatinous materials of a tree fern collected in the Solomon Islands.

CANDIDA BUTYRI Nakase

Candida butyri has oval or elongate cells, 1.5 to 7.5 by 2.5 to 17.5 μm, and forms abundant pseudohyphae with round or short-oval blastoconidia in chains or in verticils. Nakase (1971c) isolated four strains from butter and one from an onion.

CANDIDA CHILENSIS Grinbergs et Yarrow

Candida chilensis has spherical or short-oval cells, 3.5 to 6.0 by 4.0 to 7.0 μm. Slide cultures give mainly pseudohyphae but also some septate hyphae. There are dense verticils of blastospores. Grinbergs & Yarrow (1970a) isolated two strains from rotted wood.

CANDIDA CHIROPTERORUM Grose et Marinkelle

Candida chiropterorum has round and oval cells, 2.1 to 5.6 by 4.2 to 12.1 μm, and forms pseudohyphae, true hyphae and chlamydospores. Grose & Marinkelle (1968) isolated 23 strains from Colombian bats *Carollia perspicillata*, *Desmodus rotundus*, *Mormoops megallophylla* and *Natalus tumidirostris*.

CANDIDA CITREA Nakase

Candida citrea has oval cells, 2.0 to 6.5 by 2.5 to 15.0 μm. On potato–glucose agar, pseudohyphae are only rudimentary with blastoconidia in verticils. Nakase (1971a) isolated two strains from lemon and one from banana. S. A. Meyer & D. Yarrow (unpublished) have found a strain, isolated from apple juice, which produces hat-shaped ascospores.

CANDIDA DENDRONEMA van der Walt, van der Klift et Scott

Candida dendronema has polymorphic cells which are spherical, oval, cylindrical or three-cornered, 1.5 to 5.0 by 2.0 to 12.0 μm, and forms abundant pseudohyphae. Van der Walt, Scott & van der Klift (1971a) isolated five strains from beetle frass, cerambycid larvae infesting *Diospyros inhacaensis* and *Acacia karroo* in Natal.

CANDIDA EDAX van der Walt et Nel

Candida edax has spherical, oval or cylindrical cells, 2.0 to 5.5 by 3.0 to 17.5 μm, and forms pseudohyphae and true hyphae. Van der Walt & Nel (1968) isolated three strains from insect frass in *Sclerocarya caffra* in the Transvaal.

CANDIDA ENTOMAEA van der Walt, Scott et van der Klift

Candida entomaea has oval, cylindrical or elongate cells, 1.5 to 4.0 by 2.5 to 13.0 μm, and forms abundant branched pseudohyphae with clusters of round blastoconidia on maize meal agar. Van der Walt, Scott & van der Klift (1972) isolated one strain from *Pinus radiata* in south Cape Province.

CANDIDA ENTOMOPHILA Scott, van der Walt et van der Klift

Candida entomophila has cells which are spherical, oval, cylindrical, three-cornered or irregular, 1.5 to 6.5 by 2.5 to 16.0 μm. On maize meal agar there are abundant branched pseudohyphae and true hyphae. The blastoconidia may be three-cornered. Strains were isolated from material associated with beetles. Van der Walt, Scott & van der Klift (1971a) isolated three strains from the tunnels of *Crossotarsus externedentatus* in *Ficus sycomorus* and in *Tabernaemontana ventricosa* in South Africa.

CANDIDA ERGATENSIS Santa María

Candida ergatensis has oval cells, 2.6 to 8.5 by 4.5 to 13.0 μm, and forms pseudohyphae and true hyphae. One strain was isolated from beetle larvae.

Santa María (1971) isolated one strain from larvae of cerambycid beetles *Ergates faber* in pine stumps from Madrid.

CANDIDA FLAVIFICANS Nakase

The cells are round or oval, 2 to 4.5 by 2.5 to 12 μm. The abundant hyphae are often septate and there are some pseudohyphae. Round to oval blastoconidia occur singly, in pairs or sometimes in chains of three or four. Nakese (1975) isolated one strain from washings of ion-exchange resin in a guanosine monophosphate manufacturing plant.

CANDIDA FLUVIOTILIS Hedrick

The oval cells are 2 to 4 by 3 to 6 μm; pseudo and not true hyphae are formed. Hedrick (1976) isolated five strains from an industrially polluted river in Indiana.

CANDIDA FRAGICOLA Nakase

Candida fragicola has oval or elongate cells, 2.0 to 3.5 by 3.5 to 30 μm, and forms abundant pseudohyphae. Nakase (1971b) isolated a strain from strawberry in Tokyo.

CANDIDA GRAMINIS Rodrigues de Miranda et Diem

The cells are oval or elongate, 1.5 to 4 by 3 to 12 μm. Some hyphae are formed and the pseudohyphae carry oval or elongate blastoconidia. Rodrigues de Miranda & Diem (1974) isolated a strain from barley leaves.

CANDIDA HOMILENTOMA van der Walt et Nakase

The cells are round, oval or cylindrical, 1.5 to 5.0 by 1.5 to 19 μm. Abundant pseudohyphae form triangular blastoconidia, usually in chains or clusters. Van der Walt & Nakase (1973) described two strains from tunnels of (i) *Xylion adustus* in *Ficus sycomorus* in Natal and (ii) *Sinoxylon ruficorne* in *Combretum apiculatum* in the Transvaal.

CANDIDA HYDROCARBOFUMARICA Yamada, Furukawa et Nakahara ex Ramírez

The round or oval cells measure 2 to 5 by 2 to 7 μm. This yeast forms abundant pseudohyphae and true hyphae with round to oval blastoconidia, in chains, in verticils or in clusters. Yamada, Furukawa & Nakahara (1970) isolated a strain of this yeast from soil. The name was validated by Ramírez (1974) who gave it a Latin description.

CANDIDA HYLOPHILA van der Walt, van der Klift et Scott

Candida hylophila has cells that are spherical, oval, cylindrical, three-cornered or irregular, 1.0 to 3.0 by 1.5 to 9.0 μm. On maize meal agar, the much branched pseudohyphae carry three-cornered blastoconidia. Van der Walt, Scott & van der Klift (1971a) isolated one strain from tunnels of *Xyleborus aemulus* in *Rapanea melanophloeos* in Cape Province.

CANDIDA IBERICA Ramírez et González

Candida iberica has oval cells, 2.6 to 6.0 by 3.0 to 7.0 μm, and forms abundant pseudohyphae on maize meal agar. Ramírez & González (1972) isolated five strains from sausages.

CANDIDA INCOMMUNIS Ohara, Nonomura et Yamazaki

Candida incommunis has cylindrical cells, 2.0 to 3.0 by 10.0 to 17.0 μm, and forms pseudohyphae on potato agar. Ohara, Nonomura & Yamazaki (1965) isolated this yeast from grape must.

CANDIDA INOSITOPHILA Nakase

The yeast has oval or elongate cells 3 to 5 by 7 to 25 μm, and both pseudo- and true hyphae with oval blastoconidia singly or in clusters. Nakase (1975) isolated one strain from washings of ion-exchange resin in a guanosine monophosphate factory.

CANDIDA INSECTAMANS Scott, van der Walt et van der Klift

Candida insectamans has spherical, oval or cylindrical cells, 1.5 to 6.5 by 2.5 to 12.0 μm, and forms branched pseudohyphae on maize meal agar. Van der Walt, Scott & van der Klift (1972) isolated one strain from the frass of buprestid larvae in dead *Acacia nilotica* var. *kraussiana* in the Transvaal.

CANDIDA INSECTORUM Scott, van der Walt et van der Klift

Candida insectorum has spherical, oval, cylindrical or elongate cells, 1.5 to 5.0 by 2.5 to 19.0 μm, and forms branched pseudohyphae on maize meal agar. Van der Walt, Scott & van der Klift (1972) isolated three strains, (i) two from the tunnels of *Crossotarsus externedentatus* in *Euclea natalensis* and (ii) one from the frass of *Phoracantha recurva* in *Eucalyptus maculata* in Natal.

CANDIDA ISHIWADAE Sugiyama et Goto

Candida ishiwadae has spherical or oval cells, 3.0 to 6.0 by 3.0 to 7.0 μm, and forms pseudohyphae and true hyphae. Sugiyama & Goto (1969) isolated two strains from soil.

CANDIDA KRISSII Goto, Yamasato et Iizuka

The yeast has oval cells, 2.5 to 5.0 by 5 to 12μm and forms pseudohyphae. Goto, Yamasato & Iizuka (1974) isolated one strain from sea-water.

CANDIDA MALTOSA Komagata, Nakase et Katsuya

This yeast has round to oval cells, 2.5 to 7 by 3 to 7.5 μm, and pseudohyphae. Komagata, Nakase & Katsuya (1964) isolated one strain from adhesives of neutralizing tanks used in manufacturing monosodium glutamate. Van Uden & Buckley (1970) considered this species synonymous with *Candida sake*, but Meyer, Anderson, Brown, Smith, Yarrow, Mitchell & Ahearn (1975) restored *Candida maltosa* because (i) it differed in DNA base composition (%GC) and (ii) there was a low percentage binding of DNA from this species with that from *Candida sake*.

CANDIDA MILLERI Yarrow

This yeast has round to oval cells, 4.0 to 6.0 by 5.0 to 6.0 μm, and does not form pseudohyphae. Yarrow (1978) described two strains isolated from San Francisco sour dough by Sugihara, Kline & Miller (1971) and one from *alpechín* by Gómez & Rico (1956).

CANDIDA NAEODENDRA van der Walt, Johannsen et Nakase

This yeast has oval to cylindrical cells, 1.5 to 5.0 by 2.0 to 11 μm, and abundant pseudohyphae with chains or clusters of triangular blastoconidia. Van der Walt, Johannsen & Nakase (1973) described two strains, isolated from frass of buprestid larvae in *Acacia nilotica* var. *kraussiana* in the Transvaal.

CANDIDA OLEOPHILA Montrocher

Although some cells are oval, most are elongate, 1.5 to 4.2 by 5 to 17.5 μm. There are abundant pseudohyphae, which are often branched and bear elongate blastoconidia. Montrocher (1967) studied one strain which had been isolated from olives by O. Verona.

CANDIDA PODZOLICA Babjeva et Reshetova

This yeast has polymorphic cells, almost round, oval, apiculate and elongate, which characteristically bear protrusions that may be free of cytoplasm. The pseudohyphae bear uniformly arranged blastoconidia of varied shape. Babjeva & Reshetova (1975) isolated several hundred strains from soil.

CANDIDA PSEUDOINTERMEDIA Nakase, Komagata et Fukazawa

The cells of this yeast are round to oval, 2.5 to 7.0 by 3.0 to 7.5 μm, and form abundant pseudohyphae with verticils of oval blastoconidia. Nakase, Komagata & Fukazawa (1976) isolated three strains from Kamaboko, a Japanese fish paste.

CANDIDA QUERCUUM Nakase

Candida quercuum has oval, elongate or cylindrical cells, 1.5 to 4.0 by 2.0 to 7.0 μm. Pseudohyphae are abundant on potato–glucose agar. Nakase (1971c) isolated one strain from an exudate of an oak.

CANDIDA RUGOPELLICULOSA Nakase

Candida rugopelliculosa has oval or elongate cells, 2.0 to 5.0 by 2.5 to 10.0 μm. Pseudohyphae are abundant on potato–glucose agar. Nakase (1971a) isolated one strain from a soya protein factory.

CANDIDA SAVONICA Sonck

The cells are round, oval (3.0 to 4.5 by 5.0 to 6.5 μm) or cylindrical, (3.5 by 12 μm). Blastoconidia occur singly, in pairs, short chains or clusters along pseudohyphae. In Finland, Sonck (1974) isolated a strain of this yeast from the cut surface of a birch log on which it formed white spots.

CANDIDA SILVANORUM van der Walt, van der Klift et Scott

Candida silvanorum has spherical, oval, cylindrical or three-cornered cells, 1.5 to 4.5 by 2.5 to 14.0 μm. On maize meal agar, pseudohyphae are abundant and blastoconidia may be three-cornered. Van der Walt, Scott & van der Klift (1971a) isolated three strains from material associated with various beetles infesting trees in South Africa.

CANDIDA SILVICULTRIX van der Walt, Scott et van der Klift

Candida silvicultrix has oval cells, 2.5 to 5.0 by 3.0 to 6.5 μm, and forms abundant pseudohyphae on maize meal agar. Van der Walt, Scott & van der Klift (1972) isolated two strains from material associated with beetles infesting trees in South Africa.

CANDIDA SORBOXYLOSA Nakase

Candida sorboxylosa has oval or elongate cells, 2.0 to 6.0 by 2.5 to 20.0 μm, and on potato–glucose agar forms pseudohyphae. Nakase (1971a) isolated one strain from pineapple, one from banana and three from artificial food for silkworms.

CANDIDA STEATOLYTICA Yarrow

Candida steatolytica has oval or cylindrical cells, 2.0 to 4.0 by 4.5 to 10.0 μm, and on potato agar forms abundant hyphae and pseudohyphae. Yarrow (1969a) examined one strain from a mastitic bovine udder, another from grape must and a third from a human toe.

CANDIDA SUECICA Rodrigues de Miranda et Norkrans

Candida suecica has oval or cylindrical cells, 2.5 to 7.0 by 3.0 to 18.0 μm. In slide cultures, it forms pseudohyphae which are sometimes swollen at one end. Rodrigues de Miranda & Norkrans (1968) isolated one strain from a Swedish estuary.

CANDIDA TEPAE Grinbergs

Candida tepae has oval cells, 2.0 to 4.0 by 3.5 to 11.0 μm, and forms abundant pseudohyphae on potato extract agar. Grinbergs (1967) isolated one strain from *Laurelia philippiana*.

CANDIDA TEREBRA Sugiyama et Goto

Candida terebra has spherical or oval cells, 2.0 to 5.0 by 3.0 to 7.0 μm, and in slide culture forms both hyphae and pseudophyphae. Sugiyama & Goto (1969) isolated two strains from soil.

CANDIDA TSUKUBAENSIS Onishi

Candida tsukubaensis has long oval or cylindrical cells, 3.0 to 6.0 by 6.0 to 12.0 μm, and forms pseudohyphae on maize meal agar. Onishi (1972) isolated one strain from a flower.

CANDIDA VALDIVIANA Grinbergs et Yarrow

Candida valdiviana has oval cells, 3.5 to 6.0 by 2.5 to 3.5 μm, and forms pseudohyphae and septate hyphae in slide cultures. Grinbergs & Yarrow (1970a) isolated one strain from rotting wood.

CANDIDA VINARIA Ohara, Nonomura et Yunome ex Smith

The cells are oval, 2.0 to 3.0 by 4.0 to 5.0 μm, but also elongate, up to 12 μm. Pseudohyphae are sparse. Smith (1973) validated the name, given by Ohara, Nonomura & Yunome (1960), who isolated one strain from grape juice.

CRYPTOCOCCUS BHUTANENSIS Goto et Sugiyama

Cryptococcus bhutanensis has spherical cells, 4.0 to 9.0 μm. Goto & Sugiyama (1970) isolated one strain from soil.

CRYPTOCOCCUS CEREANUS Phaff, Miller, Miranda, Heed et Starmer

The cells are oval to elongate and sometimes curved, 2.1 to 3.9 by 4.0 to 10.5 μm, occasionally up to 13.5 μm long. Budding is on broad shoulders. Phaff, Miller, Miranda, Heed & Starmer (1974) examined 11 strains from 'rot pockets' in the cacti *Cereus schotti*, *Cereus giganteus* and *Cereus thurberi* in Arizona and Mexico.

CRYPTOCOCCUS HEVEANENSIS (Groenewege) Baptist et Kurtzman

Synonyms
Torula heveanensis Groenewege (1921)
Candida heveanensis (Groenewege) Diddens et Lodder (1942)
Torulopsis heveanensis (Groenewege) Mager et Aschner (1947)

Explanation
Phaff & Fell (1970) described the strains of this species as *Cryptococcus laurentii* var. *magnus*, but Baptist & Kurtzman (1976) have re-established these yeasts as a distinct species, chiefly because of differences in DNA base composition from other strains of *Cryptococcus laurentii* var. *laurentii*.

CRYPTOCOCCUS HIMALAYENSIS Goto et Sugiyama

Cryptococcus himalayensis has spherical cells, 3.5 to 9.0 μm. Goto & Sugiyama (1970) isolated two strains from soil.

CRYPTOCOCCUS MAGNUS (Lodder et Kreger-van Rij) Baptist et Kurtzman

Phaff & Fell (1970) described this species as *Cryptococcus laurentii* var. *magnus*. Baptist & Kurtzman (1976) gave this former variety specific status because they found differences (i) in the electrophoretic migrations of six enzymes and (ii) in the buoyant densities of the DNA.

DEBARYOMYCES MELISSOPHILUS (van der Walt et van der Klift) Kurtzman et Kreger-van Rij

Synonym
Pichia melissophila van der Walt et van der Klift (1972)

Description
Debaryomyces melissophilus has spherical or oval cells, 1.5 to 5.0 by 2.5 to 5.5 μm, and forms neither hyphae nor pseudohyphae on maize meal agar. The yeast is homothallic, conjugation occurring both by fusion between mother-cell and

bud and between separate cells. Asci occur on V-8 agar or malt agar; each ascus has one to four spherical or oval ascospores. Kurtzman & Kreger-van Rij (1976) transferred this species to the genus *Debaryomyces* on account of the internal wall structure of the ascospores as revealed by electron microscopy. Van der Walt & van der Klift (1972) isolated seven strains from the alimentary canals of African honey bees *Apis mellifica* var. *adonsonii*.

DEBARYOMYCES NEPALENSIS Goto et Sugiyama

Debaryomyces nepalensis has spherical cells, 2.0 to 5.0 μm. Each ascus has one spherical ascospore with a warty wall. Goto & Sugiyama (1968) isolated two strains from soil.

DEBARYOMYCES POLYMORPHUS (Klöcker) Kurtzman et Smiley

Synonyms
Pichia polymorpha Klöcker (1913)
Saccharomyces polymorpha (Klöcker) Novák et Zsolt (1961)
Debaryomyces cantarellii Capriotti (1961)

Explanation
Kreger-van Rij (1970a, c) described these yeasts as *Debaryomyces cantarellii* and *Pichia polymorpha*. Kurtzman & Smiley (1974) proposed that the two species should be combined in *Debaryomyces* for two reasons: (i) examination of the ascospores of *Pichia polymorpha* by scanning electron microscopy showed wart-like protuberances of the wall that are typical of *Debaryomyces*; (ii) the two species are indistinguishable in their other features. See also Kurtzman, Smiley & Baker (1975).

DEBARYOMYCES YARROWII Santa María et García Aser

Debaryomyces yarrowii has oval cells, 3.9 to 7.8 by 5.5 to 15.6 μm, and abundant primitive pseudohyphae. Each ascus has one to three round, wrinkled ascospores. Santa María & García Aser (1971) isolated one strain from the frass of insects.

FILOBASIDIUM Olive

Olive (1968) described this genus. Hyphae, with blastoconidia and clamp connections, form non-septate basidia which bear terminally sessile thin-walled basidiospores.

FILOBASIDIUM CAPSULIGENUM (Fell, Statzell, Hunter et Phaff) Rodrigues de Miranda

Fell & Phaff (1970) described this yeast as *Leucosporidium capsuligenum*. Rodrigues de Miranda (1972) reclassified the species in the genus *Filobasidium*, since he found *Tilletia*-like basidia growing out from dikaryotic hyphae with clamp connections.

FILOBASIDIUM FLORIFORME Olive

Filobasidium floriforme has slender non-septate basidia that bear six to eight sessile basidiospores terminally. Yeast-like colonies of budding cells develop from single basidiospore isolations. Olive (1968) isolated the organism from dead grass.

GEOTRICHUM CAPITATUM (Diddens et Lodder) von Arx

Do Carmo-Sousa (1970) described this species as *Trichosporon capitatum*. Von Arx, Rodrigues de Miranda, Smith & Yarrow (1977) indicated, without explanation, that they reclassified the species in the genus *Geotrichum*.

GEOTRICHUM FERMENTANS (Diddens et Lodder) von Arx

Do Carmo-Sousa (1970) described this species as *Trichosporon fermentans*. Von Arx, Rodrigues de Miranda, Smith & Yarrow (1977) indicated, without explanation, that they reclassified the species in the genus *Geotrichum*.

GEOTRICHUM PENICILLATUM (do Carmo-Sousa) von Arx

Do Carmo-Sousa (1970) described this species as *Trichosporon penicillatum*. Von Arx, Rodrigues de Miranda, Smith & Yarrow (1977) indicated, without explanation, that they reclassified the species in the genus *Geotrichum*.

GUILLIERMONDELLA Nadson et Krassilnikov

Nadson & Krassilnikov (1928) described this genus, which is composed of one species. There are oval budding cells and branched, septate hyphae. Asci are lateral, terminal or intercalary on the hyphae, with four ascospores per ascus.

GUILLIERMONDELLA SELENOSPORA Nadson et Krassilnikov

Kreger-van Rij (1970b) described this species as *Endomycopsis selenospora*. With the rejection of the name *Endomycopsis*, the species has reverted to its original genus (von Arx, 1972).

HANSENIASPORA GUILLIERMONDII Pijper

Synonym
Hanseniaspora melligeri Lodder (1932)

Explanation
Following Lodder & Kreger-van Rij (1952), Phaff (1970a) included this species in *Hanseniaspora valbyensis*. Pijper (1928a,b) described *Hanseniaspora guilliermondii* as producing four ascospores per ascus and fermenting D-glucose only weakly and, hence, distinguished this species from *Hanseniaspora valbyensis*. However, Lodder & Kreger-van Rij (1952) found these distinctions either unimportant or non-existent. Meyer, Brown & Smith (1977) have now restored *Hanseniaspora guilliermondii* because (i) DNA from this species does not reassociate markedly with that of *Hanseniaspora valbyensis*, and (ii) strains of the former, but not the latter, grow aerobically on 2-keto-gluconate and at 37°C. Smith, Simione & Meyer (1977) have discussed *Kloeckera apis*, the non-sexual ('imperfect') state of this yeast.

HANSENIASPORA OCCIDENTALIS Smith

Smith (1974) established this species, having found ascospores in the type strain of *Kloeckera occidentalis*. Phaff (1970c) had described *Kloeckera javanica* and included *Kloeckera occidentalis* as a synonym. Now *Kloeckera javanica* is regarded as the anascosporogenous state of *Hanseniaspora occidentalis* which has apiculate, oval asci, 3 to 7 by 6 to 11 μm, each of which contains one or two ascospores, 2 to 3.5 by 3 to 5 μm.

HANSENIASPORA VINEAE van der Walt et Tscheuschner

This species is similar to *Hanseniaspora uvarum* (Niehaus) Shehata, Mrak et Phaff (1955) and *Hanseniaspora osmophila* (Niehaus) Phaff, Miller et Shifrine (1956). Van der Walt & Tscheuschner (1957) distinguished *Hanseniaspora vineae* from *Hanseniaspora valbyensis* and *Hanseniaspora uvarum* because the ascospores of *Hanseniaspora vineae* were warty, some cells were longer, there were rudimentary pseudohyphae and the yeast could not use maltose for aerobic growth. Phaff (1970a) included *Hanseniaspora vineae* as a synonym of *Hanseniaspora osmophila* because, contrary to the findings of van der Walt & Tscheuschner (1957), (i) the ascospores were not very warty and (ii) maltose was used as a substrate for aerobic growth. However, Meyer, Smith & Simione (1978) considered *Hanseniaspora vineae* a distinct species, chiefly because the percentage of relative binding of DNA from strains of each species with that from the other species was only between 38 and 60.

HANSENULA BECKII Wickerham

Kreger-van Rij (1970b) described this yeast as *Endomycopsis bispora*. With the abolition of the generic name *Endomycopsis* (see von Arx, 1972), the species is accepted as *Hansenula beckii* (see also Wickerham, 1970, p. 254).

HANSENULA DRYADOIDES Scott et van der Walt

Hansenula dryadoides has spherical or oval cells, 3.0 to 5.0 by 3.0 to 8.0 μm, and forms branched, rudimentary pseudohyphae. On Difco Bacto Yeast Morphology agar and on V-8 agar, vegetative cells become asci without copulation and each contains one or two hat-shaped ascospores with prominent brims. Scott & van der Walt (1971a) isolated four strains from material associated with Ambrosia beetles *Platypus externedentatus* in two species of *Ficus* in Natal.

HANSENULA LYNFERDII van der Walt et Johannsen

This yeast has round or oval cells, 2.5 to 5.5 by 3.0 to 7.0 μm, and simple pseudohyphae of cylindrical cells which branch and form blastoconidia. Cells become asci on Difco YM agar, 2% malt extract agar and on V-8 agar, without first conjugating. The asci each contain one to four ascospores which are hat shaped with prominent brims. Van der Walt & Johannsen (1975a) isolated one strain from soil.

HANSENULA MUSCICOLA (Nakase et Komagata) Yarrow

Synonym
Endomycopsis muscicola Nakase et Komagata (1966)

Description
Hansenula muscicola has round, oval or elongate cells, 3.0 to 8.0 by 3.5 to 20.0 μm. In slide culture, there are abundant branched hyphae and some pseudohyphae. Each ascus contains one to four ascospores. Nakase & Komagata (1966) isolated three strains from moss. Yarrow (1972) discussed the characteristics and classification of this species.

HANSENULA OFUNAENSIS Makiguchi et Asai

The oval cells are 1.5 to 4.0 by 2.5 to 6.5 μm, the pseudohyphae form oval blastoconidia, and each ascus contains one to four hat-shaped ascospores (Asai, Makiguchi, Shimada & Kurimura, 1976).

HANSENULA PHILODENDRA van der Walt et Scott

Hansenula philodendra has round or oval cells, 1.5 to 5.0 by 1.5 to 6.5 μm, and forms neither hyphae nor pseudohyphae on Difco Bacto Yeast Morphology agar or on maize meal agar. Ascospores are formed readily on V-8 agar, malt extract agar and on YM agar. Each ascus contains one to four hat-shaped ascospores with short brims. Scott & van der Walt (1970b) isolated two strains from the frass of the beetle *Xylion adustus* in dead *Ficus sycomorus* in Natal.

HANSENULA SYDOWIORUM Scott et van der Walt

Hansenula sydowiorum has almost spherical or oval cells, 2.0 to 5.5 by 2.8 to 7.0 μm. There is no filamentous growth on Difco Bacto Yeast Morphology agar or on maize meal agar, although branching pseudohyphae occur along the margin of old streak cultures on malt agar. Asci, each containing one to four hat-shaped ascospores, are formed slowly and sparsely on malt agar in three- to four-week-old cultures. Scott & van der Walt (1970a) isolated three strains from the frass of the beetle *Sinoxylon ruficorne* in dead *Combretum apiculatum* in the Transvaal.

HORMOASCUS von Arx

Von Arx (1972) invented *Hormoascus* for the single species, *Endomycopsis platypodis*. He defined the genus as having branched, regularly septate hyphae, the septa thickened with plugged septal pores; the hyphae carry chains of blastoconidia. The asci are borne apically or laterally on ascophores, often in short chains, oval or club-shaped, each ascus containing two to four hat-shaped ascospores. Nitrate is utilized. The yeasts are associated with bark beetles.

HORMOASCUS PLATYPODIS (Baker et Kreger-van Rij) von Arx

Kreger-van Rij (1970b) described this yeast as *Endomycopsis platypodis*. With the abolition of the genus *Endomycopsis*, von Arx (1972) renamed the species *Hormoascus platypodis*.

HYPHOPICHIA von Arx et van der Walt

Von Arx & van der Walt (1976) created this genus in order to accommodate a single species, formerly *Endomycopsis burtonii*. The following characteristics feature in the definition of *Hyphopichia*. There are broad, septate hyphae, which include intercalary or terminal conidiogenous cells; on these the conidia are produced by budding from denticles, often in chains. The strains are heterothallic and the asci, formed from conjugating cells, each contain one to four hat-shaped ascospores.

HYPHOPICHIA BURTONII (Boidin, Pigal, Lehodey, Vey et Abadie) von Arx et van der Walt

Kreger-van Rij (1970b) described this yeast as *Endomycopsis burtonii*. With the abolition of the genus *Endomycopsis*, von Arx & van der Walt (1976) have transferred the species to *Hyphopichia*.

KLUYVEROMYCES BLATTAE Henninger et Windisch

The cells are round or oval, 5.5 to 9 by 6.5 to 11 μm, and there are no filaments. Each ascus contains up to nine round ascospores. Henninger & Windisch (1976b) isolated this yeast from the intestine of the cockroach *Blatta orientalis*.

KLUYVEROMYCES THERMOTOLERANS (Philippov) Yarrow

Synonyms

Zygosaccharomyces thermotolerans [termotolerans] Philippov (1932)
Saccharomyces veronae Lodder et Kreger-van Rij (1952)
Kluyveromyces veronae (Lodder et Kreger-van Rij) van der Walt (1970b)

Explanation

Van der Walt (1970b) described this yeast as *Kluyveromyces veronae*. However, Yarrow (1972) found it synonymous with *Zygosaccharomyces thermotolerans*, so he accepted the earlier specific name.

KLUYVEROMYCES WALTII Kodama

This yeast has oval cells, 2.5 to 6.0 by 6.0 to 10.0 μm and simple pseudohyphae. Conjugation of independent cells precedes ascus formation; there are one to four round ascospores per ascus, from which they are readily released. Kodama & Kyono (1974b) isolated one strain from an exudate of *Ilex integra* in Miyazaki.

LIPOMYCES ANOMALUS Babjeva et Gorin

Growth is slow. The round or oval cells (4.5 to 7.0 by 5.3 to 9.2 μm), which bud on a broad base, are formed in chains (simple pseudohyphae). The yeast is homothallic; asci are usually formed by conjugation of two buds on the same or different cells. Each ascus contains 4 to 30 or more round or oval ascospores. Babjeva & Gorin (1975) isolated two strains of this species from podzolic soil under a pine forest in Komi ASSR.

LIPOMYCES TETRASPORUS (Krassilnikov, Babjeva et Meavahd) Nieuwdorp, Bos et Slooff

Synonyms

Zygolipomyces tetrasporus nom. nud. Krassilnikov, Babjeva et Meavahd (1967)
Zygolipomyces lactosus nom. nud. Krassilnikov, Babjeva et Meavahd (1967)

Description

The cells are round or oval, 4 to 7 by 4 to 8.5 μm and there are no pseudohyphae. Asci are formed by conjugating buds or, more often, by young cells, each conjugating with a bud of another cell by means of a tube. Cells may form asci without previous conjugation. Each ascus contains one to four brown pigmented, ridged ascospores. Nieuwdorp, Bos & Slooff (1974) studied 12 strains isolated from soil.

METSCHNIKOWIA LUNATA Golubev

The cells are lunate, 2 to 4 by 5 to 8 μm, or rarely oval, 2.5 to 5 by 3.5 to 8 μm. Rudimentary pseudohyphae are formed. Asci are sphaeropedunculate, the spherical part 7 to 10 by 7 to 12 μm, the cylindrical peduncles 2 to 4 by 13 to 40 μm. There are one or two needle-shaped ascospores per ascus. Golubev (1977) created this species, having obtained ascospores in a yeast strain previously designated *Selenotila intestinalis* (see Barnett & Pankhurst, 1974) which had been isolated from flowers of *Vicia cracca* in the Spasski district, Primorski Territory, USSR.

NADSONIA COMMUTATA Golubev

The apically budding cells are oval, 4.2 to 7.7 by 5.6 to 11.2 μm. Budding is bipolar on a broad base. Neither hyphae nor pseudohyphae are formed. Asci are usually formed from mother-cells, sometimes from daughter-cells or from combined mother- and daughter-cells. Each ascus contains a single ascospore or sometimes two. Golubev (1973) isolated this yeast from Antarctic soil.

PACHYTICHOSPORA van der Walt

Van der Walt (1978) created this genus to accommodate the species *Saccharomyces transvaalensis*. Budding vegetative cells are diploid. Haploid cells of opposite mating type agglutinate. Each ascus contains one or two ascospores, which are round or oval, with stout, eccentrically thickened walls. The yeasts ferment sugars anaerobically, but do not utilize nitrate.

PACHYTICHOSPORA TRANSVAALENSIS (van der Walt) van der Walt

Van der Walt (1970c) described this yeast as *Saccharomyces transvaalensis*. Van der Walt (1978) has now transferred it to the genus *Pachytichospora* because electron micrographs show that the ascospores have walls which are thickened eccentrically.

PHAFFIA Miller, Yoneyama et Soneda

Vegetative cells are oval. Carotenoid pigments, mainly of astacanthin and some β-carotene, make the yeast orange to red. There are pseudohyphae but no true hyphae. Chlamydospores are formed. Sugars are fermented anaerobically and starch-like compounds are produced. The genus was described by Miller, Yoneyama & Soneda (1976).

PHAFFIA RHODOZYMA Miller, Yoneyama et Soneda

The cells are oval, 3.8 to 7.5 by 5.5 to 10.5 μm. Budding often recurs at a previous bud site. Round chlamydospores may be formed. There are pseudohyphae but no true hyphae. No sexual reproduction was observed. Miller, Yoneyama & Soneda (1976) described 10 strains of this yeast, which was isolated from exudates of deciduous trees in Japan and North America.

Previously, Phaff, Miller, Yoneyama & Soneda (1972) had described this species invalidly, as *Rhodozyma montanae*.

PICHIA ABADIEAE Jacob

The cells are oval, 2 by 3.5 to 2.5 by 6 μm. Pseudohyphae form a few oval blastoconidia. Oval asci each contain one to four round ascospores, 2 to 3.2 μm. Jacob (1969) isolated *Pichia abadieae* from tanning liquors.

PICHIA AMBROSIAE van der Walt et Scott

Synonym
Ambrosiozyma ambrosiae van der Walt (1972)

Description
Pichia ambrosiae has round or oval cells, 2.0 to 6.0 by 4.0 by 12.0 μm, and on maize meal agar forms both pseudohyphae and true hyphae which have dolipore-like bodies in the septa. Copulation between vegetative cells usually

produces simple or terminally branched hyphae with one or several successive asci at the apex or apices. Each ascus usually has four hat-shaped ascospores. Van der Walt & Scott (1971a) isolated the species from material associated with Ambrosia beetles infesting trees in Natal.

PICHIA BESSEYI Kurtzman et Wickerham

Pichia besseyi has spherical or oval cells, 4.3 by 4.9 to 7.0 by 8.2 μm, and some pseudohyphae on maize meal agar after two to three weeks at 25°C. Ascosporulation occurs after one week at 25°C on Difco Bacto YM agar, malt extract agar and on Difco Bacto Yeast Morphology agar. Most asci contain four hemispherical ascospores. Kurtzman & Wickerham (1972) isolated one strain from marsh water in Quebec.

PICHIA CASTILLAE Santa María et García Aser

Pichia castillae has cylindrical or oval cells, 1.5 to 4.0 by 2.5 to 8.0 μm, but no pseudohyphae or true hyphae. The ascospores are hat-shaped. Santa María & García Aser (1970) isolated a strain from insect frass in *Gymnocladus canadiensis*.

PICHIA HEIMII Pignal

Pichia heimii has round or oval cells, 4.5 to 5.0 by 5.0 to 7.0 μm, and forms pseudo- but not true hyphae. Each ascus has one to four hat-shaped ascospores. Pignal (1970) isolated one strain from decaying insect-invaded wood from equatorial Africa.

PICHIA HUMBOLDTII Rodrigues de Miranda et Török

This species is the ascosporogenous state of *Candida ingens*, which was described by van Uden & Buckley (1970). Rodrigues de Miranda & Török (1976) treated the type strain of *Candida ingens* with the mutagen N-methyl-N'-nitro-N-nitrosoguanidine and obtained a stable, homothallic, ascosporogenous yeast. Each ascus contains one to four large, round to oval ascospores with smooth walls. The ascospores are liberated easily from the ascus; they are formed on acetate agar.

PICHIA LINDNERII Henninger et Windisch

The cells are round or oval, 1.0 to 2.7 by 2.0 to 5.0 μm; there are no pseudo- or true hyphae. The yeast is homothallic and each ascus contains four cap-shaped ascospores. Henninger & Windisch (1975a) isolated *Pichia lindnerii* from forest soil near Berlin.

PICHIA METHANOLICA Makiguchi

The cells are round or oval, 2 to 7 by 2.5 to 10 μm, and form neither septate hyphae nor pseudohyphae. Asci are formed after conjugation between mother-cell and bud; each ascus contains one to four (usually four) round or hemispherical ascospores, which are liberated easily. Kato, Kurimura, Makiguchi & Asai (1974) isolated five strains of *Pichia methanolica* from soil in Japan.

PICHIA MUCOSA Wickerham et Kurtzman

Pichia mucosa has round or oval cells, 2.5 by 4.5 to 5.1 by 8.0 μm, and forms pseudohyphae, but not true hyphae, on Difco Bacto Yeast Morphology agar. After one week on Difco Bacto YM agar or malt extract agar there are asci, each with one or two round or oval, ridged, ascospores. Wickerham & Kurtzman (1971) isolated one strain from soil in Illinois.

PICHIA NAGANISHII Kodama

Pichia naganishii has round or oval cells, 2.5 to 5.0 by 3.0 to 7.0 μm, and forms no pseudohyphae in slide culture on potato agar. Each ascus has two or four ascospores which are hat-shaped with a narrow brim. Kodama (1972) isolated one strain from an exudate of a camellia in Nagasaki and Kodama & Kyono (1974a) described the species.

PICHIA NAKAZAWAE Kodama

The cells are oval to cylindrical, 2.0 to 5.0 by 4.0 to 15.0 μm. Pseudohyphae bear short chains of oval blastoconidia. The asci are formed after conjugation; each ascus contains two to four hat-shaped ascospores, which are released easily. Kodama (1975) described a strain of this yeast, which had been isolated from an exudate of *Quercus myrsinaeforia* in Yamaguchi.

PICHIA NORVEGENSIS Leask et Yarrow

Leask & Yarrow (1976) reported this homothallic, ascosporogenous state of *Candida norvegensis*, which is described by van Uden & Buckley (1970). Asci are formed on acetate agar (Fowell, 1952), each containing one to four ascospores.

PICHIA PHILOGAEA van der Walt et Johannsen

The cells are round, oval or sausage-shaped to cylindrical, 3.0 to 8.0 by 3.0 to 11.0 μm. Pseudohyphae are abundant, with cells up to 20 μm long and 2.5 to 3.0 μm across. Asci are formed, on 2% malt, V-8, maize meal and Gorodkowa agars, without immediately previous conjugation. Each ascus contains one to four hat-shaped ascospores. The yeast is probably homothallic. Van der Walt & Johannsen (1975b) isolated two strains from soil near Pretoria.

PICHIA RABAULENSIS Soneda et Uchida

The cells are oval or elongate, 3.2 to 5.5 by 3.4 to 8.8 μm. There are abundant, branched pseudohyphae with oval blastoconidia in short chains. Asci, formed without immediately preceding conjugation, each contain one to four hat-shaped ascospores, which are easily liberated. Soneda & Uchida (1971) isolated two strains of this homothallic yeast in New Britain, New Guinea, one strain from the faeces of a snail and the other from the slime flux of a mango tree.

PICHIA SARGENTENSIS Wickerham et Kurtzman

Pichia sargentensis has round or oval cells, 2.8 to 5.0 by 5.0 to 9.5 μm, and forms some pseudohyphae on Difco Bacto Yeast Morphology agar. After one week on Difco Bacto YM agar or on malt extract agar, there are asci, each with one to four spherical, ringed ascospores. Wickerham & Kurtzman (1971) isolated one strain from a lake in New Hampshire.

PICHIA SCUTULATA Phaff, Miller et Miranda

The heterothallic, diploid, vegetative cells are round or oval and sometimes elongate, 2.6 to 6.6 by 3.9 to 11.8 μm. Some strains form simple pseudohyphae. The ascospores, which are formed on many solid media, are round. There are usually four ascospores per ascus, two of mating type a and two of mating type α. Phaff, Miller & Miranda (1976) described 11 strains, isolated from trees in Hawaii and British Columbia.

PICHIA SPARTINAE Ahearn, Yarrow et Meyers

Pichia spartinae has round or oval cells, 3.0 to 5.0 by 3.5 to 6.0 μm, and forms pseudohyphae in slide culture. After two to four days at 25°C on 2% Diamalt agar, asci were formed, each with one to four hat-shaped ascospores. Ahearn, Yarrow & Meyers (1970) isolated *Pichia spartinae* from the rhizosphere of *Spartina alterniflora* in Louisiana marshland.

PICHIA VERONAE Kodama

The cells are round to oval, 3.0 to 7.0 by 5.0 to 8.5 μm, and pseudohyphae are formed. Each ascus contains two to four hat-shaped ascospores. Kodama & Kyono (1974b) isolated this yeast from an exudate of *Betula grossa* in Ehima, Japan.

RHODOSPORIDIUM BISPORIDIIS Fell, Hunter et Tallman

The cells are almost round, 4.0 by 4.7 to 4.7 by 8.7 μm, to elongate, 2.7 by 8.0 to 3.4 by 11.4 μm. Budding may occur on a short neck. Hyphae have hook-to-peg clamp connections, but no conidia. The teliospores are round, 9.4 to 11.4 μm, or almost so, 6.0 by 6.7 to 8.7 by 11.4 μm, intercalary, single, although occasionally in pairs. Hook-to-peg clamps connect the teliospores to the hyphae. Each teliospore germinates to produce a basidium, 1.5 to 2 by 19 to 140 μm. There may be a few endospores in the basidia and in hyphae. Fell, Hunter & Tallman (1973) isolated nine strains from sea-water.

RHODOSPORIDIUM CAPITATUM Fell, Hunter et Tallman

The cells are almost round, 1.3 by 2.2 to 4.7 by 8.7 μm, to oval, 2.0 by 4.0 to 4.0 by 8.0 μm, or elongate, 1.3 by 4.0 to 4.0 by 14.1 μm. Teliospores form on short hyphae and produce large groups of teliospores by budding. True hyphae, without conidia, have intercalary, terminal and lateral teliospores and may have endospores. Fell, Hunter & Tallman (1973) described 14 strains; these were isolated from sea-water.

RHODOSPORIDIUM DACRYOIDUM Fell, Hunter et Tallman

The cells are oval, 1.3 by 3.4 to 5.4 by 8.7 μm. There may be true hyphae, without conidia, or simple pseudohyphae. The hyphae bear lateral, or sometimes terminal, teliospores, which may be pear- or tear-shaped (dacryoid). A basal clamp connection joins the teliospore to a hypha or to a sporophore. Fell, Hunter & Tallman (1973) isolated 10 strains from sea-water.

RHODOSPORIDIUM INFIRMO-MINIATUM Fell, Hunter et Tallman

Phaff & Fell (1970) described the asexual state of this yeast as *Cryptococcus infirmo-miniatus*. Fell, Hunter & Tallman (1973) found three mating types. Hyphae are formed after conjugation and there are occasional hook-to-peg clamp connections. There are two kinds of teliospore: (i) terminal and intercalary, single and double, round (6.5 by 8.1 μm) to almost round (5.4 by 6.5 to 8.6 by 10.8 μm); (ii) probably older teliospores, brown-pigmented with a thick cell wall, intercalary, oval (7.6 by 8.1 to 17.3 by 18.4 μm), in groups of up to 12. In germination, teliospores form basidia, usually pear-shaped, the large base attached to the teliospore. Endospores are sometimes formed in hyphae.

RHODOSPORIDIUM MALVINELLUM Fell et Hunter

The cells measure 2.0 to 6.7 by 2.7 to 7.4 μm; they bud on short necks. There are no pseudohyphae. Hyphae with clamp connections form after conjugation between compatible mating-types. Teliospores, almost round (5.1 to 10 by 7.1 to 12.0 μm) are rare. Seventeen strains were isolated and the species was described by Fell (1970).

RHODOTORULA ACHENIORUM (Buhagiar et Barnett) Rodrigues de Miranda

Synonym
Sterigmatomyces acheniorum Buhagiar et Barnett (1973)

Description
The cells are oval, elongate or pear-shaped, 1.6 to 4.0 by 2.6 to 7.9 μm. Usually each bud is formed on a narrow stalk, about 1 μm long. A few simple pseudohyphae are produced. Buhagiar & Barnett (1973) isolated this yeast from fresh strawberries. Rodrigues de Miranda (1975) moved the species from *Sterigmatomyces* to *Rhodotorula* because, contrary to the findings of Buhagiar & Barnett (1973), (i) red cultures occurred and (ii) the bud stalks did not appear to be sterigmata, although no photographic or other evidence was given.

RHODOTORULA ARAUCARIAE Grinbergs et Yarrow

Rhodotorula araucariae has round or oval cells 3.5 to 5.5 by 5.0 to 11.0 μm. It does not form pseudohyphae on either rice or potato agar. Grinbergs & Yarrow

(1970b) isolated one strain from a rotting monkey puzzle tree (*Araucaria araucana*) in Chile.

SACCHAROMYCES SERVAZZII Capriotti

The cells are round or oval, 3 to 5 by 3.2 to 5.5 µm, and do not form pseudohyphae. Each ascus contains one to four oval ascospores. Capriotti (1967) described *Saccharomyces servazzii*, 15 strains of which had been isolated from soil in Finland.

SACCHAROMYCOPSIS Schiönning

Synonyms
Endomycopsis Stelling-Dekker (1931) *Nomen illegitimum* (*pro parte*)
Prosaccharomyces Novák et Zsolt (1961)
?*Endomycopsella* Boedijn (1960)
[*non Saccharomycopsis sensu* Guilliermond (1912)]

Description
Vegetative reproduction is both by budding and by forming septa. There are abundant true hyphae and pseudohyphae. Asci are terminal, intercalary, lateral, single, in small clusters or short chains on hyphae, or occurring free. Ascospores are round, oval or hemispherical with equatorial or sub-equatorial ledges. Nitrate is not utilized. (See van der Walt & Scott, 1971b.)

SACCHAROMYCOPSIS CAPSULARIS Schiönning

Kreger-van Rij (1970b) described this species as *Endomycopsis capsularis*. With the abolition of the genus *Endomycopsis*, van der Walt & Scott (1971b) reinstated the name *Saccharomycopsis capsularis* of Schiönning (1903).

SACCHAROMYCOPSIS CRATAEGENSIS Kurtzman et Wickerham

The cells are oval or elongate, 2.0 to 3.0 by 4.0 to 9.5 µm. Abundant true hyphae with blastoconidia and some pseudohyphae are formed. *Saccharomycopsis crataegensis* is heterothallic; asci are produced on YM agar after conjugation. Oval asci, 4.0 to 6.0 by 7.5 to 10.0 µm, are borne either directly on the hyphae or on short stalks. Each ascus contains two elongate or oval ascospores, each with a median, longitudinal ledge. Kurtzman & Wickerham (1973) described this yeast, of which one strain was isolated in Illinois from grapes (*Vitis labrusca*) and four strains from fallen hawthorn fruits (*Crataegus* sp.).

SACCHAROMYCOPSIS FIBULIGERA (Lindner) Klöcker

Kreger-van Rij (1970b) described this species as *Endomycopsis fibuligera*. With the abolition of the genus *Endomycopsis*, van der Walt & Scott (1971b) reinstated the name *Saccharomycopsis fibuligera* of Klöcker (1924), who wrote it *Saccharomycopsis fibuliger*.

SACCHAROMYCOPSIS LIPOLYTICA (Wickerham, Kurtzman et Herman) Yarrow

Van Uden & Buckley (1970) described the anascosporogenous state of this species as *Candida lipolytica*. Wickerham, Kurtzman & Herman (1970) found an ascosporogenous state, *Endomycopsis lipolytica*. With the abolition of that genus, Yarrow (1972) gave this yeast the name *Saccharomycopsis lipolytica*. The yeast is usually heterothallic and haploid. Each ascus contains one to four ascospores.

SACCHAROMYCOPSIS MALANGA (Dwidjoseputro) Kurtzman, Vesonder et Smiley

Synonym
Hansenula malanga Dwidjoseputro [Dwidjoseputro & Wolf (1970)]

Description
The cells are round, 3.2 to 6 µm, or oval to elongate, 3 to 6 by 8 to 13 µm, with multilateral budding on a narrow base. There are a few pseudohyphae and

abundant true hyphae with blastoconidia, singly or in clusters, either on the sides of hyphae or terminally. The asci are round to oval, 7 to 8 by 8 to 10 μm, and each contains two hat-shaped ascospores, 2.5 to 3.5 by 4 to 6 μm. Asci are lateral or terminal on the hyphae and may be single or in clusters of up to 10 in number. Dwidjoseputro & Wolf (1970) isolated this yeast from an Indonesian food, ragi-tape, and Kurtzman, Vesonder & Smiley (1974) reported seven strains from Chinese yeast ball (*Chiu-yüeh* or *peh-yüeh*), which is the commercial starter for making *lao-chao*, a fermented rice preparation. The latter authors transferred the yeast from the genus *Hansenula*, because the species does not utilize nitrate.

SACCHAROMYCOPSIS VINI (Kreger-van Rij) van der Walt et Scott

Kreger-van Rij (1970b) described this yeast as *Endomycopsis vini*. With the abolition of the genus *Endomycopsis*, van der Walt & Scott (1971b) transferred the species to *Saccharomycopsis*.

SARCINOSPORON King et Jong

King & Jong (1975) created this genus to accommodate *Trichosporon inkin*. Vegetative cells are unicellular or filamentous with true septa, multiplying by budding singly or successively in chains. Sporangia are formed by septation of individual blastoconidia or of hyphal cells and contain one to several endospores. Chlamydospores are formed.

SARCINOSPORON INKIN (Oho) King et Jong

Do Carmo-Sousa (1970) described this yeast as *Trichosporon inkin*. King & Jong (1975) transferred it to the genus *Sarcinosporon*. They observed no arthroconidia, deciding that the septation of the hyphae occurred in sporangia.

SCHIZOSACCHAROMYCES SLOOFFIAE Kumbhojkar

The cells are round to long-oval, 7.2 to 8.1 by 7.2 to 9.0 μm. There are no hyphae or pseudohyphae. Each ascus contains four to eight round to oval ascospores. Kumbhojkar (1972) examined two strains which had been isolated from Indian honey.

SELENOZYMA Yarrow

The cells are hemispherical, or sometimes round or oval. Vegetative reproduction is by budding, usually on the convex face or ends of the cells. Hyphae are not formed; the cells may adhere in short chains. There is no visible carotenoid pigmentation. The yeasts are asexual. They may ferment sugars, producing ethanol. The genus was formally described by von Arx, Rodrigues de Miranda, Smith & Yarrow (1977).

SELENOZYMA PELTATA (Yarrow) Yarrow

Synonyms
Torulopsis peltata Yarrow (1968)
Selenotila peltata (Yarrow) Yarrow (1969b)

Description
The cells are round or oval, 2.2 to 3.8 by 3.2 to 5.2 μm, or crescentic, 1.1 to 1.6 by 3.3 to 4.8 μm. There are no hyphae or pseudohyphae. Yarrow (1968) described two strains, isolated from a mastitic bovine udder.

SPOROBOLOMYCES ANTARCTICUS Goto, Sugiyama et Iizuka

Sporobolomyces antarcticus has elongate or fusiform cells, 2.6 to 3.0 by 5.0 to 15.0 μm, and forms abundant pseudohyphae. Ballistoconidia are long-ovoid or fusiform. Goto, Sugiyama & Iizuka (1969) isolated one strain from the sediment of Lake Vanda, Victoria Land, Antarctica.

SPOROBOLOMYCES PUNICEUS (Komagata et Nakase) Ahearn et Yarrow

Synonym
Candida punicea Komagata et Nakase (1965)

Description

Sporobolomyces puniceus has spherical, oval, sausage-shaped or lemon-shaped cells, 3.0 to 8.5 by 4.5 to 20.0 μm, and forms pseudohyphae and true hyphae. Komagata & Nakase (1965) described one strain, which was isolated from frozen salmon. Ballistospores were observed in this species by Ahearn, Roth & Meyers (1968) and by Yarrow (1969a). Yarrow (1972) proposed the new combination, *Sporobolomyces puniceus*.

STEPHANOASCUS Smith, van der Walt et Johannsen

There are budding cells, pseudohyphae and septate, branching hyphae. Oval, unicellular conidia are formed on denticulate cells. Spherical, thick-walled asci develop singly from the subapical cell of a short evagination from two fusing hyphae. Each ascus, crowned by a small, blister-like apical cell, contains one to four hemispherical, hat-shaped ascospores (Smith, van der Walt & Johannsen, 1976).

STEPHANOASCUS CIFERRII Smith, van der Walt et Johannsen

Van Uden & Buckley (1970) described the asexual state of this heterothallic yeast as *Candida ciferrii*. Asci develop singly from a central swollen cell after fusion of two hyphae. Each ascus contains one to four, hemispherical, hat-shaped ascospores. Smith, van der Walt & Johannsen (1976) examined six strains, including one from soil and four from mammals.

STERIGMATOMYCES APHIDIS Henninger et Windisch

The cells are elongate, 1.4 to 3.6 by 4.3 to 11.5 μm, and often pointed at one or both ends. Some cells may be 40 μm long. New cells are formed multilaterally on sterigmata. There are a few pseudohyphae in the form of small chains of cells, connected by sterigmata. Henninger & Windisch (1975b) isolated two strains from the secretions of Aphididae on leaves of *Solanum pseudocapsicum*.

STERIGMATOMYCES ELVIAE Sonck et Yarrow

Sterigmatomyces elviae has almost round or oval cells, 3.0 to 5.0 by 4.0 to 8.0 μm, which reproduce vegetatively by forming conidia on sterigmata. Neither hyphae nor pseudohyphae are formed in slide culture with potato, rice or maize meal agar. Sonck & Yarrow (1969) examined one strain which was isolated from human skin.

STERIGMATOMYCES NECTAIRII Rodrigues de Miranda

The cells are round to oval, 2.0 to 8.0 by 3.0 to 8.0 μm, and have short sterigmata. Rodrigues de Miranda (1975) examined one strain, isolated from cheese of St Nectaire.

STERIGMATOMYCES PENICILLATUS Rodrigues de Miranda

The cells are round to oval, 2.0 to 5.5 by 2.5 to 6.0 μm. Most sterigmata are short, but some may be at least 25 μm long. In slide culture, a tree-like arrangement is formed of round or elongate cells with long sterigmata, which may exceed 150 μm. Rodrigues de Miranda (1975) examined two strains that had been isolated as contaminants on agar plates.

STERIGMATOMYCES POLYBORUS Scott et van der Walt

Sterigmatomyces polyborus has round or oval cells, 1.5 to 4.0 by 2.5 to 5.5 μm, which reproduce vegetatively by forming single conidia on sterigmata. Neither true hyphae nor pseudohyphae are formed on maize meal agar. Scott & van der Walt (1970b) isolated two strains from the tunnels of the wood-boring beetles *Xyloborus torquatus* infesting *Cussonia umbellifera* in Natal.

SYMPODIOMYCES Fell et Statzell

Fell & Statzell (1971) described this genus. On malt agar, yeast-like cells develop conidiophores, each producing a terminal conidium. At the side of the **conidium**, a new growing tip of the conidiophore develops forming a new terminal conidium. In this way an elongate conidiophore is formed with successive conidia. The conidiophores possess several scars, each of which was terminal before the development of a new conidium. Conidia formation is irregular, both in distance between conidia and location on the conidiophore.

There are few branched hyphae which form conidia, singly on a short peg, sympodially on an elongate conidiophore, or as clusters. No ascospores, teliospores or ballistoconidia have been seen.

SYMPODIOMYCES PARVUS Fell et Statzell

Sympodiomyces parvus, after five days at 12 °C, has round or oval single cells, 1.0 to 1.3 by 1.3 to 3.4 μm. There are also oval cells in chains, 2.6 to 3.4 by 4.7 to 5.4 μm. After one month at 12 °C on maize meal agar, there are septate hyphae with clusters of conidia. Fell & Statzell (1971) isolated 52 strains from sea-water, mainly from the Antarctic.

TORULASPORA Lindner

The genus *Torulaspora*, accepted by Stelling-Dekker (1931), was rejected by Lodder & Kreger-van Rij (1952) because they found it indistinguishable from *Saccharomyces*. Van der Walt & Johannsen (1975d) have now restored and redefined the genus.

The vegetative cells reproduce exclusively by budding and are mainly haploid. Cell fusion does not usually immediately precede ascus formation. Each ascus contains one to four ascospores, which are round to long-oval, usually warty, without ledges. Nitrate is not utilized.

TORULASPORA DELBRUECKII Lindner

Van der Walt (1970c) described this species as *Saccharomyces delbrueckii*. Now, van der Walt & Johannsen (1975d) have restored the original name, given by Lindner (1904).

TORULASPORA GLOBOSA (Klöcker) van der Walt et Johannsen

Van der Walt (1970c) described this species as *Saccharomyces kloeckerianus*; van der Walt & Johannsen (1975d) changed the name to *Torulaspora globosa*.

TORULASPORA PRETORIENSIS (van der Walt et Tscheuschner) van der Walt et Johannsen

Van der Walt (1970c) described this species as *Saccharomyces pretoriensis*; van der Walt & Johannsen (1975d) changed the name to *Torulaspora pretoriensis*.

TORULOPSIS AURICULARIAE Nakase

Torulopsis auriculariae has oval cells, 3.0 to 4.5 by 4.0 to 7.0 μm, and a few elongated cells but no pseudohyphae. Nakase (1971b) isolated two strains from a fruiting body of the fungus *Auricularia auricula-judae*.

TORULOPSIS AUSTROMARINA Fell et Hunter

The cells are round (1.3 to 7.4 μm) to oval (3.4 by 4.0 to 4.0 by 7.4 μm) and do not form pseudohyphae. Fell & Hunter (1974) examined 85 strains from the Antarctic Ocean.

TORULOPSIS BACARUM Buhagiar

The cells are oval, 1.6 to 3.1 by 3.1 to 5.8 μm, and do not form pseudohyphae. Buhagiar (1975) isolated three strains from soft fruit.

TORULOPSIS BOMBICOLA Spencer, Gorin et Tulloch

Torulopsis bombicola has round or oval cells, 1.7 to 2.6 by 1.0 to 2.0 μm, and no pseudohyphae. Spencer, Gorin & Tulloch (1970) isolated the yeast from bumble-bee honey.

TORULOPSIS DENDRICA van der Walt, van der Klift et Scott

Torulopsis dendrica has oval or almost cylindrical cells, 2.0 to 5.5 by 3.0 to 6.5 μm, and no pseudohyphae. Van der Walt, Scott & van der Klift (1971b) isolated seven strains from material associated with insects in Natal.

TORULOPSIS FRAGARIA Barnett et Buhagiar

Torulopsis fragaria has polymorphic cells, 3.4 to 12.9 by 3.2 to 6.1 μm, and no pseudohyphae. Barnett & Buhagiar (1971) isolated 35 strains from strawberries and 5 from blackcurrants.

TORULOPSIS FRUCTUS Nakase

Torulopsis fructus has nearly round or oval cells, 1.5 to 6.5 by 2.0 to 7.0 μm, and no pseudohyphae. Nakase (1971b) isolated two strains from banana.

TORULOPSIS HALOPHILUS Onishi

Torulopsis halophilus has oval cells, 2.0 to 4.0 by 3.0 to 5.0 μm, and no pseudohyphae. Van Uden & Vidal-Leiria (1970) did not validate *Torulopsis halophilus*, *nomen nudum*, because they found its characteristics differed from those described by Onishi (1957) who, having isolated one strain from soy mash, gave no Latin description. Yarrow & Meyer (1978) published a Latin description on transferring *Torulopsis* species to *Candida*.

TORULOPSIS HUMILIS Nel et van der Walt

Torulopsis humilis has round or oval cells, 2.0 to 5.5 by 3.5 to 7.0 μm, and no pseudohyphae. Nel & van der Walt (1968) isolated one strain from Bantu beer.

TORULOPSIS INSECTALENS Scott, van der Walt et van der Klift

Torulopsis insectalens has round or oval cells, 2.0 to 2.5 by 2.5 to 3.0 μm, and no pseudohyphae. Van der Walt, Scott & van der Klift (1971b) isolated 10 strains from material associated with insects in Natal.

TORULOPSIS KARAWAIEWI Jurzitza *nom. nud.*

Torulopsis karawaiewi has round or oval cells, 2.0 to 4.5 by 2.2 to 4.8 μm, and no pseudohyphae. Jurzitza (1970) isolated three strains from anobiid beetles.

TORULOPSIS KRUISII Kocková-Kratochvílová et Ondrušová

The cells are oval, 2.0 to 4.0 by 3.0 to 5.0 μm. Kocková-Kratochvílová & Ondrušová (1971) examined 14 strains isolated from the pilei of higher fungi.

TORULOPSIS MANNITOFACIENS Onishi et Suzuki

Torulopsis mannitofaciens has oval or spherical cells, 2.3 to 4.6 μm, and no pseudohyphae. Onishi & Suzuki (1969) isolated one strain from soy mash.

TORULOPSIS METHANOLOVESCENS Oki et Kouno

The cells are round or short-oval, 1.2 to 2.0 by 1.5 to 3.0 μm. There are no pseudohyphae. Oki, Kouno, Kitai & Ozaki (1972) isolated this species from a flower of *Rhododendron indici* in Tsujido, Japan.

TORULOPSIS MULTIS-GEMMIS Buhagiar

The cells are mainly round to oval, 1.6 to 3.9 by 1.8 to 4.7 μm, and form some simple pseudohyphae. Buhagiar (1975) examined 20 strains isolated from raspberries *Rubus idaeus*.

TORULOPSIS MUSAE Nakase

Torulopsis musae has nearly round or oval cells, 2.0 to 5.5 by 2.5 to 7.0 μm, and no pseudohyphae. Nakase (1971b) isolated one strain from banana.

TORULOPSIS NAGOYAENSIS Asai et Makiguchi

The cells are round or oval, 2.0 to 5.0 by 2.0 to 5.0 μm, and do not form pseudohyphae. Asai, Makiguchi, Shimada & Kurimura (1976) isolated two strains from sewage in Nagoya, Japan.

TORULOPSIS NAVARRENSIS Moriyon et Ramírez

The cells are oval, 1.5 to 3.5 by 2.5 to 5.0 μm, and do not form pseudohyphae. Seven strains were isolated from forest soil in Navarra and described by Moriyon & Ramírez (1974) and Ramírez & Moriyon (1978).

TORULOPSIS NEMODENDRA van der Walt, van der Klift et Scott

Torulopsis nemodendra has round or oval cells, 1.5 to 4.5 μm, and no pseudohyphae. Van der Walt, Scott & van der Klift (1971b) isolated five strains from the tunnels of *Xyleborus aemulus* in *Rapanea melanophloeos* in the Cape Province.

TORULOPSIS NODAENSIS Onishi

Torulopsis nodaensis has oval cells, 2.0 to 4.0 by 3.0 to 5.0 μm, and no pseudohyphae. One strain was isolated from soy mash. Van Uden & Vidal-Leiria (1970) did not validate *Torulopsis nodaensis*, *nomen nudum*, because the characteristics of the original strain differed from those described by Onishi (1957). Yarrow & Meyer (1978) published a Latin description on transferring *Torulopsis* species to *Candida*.

TORULOPSIS PAMPELONENSIS Ramírez et Martinez

The cells are oval, 2.0 to 3.5 by 2.5 to 5.0 μm, and do not form pseudohyphae. Ramírez & Martinez (1978) isolated eight strains from soil and leaves in a beech forest in Navarra.

TORULOPSIS PHILYLA van der Walt, van der Klift et Scott

Torulopsis philyla has oval cells, 2.5 to 5.0 by 3.0 to 8.0 μm, and no pseudohyphae. Van der Walt, Scott & van der Klift (1971b) isolated one strain from the tunnels of *Xyleborus ferrugineus* infesting *Harpephyllum caffrum* in Natal.

TORULOPSIS PIGNALIAE Jacob

Torulopsis pignaliae has almost round cells, 2.9 to 5.0 μm, and no pseudohyphae. Jacob (1970) isolated six strains from tanning liquor.

TORULOPSIS PSYCHROPHILA Goto, Sugiyama et Iizuka

Torulopsis psychrophila has round or oval cells, 2.6 to 4.0 by 2.6 to 4.5 μm, and simple pseudohyphae. Goto, Sugiyama & Iizuka (1969) isolated two strains from penguin dung from Ross Island.

TORULOPSIS PUSTULA Buhagiar

The cells, 2.4 to 5.2 by 4.5 to 7.8 μm, are oval or pear-shaped, with one or two buds on or near one pole. There are no pseudohyphae. Buhagiar (1975) isolated four strains from blackcurrants *Ribes nigrum*.

TORULOPSIS SCHATAVII Kocková-Kratochvílová et Ondrušová

The cells are oval, 3.3 to 4.0 by 4.5 to 5.0 μm, and do not form pseudohyphae. Kocková-Kratochvílová & Ondrušová (1971) examined 10 strains, isolated from the pilei of higher fungi.

TORULOPSIS SILVATICA van der Walt, van der Klift et Scott

Torulopsis silvatica has round or oval cells, 1.5 to 3.0 by 2.5 to 4.0 μm, and no pseudohyphae. Van der Walt, Scott & van der Klift (1971b) isolated two strains from tunnels of *Crossotarsus externedentatus* in *Ficus* species in Natal.

TORULOPSIS SONORENSIS Miller, Phaff, Miranda, Heed et Starmer

The cells are round to elongate, 2.4 to 5.4 by 2.8 to 7.0 μm, and do not form pseudohyphae. Miller, Phaff, Miranda, Heed & Starmer (1976) examined 35 strains from *Drosophila mojavensis* and from cacti in southern Arizona and northern Mexico.

TORULOPSIS SORBOPHILA Nakase

The cells are round or oval, 2 to 4.5 by 2 to 5 μm, and do not form pseudohyphae. Nakase (1975) examined one strain from ion-exchange resin in a guanosine monophosphate manufacturing plant.

TORULOPSIS SPANDOVENSIS Henninger et Windisch

The cells are round or oval, 2.8 to 6.0 by 2.8 to 6.0 μm, and do not form pseudohyphae. Henninger & Windisch (1976a) isolated two strains from beer.

TORULOPSIS TANNOTOLERANS Jacob

The cells are oval, 2 to 4.3 by 3.1 to 7.3 μm, and do not form pseudohyphae. Jacob (1970) isolated this species from tanning liquors.

TORULOPSIS XESTOBII Jurzitza *nom. nud.*

Torulopsis xestobii has oval cells, 2.9 to 4.5 by 3.3 to 6.1 μm, and no pseudohyphae. Jurzitza (1970) isolated one strain from the beetle *Xestobium plumbeum*.

TRICHOSPORON AQUATILE Hedrick et Dupont

Trichosporon aquatile has round, oval or elongate cells, 4.0 to 5.0 by 5.0 to 12.0 μm, and forms hyphae and arthroconidia on potato–glucose agar or on Difco Bacto Yeast Morphology agar. Hedrick & Dupont (1968) isolated two strains from fresh water.

TRICHOSPORON BRASSICAE Nakase

Trichosporon brassicae has round or oval cells, 3.0 to 7.5 by 3.5 to 13.0 μm, and hyphae and arthroconidia in slide culture. Nakase (1971b) isolated one strain from cabbage.

TRICHOSPORON ERIENSE Hedrick et Dupont

Trichosporon eriense has oval or elongate cells, 4.0 by 4.0 to 10.0 μm, and forms hyphae and arthroconidia on potato–glucose agar or on Difco Bacto Yeast Morphology agar. Hedrick & Dupont (1968) isolated one strain from Lake Erie, near Cleveland.

TRICHOSPORON FENNICUM Sonck et Yarrow

Trichosporon fennicum has round or oval cells, 3.0 to 5.0 by 3.0 to 8.0 μm, and in slide culture on potato agar forms abundant hyphae with arthroconidia. Sonck & Yarrow (1969) examined one strain, isolated from a cat *Felis ocreata*. A further 15 strains have been isolated in Finland from hens *Gallus domesticus* and birch trees *Betula alba* (D. Yarrow, unpublished observations).

TRICHOSPORON MELIBIOSACEUM Scott et. van der Walt

Trichosporon melibiosaceum has round, oval or elongate cells, 1.5 to 6.5 by 2.5 to 27.0 μm, and, on maize meal agar, forms pseudohyphae and hyphae with arthroconidia. Scott & van der Walt (1970b) isolated this species from the frass of *Bostrichioplites cornutus* infesting *Acacia karroo* in the Transvaal.

TRICHOSPORON TERRESTRE van der Walt et Johannsen

The budding yeast cells are oval or cylindrical, 2.5 to 6.0 by 3.0 to 11.0 μm, and there are abundant pesudohyphae, hyphae and arthroconidia. Endospores occur in old cultures. Van der Walt & Johannsen (1975c) isolated one strain from soil near Pretoria.

WICKERHAMIELLA DOMERCQII van der Walt

Van Uden & Vidal-Leiria (1970) described the asexual state of this yeast as *Torulopsis domercqii*. Van der Walt & Liebenberg (1973) found that two strains produce elongate and rugose ascospores, one per ascus, after cell fusion.

ZENDERA Redhead et Malloch

Branched, septate hyphae form arthroconidia. Asci, formed after fusion of gametangia from two cells, each contain two, four or eight smooth, oval ascospores, gelatinously sheathed (Redhead & Malloch, 1977).

ZENDERA OVETENSIS (Peláez et Ramírez) Redhead et Malloch

Kreger-van Rij (1970b) described this yeast as *Endomycopsis ovetensis*. With the abolition of the genus *Endomycopsis*, Redhead & Malloch (1977) have transferred the species to the genus *Zendera*.

ZYGOSACCHAROMYCES Barker

Yeasts of this genus have the same characteristics as those of *Saccharomyces*, except that conjugation occurs immediately before ascospore formation. The genus *Zygosaccharomyces* was described by Barker (1901a,b) and accepted by Stelling-Dekker (1931). However, following the genetical work of Öjvind Winge and Carl Lindegren, Lodder & Kreger-van Rij (1952) rejected the genus as they considered it to be composed of haploid *Saccharomyces* species and van der Walt (1970c) perpetuated this rejection. Nonetheless, there has been a tendency to restore *Zygosaccharomyces* as an accepted genus, for example Kudriavzev (1954), Capriotti (1958), Novák & Zsolt (1961), van der Walt & Johannsen (1975d) and von Arx, Rodrigues de Miranda, Smith & Yarrow (1977).

ZYGOSACCHAROMYCES BAILII Lindner

Van der Walt (1970c) described yeasts of this species as *Saccharomyces bailii*.

ZYGOSACCHAROMYCES BISPORUS Naganishi

Van der Walt (1970c) described yeasts of this species as *Saccharomyces bisporus*.

ZYGOSACCHAROMYCES CIDRI (Legakis) Yarrow

Van der Walt (1970c) described yeasts of this species as *Saccharomyces cidri*.

ZYGOSACCHAROMYCES FERMENTATI Naganishi

Van der Walt (1970c) described yeasts of this species as *Saccharomyces montanus*. [N.B. *Saccharomyces fermentati*, synonym *Zygosaccharomyces fermentati* (Saito) Krumbholz, is here listed as *Torulaspora delbrueckii*.]

ZYGOSACCHAROMYCES FLORENTINUS Castelli ex Kudrjawzew

Van der Walt (1970c) described yeasts of this species as *Saccharomyces florentinus*. Kudrjawzew [Kudriavzev] (1960) validated this species of Castelli (1938) by giving it a Latin description.

ZYGOSACCHAROMYCES MICROELLIPSODES (Osterwalder) Yarrow

Van der Walt (1970c) described yeasts of this species as *Saccharomyces microellipsodes*.

ZYGOSACCHAROMYCES MRAKII Capriotti

Van de Walt (1970c) described yeasts of this species as *Saccharomyces mrakii*.

ZYGOSACCHAROMYCES ROUXII (Boutroux) Yarrow

Van der Walt (1970c) described yeasts of this species as *Saccharomyces rouxii*.

5
A simple guide to laboratory methods for identifying yeasts

Procedures for identifying yeasts may differ greatly from those for classifying them. For example, when comparing *Debaryomyces* with *Pichia*, Kreger-van Rij (1970c) made electron micrographs to study the walls of the ascospores. This was perfectly proper for a taxonomist engaged in describing the characteristics of genera and species, although electron microscopy would be impracticable for most people engaged in identifying yeasts. Whilst taxonomic descriptions should be as complete as can be, the nature and number of criteria used for identification must depend on such factors as the number of identifications to be done, the purpose to which they will be put and whether there is a particular, restricted range of possible species in the niche that is under study. When identifying hundreds or thousands of yeasts isolated in a survey, it is usually acceptable if a few are identified incorrectly, because it is the total picture rather than the detail which is important. This situation differs greatly from that of having to identify a single or only a few strains, for which purpose anything short of the greatest reliability attainable is futile.

Van der Walt (1970d) has written an exhaustive and authoritative account of the methods used when classifying yeasts. The present chapter assumes a general knowledge of microbial techniques and describes only those well-tried methods that the authors of this *Guide* consider most helpful for identifications or those methods most commonly used. In order to clinch an identification, it is desirable to compare the yeast at issue with at least one authentic strain of the supposed species. It is also essential to consult the full description of this species in the latest edition of *The Yeasts. A Taxonomic Study* and, perhaps, in more than one edition (first, Lodder & Kreger-van Rij, 1952; second, Lodder, 1970; third, Kreger-van Rij, in preparation). Whilst Lodder & Kreger-van Rij (1952) included excellent drawings of every species described, this was not done for every species in the second edition. Consequently, it may often prove valuable to refer also to the original communication in which a given yeast was first described, particularly if published after the first edition of *The Yeasts* was written.

Media for cultivating yeasts. Yeasts can be maintained on agar slopes of malt--yeast-glucose-peptone, yeast-glucose-peptone or malt extract. Details of culture and test media are given on pp. 45-46.

Incubation temperatures. For taxonomic tests, yeasts are usually incubated at 25 °C, although optimum temperatures for growth are higher for some yeasts and lower for others (reviewed by Stokes, 1971). Certain yeasts, such as five species of *Leucosporidium* (Fell & Phaff, 1970), only grow at temperatures below 25 °C. Long-established laboratory strains of, for example, *Candida utilis* or *Saccharomyces cerevisiae*, may grow better at 30 °C than at 25 °C. But many yeasts that grow poorly at 30 °C grow well at 25 °C. Growth tests may often be done successfully at 25 °C with yeasts that under certain conditions could not be isolated at that temperature but were isolated only at a lower temperature. This is probably because a minority of the cells grow at 25 °C (Buhagiar & Barnett, 1971).

Microscopical appearance of non-filamentous vegetative cells
Purpose. The purpose of examining vegetative yeast cells microscopically is chiefly to answer the following questions. (i) Does the yeast reproduce by budding, splitting or both? (ii) If the yeast forms buds, where do they occur on

the mother-cells and how wide is the isthmus between mother and daughter? (iii) What are the shapes and sizes of the vegetative cells?

Procedure. Yeast from a young, growing culture (say, one day old, if practicable) is inoculated into 30 ml of sterile liquid culture medium in a 100-ml Erlenmeyer flask. The medium should be malt extract, yeast–glucose–peptone or malt–yeast–glucose–peptone, since most yeasts can grow on these media. The culture is examined microscopically after incubation for two or three days at 25°C or 28°C.

Comment. This technique is a survival from the days before microbiologists appreciated the value, for making reproducible observations, of standardizing conditions wherever practicable by (i) using chemically defined media and (ii) studying cells from exponentially growing cultures.

Microscopical examination for filamentous growth
Purpose. Although all species form separate cells under some conditions, yeasts vary in their capacity to produce filaments. For example, *Cryptococcus albidus* appears to produce no filaments, *Saccharomyces cerevisiae* may form pseudohyphae, *Candida albicans* forms both pseudohyphae and true hyphae, whilst *Trichosporon aquatile* produces true hyphae only. The purpose of this examination is to ascertain whether or not there is filamentous growth; if so, what kind of filaments, and what kinds of cells (if any) grow from the filaments. The procedure described below has been developed empirically to provide conditions that favour filament production by yeasts.

Procedure. Slide cultures are made as follows. Petri dishes are autoclaved, each dish containing a piece of filter paper, a U-shaped glass rod support, two clean microscope slides and clean coverslips. Working aseptically, an agar medium (one or more of those given below) is melted and poured into a boiling-tube, wide and deep enough to hold a microscope slide. Each slide is dipped into the agar and replaced on its glass rod support in the Petri dish. Yeast from an actively growing culture is lightly inoculated with a straight wire along the length of each slide and a coverslip placed over a portion of the inoculated agar. The filter paper, wetted with sterile water, prevents drying. Incubation is at 25 °C. After wiping the agar from the back of each slide, the cultures are examined microscopically every one or two days for up to about two weeks. Either maize (corn) meal agar or potato–glucose agar is generally used.

Microscopical examination for ballistoconidia
Purpose. Finding ballistoconidia gives decisive identification of a yeast as a member of the family Sporobolomycetaceae, that is, of one of the genera *Bullera*, *Sporobolomyces*, *Sporidiobolus* or *Aessosporon*.

Procedure. Agar medium, 10 to 15 ml of malt extract, potato–glucose or maize meal, is poured into a Petri dish. When the agar is set, it may be necessary to dry the surface. The medium is then inoculated with the yeast to be tested in lines along two diameters at right angles. This inoculated dish is inverted over another Petri dish bottom, also containing agar medium on which is placed a sterile microscope slide; the two dishes are taped together all round the circumference. The preparation is incubated for up to three weeks at about 20 °C. Discharged ballistoconidia form colonies on the lower dish and the slide collects ballistoconidia which can be examined microscopically.

Microscopical examination for ascospores
Purpose. The intentions are threefold: (i) to establish whether or not the yeast can form ascospores and, if so, (ii) whether they are produced from ordinary vegetative cells, without prior conjugation, after conjugation of two vegetative

cells, or after conjugation between a mother-cell and its bud; and (iii) to investigate the shapes and sizes of the ascospores and asci.

Comment. It is most helpful for identification if ascospores are formed readily. However, they may not be observed for the following reasons: (i) the media chosen are unsuitable for the yeast in question; (ii) the strain is heterothallic and haploid; (iii) the asci are difficult to observe and the observer insufficiently experienced; (iv) the yeast is anascosporogenous. For the first reason it is usual to try several media when testing for ascospore formation.

Procedure. A young culture, actively growing overnight or up to about two days at 25 °C on, say, malt–yeast–glucose–peptone or malt extract (usually agar, but sometimes broth), is used to inoculate each ascosporulation medium. These inoculated media are incubated at 25 °C for three days and examined microscopically. Those in which ascospores are not seen are incubated further at about 20 °C and examined every week for at least six weeks.

Van der Walt (1970*d*) lists 14 media on which ascospores are formed. The following five media seem to be among the most effective: V-8 agar, Gorodkowa agar, acetate agar, malt–yeast–glucose–peptone agar and malt extract agar. Ascospores may also be seen near the edge of the coverslips in slide cultures with maize meal agar.

Assessing the ability to use organic compounds as sole source of carbon for aerobic growth

Purpose. It is practicable to distinguish between many yeast species by their differing abilities to utilize certain organic compounds, each as the sole major source of carbon. These compounds, listed on p. 47, include sugars, alditols and organic acids. The tests described below, often called 'assimilation' tests, are used to assess the ability of each yeast to use the compounds for aerobic growth.

Procedures. Two alternative methods of testing are most commonly used, tubes of liquid media or auxanograms. A third method, replica plating, is also convenient.

(i) *Liquid medium.* The tests are done in test-tubes (e.g. 180 mm by 16 mm), rimless if with caps. Each tube contains a standard amount of liquid medium, usually 5 to 10 ml. The medium is a nitrogen base which, except those for negative controls with no carbon source, contains a standard amount of one test substrate, such as D-glucose (a positive control), sucrose or citrate, at a convenient concentration of, say, 50 mM. Without being inhibitory, this concentration allows for some yeasts having a low affinity for the substrate (K_m of, say, 10 or 20 mM). However, such a high substrate concentration entails the risk of false positive results from the presence of significant concentrations of impurities, particularly with sugars, such as D-glucose in maltose or D-galactose in L-arabinose.

For inoculation, the yeast to be tested is grown on some convenient medium, say malt–yeast–glucose–peptone agar, and a young, actively growing culture (typically grown overnight, but longer with slow-growing yeasts) is suspended in nitrogen base to give $\sim 25 \times 10^6$ cells per ml, that is very roughly 500 μg dry weight of yeast per ml ($A_{640} \approx 1.0$). Each tube is inoculated with 100 μl of standardized suspension at 25 °C.

It is common practice for the tubes to be stationary, perhaps tilted to improve aeration by exposing maximum surface of the liquid, and shaken by hand from day to day. However, full aeration and better mixing are obtained if the tubes are rocked or rotated continuously, so the tests are quicker and better standardized and the results more reliable. The angle of rocking used (say, from 15° to 40° from the horizontal) should be maximal, without causing loss of material or wetting of plugs. On the other hand, even the most energetic rocking of test-tubes is insufficient for certain yeasts which sediment rapidly. For such yeasts, T-tubes

(Monod tubes) are useful (Monod, Cohen-Bazire & Cohn, 1951; Barnett & Ingram, 1955); these can be rocked to an angle of up to 90° both sides of the horizontal or shaken on a reciprocating shaker. The disadvantages of T-tubes are that they are expensive to buy and troublesome to clean and handle.

Commonly, growth is assessed by eye, relative to the negative and positive control tubes, sometimes using as a criterion the degree of obscuring by the grown yeast of black lines drawn on a card and placed behind each tube. However, a photoelectric assessment of growth is better in giving objective measurements. A nephelometer measures scattered light (see, for example, Powell, 1963; Mallette, 1969; **Meynell & Meynell,** 1970) and is suitable for measuring growth in test-tubes, as it is fairly insensitive to the unevenness of the tubes. For most identifications, it should be unnecessary to assess growth for longer than one week, although it is quite usual to do so for up to four weeks.

For very large numbers of identifications, it is practicable to construct an automatically recording nephelometer. Wentink & la Rivière (1962) made a crude form of such an instrument; with modern photoreceptors and electronics, a highly sophisticated design could be constructed to record growth in hundreds of tubes, reliably and simultaneously.

(ii) *Auxanograms*. Yeast is inoculated into an agar medium containing all growth requirements save a source of carbon. Various compounds are placed on the agar surface and growth of the yeast may occur in the region of those compounds that the yeast can use as a carbon source.

Tubes, each containing about 15 ml of molten nitrogen base agar medium at 45 °C, are inoculated with 500 μl of yeast suspension, with roughly 25×10^6 cells per ml, prepared as for the tests in liquid medium (p. 41). The concentration of yeast may be critical for the success of the auxanograms. For each tube, the inoculum is well mixed with the medium, care being taken to avoid bubbles forming, and the contents are poured immediately into a Petri dish. It is necessary to work quickly to avoid (i) killing the yeast at the high temperature and (ii) the agar setting prematurely. The Petri dishes containing inoculated medium are left undisturbed and horizontal for about 30 min. Then the surface of the agar is dried by incubating the dishes, open and inverted, at 37 °C for about 90 min (although such treatment could destroy psychrophilic yeasts, such as *Leucosporidium antarcticum*). Special care is necessary at this stage to avoid microbial contamination.

About 5 mg of each test compound are placed with a small spatula as a little pile on the agar surface, close to the edge of the Petri dish. Three compounds can be tested in each dish, two placed opposite each other and the third opposite a negative control area without added substrate. The dishes are incubated and examined for growth every two days for about a week.

A washed agar gives best results for auxanograms (for example, Special Agar (Noble), marketed by Difco Laboratories Inc., Detroit, Michigan, USA).

(iii) *Replica plating*. Up to 25 yeast colonies, growing on the surface of agar medium in a 'master' Petri dish, are 'printed', that is, inoculated simultaneously, on other dishes with different sources of carbon. The printing is done with a velveteen pad.

The surface of a nutrient agar (e.g. malt–yeast–glucose–peptone) in a Petri dish is inoculated by means of a straight wire with up to 25 actively growing strains. These inoculations are arranged as a pattern on the agar. The yeasts are allowed to grow overnight into small discrete colonies. Every dish must be oriented with a mark. The test dishes on which the printing is done each contain nitrogen base, about 25 mM carbon source (none for the negative control), and 2% (w/v) of washed agar. The velveteen (cotton) cloth can be wrapped in aluminium foil and autoclaved. A cylindrical wooden block, which fits inside the Petri dish, is covered with the sterile velveteen (i) by pressing it down on the block with a sterile Petri dish bottom and (ii) by fixing the velveteen with a metal ring or rubber band. The master plate is inverted and pressed against the mounted velveteen. Then the test dishes are similarly pressed on the velveteen, negative control first, D-glucose (positive control) last. Up to 10 printings can be done from one master plate.

Table 5.1 *A summary of the methods of assessing aerobic growth discussed above*

Method	Comments	References
Liquid media	Easy to measure growth. Tests for starch production can be done in these tubes	
Stationary tubes	Undefined (limiting) oxygen concentration in the medium and lack of uniform conditions	Wickerham & Burton (1948); Wickerham (1951)
Agitated tubes	Oxygen concentration not limiting during crucial part of test; results obtained more quickly than with stationary tubes	Barnett & Ingram (1955); Ahearn, Roth, Fell & Meyers (1960); Buhagiar & Barnett (1971)
Solid media	Objective assessment of growth impracticable. Microbial contaminants easily observed	
Auxanograms	Agar N-base + inoculum in Petri dishes; crystals of test substrates placed on agar surface. By contrast with liquid media, it is unnecessary to sterilize most substrates; can observe responses to varying substrate concentration and test substrates of low solubility	Lodder (1934); Pontecorvo (1949); Barnett & Ingram (1955)
Replica plates	Each Petri dish contains medium with a single carbon source. Inoculations from colonies growing in another Petri dish, 'printed' with a velveteen pad	Lederberg & Lederberg (1952); Shifrine, Phaff & Demain (1954); Beech, Carr & Codner (1955)

Assessing the ability to use nitrogen compounds for aerobic growth
Purpose. The test of ability to use nitrate as sole source of nitrogen is a valuable aid to identifying yeasts, since roughly one quarter of all species utilize nitrate and this ability is usually a uniform feature of all the strains within the species. Certain other test compounds, such as nitrite, ethylamine and L-lysine, have also been found to be helpful.

Procedures. The methods, using liquid or solid media, for assessing growth are closely similar to those described for using organic compounds. Carbon base medium is used in place of nitrogen base. Nitrite may prove toxic to the yeast because nitrous acid is formed at pH values of less than 6, so media are best adjusted to pH 6.5. Because of this toxicity, auxanograms provide one of the more effective methods for testing utilization of nitrite and of ethylamine, which can also be inhibitory. In solution, nitrogenous test substrates should be added to give a concentration of from 2 to 5 mM. Growth may occur in negative control tubes because ammonia from the atmosphere dissolves in the medium.

Assessing the ability to use certain sugars anaerobically
Purpose. Just over half the yeast species 'ferment' at least D-glucose semi-anaerobically. The ability to use up to 13 compounds (sugars or certain of their derivatives, see p. 47) is assessed by looking for the formation of gas (CO_2).

Procedures. Large Durham tubes are used, say, test-tubes of about 150 mm by 12 mm with insert tubes of about 50 mm by 6 mm. The efficiency of collecting gas in the inner tube is improved by fusing a small globule of glass to the inner tube's rim, to keep it from touching the bottom of the outer tube. The tubes are filled with 10 to 15 ml of yeast extract medium (0.5%(w/v) of a commercial dried yeast extract) with 50 mM test sugar. Negative control tubes contain no sugar. Further controls can include: (i) tubes inoculated with yeasts known to ferment all the test substrates; (ii) a yeast that ferments D-glucose only. Each tube is inoculated with 100 µl of a fresh yeast suspension of about 10^7 cells per ml of yeast extract broth, taken from an actively growing culture.

If it uses the sugar in the Durham tube aerobically, the yeast will probably grow in the upper part of the tube and adapt to that sugar. The adapted (induced or derepressed) yeast cells fall to the bottom of the tube, where the oxygen concentration quickly decreases. Here the yeast may catabolize the sugar anaerobically, liberating bubbles of CO_2 which are trapped in the inner inverted tube. The tubes are incubated at 25 °C for about one week; each day they are shaken to bring down the yeast and examined for bubbles of gas. Some workers prolong the incubation for a month.

Assessing vitamin requirements

Purpose. Yeasts differ from each other in their requirements for the exogenous supply of certain growth factors (listed below). Accordingly, tests are made of the ability to grow without each in turn, or all, of these growth factors.

Procedures. These growth tests are done in liquid medium as for organic compounds with a chemically defined growth medium, complete except for the growth factors. A series of tubes is prepared, with batches lacking (i) all the growth factors and (ii) each growth factor in turn. The following list of the factors gives the amount of each to add per litre of medium.

p-Aminobenzoic acid	200 µg
Biotin	20 µg
Folic acid	2 µg
myo-Inositol*	10 mg
Nicotinic acid	400 µg
Pantothenate (Ca)	2 mg
Pyridoxine HCl	400 µg
Riboflavin	200 µg
Thiamin HCl	400 µg

*[N.B. *myo*-Inositol is also used for aerobic growth tests as a major source of carbon.]

Detecting production of extracellular starch-like compounds

Purpose. This test helps to identify certain species, especially those of the genus *Cryptococcus*, which characteristically form extracellular starch-like polysaccharide, giving a blue colour with iodine.

Procedure. After growth in liquid medium on a sugar or an alditol, one drop of Lugol's iodine solution is shaken with the medium and yeast in the tube. A blue, purple or green colour indicates that the test is positive. Lugol's iodine solution (Cowan & Steel, 1966) is made up with 5 g iodine, 10 g potassium iodide and water to 100 ml. The potassium iodide and iodine are dissolved in 10 ml of water and made up to volume with more water. For use, the solution is diluted further, 1 : 5, with water.

Assessing growth at high concentrations of D-glucose

Purpose. Certain yeasts are well known to be able to grow in media with high concentrations of sugar. The following procedure tests this ability.

Procedure. Slopes are prepared of yeast extract agar, some containing 50% (w/w) D-glucose and others 60% (w/w) D-glucose. The slopes are inoculated lightly with streaks and examined for growth at 25 °C for up to four weeks. It may be necessary to take precautions against drying of the medium, for example sealing with wax paper.

The media

Acetate agar (McClary, Nulty & Miller, 1959). One litre of medium contains 9.8 g potassium acetate, 1.0 g D-glucose, 1.2 g NaCl, 0.7 g $MgSO_4.7H_2O$ and 2.5 g dried yeast extract and is formed into slopes with 20 g agar after autoclaving at 120 °C for 15 min.

Gorodkowa agar (van der Walt, 1970d). This contains D-glucose 0.1% (w/v), peptone 1% (w/v), NaCl 0.5% (w/v) and agar 2% (w/v).

Maize (corn) meal agar (van der Walt, 1970d). Yellow maize meal (12.5 g) is stirred with 300 ml water at 60 °C for 1 h and filtered. The volume of the filtrate is made up to 300 ml, 3.8 g agar added and the medium autoclaved at 120 °C for 15 min. This medium is marketed widely, for example by Oxoid Ltd (Basingstoke, Hampshire, England) or Difco Laboratories Inc. (Detroit, Michigan, USA).

Malt extract (Lodder & Kreger-van Rij, 1952). One kilogram of ground malt mixed with 2.6 litres of water is stirred at 45 °C for 3 h; the temperature is then raised to 63 °C for 1 h. After filtering, the mixture is sterilized for 15 min at 120 °C, refiltered and diluted to 15° Balling (specific gravity 1.06). The pH may be adjusted to ~ 5.4. For malt extract agar, the broth is adjusted to 10° Balling (specific gravity 1.04), 2% (w/v) agar is added, and the mixture is sterilized at 110 °C for 15 min. These media are marketed by several manufacturers including BBL (Becton, Dickinson & Co, Cockeysville, Maryland, USA) and Oxoid Ltd (Basingstoke, Hampshire, England).

Malt–yeast–glucose–peptone (Wickerham, 1951). The broth is made with 3 g dried malt extract, 3 g dried yeast extract, 5 g peptone, 10 g D-glucose and 1000 ml water. Unadjusted, the pH $\approx$ 5.5. Agar (20 g) is added for the solid medium. Difco Laboratories Inc. (Detroit, Michigan, USA) market these media as YM broth or YM agar.

Potato–glucose agar (van der Walt, 1970d). Washed potatoes are peeled and grated finely; 100 g are soaked overnight in 300 ml water in a refrigerator, filtered through cloth and the filtrate autoclaved for 1 h at 120 °C. Then 230 ml of the autoclaved extract is added to 770 ml of water, 20 g D-glucose and 20 g agar. After dissolving the agar, the medium is sterilized at 120 °C for 15 min. This medium is widely marketed, for example by Difco Laboratories Inc. (Detroit, Michigan, USA) or Oxoid Ltd (Basingstoke, Hampshire, England).

V-8 agar (van der Walt, 1970d). This contains a mixture of juices from several vegetables, and baker's yeast. In vessel A, 14 g agar are dissolved in 340 ml water. Vessel B contains 350 ml of V-8 juice (Campbell Soup Company, Camden, NJ, USA) well mixed with 5 g of compressed yeast dispersed in 10 ml water. Vessel B and its contents are heated for ~ 10 min in flowing steam and adjusted to pH 6.8 at ~ 20 °C. The hot contents of A and B are mixed, distributed into tubes, autoclaved at 120 °C for 15 min and cooled as slopes.

Yeast–glucose–peptone (van der Walt, 1970d). Broth is made with 5 g dried yeast extract, 20 g D-glucose and 10 g peptone with 1000 ml water, without adjusting the pH. The medium may be autoclaved at 120 °C for 15 min.

Table 5.2. *Composition of certain chemically defined media.* (Quantities are per litre of medium)

Ingredient	Yeast morphology agar	Nitrogen base	Carbon base	Vitamin-free medium
Nitrogen sources				
$(NH_4)_2SO_4$	3.5 g	5 g	None	5 g
L-Asparagine	1.5 g	None	None	None
Carbon source				
D-Glucose	10 g	None	10 g	10 g
Amino acids				
L-Histidine	10 mg	10 mg	1 mg	10 mg
DL-Methionine	20 mg	20 mg	2 mg	20 mg
DL-Tryptophan	20 mg	20 mg	2 mg	20 mg
Growth factors				
p-Aminobenzoic acid		200 µg		None
Biotin		20 µg		None
Folic acid		2 µg		None
myo-Inositol		10 mg		None
Nicotinic acid		400 µg		None
Pantothenate (Ca)		2 mg		None
Pyridoxine HCl		400 µg		None
Riboflavin		200 µg		None
Thiamin HCl		400 µg		None
Trace element sources				
H_3BO_3		500 µg		
$CuSO_4.5H_2O$		40 µg		
KI		100 µg		
$FeCl_3.6H_2O$		200 µg		
$MnSO_4.4H_2O$		400 µg		
$Na_2MoO_4.2H_2O$		200 µg		
$ZnSO_4.7H_2O$		400 µg		
Salts				
KH_2PO_4		850 mg		
K_2HPO_4		150 mg		
$MgSO_4.7H_2O$		500 mg		
NaCl		100 mg		
$CaCl_2.6H_2O$		100 mg		

Notes. (i) These media, which may be solidified with ~ 2% (w/v) washed agar, are marketed, dehydrated, by Difco Laboratories Inc. (Detroit, Michigan, USA). (ii) The amino acids may be omitted with advantage from the nitrogen base in order to reduce growth of some yeasts in the negative controls. (Difco markets Yeast Nitrogen Base Without Amino Acids.) (iii) References: Wickerham & Burton (1948); Wickerham (1951); Barnett & Ingram (1955); van der Walt (1970*d*); *Difco Manual of Dehydrated Culture Media and Reagents*, ninth edition (1972, Detroit, Difco Laboratories Inc.).

6
Characteristics of the species

This chapter tabulates the characteristics of 439 species as responses to taxonomic tests or results of observations, both derived from the records of the Yeast Division of the Centraalbureau voor Schimmelcultures. The way the tests and observations are made is described in Chapter 5.

In the table, a single asterisk after a specific name indicates that the species is not recognized in *The Yeasts* (Lodder, 1970); two asterisks mean that the yeast is included in that work, but under another name. Chapter 4 gives information about both kinds of species, with references.

Numbered characteristics used in Chapters 6, 7 and 8

Anaerobic fermentation
1   D-Glucose
2   D-Galactose
3   Maltose
4   Methyl α-D-glucopyranoside
5   Sucrose
6   α, α-Trehalose
7   Melibiose
8   Lactose
9   Cellobiose
10 Melezitose
11 Raffinose

Aerobic utilization and growth
12 D-Galactose
13 L-Sorbose
14 D-Ribose
15 D-Xylose
16 L-Arabinose
17 D-Arabinose
18 L-Rhamnose
19 Sucrose
20 Maltose
21 α, α-Trehalose
22 Methyl α-D-glucopyranoside
23 Cellobiose
24 Salicin
25 Arbutin
26 Melibiose
27 Lactose
28 Raffinose
29 Melezitose
30 Inulin
31 Starch
32 Glycerol
33 Erythritol
34 Ribitol
35 Galactitol
36 D-Mannitol
37 D-Glucitol
38 *myo*-Inositol
39 D-δ-Gluconolactone
40 2-Ketogluconate
41 5-Ketogluconate
42 DL-Lactate
43 Succinate
44 Citrate
45 Methanol

46 Ethanol
47 Ethylamine
48 D-Glucosamine
49 Nitrate
50 Nitrite
51 Without vitamins
52 Without inositol
53 Without pantothenate
54 Without biotin
55 Without thiamin
56 Without pyridoxine
57 Without niacin
58 Without folic acid
59 Without *p*-aminobenzoate
60 50% D-Glucose
61 60% D-Glucose
62 37°C
63 0.01% Cycloheximide
64 0.1% Cycloheximide
65 Starch formation

Appearance and reproduction
66 Pink colonies
67 Budding cells
68 Splitting cells
69 Apical budding
70 Cells spherical, oval or cylindrical
71 Cells of other shapes
72 Filamentous
73 Pseudohyphae
74 Septate hyphae
75 Ascosporogenous
76 Ascospores spherical, oval or reniform
77 Ascospores hat- or Saturn-shaped
78 Ascospores of other shapes

Table 6.1. *Characteristics of the species*

		1 D–Glucose fermentation	2 Galactose fermentation	3 Maltose fermentation	4 Me α–D–glucoside fermn	5 Sucrose fermentation	6 Trehalose fermentation	7 Melibiose fermentation	8 Lactose fermentation	9 Cellobiose fermentn	10 Melezitose fermentn	11 Raffinose fermentation	12 D–Galactose growth	13 L–Sorbose growth	14 D–Ribose growth	15 D–Xylose growth	16 L–Arabinose growth	17 D–Arabinose growth	18 L–Rhamnose growth	19 Sucrose growth	20 Maltose growth	21 Trehalose growth	22 Me α–D–glucoside grth	23 Cellobiose growth	24 Salicin growth	25 Arbutin growth	26 Melibiose growth	27 Lactose growth	28 Raffinose growth	29 Melezitose growth	30 Inulin growth	31 Starch growth
1	Aciculoconidium aculeatum**	V	–	–	?	–	V	–	–	?	?	–	–	–	–	–	–	–	–	+	+	+	+	+	+	+	–	–	–	V	–	+
2	Aessosporon dendrophilum*	–	–	–	–	–	–	–	–	–	–	–	+	V	+	+	+	+	V	+	+	+	D	+	V	+	–	D	–	+	–	+
3	Aessosporon salmonicolor**	–	–	–	–	–	–	–	–	–	–	–	V	V	V	V	V	V	–	+	–	+	–	–	–	–	–	–	V	–	–	–
4	Ambrosiozyma cicatricosa*	+	–	S	?	+	?	–	–	?	?	–	–	–	+	+	+	–	–	+	+	+	+	+	+	+	–	–	–	+	–	–
5	Ambrosiozyma monospora**	S	–	V	?	V	V	–	–	?	?	–	–	–	V	+	+	–	–	+	+	+	+	+	+	+	–	–	–	+	–	–
6	Ambrosiozyma philentoma*	S	–	–	?	V	?	–	–	?	?	–	–	–	+	+	–	–	+	+	+	+	+	+	+	+	–	–	–	+	–	–
7	Arthroascus javanensis**	–	–	–	–	–	–	–	–	–	–	–	–	S	–	–	–	–	–	–	V	V	–	–	–	–	–	–	–	–	–	V
8	Botryoascus synnaedendrus*	–	–	–	–	–	–	–	–	–	–	–	–	–	D	V	–	D	+	–	D	–	–	–	–	–	–	–	–	–	–	D
9	Brettanomyces abstinens*	+	D	–	–	V	–	–	–	?	–	–	+	–	–	–	–	–	–	–	V	–	–	D	D	D	–	–	–	–	–	–
10	Brettanomyces anomalus	+	+	V	V	V	V	–	+	+	V	V	+	–	V	–	–	–	–	+	V	+	V	+	+	+	–	+	D	V	V	–
11	Brettanomyces claussenii	+	+	+	+	+	+	–	+	+	+	V	+	–	V	–	–	–	–	+	+	+	+	+	+	+	–	+	+	+	V	–
12	Brettanomyces custersianus	S	–	–	–	–	S	–	–	–	–	–	V	–	V	–	–	–	–	S	V	+	–	–	–	–	–	–	–	–	–	–
13	Brettanomyces custersii	+	S	+	+	+	+	–	–	+	+	S	+	–	S	–	–	–	–	+	+	+	+	+	+	+	–	S	S	+	S	–
14	Brettanomyces lambicus	+	V	+	+	+	+	–	–	–	+	–	S	–	S	–	–	–	–	+	+	+	+	–	–	–	–	–	–	+	V	–
15	Brettanomyces naardenensis*	S	V	–	?	–	S	–	–	?	?	–	V	V	–	+	–	–	V	–	V	+	V	+	V	–	–	–	–	V	–	V
16	Bullera alba	–	–	–	–	–	–	–	–	–	–	–	V	V	V	+	+	V	+	+	+	+	+	+	S	S	V	+	+	+	–	V
17	Bullera piricola*	–	–	–	–	–	–	–	–	–	–	–	+	D	V	+	+	+	–	+	+	+	–	+	+	+	+	+	+	+	–	+
18	Bullera tsugae	–	–	–	–	–	–	–	–	–	–	–	–	V	–	–	–	–	–	+	+	+	–	+	+	S	–	+	–	+	–	–
19	Candida aaseri	–	–	–	–	–	–	–	–	–	–	–	+	+	+	+	+	–	–	+	+	+	+	+	+	+	–	D	–	+	–	–
20	Candida albicans	+	V	+	?	V	V	–	–	–	–	–	+	V	–	+	V	–	–	+	+	+	V	V	V	V	–	–	–	V	–	+
21	Candida amylolenta*	–	–	–	–	–	–	–	–	–	–	–	+	V	V	+	+	S	V	+	+	+	+	+	+	+	+	–	+	+	–	+
22	Candida aquatica	V	–	V	?	V	V	V	–	V	–	V	+	+	–	+	+	–	–	+	+	+	–	+	+	+	+	+	+	+	–	+
23	Candida beechii	+	–	–	–	–	+	–	–	–	–	–	–	+	–	–	–	–	–	–	–	+	–	+	+	+	–	–	–	–	–	–
24	Candida berthetii	S	–	–	–	–	–	–	–	–	–	–	–	–	–	–	–	–	–	–	–	–	–	+	+	+	–	–	–	–	–	–
25	Candida blankii	S	S	–	?	S	?	–	–	?	?	–	+	+	–	+	+	–	+	+	+	+	+	+	+	+	–	+	+	+	–	+
26	Candida bogoriensis	–	–	–	–	–	–	–	–	–	–	–	D	D	D	+	+	D	–	–	+	+	–	+	D	+	–	–	–	+	–	+
27	Candida boidinii	+	–	–	–	–	–	–	–	–	–	–	–	V	+	+	V	–	–	–	–	–	–	–	–	–	–	–	–	–	–	–
28	Candida boleticola*	+	–	–	–	–	?	–	–	–	?	–	+	D	+	+	–	D	–	–	–	+	–	V	D	+	–	–	–	–	–	–
29	Candida bombi*	+	–	–	–	+	S	–	–	–	–	–	+	–	+	–	–	–	–	+	–	V	–	–	–	–	–	–	+	–	–	–
30	Candida brassicae*	+	–	–	?	–	?	?	–	?	?	–	+	D	D	+	+	D	–	+	+	+	+	+	+	D	+	–	+	+	+	–
31	Candida brumptii	V	–	–	–	–	–	–	–	–	–	–	+	–	–	+	–	–	–	–	+	+	–	–	–	–	–	–	–	–	–	+
32	Candida buffonii	–	–	–	–	–	–	–	–	–	–	–	–	D	–	D	D	–	–	–	+	+	–	+	D	D	–	–	–	+	–	+
33	Candida buinensis*	+	+	S	?	–	?	–	–	?	?	–	+	+	D	+	D	+	–	+	+	+	+	+	+	+	–	–	–	+	–	+
34	Candida butyri*	+	S	–	?	–	?	–	–	?	?	–	+	D	+	+	+	+	–	+	+	+	V	+	+	+	–	D	–	D	–	–
35	Candida cacaoi	+	–	–	–	–	+	–	–	+	–	–	+	–	D	+	V	–	–	–	–	+	–	+	+	+	–	–	–	–	–	–
36	Candida catenulata	S	–	–	–	–	–	–	–	–	–	–	+	–	D	D	–	–	–	–	D	D	–	–	–	–	–	–	–	–	–	V
37	Candida chilensis*	+	–	–	?	–	–	–	–	–	?	?	–	+	+	+	+	+	+	+	+	+	+	+	+	+	–	+	–	+	–	–
38	Candida chiropterorum*	–	–	–	–	–	–	–	–	–	–	–	+	+	D	+	+	D	+	+	+	+	+	+	+	+	–	–	D	D	–	+

Codes in table: + positive; – negative; D delayed for longer than 7 days;

Characteristics of the species 49

	32 Glycerol growth	33 Erythritol growth	34 Ribitol growth	35 Galactitol growth	36 D–Mannitol growth	37 D–Glucitol growth	38 myo–Inositol growth	39 Gluconolactone growth	40 2–Ketogluconate growth	41 5–Ketogluconate growth	42 D,L–Lactate growth	43 Succinate growth	44 Citrate growth	45 Methanol growth	46 Ethanol growth	47 Ethylamine growth	48 D–Glucosamine growth	49 Nitrate growth	50 Nitrite growth	51 Growth sans vitamins	52 Growth sans inositol	53 Grth sans pantothenate	54 Growth sans biotin	55 Growth sans thiamin	56 Grth sans pyridoxine	57 Growth sans niacin	58 Growth sans folic acid	59 Growth sans PABA	60 Growth in 50% glucose	61 Growth in 60% glucose	62 Growth at 37 degrees	63 0.01% cyclohex. growth	64 0.1% cyclohex. growth	65 Starch formation	66 Pink colonies	67 Budding cells	68 Splitting cells	69 Apical budding	70 Cells sph, ovl, cylind	71 Cells of other shapes	72 Filamentous	73 Pseudohyphae	74 Septate hyphae	75 Ascosporogenous	76 Ascosp sph, ovl, renfm	77 Ascosp hat or saturn	78 Ascosp of other shapes
1	V	–	V	–	+	V	–	?	?	?	V	S	S	–	+	+	+	–	–	–	+	+	–	–	+	+	+	+	–	–	–	?	?	–	–	+	–	–	+	–	+	–	+	–	N	N	N
2	D	–	+	+	D		–	?	?	?		–	+	+		–	+	+	–	–	?	–	?	?	?	?		?	?	?	?	–	–	–	–	–	+	–	+	–	+	–	–	N	N	–	N	N	N
3	V	–	V	–	+	+		–	+	–	–		–	V	V		–	D		?	?	+	+	+	+	+	+	+		+	+	+	+		–	–	V	?	?	–	+	+	–	–	+	–	+	+	+	–	N	N	N
4	+	+	+	–	+	+		–	?	?	?		+	+	+		–	+		+	–	–	–	–	?	?	?	?		?	?	?	?		?	?	+	?	?	–	–	+	–	–	+	–	+	+	+	+	–	+	–
5	+	+	+	–	+	+		–	–	–	–		V	+	+		–	+		?	–	–	?	–	?	?	?	?		?	?	?	?		–	–	V	?	?	?	–	+	–	–	+	–	+	+	+	+	–	+	–
6	+	+	+	–	+	+		–	?	?	?		+	+	+		–	?		?	?	–	?	–	?	?	?	?		?	?	?	?		?	?	?	?	–	–	+	–	–	+	–	+	+	+	–	+	–
7	+	–	–	–	–	–		–	?	?	?		–	+	–		–	+		?	?	–	?	–	?	?	?	?		?	?	?	?		–	–	–	?	?	?	–	+	–	+	+	+	–	+	–	+	–
8	+	+	+	–	V	+		+	?	?	?		–	D	D		–	+		+	–	–	?	–	?	?	?	?		?	?	?	?		–	–	–	?	?	–	–	+	–	–	+	–	+	+	+	–	+	–
9	D	–	–	–	–		–	?	?	?		–	–	–		–	+		?	–	+	?	–	?	?	?	?		?	?	?	?		–	–	+	+	?	–	–	+	–	–	+	–	+	–	–	N	N	N
10	+	–	V	–	–	V		–	?	?	?		V	V	–		–	+		+	?	+	?	–	?	?	?	?		?	?	?	?		–	–	+	+	?	?	–	+	–	–	+	–	+	+	+	–	N	N	N
11	+	–	V	–	–	V		–	?	?	?		V	V	–		–	+		+	?	V	?	–	?	?	?	?		?	?	?	?		–	–	+	+	?	?	–	+	–	–	+	–	+	+	+	–	N	N	N
12	+	–	V	–	–	V		–	?	?	?		+	+	–		–	+		V	?	–	?	–	?	?	?	?		?	?	?	?		–	–	V	+	?	?	–	+	–	–	+	–	+	+	–	–	N	N	N
13	V	–	V	–	–	V		–	?	?	?		S	S	–		–	+		+	?	+	?	–	?	?	?	?		?	?	?	?		–	–	+	+	?	?	–	+	–	–	+	–	+	+	–	–	N	N	N
14	+	–	–	–	–	V		–	?	?	?		V	V	–		–	+		+	?	+	?	–	?	?	?	?		?	?	?	?		–	–	+	+	?	?	–	+	–	–	+	–	+	+	–	–	N	N	N
15	–	–	V	–	+	+		–	?	?	?		V	+	–		–	+		?	–	–	?	–	?	?	?	?		?	?	?	?		–	–	–	–	–	?	–	+	–	–	+	–	+	+	–	–	N	N	N
16	V	V	V	V	V	V		+	V	+	V		S	+	V		–	–		?	?	–	V	–	?	?	?	?		?	?	?	?		–	–	–	?	?	V	–	+	–	–	+	–	–	N	N	–	N	N	N
17	V	–	D	D	+	+		+	S	+	+		–	+	+		–	D		–	–	+	–	–	?	?	?	?		?	?	?	?		–	–	–	–	–	–	–	+	–	–	+	–	V	V	–	–	N	N	N
18	+	–	–	–	+	+		–	+	+	+		–	S	S		–	D		?	?	+	+	+	+	+	+	+		+	+	+	+		–	–	–	?	?	–	–	+	–	–	+	–	–	N	N	–	N	N	N
19	+	+	+	–	+	+		–	?	?	?		–	+	+		–	+		+	?	–	–	V	+	+	V	V		+	+	+	+		–	–	+	?	?	–	–	+	–	–	V	–	+	+	–	–	N	N	N
20	V	–	V	–	+	V		–	?	?	?		V	V	V		–	V		?	?	–	?	S	+	+	S	V		+	+	+	+		?	?	?	?	–	–	+	–	–	+	–	+	+	+	–	N	N	N
21	+	+	+	V	+	+		+	?	?	?		–	+	+		–	+		?	?	–	–	–	?	?	?	?		?	?	?	?		S	?	–	?	?	+	–	+	–	–	+	–	+	+	–	–	N	N	N
22	V	–	+	–	+	+		–	?	?	?		–	S	S		–	+		?	?	+	?	S	+	+	S	S		+	+	+	+		–	–	–	?	?	?	–	+	–	–	+	+	+	+	–	–	N	N	N
23	+	–	+	–	+	+		–	?	?	?		S	+	+		–	+		?	?	–	?	S	+	+	S	+		+	+	+	+		?	?	–	?	?	?	–	+	–	–	+	–	+	+	–	–	N	N	N
24	+	–	–	–	–	–		–	?	?	?		–	S	S		–	+		+	–	+	?	+	+	+	+	+		+	+	+	+		?	?	+	?	?	?	–	+	–	–	+	–	+	+	–	–	N	N	N
25	+	+	+	+	+	+		S	?	?	?		–	S	+		–	+		?	+	–	?	+	+	+	+	+		+	+	+	+		?	?	+	?	?	?	–	+	–	–	+	–	+	+	–	–	N	N	N
26	D	–	+	–	+	+		–	+	+	+		–	+	+		–	+		+	D	–	+	S	+	+	+	S		+	+	+	+		?	?	–	?	?	?	–	+	–	–	+	–	+	+	–	–	N	N	N
27	+	+	+	–	+	+		–	V	–	–		S	–	–		+	+		?	?	+	?	S	+	+	S	+		+	+	+	+		?	?	V	?	?	?	–	+	–	–	+	–	+	+	–	–	N	N	N
28	+	+	+	–	+	+		–	–	+	+	D		–	+	+		–	+		+	?	–	?	–	+	+	–	+		+	+	+	+		?	?	–	?	?	?	–	+	–	–	+	–	+	+	–	–	N	N	N
29	+	–	–	–	D	+		–	?	?	?		–	+	D		–	–		+	–	–	–	+	+	+	+	+		+	+	+	+		+	?	V	?	?	?	–	+	–	–	+	–	+	+	–	–	N	N	N
30	+	–	+	+	+	+		–	D	+	–		+	+	+		–	?		?	?	–	+	+	+	+	+	+		+	+	+	+		?	?	+	–	–	?	–	+	–	–	+	–	+	+	–	–	N	N	N
31	+	–	–	–	+	+		–	?	?	?		+	D	D		–	+		+	D	–	–	V	+	+	V	V		+	+	+	+		?	?	–	?	?	?	–	+	–	–	+	–	+	+	–	–	N	N	N
32	D	–	+	–	+	+		–	?	?	?		–	–	D		–	S		+	D	+	+	–	–	+	–	–		+	–	+	+		?	?	–	?	?	?	–	+	–	–	+	–	+	+	–	–	N	N	N
33	D	–	–	–	+	D		–	+	–	–		–	+	+		–	?		?	?	–	?	–	?	?	?	?		?	?	?	?		?	?	–	?	–	?	–	+	–	–	+	–	+	+	–	–	N	N	N
34	+	+	+	+	+	+		–	?	?	?		–	+	+		–	+		?	?	–	?	–	+	+	–	–		+	+	+	+		?	?	–	?	?	?	–	+	–	–	+	–	+	+	–	–	N	N	N
35	+	+	+	–	+	+		–	D	–	–		–	+	–		–	+		+	+	+	+	+	+	+	+	+		?	?	+	?	?	–	–	+	–	–	+	–	+	+	–	–	N	N	N
36	+	–	V	–	+	+		–	?	?	?		+	+	+		–	+		?	D	–	?	S	+	+	S	S		+	+	+	+		?	?	–	?	?	?	–	+	–	–	+	–	+	+	–	–	N	N	N
37	+	+	+	–	+	+		–	?	?	?		+	+	+		–	D		+	?	+	+	–	?	?	?	?		?	?	?	?		?	?	–	?	?	?	–	+	–	–	+	–	+	+	–	–	N	N	N
38	+	+	+	+	+	+		+	?	?	?		D	–	+		–	?		?	D	–	–	–	?	?	?	?		?	?	?	?		–	–	+	?	?	–	–	+	–	–	+	–	+	–	+	–	N	N	N

N not applicable; S slow utilisation; V variable; ? result not known.

Characteristics of the species

		1 D-Glucose fermentation	2 Galactose fermentation	3 Maltose fermentation	4 Me α-D-glucoside fermn	5 Sucrose fermentation	6 Trehalose fermentation	7 Melibiose fermentation	8 Lactose fermentation	9 Cellobiose fermentn	10 Melezitose fermentn	11 Raffinose fermentation	12 D-Galactose growth	13 L-Sorbose growth	14 D-Ribose growth	15 D-Xylose growth	16 L-Arabinose growth	17 D-Arabinose growth	18 L-Rhamnose growth	19 Sucrose growth	20 Maltose growth	21 Trehalose growth	22 Me α-D-glucoside grth	23 Cellobiose growth	24 Salicin growth	25 Arbutin growth	26 Melibiose growth	27 Lactose growth	28 Raffinose growth	29 Melezitose growth	30 Inulin growth	31 Starch growth
39	Candida citrea*	+	−	−	−	−	−	−	−	−	−	−	−	−	−	−	−	−	−	−	−	−	−	−	−	−	−	−	−	−	−	−
40	Candida conglobata	+	S	−	−	−	+	−	−	V	−	−	+	+	+	+	D	−	−	−	−	+	−	+	+	+	−	−	−	−	−	−
41	Candida curiosa	+	V	−	−	+	S	V	−	−	−	V	D	D	D	D	D	V	V	+	−	+	−	+	D	+	V	V	D	−	−	−
42	Candida curvata	−	−	−	−	−	−	−	−	−	−	−	+	V	+	+	V	−	+	+	+	+	V	+	+	+	−	+	+	V	−	+
43	Candida dendronema*	+	S	V	?	V	S	−	−	?	−	−	+	+	+	+	+	+	+	+	+	+	+	+	+	+	−	−	−	−	−	V
44	Candida diddensii	V	V	V	?	V	V	−	−	−	−	−	+	V	+	+	+	V	V	+	+	+	V	+	+	+	−	−	−	V	−	−
45	Candida diffluens	−	−	−	−	−	−	−	−	−	−	−	D	D	−	−	−	−	−	+	+	+	V	−	V	V	−	−	−	+	−	−
46	Candida diversa	+	−	−	−	−	−	−	−	−	−	−	−	−	−	V	−	−	−	−	−	−	−	−	−	−	−	−	−	−	−	−
47	Candida edax*	−	−	−	−	−	↓	−	−	−	−	−	+	+	+	+	+	+	+	+	+	+	+	+	+	+	+	D	+	+	−	D
48	Candida entomaea*	+	+	V	?	S	?	−	−	?	?	−	+	−	+	+	+	+	+	+	+	+	+	+	+	+	−	+	−	+	−	−
49	Candida entomophila*	+	S	−	?	+	?	V	−	?	?	+	+	D	+	+	+	D	−	+	+	+	+	+	+	+	+	+	+	+	−	−
50	Candida ergatensis*	+	S	−	?	−	?	−	−	?	−	−	+	+	V	+	D	D	−	+	+	+	+	+	+	+	−	+	−	−	−	−
51	Candida flavificans*	S	S	−	−	−	?	−	−	?	−	−	+	−	D	D	−	−	−	−	D	−	+	+	+	−	−	−	−	−	−
52	Candida fluviotilis*	S	−	V	?	−	S	−	−	?	−	−	+	−	−	+	−	−	−	+	+	+	+	+	+	+	−	+	−	+	−	+
53	Candida foliarum	−	−	−	−	−	−	−	−	−	−	−	V	−	V	S	V	V	−	−	−	+	−	S	−	−	−	−	−	−	−	−
54	Candida fragicola*	+	+	−	−	−	−	−	−	?	−	−	+	−	D	D	D	D	−	−	−	−	−	V	V	S	−	−	−	−	−	D
55	Candida freyschussii	+	−	−	?	−	−	−	−	+	−	−	−	−	−	+	−	−	+	+	+	D	+	+	+	+	−	−	−	+	−	−
56	Candida friedrichii	+	S	−	−	−	S	S	−	−	−	−	+	+	+	+	+	+	−	+	+	+	+	+	+	+	+	−	+	+	−	−
57	Candida glaebosa	−	−	−	−	−	−	−	−	−	−	−	+	+	−	+	V	V	−	+	+	+	+	+	+	+	+	+	+	V	−	−
58	Candida graminis*	−	−	−	−	−	−	−	−	−	−	−	D	+	+	V	−	V	−	+	+	+	D	+	+	+	−	D	−	+	−	V
59	Candida homilentoma*	+	+	V	?	−	?	−	−	?	?	−	+	V	D	+	+	V	+	+	+	+	+	+	+	+	−	−	−	+	−	+
60	Candida humicola	−	−	−	−	−	−	−	−	−	−	−	V	V	+	+	V	V	V	+	+	+	+	V	+	+	+	V	V	V	−	V
61	Candida hydrocarbofumarica*	−	−	−	−	−	−	−	−	−	−	−	+	+	D	+	+	D	+	+	+	+	+	+	+	+	−	+	D	+	−	+
62	Candida hylophila*	−	−	−	−	−	−	−	−	−	−	−	−	+	−	−	−	−	−	−	−	−	−	−	−	−	−	−	−	−	−	−
63	Candida iberica*	V	−	−	−	−	?	−	−	−	−	−	D	+	−	−	−	−	−	−	−	+	−	−	V	V	−	−	−	−	−	−
64	Candida incommunis*	S	−	−	?	V	?	−	−	?	?	−	−	+	−	V	−	−	−	+	+	+	+	+	+	+	−	−	−	+	−	−
65	Candida inositophila*	+	V	S	?	S	?	−	−	?	?	V	+	+	D	+	+	−	+	+	+	+	V	+	+	+	−	−	+	+	−	+
66	Candida insectamans*	S	−	−	?	−	?	−	−	?	−	−	−	−	+	−	−	−	−	−	+	+	+	+	+	+	−	−	−	−	−	+
67	Candida insectorum*	+	S	−	?	−	?	?	−	?	?	−	+	D	+	+	+	D	+	+	+	+	+	+	+	+	+	V	V	D	+	−	V
68	Candida intermedia	+	+	V	?	+	V	−	−	−	S	V	+	+	−	+	V	−	V	+	+	+	+	+	+	+	−	+	+	+	−	+
69	Candida ishiwadae*	+	V	+	?	V	?	−	−	?	?	−	V	+	+	+	D	+	+	+	+	+	+	+	+	+	−	−	−	+	−	+
70	Candida javanica	−	−	−	−	−	−	−	−	−	−	−	+	+	+	+	+	+	−	−	+	+	−	+	+	+	+	−	+	−	−	+
71	Candida kefyr	+	+	−	−	+	−	−	−	−	−	S	+	−	−	−	V	−	−	+	−	−	−	V	V	+	−	+	+	−	+	−
72	Candida krissii*	V	−	−	−	−	−	−	−	?	−	−	−	D	−	−	−	−	−	−	−	−	−	V	+	V	−	−	−	−	−	−
73	Candida lambica	+	−	−	−	−	−	−	−	−	−	−	−	−	−	+	−	−	−	−	−	−	−	−	−	−	−	−	−	−	−	−
74	Candida lusitaniae	+	V	V	?	V	V	−	−	+	−	−	V	+	V	+	V	−	+	+	+	+	V	+	+	+	−	−	−	+	−	−
75	Candida macedoniensis	+	+	−	−	+	−	−	−	−	−	+	+	−	−	+	+	−	−	+	−	V	−	+	+	+	−	+	+	−	+	−
76	Candida maltosa*	+	V	S	?	+	+	−	−	?	?	−	+	V	−	+	−	−	−	+	+	+	+	V	+	+	−	−	−	−	−	−
77	Candida marina	−	−	−	−	−	−	−	−	−	−	−	−	+	+	+	D	+	−	−	+	−	−	D	+	−	−	−	−	−	−	−
78	Candida maritima	+	−	−	?	S	−	−	−	−	−	S	−	−	+	−	−	+	+	+	V	+	+	−	+	−	−	V
79	Candida melibiosica	+	S	−	?	−	+	S	−	+	−	S	+	+	D	+	−	−	−	+	+	V	+	+	+	+	−	+	+	−	−
80	Candida melinii	−	−	−	−	−	−	−	−	−	−	−	−	−	−	+	−	−	+	+	V	+	−	−	−	−	−	+	−	−	−	−
81	Candida membranaefaciens	+	V	−	?	V	V	−	−	−	−	V	+	+	+	+	+	+	−	+	+	+	+	+	+	+	−	−	+	+	+	−
82	Candida mesenterica	−	−	−	−	−	−	−	−	−	−	−	−	V	V	−	−	−	−	+	+	+	V	V	V	V	−	−	−	V	−	−

Codes in table: + positive; − negative; D delayed for longer than 7 days;

Characteristics of the species

	32 Glycerol growth	33 Erythritol growth	34 Ribitol growth	35 Galactitol growth	36 D–Mannitol growth	37 D–Glucitol growth	38 myo–Inositol growth	39 Gluconolactone growth	40 2–Ketogluconate growth	41 5–Ketogluconate growth	42 D,L–Lactate growth	43 Succinate growth	44 Citrate growth	45 Methanol growth	46 Ethanol growth	47 Ethylamine growth	48 D–Glucosamine growth	49 Nitrate growth	50 Nitrite growth	51 Growth sans vitamins	52 Growth sans inositol	53 Grth sans pantothenate	54 Growth sans biotin	55 Growth sans thiamin	56 Grth sans pyridoxine	57 Growth sans niacin	58 Growth sans folic acid	59 Growth sans PABA	60 Growth in 50% glucose	61 Growth in 60% glucose	62 Growth at 37 degrees	63 0.01% cyclohex. growth	64 0.1% cyclohex. growth	65 Starch formation	66 Pink colonies	67 Budding cells	68 Splitting cells	69 Apical budding	70 Cells sph, ovl, cylind	71 Cells of other shapes	72 Filamentous	73 Pseudohyphae	74 Septate hyphae	75 Ascosporogenous	76 Ascosp sph, ovl, renfm	77 Ascosp hat or saturn	78 Ascosp of other shapes
39	+	–	–	–	–	–	–	+	–	–	–	+	V	–	?	?	–	–	?	–	+	+	–	–	+	+	+	+	?	?	–	?	?	–	–	+	–	–	+	–	+	+	–	–	N	N	N
40	+	+	+	–	+	+	–	+	–	–	V	+	–	–	+	?	+	–	?	–	+	+	–	+	+	+	+	+	?	?	–	?	?	–	–	+	–	–	+	–	+	+	–	–	N	N	N
41	D	–	+	–	+	+	V	?	?	?	V	D	V	–	V	?	?	+	?	–	+	+	–	–	+	+	+	+	?	?	–	?	?	–	–	+	–	–	+	–	+	+	+	–	N	N	N
42	+	+	+	–	V	V	+	?	?	?	V	V	V	–	+	?	?	–	?	S	+	+	S	S	+	+	+	+	?	?	V	?	?	+	–	+	–	–	+	–	+	+	–	–	N	N	N
43	+	+	+	+	+	+	–	+	–	?	–	+	+	–	+	?	?	–	?	–	+	+	–	+	+	+	+	+	+	?	–	?	?	–	–	+	–	–	+	+	+	+	–	–	N	N	N
44	+	+	+	–	+	+	–	+	–	V	–	S	V	–	+	?	V	–	?	–	+	+	–	–	+	+	+	+	?	?	V	?	?	–	–	+	–	–	+	+	V	V	–	–	N	N	N
45	D	+	V	–	D	+	–	–	–	–	D	D	–	–	+	+	?	+	+	+	+	+	+	+	+	+	+	+	?	?	–	?	?	–	–	+	–	–	+	–	+	V	–	–	N	N	N
46	–	–	+	–	+	+	–	?	?	?	–	+	+	–	+	+	–	–	–	–	–	+	–	–	–	+	+	+	–	–	S	?	?	–	–	+	–	–	+	–	V	V	–	–	N	N	N
47	+	+	+	D	+	+	+	?	?	?	+	+	+	–	+	+	+	+	+	+	+	+	+	+	+	+	+	+	–	–	+	?	?	–	–	+	–	–	+	–	+	+	+	–	N	N	N
48	+	+	+	+	+	+	–	?	?	?	–	+	+	–	+	?	?	–	?	–	?	?	?	?	?	?	?	?	+	?	+	?	?	–	–	+	–	–	+	–	+	+	+	–	N	N	N
49	+	+	+	+	+	+	–	?	?	?	–	+	+	–	+	?	?	–	?	–	?	?	?	?	?	?	?	?	?	?	+	?	?	–	–	+	–	–	+	+	+	+	+	–	N	N	N
50	+	+	+	+	+	+	–	?	?	?	–	+	+	–	+	?	+	–	?	–	?	?	?	?	?	?	?	?	–	–	–	?	?	–	–	+	–	–	+	–	+	+	+	–	N	N	N
51	+	+	–	–	–	D	–	–	–	–	–	D	–	–	?	?	–	–	?	–	+	+	–	–	–	+	+	+	–	–	+	–	–	–	–	+	–	–	+	–	+	–	+	–	N	N	N
52	+	–	+	–	+	+	–	?	?	?	+	+	+	–	+	?	?	–	?	–	+	+	+	+	+	+	+	+	?	?	?	?	?	–	–	+	–	–	+	–	–	+	–	–	N	N	N
53	S	–	V	–	+	+	–	?	?	?	S	S	S	–	+	+	+	+	+	S	+	+	S	S	+	+	+	+	?	?	–	?	?	–	–	+	–	–	+	–	–	+	+	–	N	N	N
54	+	–	D	–	D	D	+	+	–	+	+	+	–	–	+	?	–	–	?	–	+	+	–	–	–	+	+	+	?	?	V	?	?	–	–	+	–	–	+	–	+	+	–	–	N	N	N
55	+	–	–	–	+	+	–	?	?	?	+	+	+	–	+	?	–	–	?	–	+	+	–	–	+	+	+	+	?	?	+	?	?	–	–	+	–	–	+	–	+	+	–	–	N	N	N
56	+	+	+	+	+	+	–	?	?	?	–	+	+	–	+	+	?	–	–	S	+	+	S	+	+	+	+	+	?	?	–	?	?	–	–	+	–	–	+	–	+	+	–	–	N	N	N
57	+	–	+	–	+	+	–	–	–	+	–	+	+	+	–	+	?	–	–	–	–	–	–	–	–	+	+	+	?	?	–	?	?	–	–	+	–	–	+	–	+	+	–	–	N	N	N
58	+	+	+	–	+	+	–	?	?	?	+	+	+	–	–	?	?	+	?	–	+	+	+	–	+	+	+	+	?	?	–	?	?	–	–	+	–	–	+	–	+	+	–	–	N	N	N
59	+	+	+	–	+	+	–	?	?	?	–	+	+	–	+	?	?	–	?	+	+	+	+	+	+	+	+	+	+	?	+	?	?	–	–	+	–	–	+	+	+	+	–	–	N	N	N
60	V	+	V	V	–	V	V	?	?	?	V	V	V	–	V	?	?	–	?	V	+	+	+	V	+	+	+	+	?	?	V	?	?	+	–	+	–	–	+	+	+	+	+	–	N	N	N
61	D	+	+	+	+	+	+	+	–	–	+	+	–	–	?	?	D	–	?	+	+	+	+	+	+	+	+	+	?	?	+	?	?	–	–	+	–	–	+	–	+	+	+	–	N	N	N
62	+	–	–	–	+	+	–	?	?	?	+	+	–	–	S	?	?	–	?	–	?	?	?	?	?	?	?	?	–	–	–	?	?	–	–	+	–	–	+	+	+	+	–	–	N	N	N
63	+	–	D	–	+	+	–	+	+	D	–	+	+	–	?	?	D	–	?	V	+	+	V	+	+	+	+	+	?	?	–	?	?	–	–	+	–	–	+	–	+	+	–	–	N	N	N
64	+	+	–	–	–	+	+	+	+	?	–	+	+	–	?	+	+	+	?	–	+	+	–	+	+	+	+	+	?	?	–	?	?	–	–	+	–	–	+	–	+	+	–	–	N	N	N
65	+	–	+	D	+	+	+	+	–	+	–	+	–	–	?	+	–	–	?	–	+	+	–	–	+	+	+	+	–	–	+	–	–	–	–	+	–	–	+	–	+	+	–	–	N	N	N
66	–	–	+	–	+	+	–	?	?	?	–	D	+	–	–	?	?	–	V	?	?	?	?	?	?	?	?	?	–	–	+	–	–	?	–	+	–	–	+	–	+	+	–	–	N	N	N
67	+	+	+	–	+	+	–	?	?	?	–	+	+	–	+	?	?	–	?	–	?	?	?	?	?	?	?	?	?	V	?	?	?	–	–	+	–	–	+	–	+	+	–	–	N	N	N
68	–	–	–	–	+	+	–	?	?	?	–	V	V	–	+	?	?	–	?	S	+	+	S	+	+	+	+	+	?	V	?	?	–	–	–	+	–	–	+	–	+	+	–	–	N	N	N
69	+	+	+	V	+	+	–	+	D	?	+	+	+	–	?	?	+	+	?	–	+	+	–	–	+	+	+	+	?	?	–	?	?	–	–	+	–	–	+	–	+	+	–	–	N	N	N
70	+	–	+	+	+	+	–	?	?	?	–	–	–	–	+	+	–	+	S	+	+	+	S	+	+	+	?	?	–	–	+	–	–	+	–	+	–	–	+	–	+	+	–	–	N	N	N
71	V	–	–	–	V	V	–	?	?	?	+	+	V	–	S	?	–	?	–	+	–	–	+	+	+	–	–	+	?	?	+	–	–	–	–	+	–	–	+	–	+	+	–	–	N	N	N
72	+	–	D	–	+	+	–	+	+	D	–	+	+	–	?	?	V	–	?	–	+	+	–	+	+	+	+	+	?	?	–	?	?	–	–	+	–	–	+	–	+	+	–	–	N	N	N
73	V	–	–	–	–	–	–	V	–	–	+	+	V	–	+	?	V	–	?	V	+	+	+	V	+	+	+	+	?	?	V	?	?	–	–	+	–	–	+	–	+	+	–	–	N	N	N
74	V	–	+	–	+	+	–	?	?	?	V	+	V	–	+	?	?	–	?	S	+	+	S	S	S	+	+	+	?	?	+	?	?	–	–	+	–	–	+	–	V	V	–	–	N	N	N
75	+	–	V	–	V	V	–	?	?	?	+	S	–	–	+	?	?	–	?	V	+	+	+	V	+	+	+	+	?	?	+	?	?	–	–	+	–	–	+	–	+	+	–	–	N	N	N
76	+	–	+	–	+	+	–	+	+	+	V	+	V	–	+	?	?	–	?	–	?	?	?	?	?	?	?	?	?	?	+	?	?	–	–	+	–	–	+	–	+	+	–	–	N	N	N
77	+	+	+	–	+	+	+	?	?	?	–	–	–	–	+	?	+	–	?	–	+	+	–	–	–	+	+	+	?	?	–	?	?	S	–	+	–	–	+	–	+	+	–	–	N	N	N
78	+	–	–	–	+	+	–	?	?	?	+	+	V	–	+	+	?	–	–	–	+	+	–	–	–	+	+	+	–	–	–	?	?	–	–	+	–	–	+	–	+	+	–	–	N	N	N
79	S	–	+	–	+	+	–	?	?	?	–	+	+	–	+	?	V	–	?	S	+	+	S	+	+	+	+	+	?	?	–	?	?	–	–	+	–	–	+	–	+	+	–	–	N	N	N
80	+	–	–	–	+	+	–	?	?	?	S	+	V	–	+	?	?	+	?	V	+	+	V	V	+	V	+	+	+	?	?	–	?	?	–	–	+	–	–	+	–	+	+	–	–	N	N	N
81	+	+	+	+	+	+	–	?	?	?	V	V	V	–	+	?	?	–	?	S	+	+	S	+	+	+	+	+	?	?	V	?	?	–	–	+	–	–	+	–	+	+	–	–	N	N	N
82	+	+	V	–	V	V	–	?	?	?	S	V	V	–	+	?	?	–	?	S	+	+	S	+	+	V	+	+	+	?	?	–	?	?	–	–	+	–	–	+	–	+	+	–	–	N	N	N

N not applicable; S slow utilisation; V variable; ? result not known.

Characteristics of the species

		1 D–Glucose fermentation	2 Galactose fermentation	3 Maltose fermentation	4 Me α–D–glucoside fermn	5 Sucrose fermentation	6 Trehalose fermentation	7 Melibiose fermentation	8 Lactose fermentation	9 Cellobiose fermentn	10 Melezitose fermentn	11 Raffinose fermentation	12 D–Galactose growth	13 L–Sorbose growth	14 D–Ribose growth	15 D–Xylose growth	16 L–Arabinose growth	17 D–Arabinose growth	18 L–Rhamnose growth	19 Sucrose growth	20 Maltose growth	21 Trehalose growth	22 Me α–D–glucoside grth	23 Cellobiose growth	24 Salicin growth	25 Arbutin growth	26 Melibiose growth	27 Lactose growth	28 Raffinose growth	29 Melezitose growth	30 Inulin growth	31 Starch growth
83	Candida milleri*	+	+	–	–	+	?	–	–	–	–	+	+	–	–	–	–	–	–	+	–	+	–	–	–	–	–	–	+	–	V	–
84	Candida mogii	+	–	+	–	+	+	–	–	–	–	V	+	+	S	+	+	–	–	+	+	+	–	–	–	–	–	–	V	–	V	+
85	Candida muscorum	–	–	–	–	–	–	–	–	–	–	–	+	+	D	+	–	D	–	+	+	+	+	+	+	+	–	+	+	+	–	–
86	Candida naeodendra*	S	S	–	?	–	?	–	–	?	?	–	+	D	+	+	+	D	+	+	+	+	+	+	+	+	–	–	–	+	–	+
87	Candida norvegensis	S	–	–	–	–	–	–	–	V	–	–	–	–	–	V	–	–	–	–	–	–	–	+	+	+	–	–	–	–	–	–
88	Candida oleophila*	+	+	S	?	S	–	–	–	–	?	–	+	+	D	+	–	–	–	+	+	+	D	+	+	+	–	–	–	+	–	–
89	Candida oregonensis	+	–	V	?	–	–	–	–	?	?	–	–	–	–	–	+	–	–	+	+	+	+	+	+	+	–	–	–	+	–	+
90	Candida parapsilosis	+	V	V	?	V	–	–	–	–	–	–	+	V	V	+	+	–	–	+	+	+	+	–	–	–	–	–	–	–	–	–
91	Candida podzolica*	–	–	–	–	–	–	–	–	–	–	–	+	+	+	+	+	+	+	+	+	+	+	D	+	+	+	+	+	+	–	+
92	Candida pseudointermedia*	+	+	S	?	+	?	–	–	?	?	–	+	+	+	+	V	V	D	+	+	+	+	+	+	+	–	–	+	+	–	+
93	Candida pseudotropicalis	+	+	–	–	+	–	–	+	–	–	V	+	–	V	–	+	–	–	+	–	–	–	+	+	+	–	+	+	–	+	–
94	Candida quercuum*	S	–	–	?	–	?	–	–	–	?	–	–	–	–	–	+	–	–	+	+	+	+	+	+	+	–	–	–	+	–	–
95	Candida ravautii	V	–	–	–	–	–	–	–	–	–	–	+	–	–	+	–	–	–	–	+	–	–	–	–	–	–	–	–	–	–	+
96	Candida rhagii	+	V	–	?	+	V	–	–	–	–	V	+	–	+	+	V	–	–	+	+	+	+	+	+	+	–	–	+	+	–	–
97	Candida rugopelliculosa*	+	–	–	–	–	–	–	–	–	–	–																				
98	Candida rugosa	–	–	–	–	–	–	–	–	–	–	–	+	V	–	V	V	V	–	–	–	–	–	–	–	–	–	–	–	–	–	–
99	Candida sake	+	V	V	?	V	V	–	–	–	–	–	+	+	V	+	–	–	–	+	+	+	V	V	V	V	–	–	–	V	–	–
100	Candida salmanticensis	+	S	+	?	+	+	+	–	–	+	+	+	+	–	+	–	+	–	+	+	+	+	+	+	+	+	V	+	+	+	D
101	Candida santamariae	+	–	–	–	+	–	–	–	–	–	–	V	+	V	V	–	–	–	–	+	–	–	+	+	+	–	–	–	–	–	–
102	Candida savonica*	S	S	–	–	–	?	–	–	?	–	–	+	–	–	D	–	D	–	–	+	–	–	+	+	+	–	–	–	–	–	–
103	Candida silvae	–	–	–	–	–	–	–	–	–	–	–	–	–	–	–	–	V	–													
104	Candida silvanorum*	+	+	S	?	S	?	?	S	?	?	S	+	D	+	+	+	V	+	+	+	+	+	+	+	+	+	–	+	+	–	+
105	Candida silvicultrix*	+	+	S	?	+	?	+	–	?	?	+	+	–	+	+	+	+	+	+	+	+	+	+	+	+	+	–	+	+	–	V
106	Candida solani	+	–	–	–	–	–	–	–	–	–	–	–	+	–	+	–	+	–	+	+	+	+	+	+	+	–	–	–	+	–	–
107	Candida sorboxylosa*	+	–	–	–	–	–	–	–	–	–	–	–	D	–	D	–	–	–	–	–	–	–	–	–	–	–	–	–	–	–	–
108	Candida steatolytica*	+	+	V	?	+	?	–	–	?	?	S	+	+	D	+	+	–	+	+	+	+	V	+	+	+	–	V	+	V	–	+
109	Candida suecica*	V	–	–	?	–	–	–	–	?	–	–	–	+	–	V	–	V	–	+	+	+	+	+	V	V	–	–	–	–	–	–
110	Candida tenuis	+	+	–	?	–	V	–	–	–	–	–	+	V	+	+	V	V	+	+	+	+	V	+	+	+	–	V	–	+	–	V
111	Candida tepae*	–	–	–	–	–	–	–	–	–	–	–	D	D	–	+	–	–	–	+	+	+	V	+	D	+	–	–	–	–	–	–
112	Candida terebra*	S	S	S	?	S	?	–	–	?	?	–	+	+	+	+	+	+	+	+	+	+	+	+	+	+	–	–	–	+	–	–
113	Candida tropicalis	+	+	+	?	V	+	–	–	?	?	–	+	V	V	+	–	–	–	V	+	+	+	+	V	V	–	–	–	+	–	+
114	Candida tsukubaensis*	–	–	–	–	–	–	–	–	–	–	–	+	+	+	+	+	D	–	+	+	+	+	+	–	+	–	+	+	+	–	+
115	Candida utilis	+	–	–	?	+	–	–	–	–	–	+	–	–	–	V	–	–	–	+	+	V	V	+	+	+	–	–	+	+	+	–
116	Candida valdiviana*	S	–	–	?	–	?	?	–	?	?	–	+	+	–	+	V	V	–	+	+	+	+	+	+	+	+	V	+	V	–	–
117	Candida valida	V	–	–	–	–	–	–	–	–	–	–	–	–	–	V	–	–	–	–	–	–	–	–	–	–	–	–	–	–	–	–
118	Candida vartiovaarai	+	–	–	?	+	–	–	–	–	–	–	–	–	–	+	–	–	–	+	+	+	+	+	+	+	–	–	–	+	–	–
119	Candida veronae	+	V	–	?	–	V	–	–	?	?	–	+	–	+	+	+	D	+	+	+	+	D	+	D	D	–	–	–	+	–	–
120	Candida vinaria*	–	–	–	–	–	–	–	–	–	–	–	+	+	–	V	–	–	–													
121	Candida vini	–	–	–	–	–	–	–	–	–	–	–																				
122	Candida viswanathii	+	S	+	?	–	+	–	–	–	–	–	+	V	–	+	V	–	–	+	+	+	+	+	+	+	–	–	–	+	–	+
123	Candida zeylanoides	S	–	–	–	–	–	–	–	–	–	–	V	V	–	V	V	–	–	–	–	+	–	V	+	+	–	–	–	–	–	–
124	Citeromyces matritensis	+	–	–	?	+	?	–	–	–	–	+	–	D	–	–	–	–	–	+	+	+	+	–	–	–	–	–	+	–	D	–
125	Cryptococcus albidus	–	–	–	–	–	–	–	–	–	–	–	V	V	V	+	+	V	V	+	+	+	V	+	+	V	+	+	+	–	V

Codes in table: + positive; – negative; D delayed for longer than 7 days;

Characteristics of the species

	32 Glycerol growth	33 Erythritol growth	34 Ribitol growth	35 Galactitol growth	36 D–Mannitol growth	37 D–Glucitol growth	38 myo–Inositol growth	39 Gluconolactone growth	40 2–Ketogluconate growth	41 5–Ketogluconate growth	42 D,L–Lactate growth	43 Succinate growth	44 Citrate growth	45 Methanol growth	46 Ethanol growth	47 Ethylamine growth	48 D–Glucosamine growth	49 Nitrate growth	50 Nitrite growth	51 Growth sans vitamins	52 Growth sans inositol	53 Grth sans pantothenate	54 Growth sans biotin	55 Growth sans thiamin	56 Grth sans pyridoxine	57 Growth sans niacin	58 Growth sans folic acid	59 Growth sans PABA	60 Growth in 50% glucose	61 Growth in 60% glucose	62 Growth at 37 degrees	63 0.01% cyclohex. growth	64 0.1% cyclohex. growth	65 Starch formation	66 Pink colonies	67 Budding cells	68 Splitting cells	69 Apical budding	70 Cells sph, ovl, cylind	71 Cells of other shapes	72 Filamentous	73 Pseudohyphae	74 Septate hyphae	75 Ascosporogenous	76 Ascosp sph, ovl, renfm	77 Ascosp hat or saturn	78 Ascosp of other shapes
83	V	–	–	–	–	–	–	V	–	–	V	V	–	–	?	V	?	–	?	–	+	S	–	V	+	V	+	+	–	–	–	V	–	–	–	+	–	–	+	–	–	N	N	–	N	N	N
84	+	–	V	V	+	V	–	+	?	?	–	V	+	–	–	+	+	–	–	+	+	+	+	+	+	+	+	+	?	?	+	–	–	–	–	+	–	–	+	–	+	+	–	–	N	N	N
85	+	–	D	D	+	+	–	+	+	?	–	+	S	–	–	?	?	+	?	–	+	+	+	–	+	+	+	+	–	–	–	?	?	–	–	+	–	–	+	–	+	+	–	–	N	N	N
86	+	+	+	–	+	+	–	?	?	?	–	+	+	–	+	?	?	–	?	–	?	?	?	?	?	?	?	?	+	?	+	–	–	–	–	+	–	–	+	+	+	+	–	–	N	N	N
87	+	–	–	–	–	–	–	–	–	–	+	+	+	–	+	?	+	–	?	–	+	+	–	–	–	+	+	+	?	?	+	?	?	–	–	+	–	–	+	–	+	+	–	–	N	N	N
88	+	–	D	–	+	+	–	+	+	D	D	+	+	–	?	+	?	–	–	?	?	?	?	?	?	?	?	?	+	?	–	–	–	–	–	+	–	–	+	–	+	+	–	–	N	N	N
89	–	–	+	–	+	+	–	?	?	?	–	S	–	–	+	+	?	–	S	+	+	S	+	+	+	+	+	+	?	?	–	?	–	–	–	+	–	–	+	–	+	+	–	–	N	N	N
90	+	–	+	–	+	+	–	?	?	?	V	V	V	–	+	?	?	–	?	S	+	+	S	+	+	+	+	+	?	?	+	?	–	–	–	+	–	–	+	–	+	+	–	–	N	N	N
91	–	–	–	+	–	D	+	?	?	?	D	+	+	–	V	?	?	–	?	+	+	+	+	+	+	+	+	+	?	?	–	?	?	+	–	+	–	–	+	–	+	+	+	–	N	N	N
92	V	–	+	V	+	+	–	+	+	?	–	+	+	–	?	?	?	–	+	–	+	+	–	+	+	+	+	+	?	?	V	?	?	–	–	+	–	–	+	–	+	+	–	–	N	N	N
93	V	–	–	–	V	V	–	?	?	?	+	V	V	–	+	?	?	–	?	–	+	–	–	–	+	–	+	–	?	?	+	?	?	–	–	+	–	–	+	–	+	+	–	–	N	N	N
94	+	–	–	–	+	+	–	?	?	?	+	+	+	–	–	?	?	–	?	–	+	+	–	–	–	+	+	+	?	?	+	?	?	–	–	+	–	–	+	–	+	+	–	–	N	N	N
95	+	–	–	–	+	+	–	?	?	?	+	V	V	–	V	?	?	–	?	S	+	+	S	S	+	+	+	+	?	?	V	?	?	–	–	+	–	–	+	–	+	+	–	–	N	N	N
96	+	+	+	–	+	+	–	?	?	?	V	V	–	–	+	?	?	–	?	V	+	+	+	V	+	+	+	+	?	?	V	?	?	–	–	+	–	–	+	–	+	+	–	–	N	N	N
97	–	–	–	–	–	–	–	–	+	–	–	+	+	–	?	?	–	?	+	+	+	+	+	+	+	+	+	+	?	?	+	?	?	–	–	+	–	–	+	–	+	+	–	–	N	N	N
98	V	–	V	–	+	+	–	V	–	–	+	+	V	–	+	?	–	–	?	–	+	+	–	V	V	+	+	+	?	?	+	?	?	?	–	+	–	–	+	–	+	+	–	–	N	N	N
99	+	–	V	–	+	+	–	V	+	V	V	+	V	–	V	?	?	–	?	V	+	+	V	+	+	+	+	+	?	?	–	?	?	–	–	+	–	–	+	–	+	+	–	–	N	N	N
100	–	–	D	–	+	+	–	D	+	+	+	+	D	–	+	?	+	–	?	–	+	+	+	–	–	+	+	+	?	?	V	?	?	–	–	+	–	–	+	–	+	+	–	–	N	N	N
101	+	–	+	–	+	+	–	V	+	V	+	+	+	–	+	?	+	–	?	–	+	–	+	+	+	+	+	+	?	?	–	?	?	–	–	+	–	–	+	–	+	+	–	–	N	N	N
102	+	–	+	–	+	+	–	D	+	–	–	+	+	–	?	?	D	–	?	S	+	+	S	+	+	+	+	+	?	?	–	?	?	–	–	+	–	–	+	–	+	+	–	–	N	N	N
103	+	–	V	–	+	+	–	?	?	?	V	V	–	–	+	?	?	–	?	S	+	+	V	S	S	+	+	+	?	?	V	?	?	–	–	+	–	–	+	–	+	+	–	–	N	N	N
104	+	+	+	–	+	+	–	?	?	?	–	+	+	–	+	?	?	–	?	–	?	?	?	?	?	?	?	?	?	?	+	?	?	–	–	+	–	–	+	+	+	+	–	–	N	N	N
105	+	+	+	–	+	+	–	?	?	?	+	+	+	–	+	?	?	–	?	–	?	?	?	?	?	?	?	?	?	?	V	?	?	–	–	+	–	–	+	–	+	+	–	–	N	N	N
106	+	–	–	–	–	–	–	?	?	?	+	+	–	–	+	?	?	–	?	S	+	+	S	S	S	+	+	+	?	?	–	?	?	–	–	+	–	–	+	–	+	+	–	–	N	N	N
107	D	–	–	–	–	–	–	–	–	–	+	+	V	–	?	?	–	–	?	–	+	+	+	–	–	+	+	+	?	?	V	?	?	–	–	+	–	–	+	–	+	+	–	–	N	N	N
108	+	–	+	+	+	+	+	+	–	+	D	+	+	–	+	+	+	–	–	–	+	+	–	–	+	+	+	+	?	?	+	?	?	–	–	+	–	–	+	+	+	+	–	–	N	N	N
109	–	–	+	–	+	+	–	?	?	?	–	–	D	–	–	?	+	–	?	–	?	?	?	?	?	?	?	?	?	?	?	?	?	–	–	+	–	–	+	–	+	+	–	–	N	N	N
110	+	+	+	–	+	+	–	?	?	?	V	V	V	–	+	?	?	–	?	S	+	+	S	V	+	+	+	+	?	?	V	?	?	–	–	+	–	–	+	–	+	+	–	–	N	N	N
111	+	–	V	–	+	+	–	–	–	?	+	+	–	–	+	?	–	–	?	–	?	?	?	?	?	?	?	?	?	?	–	?	?	–	–	+	–	–	+	–	+	+	–	–	N	N	N
112	+	+	+	D	+	+	–	+	?	?	–	+	+	–	?	?	+	–	–	?	–	+	+	+	–	–	+	+	+	?	?	+	?	?	–	–	+	–	–	+	–	+	+	–	–	N	N	N
113	V	–	V	–	+	+	–	+	+	+	V	+	V	–	+	?	?	–	?	–	+	+	–	V	+	+	+	+	?	?	+	?	?	–	–	+	–	–	+	–	+	+	+	–	N	N	N
114	+	+	–	–	–	–	+	?	?	?	+	+	+	–	+	?	–	+	+ +	+	+	+	+	+	+	+	?	?	–	–	+	–	–	+	–	+	–	–	N	N	N
115	+	–	–	–	V	–	–	?	?	?	+	V	+	–	V	?	?	+	?	V	+	+	+	V	+	+	+	+	?	?	–	?	?	–	–	+	–	–	+	–	+	+	–	–	N	N	N
116	D	–	V	V	+	+	+	+	–	?	V	+	+	–	?	?	?	+	–	?	?	?	?	?	?	?	?	?	?	V	+	–	–	–	–	+	–	–	+	–	+	+	–	–	N	N	N
117	+	–	–	–	–	–	–	V	–	–	V	V	–	–	+	?	–	–	V	+	+	V	?	?	?	?	–	–	+	–	–	+	–	+	+	–	–	N	N	N
118	+	–	–	–	+	+	–	?	?	?	+	+	–	–	+	?	–	+	?	+	+	+	+	+	+	+	+	+	?	?	–	?	?	–	–	+	–	–	V	V	–	N	N	N
119	+	+	+	–	+	+	–	?	?	?	–	+	+	–	+	?	–	–	?	–	?	?	+	+	–	–	+	+	+	+	+	?	?	–	–	+	–	–	+	–	+	+	–	–	N	N	N
120	+	–	–	–	V	V	–	+	–	–	–	–	–	–	?	?	D	–	?	–	?	+	+	+	–	?	?	?	?	?	?	–	–	+	–	–	+	–	+	+	–	–	N	N	N
121	V	–	V	–	+	+	–	?	?	?	V	V	–	–	+	?	?	–	?	–	+	+	–	V	+	–	+	+	?	?	–	?	?	–	–	+	–	–	+	–	+	+	–	–	N	N	N
122	+	–	+	–	+	+	–	?	?	?	V	V	+	–	+	?	?	–	?	S	+	+	S	+	+	+	+	+	?	?	+	?	?	–	–	+	–	–	+	–	+	+	–	–	N	N	N
123	+	–	V	–	+	+	–	D	+	V	–	+	+	–	+	?	V	–	?	V	+ +	V	+	+	+	?	?	–	?	?	–	–	+	–	–	+	–	+	+	–	–	N	N	N
124	D	–	D	–	+	+	–	?	?	?	S	–	S	–	S	?	?	?	+	S	?	?	?	?	?	?	+	+	–	?	?	–	–	+	–	–	+	–	–	N	N	+	+	–	–
125	V	V	V	V	V	V	–	+	V	+	V	V	V	–	V	?	?	+	+	V	+	+	+	V	+	+	+	+	–	–	–	?	?	+	–	+	–	–	+	–	–	N	N	–	N	N	N

N not applicable; S slow utilisation; V variable; ? result not known.

Characteristics of the species

		1 D–Glucose fermentation	2 Galactose fermentation	3 Maltose fermentation	4 Me α–D–glucoside fermn	5 Sucrose fermentation	6 Trehalose fermentation	7 Melibiose fermentation	8 Lactose fermentation	9 Cellobiose fermentn	10 Melezitose fermentn	11 Raffinose fermentation	12 D–Galactose growth	13 L–Sorbose growth	14 D–Ribose growth	15 D–Xylose growth	16 L–Arabinose growth	17 D–Arabinose growth	18 L–Rhamnose growth	19 Sucrose growth	20 Maltose growth	21 Trehalose growth	22 Me α–D–glucoside grth	23 Cellobiose growth	24 Salicin growth	25 Arbutin growth	26 Melibiose growth	27 Lactose growth	28 Raffinose growth	29 Melezitose growth	30 Inulin growth	31 Starch growth
126	Cryptococcus ater	−	−	−	−	−	−	−	−	−	−	−	V	V	V	+	+	V	+	+	+	+	+	+	+	+	−	+	V	V	−	−
127	Cryptococcus bhutanensis*	−	−	−	−	−	−	−	−	−	−	−	D	+	V	+	+	D	−	+	+	+	+	+	+	+	−	V	+	+	−	+
128	Cryptococcus cereanus*	−	−	−	−	−	−	−	−	−	−	−	−	+	D	+	−	−	−	−	−	+	−	+	+	+	−	−	−	−	−	−
129	Cryptococcus dimennae	−	−	−	−	−	−	−	−	−	−	−	+	V	+	+	+	+	+	+	−	+	−	+	+	+	−	+	+	−	−	−
130	Cryptococcus flavus	−	−	−	−	−	−	−	−	−	−	−	+	−	+	+	+	V	+	+	+	+	+	+	+	+	+	+	+	+	V	+
131	Cryptococcus gastricus	−	−	−	−	−	−	−	−	−	−	−	+	−	−	+	+	−	V	V	+	+	−	+	+	+	−	V	−	+	−	V
132	Cryptococcus heveanensis*	−	−	−	−	−	−	−	−	−	−	−	+	V	V	+	+	V	V	+	+	V	V	+	V	V	+	+	+	+	−	V
133	Cryptococcus himalayensis*	−	−	−	−	−	−	−	−	−	−	−	+	+	+	+	+	D	+	−	−	+	−	+	+	+	−	+	−	−	−	−
134	Cryptococcus hungaricus	−	−	−	−	−	−	−	−	−	−	−	+	V	V	+	+	V	V	+	+	+	V	+	+	+	V	V	+	+	−	V
135	Cryptococcus kuetzingii	−	−	−	−	−	−	−	−	−	−	−	−	−	V	+	+	V	−	+	−	+	−	+	+	+	−	+	−	−	−	−
136	Cryptococcus lactativorus	−	−	−	−	−	−	−	−	−	−	−	−	+	−	+	−	−	−	−	−	−	−	−	−	−	−	−	−	−	−	−
137	Cryptococcus laurentii	−	−	−	−	−	−	−	−	−	−	−	+	V	V	+	+	V	V	+	+	V	V	+	V	V	+	+	+	+	−	V
138	Cryptococcus luteolus	−	−	−	−	−	−	−	−	−	−	−	+	V	V	+	+	V	+	+	+	+	+	+	+	+	V	V	+	+	−	−
139	Cryptococcus macerans	−	−	−	−	−	−	−	−	−	−	−	V	V	V	+	+	+	V	+	+	+	−	+	V	+	−	V	+	V	−	D
140	Cryptococcus magnus**	−	−	−	−	−	−	−	−	−	−	−	+	V	V	+	+	V	V	+	+	V	V	+	V	V	+	+	+	+	−	V
141	Cryptococcus melibiosum	−	−	−	−	−	−	−	−	−	−	−	+	V	V	−	+	V	−	−	−	−	−	D	+	+	D	−	−	−	−	−
142	Cryptococcus neoformans	−	−	−	−	−	−	−	−	−	−	−	+	V	V	+	V	V	+	+	+	+	+	V	V	+	−	−	V	+	−	V
143	Cryptococcus skinneri	−	−	−	−	−	−	−	−	−	−	−	V	V	V	+	V	D	+	−	−	+	V	+	V	+	−	−	−	−	−	−
144	Cryptococcus terreus	−	−	−	−	−	−	−	−	−	−	−	V	+	+	+	V	V	−	−	V	V	−	+	+	+	−	V	−	V	−	V
145	Cryptococcus uniguttulatus	−	−	−	−	−	−	−	−	−	−	−	V	−	−	+	+	−	−	+	+	V	V	V	V	V	−	−	V	+	−	V
146	Debaryomyces castellii	+	−	V	?	+	?	+	−	?	?	+	+	+	−	+	+	−	+	+	+	+	+	+	+	+	+	+	+	+	+	+
147	Debaryomyces coudertii	−	−	−	−	−	−	−	−	−	−	−	+	V	+	+	+	−	−	−	+	+	−	+	−	+	−	−	−	−	−	−
148	Debaryomyces hansenii	V	V	V	?	V	−	−	−	−	−	V	+	V	V	+	+	V	V	+	+	+	+	+	+	+	V	V	+	+	V	+
149	Debaryomyces marama	V	−	V	?	V	?	?	−	?	?	−	+	V	V	+	+	−	−	+	+	+	+	+	+	+	V	V	+	+	V	+
150	Debaryomyces melissophila*	−	−	−	−	−	−	−	−	−	−	−	+	−	−	−	−	−	−	+	+	−	+	V	−	−	−	−	−	+	−	−
151	Debaryomyces nepalensis*	S	V	−	?	S	?	S	−	?	?	S	+	+	+	+	+	−	−	+	+	+	+	+	+	+	+	−	+	+	−	+
152	Debaryomyces phaffii	V	V	V	?	V	?	V	−	?	?	S	+	+	+	+	V	V	−	+	+	+	+	+	+	+	+	V	+	+	+	+
153	Debaryomyces polymorpha**	+	V	V	−	+	?	V	−	?	?	+	+	+	+	+	V	V	−	+	+	+	−	+	+	+	V	V	+	+	+	+
154	Debaryomyces tamarii	S	S	−	?	V	?	?	V	?	?	−	+	−	−	−	−	−	−	+	−	+	−	+	+	+	+	+	+	−	−	−
155	Debaryomyces vanriji	V	V	V	?	V	?	V	−	?	?	V	+	+	+	+	V	−	V	+	+	+	+	+	+	+	−	+	+	+	V	+
156	Debaryomyces yarrowii*	−	−	−	−	−	−	−	−	−	−	−	+	+	−	+	−	−	−	+	+	−	D	V	V	+	−	−	+	+	−	−
157	Dekkera bruxellensis	+	−	+	S	+	S	−	−	−	+	−	−	−	V	−	−	−	−	+	+	+	+	−	−	−	−	−	S	+	S	−
158	Dekkera intermedia	+	+	+	V	+	+	−	−	+	V	V	+	−	V	−	−	−	−	+	+	+	+	+	+	+	−	−	V	+	V	−
159	Filobasidium capsuligenum**	V	−	V	?	−	?	−	−	?	−	−	V	−	V	V	V	V	−	D	+	+	+	D	V	S	−	−	−	−	−	+
160	Filobasidium floriforme*	−	−	−	−	−	−	−	−	−	−	−	+	+	+	+	+	+	+	+	+	+	+	+	+	+	−	+	+	+	−	+
161	Geotrichum capitatum**	−	−	−	−	−	−	−	−	−	−	−	V	V	−	−	−	−	−	−	−	−	−	−	−	−	−	−	−	−	−	−
162	Geotrichum fermentans**	S	S	−	−	−	−	−	−	−	?	−	+	+	−	+	V	−	−	−	−	−	−	+	+	+	−	−	−	−	−	−
163	Geotrichum penicillatum**	S	S	−	−	−	−	−	−	−	−	−	+	+	−	+	−	−	−	−	−	−	−	−	−	−	−	−	−	−	−	−
164	Guilliermondella selenospora**	S	−	−	−	−	−	−	−	−	−	−	+	−	S	+	+	V	−	−	−	−	−	−	−	−	−	−	−	−	−	V

Codes in table: + positive; − negative; D delayed for longer than 7 days;

Characteristics of the species

Column headers (32–78):
32 Glycerol growth
33 Erythritol growth
34 Ribitol growth
35 Galactitol growth
36 D–Mannitol growth
37 D–Glucitol growth
38 myo–Inositol growth
39 Gluconolactone growth
40 2–Ketogluconate growth
41 5–Ketogluconate growth
42 D,L–Lactate growth
43 Succinate growth
44 Citrate growth
45 Methanol growth
46 Ethanol growth
47 Ethylamine growth
48 D–Glucosamine growth
49 Nitrate growth
50 Nitrite growth
51 Growth sans vitamins
52 Growth sans inositol
53 Grth sans pantothenate
54 Growth sans biotin
55 Growth sans thiamin
56 Grth sans pyridoxine
57 Growth sans niacin
58 Growth sans folic acid
59 Growth sans PABA
60 Growth in 50% glucose
61 Growth in 60% glucose
62 Growth at 37 degrees
63 0.01% cyclohex. growth
64 0.1% cyclohex. growth
65 Starch formation
66 Pink colonies
67 Budding cells
68 Splitting cells
69 Apical budding
70 Cells sph, ovl, cylind
71 Cells of other shapes
72 Filamentous
73 Pseudohyphae
74 Septate hyphae
75 Ascosporogenous
76 Ascosp sph, ovl, renfm
77 Ascosp hat or saturn
78 Ascosp of other shapes

#	32-37	38-41	42-44	45-46	47-50	51-55	56-59	60-64	65-66	67-71	72-74	75	76-78
126	– – – – + +	+ – + +	– + V	– –	? ? – –	+ + + –	+ + + +	– – – ? ?	+ –	+ – – + –	– N N	–	N N N
127	– – – – + +	– ? ? ?	– + D	– –	? ? + ?	– ? ? ?	? ? ? ?	? ? – ? ?	+ –	+ – – + –	– N N	–	N N N
128	+ + + – + +	+ V – –	– – –	– +	+ ? – ?	– + + –	+ + + +	– – + + ?	– –	+ – – + +	– N N	–	N N N
129	V – + D + +	+ + + +	V + V	– +	? ? – +	– + + +	+ + + +	– – – ? ?	+ –	+ – – + –	– N N	–	N N N
130	– + V V + +	+ + + +	V + V	– –	? ? – –	– + + +	+ + + +	– – – ? ?	– –	+ – – + –	– N N	–	N N N
131	– – – – D V	+ + + V	– V –	– V	? ? – –	– + + +	+ + + +	– – – ? ?	V –	+ – – + –	– N N	–	N N N
132	V – V V V +	+ V + +	+ V V	– +	? ? + ?	– + + +	+ + + +	– – – ? ?	+ –	+ – – + –	– N N	–	N N N
133	– – – + + +	+ ? ? ?	– + +	– –	? ? + ?	+ + + +	+ + + +	? ? – ? ?	+ –	+ – – + –	– N N	–	N N N
134	V – V + + +	+ + + +	+ + +	– –	? ? – V	– + + +	+ + + +	– – – ? ?	+ +	+ – – + –	– N N	–	N N N
135	V – V + + +	+ – + +	V + +	– +	? ? – +	+ + + +	+ + + +	– – – ? ?	+ –	+ – – + –	– N N	–	N N N
136	+ – D – – –	+ – – –	+ + –	– +	? – – –	– + + +	+ + + +	– – + ? ?	– –	+ – – + –	– N N	–	N N N
137	V + V V V +	+ V + +	V V V	– V	? ? – V	V + + + V	+ + + +	– – V ? ?	+ V	+ – – + –	V V V	–	N N N
138	V V V + + V	+ + + +	V + +	– V	? ? – V	V + + + V	+ + + +	– – – ? ?	+ –	+ – – + –	– N N	–	N N N
139	V + – – + +	+ V + +	V + +	– V	? ? + +	– + + –	+ + + +	– – – ? ?	+ +	+ – – + –	– N N	–	N N N
140	V – V V V +	+ V + +	– V V	– V	? ? – V	V + + + V	+ + + +	– – – ? ?	+ –	+ – – + –	– N N	–	N N N
141	+ – + – – –	+ + + +	– – –	– +	? ? – –	– + + – S	+ + + +	– – – ? ?	V –	+ – – + –	– N N	–	N N N
142	V V + + + +	+ + + +	V V V	– V	? ? – –	– + + +	+ + + +	– – + ? ?	V –	+ – – + –	V – V	–	N N N
143	V – + V + +	+ + + +	V + +	– V	? ? – –	– – + +	+ + + +	– – – ? ?	+ –	+ – – + –	– N N	–	N N N
144	– – V V + +	+ V + +	– V –	– –	? ? + +	+ + + +	+ + + +	– – – ? ?	+ –	+ – – + –	– N N	–	N N N
145	V – – – V V	+ – + V	– V V	– V	? ? – –	V + + + V	+ + + +	– – – ? ?	V –	+ – – + –	– N N	–	N N N
146	D – + + + +	– ? ? ?	V + +	– +	? ? – ?	– ? ? ? ?	? ? ? ?	+ ? – ? ?	– –	+ – – + –	V V –	+	+ – –
147	+ + + – + +	– ? ? ?	– + –	– +	? ? – ?	– ? ? ? ?	? ? ? ?	+ ? – ? ?	– –	+ – – + –	– N N	+	+ – –
148	+ V + V + +	– ? ? ?	+ + V	– +	? ? – V	V + + V +	+ + + +	+ ? V ? ?	– –	+ – – + –	V V –	+	+ – –
149	+ + + – + +	– ? ? ?	V + +	– +	? ? – ?	– ? ? ? ?	? ? ? ?	V ? – ? ?	– –	+ – – + –	– N N	+	+ – –
150	+ + + – + +	– ? ? ?	– – +	– S	? ? – ?	– + + – +	+ + + +	+ S – ? ?	– –	+ – – + –	– N N	+	+ – –
151	+ + + – + +	– ? ? ?	D + +	– ?	? + – ?	? ? ? ? ?	? ? ? ?	? ? ? ? – –	+ – + –	– N N	+	+ – –
152	+ + + + + +	– + + ?	– + +	– +	? ? – ?	+ + + + +	+ + + +	+ ? + ? ? – –	+ – – + –	V V –	+	+ – –
153	+ + + + + +	– + + ?	V V +	– +	? ? – –	+ + + + +	+ + + +	+ ? V ? ? – –	+ – – + –	V V –	+	+ – –
154	+ – – – + –	– ? ? ?	– + –	– +	? ? – ?	– ? ? ? ?	? ? ? ?	V ? – ? ? – –	+ – – + –	– N N	+	– – –
155	+ + + + + +	– ? ? ?	V + +	– +	? ? – ?	V ? ? ? ?	? ? ? ?	V ? + ? ? – –	+ – – + –	V V –	+	+ – –
156	D – + + + +	– ? ? ?	– D –	– +	? – – – ?	+ + + + +	+ + + +	+ ? – ? ? – –	+ – – + –	+ + –	+	+ – –
157	S – V – – V	– ? ? ?	V V –	– +	+ ? V ? – ?	? ? ? ? ?	? ? ? ?	– – + + ? – –	+ – – + –	+ + –	+	– + –
158	S – V – – V	– ? ? ?	V V –	– +	+ ? V ? – ?	? ? ? ? ?	? ? ? ?	– – + + ? – –	+ – – + –	+ + –	+	– + –
159	+ – + V + +	+ – + –	– S S	– +	? ? – – – + V + –	V + + +	– – – ? ?	+ –	+ – – + –	V V V	–	N N N
160	+ – + D + +	+ ? ? ?	– + +	– ?	? + ? ? – ? ? ? ?	? ? ? ?	? ? ? ? ? + –	+ – – + –	V V V	–	N N N
161	+ – – – – –	– ? ? ?	V V –	– +	? ? – – + + + –	+ + + +	– – + ? ? – –	– + N + –	+ – +	–	N N N
162	+ – + – + +	– ? ? ?	+ V V	– +	? ? – – + + + + +	+ + + +	– – V ? ? – –	– + N + –	+ – +	–	N N N
163	+ – – – – V	– ? ? ?	V V –	– +	? ? – – + + + + +	+ + + +	– – – ? ? – –	– + N + –	+ – +	–	N N N
164	V – + – – –	– ? ? ?	+ + –	– +	? ? – ? – ? ? ? ?	? ? ? ?	– – V ? ? – –	+ – – + –	+ – +	+	+ – –

N not applicable; S slow utilisation; V variable; ? result not known.

Characteristics of the species

		1 D–Glucose fermentation	2 Galactose fermentation	3 Maltose fermentation	4 Me α–D–glucoside fermn	5 Sucrose fermentation	6 Trehalose fermentation	7 Melibiose fermentation	8 Lactose fermentation	9 Cellobiose fermentn	10 Melezitose fermentn	11 Raffinose fermentation	12 D–Galactose growth	13 L–Sorbose growth	14 D–Ribose growth	15 D–Xylose growth	16 L–Arabinose growth	17 D–Arabinose growth	18 L–Rhamnose growth	19 Sucrose growth	20 Maltose growth	21 Trehalose growth	22 Me α–D–glucoside grth	23 Cellobiose growth	24 Salicin growth	25 Arbutin growth	26 Melibiose growth	27 Lactose growth	28 Raffinose growth	29 Melezitose growth	30 Inulin growth	31 Starch growth
165	Hanseniaspora guilliermondii*	+	–	–	–	–	–	–	–	?	–	–	–	–	–	–	–	–	–	–	–	–	–	+	+	+	–	–	–	–	–	–
166	Hanseniaspora occidentalis*	+	–	–	–	+	–	–	–	?	–	–	–	–	–	–	–	–	–	+	–	V	–	+	+	+	–	–	–	–	–	–
167	Hanseniaspora osmophila	+	–	V	–	–	?	–	–	?	–	–	–	–	–	–	–	–	–	V	+	V	–	+	+	+	–	–	–	–	–	–
168	Hanseniaspora uvarum	+	–	–	–	–	?	–	–	V	–	–	–	–	–	–	–	–	–	–	–	V	–	+	+	+	–	–	–	–	–	–
169	Hanseniaspora valbyensis	+	–	–	–	–	?	–	–	V	–	–	–	–	–	–	–	–	–	–	–	V	–	+	+	+	–	–	–	–	–	–
170	Hanseniaspora vineae*	+	–	V	–	–	?	–	–	?	–	–	–	–	–	–	–	–	–	V	+	V	–	+	+	+	–	–	–	–	–	–
171	Hansenula anomala	+	V	V	?	+	?	–	–	?	?	+	V	–	V	V	V	–	–	+	+	+	+	+	+	+	–	–	+	+	–	+
172	Hansenula beckii**	V	–	–	?	–	?	–	–	?	?	–	–	–	V	+	–	–	+	+	+	+	+	+	+	+	–	–	+	+	–	–
173	Hansenula beijerinckii	+	–	–	?	+	?	–	–	?	?	+	–	–	–	+	–	–	V	+	+	V	V	+	V	+	–	–	+	+	+	–
174	Hansenula bimundalis	D	–	–	?	–	?	–	–	?	?	–	–	–	–	+	V	V	+	+	+	+	V	+	+	+	–	–	–	–	–	–
175	Hansenula californica	+	–	–	?	–	?	–	–	–	–	–	–	+	–	+	–	–	V	+	V	V	+	+	V	+	–	–	–	–	–	–
176	Hansenula canadensis	–	–	–	–	–	–	–	–	–	–	–	–	–	–	+	–	–	+	+	+	S	V	+	+	+	–	–	–	+	–	–
177	Hansenula capsulata	+	–	V	?	–	?	–	–	?	?	–	–	–	+	+	V	V	V	–	+	+	V	+	V	+	–	V	–	V	–	+
178	Hansenula ciferrii	+	S	V	?	+	?	–	–	?	?	+	+	–	+	S	S	–	S	+	+	+	+	S	+	+	–	–	+	+	–	+
179	Hansenula dimennae	+	–	–	?	–	–	–	–	?	–	–	–	+	–	+	–	–	V	–	–	–	V	+	+	+	–	–	–	–	–	–
180	Hansenula dryadoides*	–	–	–	–	–	–	–	–	–	–	–	–	–	–	–	–	–	–	–	–	–	–	+	+	+	–	–	–	–	–	–
181	Hansenula fabianii	+	–	S	?	+	?	–	–	?	?	+	–	–	–	+	–	V	–	+	+	+	+	+	+	+	–	–	+	+	–	+
182	Hansenula glucozyma	S	–	–	–	–	?	–	–	?	–	–	–	–	+	V	V	V	V	–	–	+	–	+	+	+	–	–	–	–	–	–
183	Hansenula henricii	S	–	–	–	–	V	–	–	V	–	–	–	–	+	+	V	–	+	–	–	V	–	+	+	+	–	–	–	–	–	–
184	Hansenula holstii	+	V	–	?	–	?	–	–	?	?	–	+	+	+	+	+	+	+	+	+	+	+	+	+	+	–	–	–	+	–	+
185	Hansenula jadinii	S	–	–	?	S	?	–	–	?	?	+	–	–	–	+	–	–	–	+	+	+	+	+	+	+	–	–	+	+	+	–
186	Hansenula lynferdii*	+	S	–	?	+	?	–	–	?	?	+	+	–	V	–	–	–	–	+	+	S	+	S	+	+	–	–	+	+	+	–
187	Hansenula minuta	S	–	–	–	–	?	–	–	?	–	–	–	–	+	+	–	V	–	–	–	+	–	+	+	+	–	–	–	–	–	–
188	Hansenula mrakii	+	–	–	–	–	–	–	–	?	–	–	–	–	–	+	–	–	V	–	–	–	–	+	+	+	–	–	–	–	–	–
189	Hansenula muscicola*	+	V	–	?	–	?	–	–	?	–	–	+	–	V	+	+	–	+	–	V	V	+	+	+	+	–	–	–	–	–	–
190	Hansenula nonfermentans	–	–	–	–	–	–	–	–	–	–	–	–	–	D	–	–	–	–	–	–	+	–	+	+	+	–	–	–	–	–	–
191	Hansenula ofunaensis*	–	–	–	–	–	–	–	–	–	–	–	+	+	S	+	+	S	+	–	–	+	–	–	–	–	+	+	–	–	–	–
192	Hansenula petersonii	+	–	–	?	+	?	–	–	?	?	+	–	–	–	+	–	–	+	+	+	+	+	+	+	+	–	–	+	+	+	–
193	Hansenula philodendra*	V	–	–	–	–	–	–	–	–	–	–	–	D	S	+	V	–	–	–	–	+	–	–	–	–	–	–	–	–	–	–
194	Hansenula polymorpha	+	–	V	?	V	?	–	–	?	?	–	V	V	+	V	–	V	V	+	+	+	V	V	V	V	–	–	–	V	–	–
195	Hansenula saturnus	+	–	–	–	+	–	–	–	?	–	+	–	–	–	+	–	–	V	+	–	–	–	+	+	+	–	–	+	–	V	–
196	Hansenula silvicola	+	S	–	?	–	?	–	–	?	?	–	+	+	V	+	+	–	+	+	V	+	+	+	+	+	–	–	–	V	–	–
197	Hansenula subpelliculosa	+	–	V	?	+	?	–	–	?	?	V	V	–	V	V	V	V	–	+	+	+	+	V	+	+	–	–	–	V	–	V
198	Hansenula sydowiorum*	+	+	S	?	+	?	S	–	?	?	S	+	–	S	S	S	–	S	+	+	S	+	S	+	+	+	–	+	+	–	S
199	Hansenula wickerhamii	–	–	–	–	–	–	–	–	–	–	–	–	+	+	+	V	+	+	–	–	+	–	–	–	–	–	–	–	–	–	–
200	Hansenula wingei	–	–	–	–	–	–	–	–	–	–	–	–	–	–	+	–	–	+	+	+	V	V	+	+	+	–	–	–	+	–	–
201	Hormoascus platypodis**	S	–	–	?	V	?	–	–	?	?	–	–	–	+	+	–	–	+	+	+	+	+	+	+	+	–	–	–	+	–	V
202	Hyphopichia burtonii**	S	V	S	?	S	?	–	–	?	?	S	+	V	+	+	V	–	–	+	+	+	+	+	+	+	–	–	+	V	–	+
203	Kluyveromyces aestuarii	+	–	–	–	–	+	–	–	–	–	+	+	–	V	–	–	–	+	V	V	+	+	+	–	+	+	V	–	–	–	–
204	Kluyveromyces africanus	+	+	–	–	–	–	–	–	–	–	–	+	–	–	–	–	–	–	V	–	V	–	–	–	–	–	–	–	–	–	–
205	Kluyveromyces blattae*	+	+	–	–	–	–	–	–	–	–	–	+	–	+	–	–	–	–	–	–	–	–	–	–	–	–	–	–	–	–	–
206	Kluyveromyces bulgaricus	+	+	–	–	+	–	–	V	–	–	+	+	V	V	V	V	–	–	+	–	V	–	V	V	V	–	V	+	–	+	–

Codes in table: + positive; – negative; D delayed for longer than 7 days;

Characteristics of the species

	32 Glycerol growth	33 Erythritol growth	34 Ribitol growth	35 Galactitol growth	36 D–Mannitol growth	37 D–Glucitol growth	38 myo–Inositol growth	39 Gluconolactone growth	40 2–Ketogluconate growth	41 5–Ketogluconate growth	42 D,L–Lactate growth	43 Succinate growth	44 Citrate growth	45 Methanol growth	46 Ethanol growth	47 Ethylamine growth	48 D–Glucosamine growth	49 Nitrate growth	50 Nitrite growth	51 Growth sans vitamins	52 Growth sans inositol	53 Grth sans pantothenate	54 Growth sans biotin	55 Growth sans thiamin	56 Grth sans pyridoxine	57 Growth sans niacin	58 Growth sans folic acid	59 Growth sans PABA	60 Growth in 50% glucose	61 Growth in 60% glucose	62 Growth at 37 degrees	63 0.01% cyclohex. growth	64 0.1% cyclohex. growth	65 Starch formation	66 Pink colonies	67 Budding cells	68 Splitting cells	69 Apical budding	70 Cells sph, ovl, cylind	71 Cells of other shapes	72 Filamentous	73 Pseudohyphae	74 Septate hyphae	75 Ascosporogenous	76 Ascosp sph, ovl, renfm	77 Ascosp hat or saturn	78 Ascosp of other shapes
165	−	−	−	−	−	−	−	+	+	−	−	−	−	−	−	?	?	−	−	−	−	−	S	S	S	S	+	+	+	−	+	?	?	−	−	+	−	+	+	−	−	N	N	+	−	+	−
166	+	−	−	−	−	−	−	+	−	−	−	−	−	−	−	?	?	−	−	−	−	−	−	−	−	−	+	+	−	−	−	?	?	−	−	+	−	+	+	−	−	N	N	+	−	+	−
167	−	−	−	−	−	−	−	+	+	−	−	−	−	−	−	?	?	−	−	−	−	−	S	S	S	S	+	+	+	−	−	?	?	−	−	+	−	+	+	−	V	V	−	+	+	−	−
168	−	−	−	−	−	−	−	+	+	−	−	−	−	−	−	?	?	−	−	−	−	−	S	S	S	S	+	+	+	−	−	?	?	−	−	+	−	+	+	−	V	V	−	+	V	V	−
169	−	−	−	−	−	−	−	+	+	−	−	−	−	−	−	?	?	−	−	−	−	−	S	S	S	S	+	+	V	−	−	?	?	−	−	+	−	+	+	−	V	V	−	+	−	+	−
170	−	−	−	−	−	−	−	+	+	−	−	−	−	−	−	?	?	−	−	S	S	S	S	S	S	+	+	+	−	−	?	?	−	−	+	−	+	+	−	V	V	−	+	+	−	−
171	+	+	V	−	+	+	−	?	?	?	+	+	+	−	+	?	?	+	?	+	+	+	+	+	+	+	+	+	?	?	V	?	?	−	−	+	−	+	+	−	V	V	−	+	−	+	−
172	+	−	D	−	+	+	−	?	?	?	V	+	+	−	+	?	?	+	?	−	?	?	?	?	?	?	?	?	−	−	+	?	?	−	−	+	−	+	+	−	+	−	+	+	−	+	−
173	+	−	−	−	V	+	−	?	?	?	+	+	−	−	−	?	?	+	?	+	+	+	+	+	+	+	+	+	?	?	V	?	?	−	−	+	−	+	+	−	V	V	−	+	−	+	−
174	+	−	−	−	−	+	+	−	?	?	?	+	+	+	−	+	?	?	+	?	−	?	?	?	?	?	?	?	?	?	?	+	?	?	−	−	+	−	+	+	−	+	−	+	+	−	+	−
175	+	−	−	−	+	V	−	+	−	?	+	+	V	−	+	?	−	+	?	+	?	?	?	?	+	+	+	+	?	?	−	?	?	−	−	+	−	−	+	−	−	N	N	+	−	+	−
176	+	−	−	−	S	+	−	?	?	?	+	+	+	−	+	?	?	+	?	−	?	?	?	?	?	?	?	?	?	?	+	?	?	−	−	+	−	−	+	−	+	−	−	+	−	+	−
177	+	+	+	−	+	+	−	V	+	−	−	V	−	+	+	?	?	+	?	−	+	−	−	−	+	+	+	+	−	−	+	−	−	−	−	+	−	−	+	−	−	N	N	+	−	+	−
178	+	+	+	−	+	+	−	?	?	?	+	+	+	−	+	?	?	+	?	+	+	+	+	+	+	+	+	+	?	?	V	?	?	−	−	+	−	−	+	−	V	V	V	+	−	+	−
179	+	−	−	−	+	V	−	?	?	?	+	+	V	−	+	?	?	+	?	−	?	?	?	?	?	?	?	?	?	?	−	?	?	−	−	+	−	−	+	−	−	N	N	+	−	+	−
180	+	−	−	−	+	+	−	+	−	−	+	+	+	−	?	?	−	+	?	+	+	+	+	+	+	+	+	+	−	−	+	?	?	−	−	+	−	−	+	−	V	V	−	+	−	+	−
181	+	−	−	−	+	+	−	?	?	?	+	+	+	−	+	?	?	+	?	−	?	?	?	?	?	?	?	?	?	?	+	?	?	−	−	+	−	−	+	−	+	+	−	+	−	+	−
182	+	+	+	−	+	+	−	+	−	−	−	V	+	+	+	+	?	+	?	−	+	+	−	−	+	+	+	+	+	?	?	+	?	?	−	−	+	−	−	+	−	−	N	N	+	−	+	−
183	+	+	+	−	+	+	−	−	+	−	−	V	V	+	+	+	?	+	?	−	+	−	+	+	+	+	+	+	+	?	?	+	?	?	−	−	+	−	−	+	−	V	V	−	+	−	+	−
184	+	V	+	V	+	+	−	V	−	?	−	+	+	−	+	?	+	+	?	−	?	?	?	?	?	?	?	?	V	?	?	−	−	−	−	+	−	−	−	−	+	+	+	+	−	+	−
185	+	−	−	−	+	+	−	?	?	?	+	+	+	−	+	?	?	+	?	+	+	+	+	+	+	+	+	+	?	?	?	?	?	−	−	+	−	−	−	−	+	+	−	+	−	+	−
186	+	+	+	−	+	D	−	D	−	−	+	+	+	−	+	?	?	+	?	+	+	+	+	+	+	+	+	+	+	?	−	?	?	−	−	+	−	−	−	+	+	+	−	+	−	+	−
187	+	−	+	−	+	+	−	D	−	−	−	V	V	+	S	+	?	+	?	+	+	−	−	+	+	+	+	+	?	?	V	+	?	−	−	+	−	−	−	−	−	N	N	+	−	+	−
188	+	−	−	−	V	V	−	?	?	?	+	+	−	−	+	?	−	+	?	+	+	+	+	+	+	+	+	−	−	V	?	?	−	−	+	−	−	+	−	+	+	−	+	−	+	−
189	+	D	+	−	+	+	−	?	?	?	+	+	+	−	+	?	−	+	?	−	?	?	?	?	?	?	?	?	−	−	+	?	?	−	−	+	−	−	+	−	+	+	+	+	−	+	−
190	+	−	+	−	+	+	−	+	−	−	−	+	V	+	S	+	?	+	?	+	+	+	+	+	+	+	+	+	?	?	+	+	−	−	−	+	−	−	+	−	−	N	N	+	−	+	−
191	+	−	+	+	+	+	+	+	−	−	−	+	+	+	?	+	?	−	− −	+	+	+	+	+	+	+	?	?	+	−	−	−	−	+	−	−	+	−	+	+	−	+	−	+	−
192	+	−	−	−	−	+	−	?	?	?	+	+	+	−	+	?	?	+	?	−	?	?	?	?	?	?	?	?	?	+	?	?	−	−	+	−	−	+	−	+	+	+	+	−	+	−
193	+	+	+	−	+	+	−	V	−	−	−	+	+	+	+	+	?	+	−	−	+	+	−	−	+	+	+	+	?	?	+	+	+	−	−	+	−	−	+	−	−	N	N	+	−	+	−
194	+	+	+	V	−	−	−	V	−	−	−	V	V	+	+	?	?	+	+	−	+	+	−	+	+	+	+	+	?	?	+	?	?	−	−	+	−	−	+	−	−	N	N	+	−	+	−
195	+	−	−	−	V	V	−	?	?	?	+	+	V	−	+	?	?	+	V	?	?	?	?	?	?	?	?	V	?	−	−	+	−	+	−	V	V	−	+	−	+	−
196	+	−	+	−	V	+	−	?	?	?	V	+	V	−	+	?	?	+	?	−	?	?	?	?	?	?	?	?	?	V	?	?	−	−	+	−	−	+	−	V	V	V	+	−	+	−
197	+	+	V	−	+	+	−	V	−	?	+	+	+	−	+	?	?	+	?	−	+	+	−	+	+	+	+	?	?	V	?	?	−	−	+	−	−	+	−	V	V	V	+	−	+	−
198	+	+	S	−	+	S	−	?	?	?	S	+	S	−	+	?	−	+	?	+	+	+	+	+	+	+	?	?	−	?	?	−	−	+	−	−	+	+	−	+	−	+	−
199	+	+	+	−	+	+	−	−	V	−	−	V	−	+	+	?	?	+	−	+	?	+	−	+	+	+	+	+	?	?	V	+	+	−	−	+	−	−	−	−	N	N	+	−	+	−
200	+	−	V	−	+	+	−	?	?	?	+	+	+	−	+	?	?	+	?	−	?	?	?	?	?	?	?	?	?	+	?	?	−	−	+	−	−	+	−	+	+	+	+	−	+	−
201	+	+	+	−	+	+	−	?	?	?	+	+	+	−	+	?	?	+	?	−	?	?	?	?	?	?	?	?	−	−	−	?	?	−	−	+	−	−	+	−	+	+	+	+	−	+	−
202	+	+	+	−	+	+	−	?	?	?	−	+	+	−	+	?	?	−	?	+	+	+	+	+	+	+	+	+	?	V	?	?	−	−	+	−	−	+	−	+	−	+	+	−	+	−
203	+	−	−	−	+	+	−	?	?	?	+	+	−	−	+	?	+	+	+	−	−	−	−	−	−	−	+	−	+	−	V	V	−	+	+	−	−
204	V	−	−	−	−	−	−	?	?	?	−	−	−	−	−	?	−	?	−	−	+	+	−	−	+	+	−	−	−	V	−	−	−	−	+	−	V	V	−	+	+	−	−
205	+	−	−	−	−	−	−	?	?	?	−	−	−	−	+	−	?	−	−	?	?	?	?	?	?	?	−	−	+	−	−	−	−	+	−	−	N	N	+	+	−	−
206	+	−	S	−	V	+	−	?	?	?	+	+	V	−	+	?	−	?	−	+	+	+	−	+	−	+	+	−	−	+	+	?	−	−	+	−	−	+	−	+	+	−	+	+	−	−

N not applicable; S slow utilisation; V variable; ? result not known.

Characteristics of the species

		1 D–Glucose fermentation	2 Galactose fermentation	3 Maltose fermentation	4 Me α–D–glucoside fermn	5 Sucrose fermentation	6 Trehalose fermentation	7 Melibiose fermentation	8 Lactose fermentation	9 Cellobiose fermentn	10 Melezitose fermentn	11 Raffinose fermentation	12 D–Galactose growth	13 L–Sorbose growth	14 D–Ribose growth	15 D–Xylose growth	16 L–Arabinose growth	17 D–Arabinose growth	18 L–Rhamnose growth	19 Sucrose growth	20 Maltose growth	21 Trehalose growth	22 Me α–D–glucoside grth	23 Cellobiose growth	24 Salicin growth	25 Arbutin growth	26 Melibiose growth	27 Lactose growth	28 Raffinose growth	29 Melezitose growth	30 Inulin growth	31 Starch growth
207	Kluyveromyces delphensis	+	–	–	–	–	S	–	–	–	–	–	–	–	–	–	–	–	–	–	–	D	–	–	–	–	–	–	–	–	–	–
208	Kluyveromyces dobzhanskii	+	S	+	+	+	+	–	–	–	V	+	+	V	–	V	V	–	–	+	+	+	+	+	+	+	–	–	+	+	–	–
209	Kluyveromyces drosophilarum	+	+	–	–	+	+	–	–	–	–	+	+	+	–	S	S	S	–	+	+	+	+	V	V	V	–	–	+	+	–	–
210	Kluyveromyces lactis	+	+	V	–	V	V	–	V	–	V	V	+	V	–	V	V	–	–	+	V	V	V	+	+	+	–	V	V	V	V	–
211	Kluyveromyces lodderi	+	+	–	–	+	–	–	–	–	–	+	+	–	–	–	–	–	–	+	–	+	–	–	–	–	–	–	+	–	–	–
212	Kluyveromyces marxianus	+	+	–	–	+	–	–	V	–	–	+	+	V	V	+	+	–	–	+	–	–	–	+	+	+	–	+	+	–	+	–
213	Kluyveromyces phaffii	+	+	–	–	–	–	–	–	–	–	–	–	+	–	–	–	–	–	–	S	–	S	–	–	–	–	–	–	–	–	–
214	Kluyveromyces phaseolosporus	+	S	–	–	+	–	–	–	–	–	+	+	+	–	+	–	–	–	+	–	V	–	S	S	+	–	–	+	–	V	–
215	Kluyveromyces polysporus	+	+	–	–	+	S	–	–	–	–	+	+	–	–	–	–	–	–	+	–	+	–	–	–	–	–	–	+	–	V	–
216	Kluyveromyces thermotolerans**	+	V	V	S	+	S	–	–	–	V	+	+	V	–	–	–	–	–	+	+	+	+	–	–	–	–	–	–	+	V	–
217	Kluyveromyces waltii*	+	–	–	–	+	–	–	–	–	–	+	–	+	–	D	–	D	–	+	–	–	–	–	–	–	–	–	+	–	–	–
218	Kluyveromyces wickerhamii	+	S	–	–	S	–	–	–	–	–	–	+	S	–	+	–	–	–	+	–	V	–	+	+	+	–	–	–	–	–	–
219	Leucosporidium antarcticum	–	–	–	–	–	–	–	–	–	–	–	V	–	–	V	–	–	–	V	V	V	–	–	–	–	–	–	V	–	–	–
220	Leucosporidium frigidum	S	S	–	–	S	?	–	–	?	–	V	+	D	–	+	+	D	–	+	–	D	–	+	+	+	–	+	+	–	–	–
221	Leucosporidium gelidum	S	S	V	?	S	?	?	–	?	?	V	+	D	V	+	+	V	+	+	+	+	V	+	+	+	+	–	+	+	–	+
222	Leucosporidium nivalis	S	S	–	–	S	?	?	–	?	–	S	+	D	D	+	+	D	–	+	–	S	–	+	+	+	+	–	+	–	–	–
223	Leucosporidium scottii	–	–	–	–	–	–	–	–	–	–	–	–	V	+	V	+	V	V	+	+	+	+	+	+	+	+	–	V	+	+	–	–
224	Leucosporidium stokesii	S	S	S	–	S	?	–	–	?	?	S	+	D	D	+	+	D	+	+	+	+	–	+	+	+	–	–	+	+	–	+
225	Lipomyces anomalus*	–	–	–	–	–	–	–	–	–	–	–	+	D	–	+	+	D	–	D	+	–	–	+	–	–	–	D	–	?	–	V
226	Lipomyces kononenkoae	–	–	–	–	–	–	–	–	–	–	–	+	+	–	V	–	–	–	+	+	+	+	V	V	V	+	–	+	+	+	+
227	Lipomyces lipofer	–	–	–	–	–	–	–	–	–	–	–	+	+	–	+	V	–	–	+	+	V	V	+	V	V	+	V	+	+	V	+
228	Lipomyces starkeyi	–	–	–	–	–	–	–	–	–	–	–	+	+	V	+	V	V	V	+	+	+	+	V	V	V	+	V	+	+	+	+
229	Lipomyces tetrasporus*	–	–	–	–	–	–	–	–	–	–	–	+	+	V	+	V	V	V	+	+	V	+	V	V	+	V	V	+	+	+	+
230	Lodderomyces elongisporus	S	V	V	–	V	S	–	–	–	–	–	+	+	–	V	–	–	–	+	+	+	+	–	–	–	–	–	–	+	–	–
231	Metschnikowia bicuspidata	+	–	–	?	–	–	–	–	–	?	?	–	+	+	V	V	–	–	–	+	+	+	V	+	V	?	–	–	–	V	–	–
232	Metschnikowia krissii	–	–	–	–	–	–	–	–	–	–	–	–	–	–	–	–	–	–	+	+	+	+	+	+	+	–	–	–	+	–	–
233	Metschnikowia lunata*	S	–	–	?	–	?	–	–	?	?	–	+	D	–	D	–	–	–	+	+	+	S	+	+	+	–	–	–	+	–	–
234	Metschnikowia pulcherrima	+	V	–	–	–	–	–	–	–	–	–	+	+	V	+	–	–	–	+	+	+	+	+	+	+	–	–	–	+	–	–
235	Metschnikowia reukaufii	+	V	–	–	–	–	–	–	–	–	–	V	V	V	V	–	–	–	+	+	+	V	+	+	+	–	–	–	+	–	–
236	Metschnikowia zobellii	V	–	–	–	–	–	–	–	–	–	–	+	+	–	V	–	–	–	+	+	+	+	+	+	+	–	–	–	+	–	–
237	Nadsonia commutata*	–	–	–	–	–	–	–	–	–	–	–	V	–	–	–	–	–	–	S	+	V	–	–	–	–	–	–	V	–	–	–
238	Nadsonia elongata	+	–	–	–	–	–	–	–	–	–	–	–	+	–	–	–	–	–	–	–	–	–	–	–	–	–	–	–	–	–	–
239	Nadsonia fulvescens	+	+	S	?	+	–	–	–	–	–	–	+	+	–	–	–	–	–	+	+	–	+	–	–	–	–	–	–	–	–	–
240	Nematospora coryli	V	–	V	–	V	V	–	–	–	–	S	V	–	–	–	–	–	–	V	V	+	–	–	–	–	–	–	V	–	–	V
241	Pachysolen tannophilus	+	–	–	–	–	–	–	–	?	–	–	+	–	V	+	+	–	–	–	–	–	–	D	V	V	–	–	–	–	–	–
242	Pachytichospora transvaalensis	+	+	–	–	–	V	–	–	–	–	–	+	–	–	–	–	–	–	–	–	V	–	–	–	–	–	–	–	–	–	–

Codes in table: + positive;   – negative;   D delayed for longer than 7 days;

	32 Glycerol growth; 33 Erythritol growth; 34 Ribitol growth; 35 Galactitol growth; 36 D−Mannitol growth; 37 D−Glucitol growth	38 myo−Inositol growth; 39 Gluconolactone growth; 40 2−Ketogluconate growth; 41 5−Ketogluconate growth	42 D,L−Lactate growth; 43 Succinate growth; 44 Citrate growth	45 Methanol growth; 46 Ethanol growth	47 Ethylamine growth; 48 D−Glucosamine growth; 49 Nitrate growth; 50 Nitrite growth; 51 Growth sans vitamins; 52 Growth sans inositol; 53 Grth sans pantothenate; 54 Growth sans biotin; 55 Growth sans thiamin	56 Grth sans pyridoxine; 57 Growth sans niacin; 58 Growth sans folic acid; 59 Growth sans PABA	60 Growth in 50% glucose; 61 Growth in 60% glucose; 62 Growth at 37 degrees; 63 0.01% cyclohex. growth; 64 0.1% cyclohex. growth; 65 Starch formation; 66 Pink colonies	67 Budding cells; 68 Splitting cells; 69 Apical budding; 70 Cells sph, ovl, cylind; 71 Cells of other shapes	72 Filamentous; 73 Pseudohyphae; 74 Septate hyphae	75 Ascosporogenous; 76 Ascosp sph, ovl, renfm; 77 Ascosp hat or saturn; 78 Ascosp of other shapes
207	+ − − − − −	− − − −	− − −	− +	− ? − ? − + + − +	− + + +	S ? + + ? − −	+ − − + −	− N N	+ + − −
208	+ − V − + +	− ? ? ?	+ S S	− +	+ ? − ? − + + − +	+ − + +	S ? − + ? − −	+ − − + −	V V −	+ + − −
209	+ − V − + +	− ? ? ?	+ + V	− +	+ ? − ? − + + − +	+ − + +	− − + + ? − −	+ − − + −	V V −	+ + − −
210	+ − V − V +	− ? ? ?	V + −	− +	+ ? − ? − + + − +	+ − + +	V ? V + ? − −	+ − − + −	V V −	+ + − −
211	+ − − − − −	− + − −	S V −	− +	S ? − ? − + + − +	+ + + +	− − − + ? − −	+ − − + −	V V −	+ + − −
212	+ − V − + +	− ? ? ?	+ + V	− +	+ ? − ? − + V − +	V − + +	− − + + ? − −	+ − − + −	+ + −	+ + − −
213	+ − − − − −	− + − −	− − −	− −	− ? − ? − + − − −	+ + + +	− − − − − − −	+ − − + −	− N N	+ + − −
214	+ − V − + +	− + − ?	+ S −	− +	+ ? − ? − + + − +	+ − + +	− − + + ? − −	+ − − + −	V V −	+ + − −
215	+ − − − − −	− + − ?	− S S	− V	− ? − ? − + + − +	+ + + +	− − − V ? − −	+ − − + −	V V −	+ + − −
216	V − V − + V	− ? ? ?	− V −	− V	+ ? − ? − ? ? − ?	? − ? ?	+ ? V − − − −	+ − − + −	+ + −	+ + − −
217	− − D − + +	− ? ? ?	− − −	− ?	? ? − ? − ? ? ? ?	? ? ? ?	? ? ? ? ? − −	+ − − + −	+ + −	+ + − −
218	+ − − − − S	− ? ? ?	+ + −	− +	+ ? − ? − + + − +	+ − + +	− − V + ? − −	+ − − + −	+ + −	+ + − −
219	V − − − V −	− ? + −	− − −	− V	? ? + ? + + + + +	+ + + +	− − − ? ? − −	+ − + + −	+ + +	− N N N
220	− − + V + +	V ? + +	− + +	− +	? ? + ? − + + − −	+ + + +	− − − ? ? + −	+ − + + −	+ − +	− N N N
221	− − + − + +	V ? + +	− S S	− D	? ? + ? − + + − −	+ + + +	− − − ? ? + −	+ − + + −	+ + +	− N N N
222	D − + − + +	+ ? + +	− S V	− +	? ? + ? − + + − −	+ + + +	− − − ? ? + −	+ − + + −	+ + +	− N N N
223	+ − V + + V	− ? + +	− − −	− +	? ? + ? + + + + +	+ + + +	− − − ? ? − −	+ − + + −	+ + +	− N N N
224	− − + − + +	V ? + +	− − −	− +	? ? + ? − + + − −	+ + + +	− − − ? ? + −	+ − + + −	+ + +	− N N N
225	− − − − − V	− ? ? ?	− − −	− −	? ? − ? − ? ? ? ?	? ? ? ?	− − − ? ? S −	+ − − + −	V V −	+ + − −
226	V − − V + +	− ? ? ?	− − V	− V	? ? − ? V + + V +	+ + + +	− − V ? ? − −	+ − − + −	− N N	+ + − −
227	− + − V + +	− ? ? ?	− − V	− V	? ? − ? V + + V +	+ + + +	− − − ? ? + −	+ − − + −	− N N	+ + − −
228	V + V + + +	V ? ? ?	− V V	− V	? ? − ? V + + V +	+ + + +	− − V ? ? + −	+ − − + −	− N N	+ + − −
229	V + V V + +	V ? ? ?	V V V	− +	? ? − ? V ? ? ? ?	? ? ? ?	− − V ? ? V −	+ − − + −	− N N	+ + − −
230	+ − + − + +	− ? ? ?	− + V	− +	+ ? − ? V ? ? ? ?	? ? ? ?	V − + + ? − −	+ − − + −	+ + −	+ + − −
231	V − V − + +	− V + ?	− V −	− V	? ? − ? − + + − S	+ + + +	− − − ? ? − −	+ − − + −	+ + −	+ − − +
232	+ − − − + −	− − ? ?	− V −	− V	? ? − ? − + + − −	+ + + +	? ? − ? ? − −	+ − − + −	+ + −	+ − − +
233	+ − S − + +	− + + −	− D D	− D	+ S − ? − ? ? ? ?	? ? ? ?	− − + ? − − −	+ − − − +	V V −	+ − − +
234	+ − + − + +	− + + −	− + V	− +	? ? − − − + + − +	+ + + +	V − V ? ? − −	+ − − + −	V V −	+ − − +
235	+ − + − + +	− + + −	− + −	− +	? ? − − − + + − +	+ + + +	V − − ? ? − −	+ − − + −	V V −	+ − − +
236	+ − + − + +	− + ? ?	− + −	− +	? ? − − − + + − −	+ + + +	? ? − ? ? − −	+ − − + −	V V −	+ − − +
237	+ − − − V V	− ? ? ?	− + −	? V	? ? − ? − ? ? ? ?	? ? ? ?	− − − ? ? ? −	+ − + + −	V V −	+ + − −
238	V − − − − V	− V − −	− V −	− +	? ? − − + + + + +	+ + + +	− − − ? ? − −	+ − + + −	V V −	+ + − −
239	+ − + − + +	− + − −	− + +	− +	? ? − − − + + − +	− + + +	− − − ? ? − −	+ − + + −	V V −	+ + − −
240	+ − − − − −	− ? ? ?	− + −	− V	? ? − ? − − V V V	+ V + +	− − + ? ? − −	+ − − + −	+ + +	+ − − +
241	+ − + − + +	− ? ? ?	− + −	− +	? ? + ? − ? ? ? ?	? ? ? ?	? ? + ? ? − −	+ − − + −	V V −	+ − + −
242	− − − − − −	− V V −	− − −	− V	− ? − − − − − − −	− + + +	− − V − − − −	+ − − + −	V V −	+ + − −

N not applicable; S slow utilisation; V variable; ? result not known.

		1 D–Glucose fermentation	2 Galactose fermentation	3 Maltose fermentation	4 Me α–D–glucoside fermn	5 Sucrose fermentation	6 Trehalose fermentation	7 Melibiose fermentation	8 Lactose fermentation	9 Cellobiose fermentn	10 Melezitose fermentn	11 Raffinose fermentation	12 D–Galactose growth	13 L–Sorbose growth	14 D–Ribose growth	15 D–Xylose growth	16 L–Arabinose growth	17 D–Arabinose growth	18 L–Rhamnose growth	19 Sucrose growth	20 Maltose growth	21 Trehalose growth	22 Me α–D–glucoside grth	23 Cellobiose growth	24 Salicin growth	25 Arbutin growth	26 Melibiose growth	27 Lactose growth	28 Raffinose growth	29 Melezitose growth	30 Inulin growth	31 Starch growth
243	Phaffia rhodozyma*	S	–	S	?	S	?	–	–	?		?	S	–	–	–	D	+	–	–	+	+	+	V		+	S	S	–	–	+	+	–	V
244	Pichia abadieae*	S	S	–	–	–	?	?	–	–		–	–	+	–	–	+	+	–	+		–	–	+	–		–	–	–	+	+	–	–		–	–
245	Pichia acaciae	+	–	+	?	–	?	–	–	?		–	–	+	+	+	–	S	–	–		–	+	+	+		+	+	+	–	–	–	–		–	+
246	Pichia ambrosiae*	S	–	–	?	–	?	–	–	?		?	–	–	–	+	–	–	–	–		+	+	+	+		+	+	+	–	–	–	–		–	–
247	Pichia angophorae	+	–	+	?	+	?	–	–	?		?	–	–	–	–	–	+	–	–		+	+	D	+		+	+	+	–	–	–	+		V	–
248	Pichia besseyi*	+	–	–	–	–	–	–	–	–		–	–	–	–	–	–	–	–	–		–	–	–	–		–	–	–	–	–	–	–		–	–
249	Pichia bovis	+	–	V	?	V	–	–	–	?		?	–	–	–	–	–	–	–	–		+	+	+	+		+	+	+	–	–	–	+		–	V
250	Pichia castillae*	–	–	–	–	–	–	–	–	–		–	–	+	+	+	+	+	–	–		–	+	+	–		+	V	V	+	–	+	–		–	D
251	Pichia chambardii	–	–	–	–	–	–	–	–	–		–	–	+	–	–	–	–	–	–		–	–	–	–		–	–	–	–	–	–	–		–	–
252	Pichia delftensis	S	–	–	–	–	–	–	–	–		–	–	–	–	–	–	–	–	–		–	–	–	–		–	–	–	–	–	–	–		–	–
253	Pichia dispora	+	–	–	–	–	V	–	–	–		–	–	–	–	–	–	–	–	–		–	–	–	–		–	–	+	–	–	–	–		–	–
254	Pichia etchellsii	S	–	V	?	V	?	–	–	?		?	–	+	+	V	+	+	V	–		+	+	+	+		+	+	+	–	–	–	+		–	–
255	Pichia farinosa	+	S	–	–	–	V	–	–	?		–	–	+	V	V	V	V	–	–		–	–	V	–		V	V	V	–	–	–	–		–	–
256	Pichia fermentans	+	–	–	–	–	–	–	–	–		–	–	–	–	–	+	–	–	–		–	–	–	–		–	–	–	–	–	–	–		–	–
257	Pichia fluxuum	S	–	–	–	–	–	–	–	–		–	–	–	–	–	–	–	–	–		–	–	–	–		–	–	–	–	–	–	–		–	–
258	Pichia guilliermondii	+	V	V	?	+	V	V	–	–		–	+	+	V	+	+	+	+	V		+	+	+	+		+	+	+	+	–	+	+		+	–
259	Pichia haplophila	V	–	–	–	–	–	–	–	–		–	–	+	V	+	+	+	–	–		–	–	–	–		–	–	–	–	–	–	–		–	–
260	Pichia heimii*	+	+	–	?	+	?	–	–	?		?	–	+	–	+	+	+	–	+		+	+	+	+		+	+	+	–	–	+	+		–	–
261	Pichia humboldtii*	–	–	–	–	–	–	–	–	–		–	–	+	V	–	–	–	–	–		–	–	–	–		–	–	–	–	–	–	–		–	–
262	Pichia kluyveri	+	–	–	–	–	–	–	–	–		–	–	–	–	–	–	–	–	–		–	–	–	–		–	–	–	–	–	–	–		–	–
263	Pichia kudriavzevii	+	–	–	–	–	–	–	–	–		–	–	–	V	V	–	–	–	–		–	–	–	–		–	–	–	–	–	–	–		–	–
264	Pichia lindnerii*	S	–	–	–	–	–	–	–	–		–	–	–	–	V	+	–	–	+		–	–	+	–		+	+	+	–	–	–	–		–	–
265	Pichia media	–	–	–	–	–	–	–	–	–		–	–	+	+	–	+	V	V	–		–	+	+	–		+	+	+	–	–	–	–		–	+
266	Pichia membranaefaciens	V	–	–	–	–	–	–	–	–		–	–	–	V	–	V	–	–	–		–	–	–	–		–	–	–	–	–	–	–		–	–
267	Pichia methanolica*	+	–	–	–	–	S	–	–	S		–	–	+	V	+	+	+	V	–		–	+	–	–		+	+	+	–	–	–	–		–	–
268	Pichia mucosa*	+	–	–	?	–	?	–	–	?		?	–	–	+	–	D	–	D	–		+	+	+	+		+	+	+	–	–	–	+		–	–
269	Pichia naganishii*	+	–	–	–	–	?	–	–	?		–	–	V	V	–	+	+	–	–		+	+	+	–		+	+	+	–	–	+	–		–	–
270	Pichia nakazawae*	+	V	V	?	–	?	–	–	?		?	–	+	+	D	+	D	V	V		+	+	+	+		+	+	D	–	–	–	+		–	+
271	Pichia norvegensis*	S	–	–	–	–	S	–	–	V		–	–	–	–	–	V	–	–	–		–	–	–	–		+	V	V	–	–	–	–		–	–
272	Pichia ohmeri	+	S	V	?	+	V	–	–	?		–	+	+	+	–	–	–	–	–		+	+	+	+		+	+	+	–	–	+	–		V	–
273	Pichia onychis	+	–	–	?	+	?	–	–	?		?	S	–	–	–	+	+	–	–		+	+	+	+		+	+	+	–	–	+	+		–	–
274	Pichia pastoris	+	–	–	–	–	?	–	–	–		–	–	–	–	–	–	–	–	+		–	+	–	–		–	–	–	–	–	–	–		–	–
275	Pichia philogaea*	+	V	V	?	S	?	–	–	?		?	–	+	+	V	+	+	S	–		+	+	+	S		+	+	+	–	–	–	+		–	–
276	Pichia pijperi	+	–	–	–	–	–	–	–	?		–	–	–	+	–	+	–	–	–		–	–	–	–		+	+	+	–	–	–	–		–	–
277	Pichia pinus	V	–	–	–	–	V	–	–	?		–	–	–	V	+	+	V	V	V		–	+	–	–		+	+	+	–	–	–	–		–	–
278	Pichia pseudopolymorpha	S	V	V	?	S	?	V	–	?		?	V	+	+	+	+	V	–	+		+	+	+	+		+	+	+	+	+	+	+		–	+
279	Pichia quercuum	S	–	–	–	–	–	–	–	?		–	–	–	–	–	–	–	–	–		–	–	–	–		+	+	+	–	–	–	–		–	–
280	Pichia rabaulensis*	+	–	–	?	+	?	–	–	?		?	+	–	–	–	+	–	–	+		+	+	D	+		+	+	+	–	–	+	+		–	–
281	Pichia rhodanensis	+	–	–	?	V	?	–	–	?		?	–	–	–	–	+	–	–	+		+	+	+	V		+	+	+	–	–	–	+		–	–
282	Pichia saitoi	+	–	–	–	–	?	–	–	–		–	–	–	–	–	–	–	–	–		–	–	V	–		–	–	–	–	–	–	–		–	–
283	Pichia salictaria	–	–	–	–	–	–	–	–	–		–	–	–	–	+	–	–	–	+		–	–	–	–		+	+	+	–	–	–	–		–	–
284	Pichia sargentensis*	+	–	–	–	–	–	–	–	?		–	–	–	–	–	+	–	–	+		–	–	–	–		+	+	+	–	–	–	–		–	–

Codes in table: + positive; – negative; D delayed for longer than 7 days;

Characteristics of the species

	32 Glycerol growth	33 Erythritol growth	34 Ribitol growth	35 Galactitol growth	36 D–Mannitol growth	37 D–Glucitol growth	38 myo–Inositol growth	39 Gluconolactone growth	40 2–Ketogluconate growth	41 5–Ketogluconate growth	42 D,L–Lactate growth	43 Succinate growth	44 Citrate growth	45 Methanol growth	46 Ethanol growth	47 Ethylamine growth	48 D–Glucosamine growth	49 Nitrate growth	50 Nitrite growth	51 Growth sans vitamins	52 Growth sans inositol	53 Grth sans pantothenate	54 Growth sans biotin	55 Growth sans thiamin	56 Grth sans pyridoxine	57 Growth sans niacin	58 Growth sans folic acid	59 Growth sans PABA	60 Growth in 50% glucose	61 Growth in 60% glucose	62 Growth at 37 degrees	63 0.01% cyclohex. growth	64 0.1% cyclohex. growth	65 Starch formation	66 Pink colonies	67 Budding cells	68 Splitting cells	69 Apical budding	70 Cells sph, ovl, cylind	71 Cells of other shapes	72 Filamentous	73 Pseudohyphae	74 Septate hyphae	75 Ascosporogenous	76 Ascosp sph, ovl, renfm	77 Ascosp hat or saturn	78 Ascosp of other shapes
243	V	–	–	–	D	V	–	+	+	D	V	+	V	–	V			–	–	–	?	–	+	+	–	+			+	+	+	+	–	–	–	–	–	+	+	+	–	–	+	–	V	V	–	–	N	N	N
244	+	–	+	D	+	+	+	?	?	?	+	+	+		–	?		?	?	–	?	–	?	?	?	?			?	?	?	?				?	?	?	–	?	–	–	+	–	–	+	+	+	–	+	?	?	?
245	+	+	+	–	+	+	–	?	?	?	–	+	+		–	+		?	?	–	?	–	+	+	–	+			+	+	+	+	S	?	+	?	?	–	–	+	–	–	+	–			+	+	–	+	–	–
246	+	+	+	–	+	+	–	?	?	?	V	+	D		–	S		?	S	–	?	–	+	+	–	–			–	+	+	+	?	?	V	?	?	–	–	+	–	–	+	–			+	+	–	+	–	+	–
247	–	–	+	–	+	+	–	?	?	?	+	+	+		–	+		?	?	–	?	–	+	+	–	–			+	+	+	+	–	–	–	?	?	–	–	+	–	–	+	–			+	+	–	+	–	+	–
248	–	–	D	–	+	+	–	–	+	–	+	+	–		–	?		?	–	–	?	–	+	+	+	–			–	+	+	+	?	?	–	?	?	–	–	+	–	–	+	–			–	N	N	+	–	+	–
249	+	–	–	–	+	+	–	?	?	?	+	+	+		–	+		+	?	–	–	–	–	+	+	–			–	+	+	+	–	–	V	?	?	–	–	+	–	–	+	–			V	V	–	+	–	+	–
250	+	+	+	D	+	+	–	?	?	?	–	+	+		–	+		?	?	–	?	–	–	+	+	S			+	+	+	+	+	?	+	?	?	–	–	+	–	–	+	–			+	+	–	+	–	+	–
251	+	–	–	–	–	–	–	?	?	?	+	+	+		–	+		?	?	–	?	–	–	+	+	+			–	+	+	+	–	–	–	?	?	–	–	+	–	–	+	–			–	N	N	+	–	+	–
252	–	–	+	–	+	+	–	+	V	–	–	+	–		–	+		+	V	–	–	–	–	+	+	–			–	+	+	+	–	–	V	?	?	–	–	+	–	–	+	–			V	V	–	+	–	+	–
253	–	–	+	–	+	+	–	–	+	–	V	+	–		–	+		?	–	–	?	–	–	+	+	+			–	+	+	+	–	–	–	?	?	–	–	+	–	–	+	–			–	N	N	+	–	+	–
254	V	–	+	–	+	+	–	D	+	–	V	+	+		–	+		?	?	–	?	V	+	+	V	+			+	+	+	+	V	?	+	?	?	–	–	+	–	–	+	–			+	+	–	+	+	–	–
255	+	+	+	–	+	+	–	?	?	?	–	V	+		–	+		?	?	–	?	V	?	?	?	?			?	?	?	?	+	?	+	?	?	–	–	+	–	–	+	–			+	+	–	+	+	–	–
256	+	–	–	–	–	–	–	V	–	–	+	+	+		–	+		?	V	–	?	–	–	+	+	–			–	–	+	+			–	+	?	–	–	+	–	–	+	–			+	+	–	+	–	+	–
257	–	–	V	–	+	+	–	–	–	–	V	+	–		–	+		?	–	–	?	–	–	+	+	–	+		+	+	+	+	V	?	+	?	?	–	–	+	–	–	+	–			V	V	–	+	–	+	–
258	+	–	+	V	+	+	–	?	?	?	V	+	+		–	+		?	?	–	?	–	–	+	+	–	V		V	+	+	+	V	?	+	?	?	–	–	+	–	–	+	–			V	V	–	+	–	+	–
259	+	+	+	+	+	+	–	D	+	–	–	–	–		–	+		?	?	–	?	–	–	+	+	+	–		+	+	+	+	–	–	V	?	?	–	–	+	–	–	+	–			V	V	–	+	–	+	–
260	+	+	+	+	+	+	–	?	?	?	–	+	+		–	+		?	?	–	–	+	+	+	+	+			+	+	+	+	?	?	–	?	?	–	–	+	–	–	+	–			+	+	–	+	–	+	–
261	+	–	–	–	–	–	–	?	?	?	V	V	–		–	+		?	?	–	?	+	+	+	+	+			+	+	+	+	?	?	V	?	?	–	–	+	–	–	+	–			+	+	–	+	+	–	–
262	+	–	–	–	–	–	–	V	–	–	+	+	+		–	+		?	+	–	?	–	+	+	–	+			–	+	+	+	–	–	+	?	?	–	–	+	–	–	+	–			+	+	–	+	+	–	–
263	+	–	–	–	–	–	–	V	–	–	+	+	V		–	+		?	V	–	?	+	+	+	+	+			+	+	+	+	V	?	+	?	?	–	–	+	–	–	+	–			+	+	–	+	+	–	–
264	+	–	+	–	+	+	–	+	–	–	V	D	+		+	+		?	–	–	–	–	+	+	–	+			+	+	+	+	–	–	+	?	?	–	–	+	–	–	+	–			–	N	N	+	–	+	–
265	+	+	+	D	+	+	–	V	V	V	–	+	+		–	+		?	?	–	?	–	–	+	+	–	–		+	+	+	+	–	–	–	?	?	–	–	+	–	–	+	–			V	V	–	+	–	+	–
266	V	–	–	–	–	–	–	?	?	?	V	V	V		–	+		?	?	–	?	V	?	?	?	?			?	?	?	?	V	?	V	?	?	–	–	+	–	–	+	–			V	V	–	+	V	V	–
267	+	+	+	V	+	+	–	+	–	–	–	+	V		+	?		?	–	–	?	S	+	+	S	+			+	+	+	+	–	–	+	?	?	–	–	+	–	–	+	–			–	N	N	+	–	+	–
268	+	–	–	–	D	+	–	+	+	–	–	V	–		–	?		?	?	–	?	–	+	+	–	–			–	+	+	+	?	?	–	?	–	–	–	+	–	–	+	–			–	N	N	+	–	+	–
269	+	+	+	–	+	+	–	+	–	–	D	+	D		+	?		?	?	–	?	–	+	+	–	–			–	+	+	+	?	?	+	?	?	–	–	+	–	–	+	–			V	V	–	+	–	+	–
270	+	+	+	–	+	+	–	–	+	–	–	+	+		–	?		?	?	–	?	–	+	+	–	–			+	+	+	+	?	?	V	?	?	–	–	+	–	–	+	–			+	+	+	+	–	+	–
271	+	–	–	–	–	–	–	–	–	–	+	+	V		–	?		?	+	–	?	–	+	+	–	–			–	+	+	+	?	?	+	?	?	–	–	+	–	–	+	–			V	V	–	+	–	+	–
272	+	–	V	–	+	+	–	?	?	?	V	+	+		–	+		?	?	–	?	–	–	+	+	–	+		+	+	+	+	V	?	+	?	?	–	–	+	–	–	+	–			+	+	–	+	–	+	–
273	+	–	V	–	+	+	–	?	?	?	+	+	+		–	+		?	?	–	?	–	–	+	+	–	–			–	+	+	+	–	–	+	?	?	–	–	+	–	–	+	–			V	V	–	+	–	+	–
274	+	–	–	–	+	+	–	–	–	–	V	+	–		+	+		?	–	–	?	–	–	+	+	–	+			+	+	+	+	–	–	V	?	?	–	–	+	–	–	+	–			–	N	N	+	–	+	–
275	S	+	+	–	+	+	–	S	+	–	–	+	+		–	+		?	?	–	–	–	–	+	+	–	–			+	+	+	+	+	?	+	?	?	–	–	+	–	–	+	–			+	+	–	+	–	+	–
276	+	–	–	–	+	+	–	+	–	–	+	+	+		–	+		?	–	–	?	–	–	+	+	–	–			–	+	+	+	–	–	S	?	?	–	–	+	–	–	+	–			V	V	–	+	–	+	–
277	+	+	+	–	+	+	–	+	V	V	–	V	V		+	V		?	V	–	V	V	+	+	V	+			+	+	+	+	–	–	V	?	?	–	–	+	–	–	+	–			V	V	–	+	–	+	–
278	+	+	+	+	+	+	–	?	?	?	+	+	+		–	+		?	?	–	+	S	+	+	S	+			+	+	+	+	+	?	–	?	?	–	–	+	–	–	+	–			V	V	–	+	+	–	–
279	+	–	–	–	+	+	–	D	–	–	V	+	+		–	+		?	–	–	?	–	–	+	+	+	–			+	+	+	+	–	–	S	?	?	–	–	+	–	–	+	–			+	+	–	+	–	+	–
280	+	–	D	–	+	+	–	D	–	–	+	+	+		–	?		?	?	–	?	–	–	+	+	S	–			–	+	+	+	?	?	+	–	?	–	–	+	–	–	+	–			+	+	–	+	–	+	–
281	+	–	–	–	+	+	–	+	–	?	+	+	+		–	+		?	–	–	?	–	–	+	+	–	–			–	+	+	+	–	–	+	?	?	–	–	+	–	–	+	–			+	+	–	+	–	+	–
282	–	–	–	–	+	+	–	–	–	–	+	+	–		–	+		?	?	–	?	–	–	+	+	+	–			–	+	+	+	–	–	–	?	?	–	–	+	–	–	+	–			V	V	–	+	V	V	–
283	+	–	–	–	+	+	–	–	+	V	+	+	+		–	+		?	–	–	?	–	–	+	+	+	–			–	+	+	+	–	–	+	?	?	–	–	+	–	–	+	–			–	N	N	+	–	+	–
284	+	–	–	–	+	+	–	+	–	–	+	+	–		–	+		?	–	–	?	–	–	+	+	+	–			+	+	+	+	?	?	+	?	?	–	–	+	–	–	+	–			+	+	–	+	–	+	–

N not applicable;   S slow utilisation;   V variable;   ? result not known.

		1 D-Glucose fermentation	2 Galactose fermentation	3 Maltose fermentation	4 Me α-D-glucoside fermn	5 Sucrose fermentation	6 Trehalose fermentation	7 Melibiose fermentation	8 Lactose fermentation	9 Cellobiose fermentn	10 Melezitose fermentn	11 Raffinose fermentation	12 D-Galactose growth	13 L-Sorbose growth	14 D-Ribose growth	15 D-Xylose growth	16 L-Arabinose growth	17 D-Arabinose growth	18 L-Rhamnose growth	19 Sucrose growth	20 Maltose growth	21 Trehalose growth	22 Me α-D-glucoside grth	23 Cellobiose growth	24 Salicin growth	25 Arbutin growth	26 Melibiose growth	27 Lactose growth	28 Raffinose growth	29 Melezitose growth	30 Inulin growth	31 Starch growth
285	Pichia scolyti	S	S	V	?	V	S	?	−	?	?	S	+	−	+	+	+	V	+	+	+	+	+	+	+	+	+	+	+	+	−	−
286	Pichia scutulata*	V	−	−	−	−	−	−	−	−	−	−	−	−	−	−	−	−	−	−	−	−	−	−	−	−	−	−	−	−	−	−
287	Pichia spartinae*	+	−	V	?	S	−	−	−	?	?	−	−	+	−	−	−	−	−	+	+	+	+	+	+	+	−	−	−	+	−	−
288	Pichia stipitis	+	+	+	V	−	+	−	−	−	−	−	+	+	+	+	+	V	+	+	+	+	+	+	+	+	−	+	−	+	−	+
289	Pichia strasburgensis	+	V	−	?	+	−	−	−	?	?	S	+	−	S	+	+	−	+	+	+	+	+	+	+	+	−	−	+	+	−	−
290	Pichia terricola	S	−	−	−	−	−	−	−	−	−	−	−	−	−	−	−	−	−	−	−	−	−	−	−	−	−	−	−	−	−	−
291	Pichia toletana	S	−	−	?	−	?	−	−	?	?	−	−	−	−	+	−	−	V	+	+	+	V	+	+	+	−	−	−	V	−	−
292	Pichia trehalophila	+	−	−	−	−	S	−	−	−	−	−	−	+	+	+	+	+	−	−	−	+	−	−	−	−	−	−	−	−	−	−
293	Pichia veronae*	+	−	−	?	V	V	−	−	?	?	−	−	−	−	+	D	−	+	+	+	+	+	+	+	+	−	−	−	+	−	−
294	Pichia vini	−	−	−	−	−	−	−	−	−	−	−	+	+	−	+	V	−	−	+	+	+	+	V	+	+	V	−	V	+	−	+
295	Pichia wickerhamii	+	−	−	?	−	−	−	−	?	?	−	−	−	−	+	−	−	+	+	+	+	+	+	+	+	−	−	−	V	−	−
296	Rhodosporidium bisporidiis*	−	−	−	−	−	−	−	−	−	−	−	+	+	S	+	+	+	+	+	+	+	−	+	V	+	+	−	+	+	−	+
297	Rhodosporidium capitatum*	−	−	−	−	−	−	−	−	−	−	−	+	+	V	+	V	+	V	+	+	+	−	+	+	+	−	V	+	+	V	V
298	Rhodosporidium dacryoidum*	−	−	−	−	−	−	−	−	−	−	−	+	V	−	−	−	−	V	V	V	+	−	V	−	V	−	−	−	V	−	−
299	Rhodosporidium infirmo−miniatum**	−	−	−	−	−	−	−	−	−	−	−	V	V	S	V	+	V	+	+	+	V	−	+	+	+	−	V	+	+	−	S
300	Rhodosporidium malvinellum*	−	−	−	−	−	−	−	−	−	−	−	V	V	−	S	V	V	S	+	+	+	−	V	+	?	−	−	+	−	?	?
301	Rhodotorula acheniorum*	−	−	−	−	−	−	−	−	−	−	−	+	S	+	+	+	S	−	+	+	+	S	S	−	?	S	S	+	+	S	−
302	Rhodotorula araucariae*	−	−	−	−	−	−	−	−	−	−	−	S	S	−	S	V	V	−	−	−	+	−	V	V	V	−	−	−	−	−	−
303	Rhodotorula aurantiaca	−	−	−	−	−	−	−	−	−	−	−	+	V	V	+	V	V	−	+	+	V	−	V	+	?	−	V	−	+	−	−
304	Rhodotorula glutinis	−	−	−	−	−	−	−	−	−	−	−	V	V	V	V	V	V	V	+	+	+	V	V	V	?	−	−	V	+	−	−
305	Rhodotorula graminis	−	−	−	−	−	−	−	−	−	−	−	+	V	V	+	V	V	V	+	V	+	V	V	V	?	−	−	+	−	−	−
306	Rhodotorula lactosa	−	−	−	−	−	−	−	−	−	−	−	−	−	−	+	+	D	+	+	+	+	D	+	+	?	+	D	+	+	−	−
307	Rhodotorula marina	−	−	−	−	−	−	−	−	−	−	−	+	S	S	+	+	D	+	+	V	D	D	+	D	?	−	V	S	+	−	S
308	Rhodotorula minuta	−	−	−	−	−	−	−	−	−	−	−	V	V	V	+	+	+	−	+	−	+	−	V	V	?	−	V	−	+	−	−
309	Rhodotorula pallida	−	−	−	−	−	−	−	−	−	−	−	V	V	V	V	−	V	V	V	−	+	−	V	V	?	−	−	−	−	−	−
310	Rhodotorula pilimanae	−	−	−	−	−	−	−	−	−	−	−	+	+	+	+	+	V	−	+	−	+	−	V	V	?	−	−	+	−	−	−
311	Rhodotorula rubra	−	−	−	−	−	−	−	−	−	−	−	V	V	V	+	V	V	V	+	+	+	V	V	V	?	−	−	+	+	−	−
312	Saccharomyces cerevisiae	+	V	V	V	V	?	V	−	−	V	V	V	−	−	−	−	−	−	V	V	V	V	−	−	−	V	−	V	V	−	V
313	Saccharomyces dairensis	+	+	−	−	−	?	−	−	−	−	−	+	−	V	−	−	−	−	−	−	V	−	−	−	−	−	−	−	−	−	−
314	Saccharomyces exiguus	+	+	−	−	+	?	−	−	−	−	V	+	−	−	−	−	−	−	+	−	+	−	−	−	−	−	−	V	−	−	−
315	Saccharomyces kluyveri	+	+	V	?	+	?	+	−	?	?	+	+	V	−	V	−	−	−	+	V	V	−	V	V	V	+	−	+	V	V	−
316	Saccharomyces servazzii*	+	+	−	−	−	?	−	−	−	−	−	+	−	V	−	−	−	−	−	−	V	−	−	−	−	−	−	−	−	−	−
317	Saccharomyces telluris	+	−	−	−	−	−	−	−	−	−	−	−	−	−	−	−	−	−	−	−	−	−	−	−	−	−	−	−	−	−	−
318	Saccharomyces unisporus	+	+	−	−	−	−	−	−	−	−	−	+	−	−	−	−	−	−	−	−	−	−	−	−	−	−	−	−	−	−	−
319	Saccharomycodes ludwigii	+	−	−	−	+	−	−	−	?	−	+	−	−	−	−	−	−	−	+	−	−	−	+	+	S	−	−	+	−	−	−
320	Saccharomycopsis capsularis**	S	−	S	?	−	?	−	−	?	−	−	−	−	V	−	−	V	−	−	+	+	+	+	+	+	−	−	−	−	−	+
321	Saccharomycopsis crataegensis*	S	−	−	−	−	V	−	−	−	−	−	−	+	+	+	V	V	−	−	−	V	−	−	−	−	−	−	−	−	−	−
322	Saccharomycopsis fibuligera**	S	−	S	?	S	?	−	−	−	?	V	−	−	−	−	−	−	−	+	+	V	+	+	+	+	−	−	V	V	−	+
323	Saccharomycopsis lipolytica**	−	−	−	−	−	−	−	−	−	−	−	V	V	V	−	−	−	−	−	−	−	−	V	V	V	−	−	−	−	−	−
324	Saccharomycopsis malanga*	S	−	S	?	−	−	−	−	−	?	−	−	−	+	V	V	V	−	−	+	V	−	+	+	+	−	−	−	−	−	+

Codes in table: + positive; − negative; D delayed for longer than 7 days;

Characteristics of the species

Column headers (32–78):
32 Glycerol growth; 33 Erythritol growth; 34 Ribitol growth; 35 Galactitol growth; 36 D–Mannitol growth; 37 D–Glucitol growth; 38 myo–Inositol growth; 39 Gluconolactone growth; 40 2–Ketogluconate growth; 41 5–Ketogluconate growth; 42 D,L–Lactate growth; 43 Succinate growth; 44 Citrate growth; 45 Methanol growth; 46 Ethanol growth; 47 Ethylamine growth; 48 D–Glucosamine growth; 49 Nitrate growth; 50 Nitrite growth; 51 Growth sans vitamins; 52 Growth sans inositol; 53 Grth sans pantothenate; 54 Growth sans biotin; 55 Growth sans thiamin; 56 Grth sans pyridoxine; 57 Growth sans niacin; 58 Growth sans folic acid; 59 Growth sans PABA; 60 Growth in 50% glucose; 61 Growth in 60% glucose; 62 Growth at 37 degrees; 63 0.01% cyclohex. growth; 64 0.1% cyclohex. growth; 65 Starch formation; 66 Pink colonies; 67 Budding cells; 68 Splitting cells; 69 Apical budding; 70 Cells sph, ovl, cylind; 71 Cells of other shapes; 72 Filamentous; 73 Pseudohyphae; 74 Septate hyphae; 75 Ascosporogenous; 76 Ascosp sph, ovl, renfm; 77 Ascosp hat or saturn; 78 Ascosp of other shapes

#	32–37	38–41	42–44	45–46	47–50	51–55	56–59	60–64	65–66	67–71	72–74	75–78
285	+++−++	−???	−−+	−+	??−?	−++−+	++++	V?V??	−−	+−−+−	++−	+−+−
286	+−−−−	−−−?	++−	−+	+−−−	++++++	++++	−−V??	−−	+−−+−	VV−	++−−
287	V−+−++	−???	V++	−+	+S−?	−++−+	++++	S?+??	−−	+−−+−	++−	+−+−
288	+++−++	−++−	+++	−+	??−?	V++V+	++++	−−V??	−−	+−−++	++−	+−+−
289	+−+−++	−???	+++	−+	??−?	−++−+	−+++	−−+??	−−	+−−+−	VV−	+−+−
290	+−−−−	−−−?	V++	−+	??−?	−++−+	++++	−−+??	−−	+−−+−	VV−	++−−
291	+−−++	−+−−	V++	−+	??−?	−++−+	++++	−−−??	−−	+−−+−	++−	+−+−
292	+++−++	−+−−	−++	++	?−−?	−++−+	++++	−−−??	−−	+−−+−	−NN	+−+−
293	+−+−+−	−+−?	+++	−+	??−?	−++−−	−+++	??V??	−−	+−−+−	++−	+−+−
294	+−+−++	−???	V++	−+	??−?	−++−+	++++	V?V??	−−	+−−+−	++−	++−−
295	+−+−++	−???	+++	−+	??−?	−++−+	−+++	−−+??	−−	+−−+−	++−	+−+−
296	V−VV++	+???	V+−	−+	??+?	−++S−	++++	−−−??	++	+−−+−	VVV	−NNN
297	+−V+++	+???	−+S	−−	??+?	−++−−	++++	−−−??	++	+−−+−	+−+	−NNN
298	+−−−VV	−???	???	−V	??−?	−+++S	+++V	−−−??	−+	+−−+−	+VV	−NNN
299	V−VV+V	++++	SSS	−V	??++	−++V−	++++	−−−??	++	+−−+−	VVV	−NNN
300	+−VV+V	−???	???	−V	??+?	−++−−	+++V	??−??	−+	+−−+−	VVV	−NNN
301	S+S−++	−???	S++	−?	??+?	−????	????	−−−??	−V	+−−+−	−NN	−NNN
302	+−+V++	−???	S+S	−+	?−+?	V????	????	??−??	−+	+−−+−	−NN	−NNN
303	+−VV++	−++V	V+V	−V	??++	−+++−	+++V	−−V??	−+	+−−+−	−NN	−NNN
304	V−VVVV	−+V−	V+V	−V	??++	V+++V	++++	−−V??	−+	+−−+−	VV−	−NNN
305	V−VVVV	−++−	−+V	−V	??++	+++++	++++	−−−??	−+	+−−+−	VV−	−NNN
306	+−D−++	−+++	D+D	−D	??+−	−+++V	+++−	−−−??	−+	+−−+−	−NN	−NNN
307	+−D+++	−+−−	D++	−−	??−−	−++−−	+++−	−−−??	−+	+−−+−	−NN	−NNN
308	+−V−VV	−+++	VV−	−V	??−−	−++−−	+++−	−−V??	−+	+−−+−	−NN	−NNN
309	+VVV++	−VVV	++V	−V	??−−	−++−−	+++−	−−−??	−+	+−−+−	−NN	−NNN
310	V−+V+V	−+V−	V+V	−V	??−−	−+++−	++++	−−V??	−+	+−−+−	−NN	−NNN
311	V−VVVV	−V−−	V+V	−V	??−−	−+++−	++++	−−V??	−+	+−−+−	VV−	−NNN
312	V−−−VV	−−−−	VV−	−V	−−−?	V????	????	V−V−−−	−−	+−−+−	VV−	++−−
313	V−−−−−	−V−−	−−−	−V	−−−−	−+−V−	V−++	−−VV−−	−−	+−−+−	−NN	++−−
314	−−−−−−	−V−−	VV−	−V	−−−?	V++VV	++++	−−−V−−	−−	+−−+−	VV−	++−−
315	V−−−VV	−VV−	VV−	−+	+−−V	????	????	−−+−−	−−	+−−+−	VV−	++−−
316	D−−−−−	−−−−	−−−	−?	−−−−	−−−−	−+++	V−V++−	−−	+−−+−	−NN	++−−
317	−−−−−−	−V−−	V−−	−V	−−−−	−V−−−	−+++	−+−+−−	−−	+−−+−	VV−	++−−
318	−−−−−−	−V−−	−−−	−S	+−−?	−+−−−	−+++	−−V++−	−−	+−−+−	−NN	++−−
319	+−−−−−	−−−−	D−−	−−	??−−	−−+−+	−−++	−−−??	−−	+−++−	VV−	++−−
320	++V−++	−???	−+V	−+	??−?	−????	????	V?V??	−−	+−−+−	+++	+−+−
321	+−V−++	−?−+	VV−	−+	?V−?	−????	????	??−??	−−	+−−+−	+++	+−+−
322	+VV−VV	V???	SVV	−+	??−?	−????	????	−−+??	−−	+−−+−	+++	+−+−
323	++V−VV	−?−−	+++	−+	?−−?	−+++−	++++	??V??	−−	+−−+−	+++	+−+−
324	+V+−++	−?++	−+−	−+	?−−?	−????	????	??+??	−−	+−−+−	+++	+−+−

N not applicable; S slow utilisation; V variable; ? result not known.

Characteristics of the species

		1 D–Glucose fermentation	2 Galactose fermentation	3 Maltose fermentation	4 Me α–D–glucoside fermn	5 Sucrose fermentation	6 Trehalose fermentation	7 Melibiose fermentation	8 Lactose fermentation	9 Cellobiose fermentn	10 Melezitose fermentn	11 Raffinose fermentation	12 D–Galactose growth	13 L–Sorbose growth	14 D–Ribose growth	15 D–Xylose growth	16 L–Arabinose growth	17 D–Arabinose growth	18 L–Rhamnose growth	19 Sucrose growth	20 Maltose growth	21 Trehalose growth	22 Me α–D–glucoside grth	23 Cellobiose growth	24 Salicin growth	25 Arbutin growth	26 Melibiose growth	27 Lactose growth	28 Raffinose growth	29 Melezitose growth	30 Inulin growth	31 Starch growth
325	Saccharomycopsis vini**	S	–	–	?	V	?	–	–	?	–	V	–	V	–	–	–	–	–	V	V	V	V	V	V	V	–	–	V	–	–	–
326	Sarcinosporon inkin**	–	–	–	–	–	–	–	–	–	–	–	+	–	+	–	–	S	–	+	+	+	+	+	S	?	–	+	–	+	–	+
327	Schizoblastosporion starkeyi–henricii	–	–	–	–	–	–	–	–	–	–	–	+	V	–	–	–	–	–	–	–	V	–	–	–	–	–	–	–	–	–	–
328	Schizosaccharomyces japonicus	+	–	+	–	+	–	+	–	–	–	+	–	–	–	–	–	–	–	+	+	–	–	–	–	–	–	–	+	–	–	–
329	Schizosaccharomyces malidevorans	+	–	–	–	+	–	–	–	–	–	+	–	–	–	–	–	–	–	+	–	–	–	–	–	–	–	–	+	–	D	–
330	Schizosaccharomyces octosporus	+	–	+	–	V	–	–	–	–	–	–	–	–	–	–	–	–	–	V	+	V	–	–	–	–	–	–	–	–	–	–
331	Schizosaccharomyces pombe	+	–	+	–	+	–	+	–	–	–	+	–	–	–	–	–	–	–	+	+	–	V	–	–	–	–	–	–	+	V	–
332	Schizosaccharomyces slooffiae*	+	–	–	–	V	–	–	–	–	–	–	–	–	–	–	–	–	–	V	S	–	–	–	–	–	–	–	–	–	–	–
333	Schwanniomyces occidentalis	+	V	V	?	+	?	V	–	?	?	V	V	V	–	V	V	–	–	+	+	V	V	V	V	V	V	–	V	+	V	+
334	Selenozyma peltata*	+	S	V	?	V	?	–	–	?	?	–	+	+	+	+	+	+	+	+	+	+	+	+	+	–	–	–	+	–	D
335	Sporidibolus johnsonii	–	–	–	–	–	–	–	–	–	–	–	V	D	V	D	–	+	–	+	+	+	+	+	V	?	–	–	–	+	–	V
336	Sporidibolus ruinenii	–	–	–	–	–	–	–	–	–	–	–	D	+	V	+	D	V	V	+	V	D	–	+	S	?	–	–	+	–	–	–
337	Sporobolomyces albo–rubescens	–	–	–	–	–	–	–	–	–	–	–	+	S	V	+	+	+	–	+	+	+	V	V	V	+	–	–	+	V	–	–
338	Sporobolomyces antarcticus*	–	–	–	–	–	–	–	–	–	–	–	–	–	+	+	+	+	D	+	+	+	+	–	D	+	+	+	+	+	–	+
339	Sporobolomyces gracilis	–	–	–	–	–	–	–	–	–	–	–	–	–	D	+	–	–	–	–	–	+	–	D	D	+	–	–	–	–	–	–
340	Sporobolomyces hispanicus	–	–	–	–	–	–	–	–	–	–	–	S	S	D	S	–	V	–	–	–	+	–	–	–	–	–	–	–	–	–	–
341	Sporobolomyces holsaticus	–	–	–	–	–	–	–	–	–	–	–	V	V	V	V	V	V	–	+	+	+	V	V	S	S	–	–	V	V	–	V
342	Sporobolomyces odorus	–	–	–	–	–	–	–	–	–	–	–	–	V	V	+	–	+	–	V	–	+	–	–	–	–	–	–	–	V	–	–
343	Sporobolomyces pararoseus	–	–	–	–	–	–	–	–	–	–	–	S	S	V	V	–	S	–	+	+	+	+	+	S	+	–	–	+	+	–	+
344	Sporobolomyces puniceus*	–	–	–	–	–	–	–	–	–	–	–	–	+	D	–	D	–	–	+	+	+	–	+	+	+	–	+	+	+	–	+
345	Sporobolomyces roseus	–	–	–	–	–	–	–	–	–	–	–	V	V	V	V	V	+	–	+	+	+	V	V	S	S	–	+	+	+	–	+
346	Sporobolomyces salmonicolor	–	–	–	–	–	–	–	–	–	–	–	V	V	V	V	V	V	–	+	–	+	–	–	–	–	–	–	V	–	–	–
347	Sporobolomyces singularis	–	–	–	–	–	–	–	–	–	–	–	–	–	–	–	–	–	–	–	–	–	–	+	+	+	–	+	–	–	–	–
348	Stephanoascus ciferrii**	–	–	–	–	–	–	–	–	–	–	–	+	+	+	+	+	+	+	+	+	+	S	+	+	+	+	–	+	–	–	V
349	Sterigmatomyces aphidis*	–	–	–	–	–	–	–	–	–	–	–	+	+	+	+	+	+	+	+	+	+	+	+	D	D	+	+	+	+	–	+
350	Sterigmatomyces elviae*	–	–	–	–	–	–	–	–	–	–	–	D	+	+	+	+	+	–	+	–	+	–	+	+	+	–	+	+	–	–	–
351	Sterigmatomyces halophilus	–	–	–	–	–	–	–	–	–	–	–	+	D	+	+	+	V	–	–	–	+	–	V	+	+	–	V	–	–	–	–
352	Sterigmatomyces indicus	–	–	–	–	–	–	–	–	–	–	–	+	D	+	+	+	V	–	–	–	+	–	–	+	+	–	–	–	–	–	–
353	Sterigmatomyces nectairii*	–	–	–	–	–	–	–	–	–	–	–	–	–	–	D	–	–	–	–	–	+	–	–	–	–	–	–	–	–	–	–
354	Sterigmatomyces penicillatus*	–	–	–	–	–	–	–	–	–	–	–	+	+	+	+	+	+	+	+	+	+	V	+	+	+	D	+	–	+	–	–
355	Sterigmatomyces polyborus*	–	–	–	–	–	–	–	–	–	–	–	+	S	+	+	S	S	+	+	+	+	S	+	S	+	S	S	+	+	–	S
356	Sympodiomyces parvus*	–	–	–	–	–	–	–	–	–	–	–	+	D	V	+	D	V	V	D	D	V	+	+	–	+	–	–	–	–	–	–

Codes in table: + positive; – negative; D delayed for longer than 7 days;

Characteristics of the species

Column headers (32–78):
- 32 Glycerol growth
- 33 Erythritol growth
- 34 Ribitol growth
- 35 Galactitol growth
- 36 D–Mannitol growth
- 37 D–Glucitol growth
- 38 myo–Inositol growth
- 39 Gluconolactone growth
- 40 2–Ketogluconate growth
- 41 5–Ketogluconate growth
- 42 D,L–Lactate growth
- 43 Succinate growth
- 44 Citrate growth
- 45 Methanol growth
- 46 Ethanol growth
- 47 Ethylamine growth
- 48 D–Glucosamine growth
- 49 Nitrate growth
- 50 Nitrite growth
- 51 Growth sans vitamins
- 52 Growth sans inositol
- 53 Grth sans pantothenate
- 54 Growth sans biotin
- 55 Growth sans thiamin
- 56 Grth sans pyridoxine
- 57 Growth sans niacin
- 58 Growth sans folic acid
- 59 Growth sans PABA
- 60 Growth in 50% glucose
- 61 Growth in 60% glucose
- 62 Growth at 37 degrees
- 63 0.01% cyclohex. growth
- 64 0.1% cyclohex. growth
- 65 Starch formation
- 66 Pink colonies
- 67 Budding cells
- 68 Splitting cells
- 69 Apical budding
- 70 Cells sph, ovl, cylind
- 71 Cells of other shapes
- 72 Filamentous
- 73 Pseudohyphae
- 74 Septate hyphae
- 75 Ascosporogenous
- 76 Ascosp sph, ovl, renfm
- 77 Ascosp hat or saturn
- 78 Ascosp of other shapes

#	32	33	34	35	36	37	38	39	40	41	42	43	44	45	46	47	48	49	50	51	52	53	54	55	56	57	58	59	60	61	62	63	64	65	66	67	68	69	70	71	72	73	74	75	76	77	78
325	+	–	V	–	V	+	–	?	?	?	–	V	–	–	+	?	?	–	?	–	?	?	?	?	?	?	?	?	V	?	–	?	?	–	–	+	–	–	+	–	+	+	+	+	–	+	–
326	S	+	–	–	S	S	+	?	?	?	+	–	–	–	+	?	?	–	–	–	?	?	?	?	?	?	?	?	–	–	+	?	?	?	–	+	+	–	+	–	+	–	+	–	N	N	N
327	+	–	–	–	V	V	–	?	?	?	–	+	–	–	V	?	?	–	?	–	?	?	?	?	?	?	?	?	–	–	–	?	?	?	–	+	–	+	+	–	V	V	–	–	N	N	N
328	–	–	–	–	–	–	–	?	?	?	–	–	–	–	–	?	?	–	?	–	–	?	?	?	?	?	?	?	–	–	+	?	?	–	+	–	+	–	+	–	+	–	+	+	+	–	–
329	–	–	–	–	–	–	–	?	?	?	–	–	–	–	–	?	?	–	?	–	–	?	?	?	?	?	?	?	+	?	+	?	?	+	–	–	+	–	+	–	–	N	N	+	+	–	–
330	–	–	–	–	–	–	–	?	?	?	–	–	–	–	–	?	?	–	?	–	–	?	?	?	?	?	?	?	+	?	+	?	?	+	–	–	+	–	+	–	–	N	N	+	+	–	–
331	V	–	–	–	–	–	–	?	?	?	–	–	–	–	–	?	?	–	?	–	–	?	?	?	?	?	?	?	+	?	+	?	?	+	–	–	+	–	+	–	–	N	N	+	+	–	–
332	–	–	–	–	–	–	–	?	?	?	–	–	–	–	–	?	?	–	?	–	–	?	?	?	?	?	?	?	+	+	+	?	?	+	–	–	+	–	+	–	–	N	N	+	+	–	–
333	V	–	V	–	+	V	–	V	V	–	–	V	V	–	+	?	?	–	–	–	+	+	–	+	+	+	+	+	–	–	V	?	?	–	–	+	–	–	+	–	V	V	–	+	–	+	–
334	+	+	+	+	+	+	–	?	?	?	+	+	+	–	–	+	?	–	–	–	?	?	?	?	?	?	?	?	–	–	+	?	?	–	–	+	–	–	+	+	–	N	N	–	N	N	N
335	+	–	V	–	+	+	–	+	–	–	V	+	+	–	V	?	?	+	+	+	+	+	+	+	+	+	+	+	–	–	+	?	?	–	+	+	–	–	+	–	+	–	N	N	N
336	S	–	+	+	+	+	–	+	–	–	–	+	S	–	–	?	?	+	+	+	+	+	+	+	+	+	+	+	V	?	–	?	?	–	+	+	–	–	–	–	+	V	+	–	N	N	N
337	S	–	+	–	V	V	–	V	–	–	S	+	S	–	D	?	?	–	–	V	?	?	?	?	–	–	V	?	?	–	+	–	–	+	–	–	N	N	–	N	N	N
338	+	+	–	–	+	+	+	?	?	?	+	+	+	?	?	?	D	+	?	?	?	?	?	?	?	?	?	?	?	+	?	?	?	–	+	+	–	–	+	+	+	–	N	N	N
339	+	–	D	–	+	S	–	D	–	–	D	+	+	–	–	?	?	–	–	?	?	?	?	–	–	–	?	?	–	+	–	–	–	N	N	–	N	N	N
340	D	–	D	–	+	+	–	+	–	–	S	+	S	–	D	?	?	+	+	+	+	+	+	+	+	+	–	–	–	?	?	–	+	–	–	+	–	+	+	+	–	N	N	N
341	V	–	+	–	+	+	–	V	–	–	–	+	+	–	D	?	?	+	+	+	+	+	+	+	+	–	–	V	?	?	–	+	–	–	+	–	+	V	+	–	N	N	N
342	+	–	+	–	+	+	–	+	–	–	V	+	V	–	V	?	?	+	+	+	+	+	+	+	+	+	–	–	V	?	?	–	+	–	–	+	–	+	+	–	–	N	N	N
343	S	–	V	–	V	V	–	S	–	–	–	+	S	–	D	?	?	–	V	+	+	+	+	+	+	–	–	V	?	?	–	+	–	–	+	–	V	V	–	–	N	N	N
344	+	–	+	?	+	+	–	?	+	+	–	V	+	–	–	–	?	+	+	–	–	+	+	–	+	+	+	?	?	–	?	?	+	+	–	–	+	–	+	+	+	–	N	N	N
345	+	–	V	–	+	+	–	+	–	–	V	V	V	–	S	?	?	+	V	V	?	?	?	?	?	–	–	–	?	?	–	+	–	–	+	–	V	V	–	–	N	N	N
346	V	–	V	–	+	+	–	+	–	–	–	V	V	–	D	?	?	+	+	+	+	+	+	+	+	+	–	–	V	?	?	–	+	–	–	+	–	+	+	+	–	N	N	N
347	S	–	–	–	+	+	–	+	+	+	–	S	S	–	+	?	?	–	–	–	?	?	?	?	?	–	–	–	?	?	–	–	+	–	–	+	–	V	V	–	–	N	N	N
348	+	+	+	+	+	+	+	?	?	?	+	+	+	–	+	?	?	–	?	V	+	+	+	V	+	+	+	+	?	?	+	?	?	–	–	+	–	–	+	+	+	+	–	+	–
349	+	+	+	D	+	+	+	+	+	–	+	+	+	–	+	+	?	?	S	+	+	+	S	+	+	+	+	–	+	?	?	–	+	–	–	+	+	–	–	N	N	N
350	+	+	+	–	+	+	–	+	+	?	D	+	+	–	?	?	–	–	–	?	?	?	?	?	?	?	+	?	?	–	–	–	–	–	–	–	N	N	–	N	N	N
351	+	+	+	–	+	D	–	?	?	?	–	S	S	–	+	?	?	+	?	–	+	+	–	–	?	?	–	?	–	–	–	–	+	–	V	V	–	–	N	N	N
352	+	+	+	–	+	–	?	?	?	–	S	S	–	+	?	?	–	?	–	+	+	–	–	?	?	–	?	–	–	–	–	+	–	V	V	–	–	N	N	N
353	–	–	D	–	+	+	–	?	?	?	D	+	–	–	+	?	?	+	?	–	?	?	?	?	?	?	–	?	–	–	–	–	+	–	–	N	N	–	N	N	N
354	–	+	+	V	+	+	+	?	?	?	–	+	+	–	–	–	?	–	–	?	?	?	?	?	?	?	–	?	–	+	–	–	–	–	+	V	V	–	–	N	N	N
355	S	+	+	+	+	S	?	?	?	S	+	S	–	–	?	?	–	?	S	?	?	?	?	?	?	?	?	?	?	?	–	–	–	–	+	–	–	N	N	–	N	N	N
356	+	+	+	–	D	+	D	?	?	?	–	–	–	–	–	?	?	–	?	–	+	+	S	–	+	+	+	+	–	–	–	?	?	–	–	+	+	–	+	–	+	–	+	–	N	N	N

N not applicable; S slow utilisation; V variable; ? result not known.

Characteristics of the species

		1 D–Glucose fermentation	2 Galactose fermentation	3 Maltose fermentation	4 Me α–D–glucoside fermn	5 Sucrose fermentation	6 Trehalose fermentation	7 Melibiose fermentation	8 Lactose fermentation	9 Cellobiose fermentn	10 Melezitose fermentn	11 Raffinose fermentation	12 D–Galactose growth	13 L–Sorbose growth	14 D–Ribose growth	15 D–Xylose growth	16 L–Arabinose growth	17 D–Arabinose growth	18 L–Rhamnose growth	19 Sucrose growth	20 Maltose growth	21 Trehalose growth	22 Me α–D–glucoside grth	23 Cellobiose growth	24 Salicin growth	25 Arbutin growth	26 Melibiose growth	27 Lactose growth	28 Raffinose growth	29 Melezitose growth	30 Inulin growth	31 Starch growth
357	Torulaspora delbrueckii**	+	V	V	?	V	V	V	–	–	?	V	V	V	–	V	–	–	–	V	V	V	V	–	–	–	V	–	V	V	V	–
358	Torulaspora globosa**	+	–	–	–	+	V	–	–	–	–	+	–	–	–	–	–	–	–	+	–	V	–	–	–	–	–	–	+	–	V	–
359	Torulaspora pretoriensis**	+	V	V	?	+	?	–	–	–	?	+	+	–	–	V	–	–	–	+	+	+	+	–	–	–	–	–	+	V	V	–
360	Torulopsis anatomiae	+	–	–	–	–	–	–	–	–	–	–	–	–	V	–	–	–	+	–	–	+	V	+	+	+	–	–	–	–	–	–
361	Torulopsis apicola	+	–	–	–	+	–	–	–	–	–	V	–	+	V	–	–	–	–	+	–	–	–	–	–	–	–	–	+	–	–	–
362	Torulopsis apis	–	–	–	–	–	–	–	–	–	–	–	S	–	–	–	–	–	–	+	–	–	–	–	–	–	–	–	+	–	–	–
363	Torulopsis auriculariae*	–	–	–	–	–	–	–	–	–	–	–	–	–	–	–	–	–	–	+	V	D	–	–	–	–	–	–	–	+	–	–
364	Torulopsis austromarina*	–	–	–	–	–	–	–	–	–	–	–	+	–	–	–	–	–	–	–	–	V	–	–	–	–	–	–	–	–	–	–
365	Torulopsis bacarum*	–	–	–	–	–	–	–	–	–	–	–	–	V	+	+	+	+	–	+	+	+	+	S	S	–	–	–	+	+	S	V
366	Torulopsis bombicola*	+	–	–	–	+	–	–	–	–	–	+	D	D	–	–	–	–	–	+	–	–	–	–	–	–	–	–	+	–	–	–
367	Torulopsis cantarelii	+	–	–	–	V	–	–	–	–	–	–	–	+	–	–	–	–	–	+	–	–	–	–	–	–	–	–	–	–	–	–
368	Torulopsis castellii	+	–	–	–	–	–	–	–	–	–	–	–	–	–	–	–	–	–	–	–	–	–	–	–	–	–	–	–	–	–	–
369	Torulopsis dendrica*	S	–	–	–	–	–	–	–	?	–	–	–	–	–	–	–	–	–	–	–	–	–	V	+	+	–	–	–	–	–	–
370	Torulopsis ernobii	+	–	–	–	–	–	–	–	–	–	–	–	V	–	+	–	–	+	+	+	–	+	+	+	+	–	–	–	S	–	+
371	Torulopsis etchellsii	+	–	+	–	–	–	–	–	–	–	–	+	V	V	–	–	–	–	–	+	–	–	–	V	–	–	–	–	–	–	–
372	Torulopsis fragaria*	–	–	–	–	–	–	–	–	–	–	–	V	+	V	+	+	D	–	+	+	+	+	+	+	+	–	+	+	+	–	–
373	Torulopsis fructus*	+	–	–	–	S	–	–	–	–	–	–	–	+	D	+	–	–	–	–	–	+	–	–	–	–	–	–	–	–	–	–
374	Torulopsis fujisanensis	–	–	–	–	–	–	–	–	–	–	–	V	+	V	+	+	D	–	–	–	–	–	+	+	+	–	–	–	–	–	–
375	Torulopsis glabrata	+	–	–	–	–	V	–	–	–	–	–	–	–	–	–	–	–	–	–	–	+	–	–	–	–	–	–	–	–	–	–
376	Torulopsis gropengiesseri	+	–	–	–	+	–	–	–	–	–	–	+	–	–	–	–	–	–	+	–	–	–	+	+	+	–	–	+	–	–	–
377	Torulopsis haemulonii	+	–	–	–	+	+	–	–	–	–	–	S	+	–	–	+	+	–	+	+	+	–	–	–	–	–	–	–	S	+	–
378	Torulopsis halonitratophila	S	–	–	–	–	–	–	–	–	–	–	V	–	–	–	–	–	–	–	–	–	–	–	–	–	–	–	–	–	–	–
379	Torulopsis halophilus*	+	+	–	–	S	?	–	–	?	–	–	S	–	–	–	D	–	–	S	–	S	–	S	+	+	–	–	–	–	–	–
380	Torulopsis holmii	+	+	–	–	+	+	–	–	–	–	V	+	–	–	–	–	–	–	+	–	+	–	–	–	–	–	–	+	–	–	–
381	Torulopsis humilis*	+	+	–	–	–	+	–	–	–	–	–	+	–	–	–	–	–	–	–	–	+	–	–	–	–	–	–	–	–	–	–
382	Torulopsis inconspicua	–	–	–	–	–	–	–	–	–	–	–	–	–	–	–	–	–	–	–	–	–	–	–	–	–	–	–	–	–	–	–
383	Torulopsis ingeniosa	–	–	–	–	–	–	–	–	–	–	–	D	+	D	D	–	–	–	+	+	+	+	+	+	+	–	D	+	+	–	+
384	Torulopsis insectalens*	–	–	–	–	–	–	–	–	–	–	–	+	–	–	–	–	V	–	–	+	–	–	+	+	+	–	–	–	–	–	–
385	Torulopsis karawaiewi*	+	–	–	–	–	–	–	–	–	–	–	–	–	–	–	–	–	–	–	–	–	–	–	V	S	–	–	–	–	–	–
386	Torulopsis kruisii*	+	S	–	?	–	?	–	–	?	?	–	+	D	V	+	D	–	–	+	+	+	+	+	+	+	–	–	–	D	–	+
387	Torulopsis lactis–condensi	+	–	–	–	+	–	–	–	–	–	+	–	–	–	–	–	–	–	+	–	–	–	–	–	–	–	–	–	S	–	–
388	Torulopsis magnoliae	+	–	–	–	+	–	–	–	–	–	–	V	V	V	–	–	–	–	+	–	–	–	–	V	V	–	–	V	–	–	–
389	Torulopsis mannitofaciens*	+	S	S	–	S	S	–	–	?	–	–	+	–	D	D	+	–	–	+	+	+	–	+	D	–	D	–	D	–	–	–
390	Torulopsis maris	–	–	–	–	–	–	–	–	–	–	–	–	–	–	–	+	–	+	–	–	–	–	–	–	–	–	–	–	–	–	–
391	Torulopsis methanolovescens*	S	–	–	–	–	?	–	–	?	–	–	–	V	D	–	–	+	–	–	+	–	–	+	+	+	–	–	–	–	–	–
392	Torulopsis molischiana	+	–	–	–	–	–	+	–	–	–	+	–	–	+	+	+	+	V	–	+	–	–	+	+	+	–	–	–	+	–	+
393	Torulopsis multis–gemmis*	S	–	–	–	–	–	–	–	–	?	–	+	+	–	+	+	–	–	+	+	+	–	–	–	–	–	–	–	+	–	–
394	Torulopsis musae*	+	–	–	?	–	D	–	–	–	–	–	–	+	–	+	–	–	–	+	+	+	D	–	–	D	–	–	–	+	–	–
395	Torulopsis nagoyaensis*	+	S	–	–	–	+	–	–	?	–	–	+	+	+	+	+	–	+	–	–	+	–	+	+	+	–	–	–	–	–	–
396	Torulopsis navarrensis*	+	S	S	?	S	?	?	–	?	?	S	+	–	D	+	–	–	–	+	+	+	+	+	+	+	+	–	D	+	–	–
397	Torulopsis nemodendra*	–	–	–	–	–	–	–	–	–	–	–	+	+	+	+	+	D	–	–	–	–	+	+	+	+	–	–	–	–	–	–
398	Torulopsis nitratophila	S	–	–	–	–	?	–	–	–	–	–	+	+	+	+	+	–	+	–	–	+	–	–	–	–	–	–	–	–	–	–
399	Torulopsis nodaensis*	+	–	–	–	–	–	–	–	–	–	–	+	+	–	–	–	–	–	–	D	–	–	–	–	–	–	–	–	–	–	–
400	Torulopsis norvegica	V	–	–	–	–	–	–	–	–	–	–	–	–	–	+	V	–	V	–	–	–	–	+	+	+	–	–	–	–	–	–

Codes in table: + positive; – negative; D delayed for longer than 7 days;

Characteristics of the species

	32 Glycerol growth	33 Erythritol growth	34 Ribitol growth	35 Galactitol growth	36 D–Mannitol growth	37 D–Glucitol growth	38 myo–Inositol growth	39 Gluconolactone growth	40 2–Ketogluconate growth	41 5–Ketogluconate growth	42 D,L–Lactate growth	43 Succinate growth	44 Citrate growth	45 Methanol growth	46 Ethanol growth	47 Ethylamine growth	48 D–Glucosamine growth	49 Nitrate growth	50 Nitrite growth	51 Growth sans vitamins	52 Growth sans inositol	53 Grth sans pantothenate	54 Growth sans biotin	55 Growth sans thiamin	56 Grth sans pyridoxine	57 Growth sans niacin	58 Growth sans folic acid	59 Growth sans PABA	60 Growth in 50% glucose	61 Growth in 60% glucose	62 Growth at 37 degrees	63 0.01% cyclohex. growth	64 0.1% cyclohex. growth	65 Starch formation	66 Pink colonies	67 Budding cells	68 Splitting cells	69 Apical budding	70 Cells sph, ovl, cylind	71 Cells of other shapes	72 Filamentous	73 Pseudohyphae	74 Septate hyphae	75 Ascosporogenous	76 Ascosp sph, ovl, renfm	77 Ascosp hat or saturn	78 Ascosp of other shapes
357	V	–	V	–	V	V	–	V	V	–	V	V	–	–	V	V	?	–	V	V	+	+	V	+	+	+	+	+	+	+	V	–	–	–	–	+	–	–	+	–	–	N	N	+	+	–	–
358	V	–	–	–	V	–	–	V	V	–	+	–	–	–	+	–	?	–	?	+	+	+	+	+	+	+	+	+	V	–	+	+	+	–	–	+	–	–	+	–	–	N	N	+	+	–	–
359	V	–	–	–	+	+	–	V	+	–	+	–	–	–	?	–	–	–	?	+	+	+	+	+	+	+	+	+	+	V	+	–	–	–	–	+	–	–	+	–	–	N	N	+	+	–	–
360	–	–	–	–	+	–	–	?	?	?	–	V	V	–	S	?	–	–	?	–	+	+	–	–	+	+	+	+	?	?	–	?	?	–	–	+	–	–	+	–	–	N	N	–	N	N	N
361	V	–	–	–	+	+	–	D	–	?	–	–	–	–	–	?	–	–	?	V	+	+	V	V	+	+	+	+	?	?	V	?	?	–	–	+	–	–	+	–	–	N	N	–	N	N	N
362	+	–	–	–	+	+	–	?	?	?	–	–	–	–	–	?	?	–	?	+	+	+	+	+	+	+	+	+	?	?	–	?	?	–	–	+	–	–	+	–	–	N	N	–	N	N	N
363	–	–	–	–	+	D	–	?	?	?	–	+	–	–	+	?	?	–	?	–	+	+	+	S	–	+	+	+	?	?	–	?	?	–	–	+	–	–	+	–	–	N	N	–	N	N	N
364	+	–	–	–	–	–	–	?	?	?	–	+	–	?	V	?	?	–	?	–	+	+	–	V	V	+	+	+	–	–	–	?	?	–	–	+	–	–	+	–	–	N	N	–	N	N	N
365	+	+	V	–	+	V	–	?	?	?	S	+	V	–	?	?	?	+	?	V	?	?	?	?	?	?	?	?	–	–	–	?	?	–	–	+	–	–	+	–	–	N	N	–	N	N	N
366	+	–	–	–	+	D	–	?	?	?	–	D	–	–	+	?	–	–	–	?	?	?	?	?	?	?	?	?	+	+	–	?	?	–	–	+	–	–	+	–	–	N	N	–	N	N	N
367	+	+	+	–	+	+	–	?	?	?	S	S	–	–	+	?	?	–	?	+	+	+	+	+	+	+	+	+	?	?	V	?	?	–	–	+	–	–	+	–	–	N	N	–	N	N	N
368	+	–	–	–	–	–	–	D	+	–	–	–	–	–	–	?	–	–	?	–	–	+	–	–	V	–	+	+	?	?	+	?	?	–	–	+	–	–	+	–	–	N	N	–	N	N	N
369	V	–	–	–	–	–	–	+	–	–	–	+	+	–	V	?	–	–	?	–	+	+	–	+	+	+	+	+	?	?	–	?	?	–	–	+	–	–	+	–	–	N	N	–	N	N	N
370	+	–	+	–	+	+	–	?	?	?	–	V	V	–	S	?	?	+	?	S	+	+	S	S	+	+	+	+	?	?	–	?	?	–	–	+	–	–	+	–	–	V	V	–	N	N	N
371	+	–	–	–	V	+	–	?	?	?	V	V	–	–	V	?	?	+	?	V	+	+	V	V	+	+	+	+	+	?	–	?	?	–	–	+	–	–	+	–	–	N	N	–	N	N	N
372	+	–	+	–	+	+	–	?	?	?	–	+	+	?	V	+	+	–	?	?	+	?	+	+	+	+	+	+	+	?	?	–	?	?	–	–	+	–	–	+	–	–	N	N	–	N	N	N
373	+	–	+	–	D	+	–	?	?	?	–	+	+	–	+	?	?	–	?	–	+	+	–	+	+	+	–	+	+	?	?	V	?	?	–	–	+	–	–	+	–	–	N	N	–	N	N	N
374	D	–	+	+	+	+	–	?	?	?	–	+	V	–	+	?	?	–	?	–	?	?	?	?	?	?	?	?	–	?	?	–	V	+	–	?	+	–	V	V	V	–	N	N	N
375	V	–	–	–	–	–	–	?	?	?	–	–	–	–	V	?	?	–	?	–	+	+	V	V	–	–	+	+	?	?	+	?	?	–	–	+	–	–	+	–	–	N	N	–	N	N	N
376	+	–	–	–	+	+	–	?	?	?	–	–	–	–	–	?	?	–	?	S	+	+	S	+	+	+	+	+	?	?	–	?	?	–	–	+	–	–	+	–	–	N	N	–	N	N	N
377	+	–	+	S	+	+	–	?	?	?	–	+	+	–	+	?	?	–	?	S	+	+	S	+	+	+	+	+	?	?	+	?	?	–	–	+	–	–	+	–	–	N	N	–	N	N	N
378	+	–	–	–	+	V	–	?	?	?	V	D	–	–	V	?	?	+	?	–	+	+	–	–	+	+	+	+	+	S	–	?	?	–	–	+	–	–	+	–	–	N	N	–	N	N	N
379	S	–	–	–	+	–	–	?	?	?	–	S	S	–	?	?	–	+	?	–	?	?	?	?	?	?	?	?	?	?	?	?	–	–	+	–	–	+	–	–	N	N	–	N	N	N
380	V	–	–	–	–	–	–	+	–	–	V	–	–	–	–	–	–	–	?	–	+	+	V	V	+	+	+	+	–	–	V	V	–	–	–	+	–	–	+	–	–	N	N	–	N	N	N
381	+	–	–	–	–	–	–	+	–	–	+	–	–	–	+	–	–	?	–	+	–	+	–	+	+	+	+	+	–	–	+	?	–	–	–	+	–	–	+	–	–	N	N	–	N	N	N
382	+	–	–	–	–	–	–	?	?	?	+	V	–	–	+	?	?	–	+	+	–	?	+	+	–	–	–	+	?	?	+	?	?	–	–	+	–	–	+	–	–	V	V	–	N	N	N
383	+	–	–	–	+	+	–	+	+	S	+	+	+	–	–	?	?	+	?	+	+	+	+	+	+	+	+	+	?	?	V	?	?	–	–	+	–	–	+	–	–	N	N	–	N	N	N
384	–	–	D	–	+	+	–	–	?	?	D	+	–	–	+	?	?	–	?	–	?	?	?	?	?	?	?	?	–	–	+	?	?	–	–	+	–	–	+	–	–	N	N	–	N	N	N
385	D	–	V	–	S	S	–	?	?	?	–	V	V	–	?	?	–	–	?	–	?	?	?	?	?	?	?	?	–	–	–	?	?	–	–	+	–	–	+	–	–	N	N	–	N	N	N
386	+	–	+	–	+	+	–	?	?	?	–	+	+	–	?	?	?	–	?	+	+	+	+	+	+	+	+	+	?	?	V	?	?	–	–	+	–	–	+	+	–	N	N	–	N	N	N
387	–	–	–	–	–	–	–	?	?	?	–	–	–	–	–	?	?	+	?	V	V	+	V	V	+	+	+	+	?	?	–	?	?	–	–	+	–	–	+	–	–	N	N	–	N	N	N
388	+	–	–	–	+	+	–	?	?	?	–	V	–	–	V	?	?	+	?	V	+	+	V	V	+	+	+	+	?	?	V	?	?	–	–	+	–	–	+	–	–	N	N	–	N	N	N
389	+	–	–	–	+	–	–	?	?	?	–	D	–	–	D	?	–	+	?	–	?	?	?	?	?	?	?	?	?	?	–	?	?	–	–	+	–	–	+	–	–	N	N	–	N	N	N
390	+	–	+	–	+	+	–	+	–	–	–	+	–	–	+	?	?	–	?	S	+	+	S	+	+	+	+	+	?	?	–	?	?	–	–	+	–	–	+	–	–	N	N	–	N	N	N
391	+	–	+	–	+	+	–	+	–	–	–	D	+	+	+	?	?	–	?	–	+	+	–	+	+	+	+	+	–	–	V	?	?	–	–	+	–	–	+	–	–	N	N	–	N	N	N
392	+	+	+	–	+	+	–	D	+	–	–	V	–	+	+	?	?	+	?	–	+	+	–	–	–	+	+	+	?	?	+	?	–	–	–	+	–	–	+	–	–	N	N	–	N	N	N
393	+	–	V	–	–	+	–	?	?	?	–	+	+	–	?	?	?	–	?	–	?	?	?	?	?	?	?	?	–	–	–	?	?	–	–	+	–	–	+	–	–	V	V	–	N	N	N
394	+	–	D	–	+	+	–	+	+	–	–	+	+	–	+	?	?	–	?	–	+	+	–	+	–	+	+	+	?	?	–	?	?	–	–	+	–	–	+	–	–	N	N	–	N	N	N
395	+	–	+	–	+	+	–	+	–	–	+	+	+	+	+	?	?	+	?	–	+	+	–	+	+	+	+	+	?	?	+	?	?	–	–	+	–	–	+	–	–	N	N	–	N	N	N
396	+	–	D	–	+	+	–	?	?	?	–	+	+	–	?	?	?	–	?	–	?	?	?	?	?	?	?	?	?	?	+	?	?	–	–	+	–	–	+	–	–	V	V	–	N	N	N
397	+	+	+	+	+	+	–	D	–	–	–	+	–	+	+	?	–	–	+	–	+	+	+	+	+	+	+	+	?	?	–	?	?	–	–	+	–	–	+	–	–	N	N	–	N	N	N
398	+	–	+	–	+	+	–	+	–	–	–	D	–	+	+	?	?	+	?	–	+	+	–	–	+	+	+	+	?	?	V	?	?	–	–	+	–	–	+	–	–	V	V	–	N	N	N
399	+	–	–	–	+	D	–	?	?	?	–	D	–	?	+	?	–	+	?	–	?	?	?	?	?	?	?	?	?	?	?	?	?	–	–	+	–	–	+	–	–	N	N	–	N	N	N
400	+	–	–	–	+	+	–	?	?	?	+	V	V	–	+	?	?	+	?	S	+	+	V	V	S	+	+	+	?	?	–	?	?	–	–	+	–	–	+	–	–	V	V	–	N	N	N

N not applicable; S slow utilisation; V variable; ? result not known.

	1 D–Glucose fermentation	2 Galactose fermentation	3 Maltose fermentation	4 Me α–D–glucoside fermn	5 Sucrose fermentation	6 Trehalose fermentation	7 Melibiose fermentation	8 Lactose fermentation	9 Cellobiose fermentn	10 Melezitose fermentn	11 Raffinose fermentation	12 D–Galactose growth	13 L–Sorbose growth	14 D–Ribose growth	15 D–Xylose growth	16 L–Arabinose growth	17 D–Arabinose growth	18 L–Rhamnose growth	19 Sucrose growth	20 Maltose growth	21 Trehalose growth	22 Me α–D–glucoside grth	23 Cellobiose growth	24 Salicin growth	25 Arbutin growth	26 Melibiose growth	27 Lactose growth	28 Raffinose growth	29 Melezitose growth	30 Inulin growth	31 Starch growth
401 Torulopsis pampelonensis*	+	S	V	?	S	?	?	–	?	?	–	+	+	–	+	–	–	–	+	+	+	D	+	+	+	+	–	+	+	–	–
402 Torulopsis philyla*	–	–	–	–	–	–	–	–	–	–	–	–	–	–	–	–	–	–	–	–	+	–	–	–	–	–	–	–	–	–	–
403 Torulopsis pignaliae*	+	–	–	–	–	?	–	–	–	–	–	–	D	V	+	+	–	–	–	–	+	–	–	–	–	–	–	–	–	–	–
404 Torulopsis pinus	–	–	–	–	–	–	–	–	–	–	–	–	–	+	+	–	–	–	–	–	+	–	–	–	–	–	–	–	–	–	–
405 Torulopsis psychrophila*	–	–	–	–	–	–	–	–	–	–	–	+	V	–	+	+	–	–	–	–	–	–	–	–	–	–	–	–	–	–	–
406 Torulopsis pustula*	–	–	–	–	–	–	–	–	–	–	–	V	+	+	+	+	–	–	–	+	–	–	+	+	+	–	–	–	–	–	–
407 Torulopsis schatavii*	S	S	–	–	–	?	–	–	–	–	–	+	–	+	D	V	V	–	–	–	+	–	–	–	V	–	–	–	–	–	–
408 Torulopsis silvatica*	–	–	–	–	–	–	–	–	–	–	–	–	–	–	–	–	–	–	–	–	+	–	–	–	–	–	–	–	–	–	–
409 Torulopsis sonorensis*	+	–	–	–	–	–	–	S	–	–	–	–	–	V	+	+	–	–	+	+	+	–	–	–	–	–	–	–	–	–	–
410 Torulopsis sorbophila*	–	–	–	–	–	–	–	–	–	–	–	+	+	–	D	–	–	–	–	–	+	–	–	–	–	–	–	–	–	–	–
411 Torulopsis spandovensis*	+	S	–	–	S	–	–	–	–	–	V	+	+	–	V	V	–	–	+	–	–	–	–	–	–	–	–	+	–	V	–
412 Torulopsis stellata	+	–	–	–	+	–	–	–	–	–	V	–	–	–	–	–	–	–	+	–	–	–	–	–	–	–	–	+	–	–	–
413 Torulopsis tannotolerans*	+	+	–	–	–	–	–	–	–	–	–	+	–	–	D	–	–	–	–	–	–	–	–	–	–	–	–	–	–	–	–
414 Torulopsis torresii	+	–	–	–	–	+	–	–	–	–	–	+	+	–	+	+	–	–	–	–	+	–	+	+	+	–	–	–	–	–	–
415 Torulopsis vanderwaltii	–	–	–	–	–	–	–	–	–	–	–	+	+	V	+	+	D	–	V	–	+	–	–	–	–	–	–	–	–	–	–
416 Torulopsis versatilis	+	+	+	–	V	+	–	S	V	–	–	+	–	V	–	V	–	–	V	+	+	–	V	V	V	V	V	–	–	–	–
417 Torulopsis wickerhamii	+	–	–	–	–	–	–	–	–	+	–	+	–	+	+	+	S	+	–	–	–	–	+	+	+	–	–	–	–	–	–
418 Torulopsis xestobii*	S	–	–	?	–	?	–	–	?	?	–	+	V	D	+	D	–	–	+	+	+	+	D	D	D	D	–	D	D	–	–
419 Trichosporon aquatile*	–	–	–	–	–	–	–	–	–	–	–	+	–	+	+	+	D	–	+	+	+	+	+	–	D	–	D	–	D	–	+
420 Trichosporon brassicae*	–	–	–	–	–	–	–	–	–	–	–	+	+	+	D	+	–	–	S	+	+	–	–	–	–	–	–	–	–	–	D
421 Trichosporon cutaneum	–	–	–	–	–	–	–	–	–	–	–	V	V	V	+	V	V	V	V	V	V	V	V	V	V	V	+	V	V	–	V
422 Trichosporon eriense*	–	–	–	–	–	–	–	–	–	–	–	D	+	–	+	–	–	–	–	–	–	–	+	V	V	–	–	–	–	–	–
423 Trichosporon fennicum*	+	S	+	?	+	?	–	S	?	?	S	+	+	+	D	+	–	–	+	+	+	+	V	+	+	–	+	D	V	–	+
424 Trichosporon melibiosaceum*	+	+	+	–	+	S	S	–	?	?	S	+	S	+	+	S	S	–	+	+	+	S	+	S	+	S	S	+	+	–	S
425 Trichosporon pullulans	–	–	–	–	–	–	–	–	–	–	–	+	V	V	V	V	V	V	+	+	+	+	+	+	+	+	V	+	+	–	+
426 Trichosporon terrestre*	–	–	–	–	–	–	–	–	–	–	–	+	+	S	+	+	S	+	+	+	+	S	S	S	S	–	–	+	–	–	–
427 Trigonopsis variabilis	–	–	–	–	–	–	–	–	–	–	–	+	+	V	V	V	–	–	–	–	+	–	–	–	–	–	–	–	–	–	–
428 Wickerhamia fluorescens	+	+	–	–	+	–	–	–	–	–	+	+	–	–	–	–	–	–	+	–	S	–	–	–	–	–	–	+	–	–	–
429 Wickerhamiella domercqii**	–	–	–	–	–	–	–	–	–	–	–	D	+	V	–	–	–	–	S	–	–	–	–	–	–	–	–	–	–	–	–
430 Wingea robertsii**	+	–	+	+	+	+	–	–	–	–	+	+	+	V	+	+	–	+	+	+	+	+	+	+	+	–	–	+	+	–	S
431 Zendera ovetensis**	–	–	–	–	–	–	–	–	–	–	–	+	+	–	–	–	–	–	–	–	–	–	–	–	–	–	–	–	–	–	–
432 Zygosaccharomyces bailii**	+	–	–	–	V	V	–	–	–	–	–	V	V	–	–	–	–	–	V	–	V	–	–	–	–	–	–	–	–	–	–
433 Zygosaccharomyces bisporus**	+	–	–	–	–	–	–	–	–	–	–	V	V	–	–	–	–	–	–	–	–	–	–	–	–	–	–	–	–	–	–
434 Zygosaccharomyces cidri**	+	+	V	V	+	+	+	–	–	–	+	+	+	–	V	–	–	–	+	V	+	V	–	–	–	–	–	+	+	V	–	–
435 Zygosaccharomyces fermentati**	+	+	+	V	+	+	+	–	–	–	V	+	+	V	–	V	–	–	–	+	+	+	+	+	+	+	–	–	+	V	–
436 Zygosaccharomyces florentinus	+	V	V	V	+	V	+	–	–	–	+	V	+	–	V	–	–	–	+	V	+	V	–	–	–	–	+	–	+	V	–
437 Zygosaccharomyces microellipsodes**	+	+	–	–	+	?	+	–	–	–	+	+	V	–	V	–	–	–	+	–	V	–	–	–	–	–	+	–	+	–	–

Codes in table: + positive; – negative; D delayed for longer than 7 days;

Characteristics of the species

	32 Glycerol growth	33 Erythritol growth	34 Ribitol growth	35 Galactitol growth	36 D–Mannitol growth	37 D–Glucitol growth	38 myo–Inositol growth	39 Gluconolactone growth	40 2–Ketogluconate growth	41 5–Ketogluconate growth	42 D,L–Lactate growth	43 Succinate growth	44 Citrate growth	45 Methanol growth	46 Ethanol growth	47 Ethylamine growth	48 D–Glucosamine growth	49 Nitrate growth	50 Nitrite growth	51 Growth sans vitamins	52 Growth sans inositol	53 Grth sans pantothenate	54 Growth sans biotin	55 Growth sans thiamin	56 Grth sans pyridoxine	57 Growth sans niacin	58 Growth sans folic acid	59 Growth sans PABA	60 Growth in 50% glucose	61 Growth in 60% glucose	62 Growth at 37 degrees	63 0.01% cyclohex. growth	64 0.1% cyclohex. growth	65 Starch formation	66 Pink colonies	67 Budding cells	68 Splitting cells	69 Apical budding	70 Cells sph, ovl, cylind	71 Cells of other shapes	72 Filamentous	73 Pseudohyphae	74 Septate hyphae	75 Ascosporogenous	76 Ascosp sph, ovl, renfm	77 Ascosp hat or saturn	78 Ascosp of other shapes
01	+	–	D	–	+	+	–	?	?	?	–	+	+	–	?	?	?	–	?	–	?	?	?	?	?	?	?	?	?	?	?	?	?	–	–	+	–	–	+	–	–	N	N	–	N	N	N
02	–	–	+	–	+	+	–	?	?	?	–	+	+	–	+	?	?	–	?	+	+	+	+	+	+	+	+	+	?	?	–	?	?	–	–	+	–	–	+	–	–	N	N	–	N	N	N
03	+	–	+	–	+	+	–	–	–	?	–	+	–	+	–	?	–	+	?	–	+	+	–	+	+	+	+	+	?	?	–	+	+	–	–	+	–	–	+	–	–	N	N	–	N	N	N
04	+	+	+	–	+	+	–	+	–	–	–	–	+	+	+	?	–	–	?	–	–	+	+	–	+	+	+	+	?	?	–	?	?	–	–	+	–	–	+	–	–	N	N	–	N	N	N
05	+	+	+	–	+	+	–	?	?	?	–	–	–	–	?	?	?	–	?	?	?	?	?	?	?	?	?	?	?	?	–	?	?	–	–	+	–	–	+	–	–	N	N	–	N	N	N
06	D	–	+	–	+	+	–	+	+	?	–	+	+	–	?	?	?	+	+	?	–	?	?	?	?	?	?	?	–	–	–	?	?	–	–	+	–	–	+	–	–	N	N	–	N	N	N
07	+	+	+	–	+	+	–	+	+	V	–	+	+	–	?	?	?	–	?	+	+	+	+	+	+	+	+	+	?	?	–	?	?	–	–	+	–	–	+	–	+	+	–	–	N	N	N
08	+	–	+	–	+	+	–	–	–	–	+	+	–	–	?	?	?	–	?	–	?	?	?	?	?	?	?	+	?	?	–	–	+	–	–	+	–	–	N	N	–	N	N	N
09	D	+	–	+	–	–	–	D	–	–	D	+	–	+	+	+	?	–	–	+	+	–	+	+	+	+	–	–	+	?	?	–	–	+	–	–	N	N	–	N	N	N
10	D	–	–	–	+	+	–	–	–	–	–	+	–	–	?	?	–	?	?	–	?	–	+	–	+	–	–	+	?	+	–	– –	–	–	+	–	–	N	N	–	N	N	N
11	+	–	+	+	+	+	–	D	–	?	V	+	V	–	?	+	?	–	?	–	+	+	+	+	+	+	?	?	–	?	?	–	–	+	–	–	N	N	–	N	N	N
12	–	–	–	–	–	–	–	?	?	?	–	–	–	–	–	–	?	?	–	–	–	+	+	–	+	+	+	+	?	?	V	?	?	–	–	+	–	–	N	N	–	N	N	N
13	+	–	–	–	–	–	–	+	–	–	–	–	–	–	?	?	?	–	?	+	S	+	–	+	+	+	+	?	?	–	V	–	–	+	–	–	N	N	–	N	N	N
14	+	–	+	–	+	+	–	?	?	?	–	+	–	–	+	?	?	–	?	+	+	+	+	+	+	+	?	?	V	?	?	–	–	+	–	–	N	N	–	N	N	N
15	+	–	+	D	+	+	–	D	+	–	–	+	+	–	–	?	?	+	?	–	+	+	+	?	?	V	?	?	–	–	+	–	–	N	N	–	N	N	N
16	+	–	–	–	V	+	–	?	?	?	–	–	–	–	–	–	?	?	+	?	S	+	+	V	S	+	+	+	+	?	?	–	?	?	–	–	+	–	–	N	N	–	N	N	N
17	+	–	+	–	+	+	–	?	?	?	V	S	–	–	+	?	?	+	?	S	+	+	+	S	+	+	+	+	?	?	–	?	?	–	–	+	–	–	N	N	–	N	N	N
18	+	–	+	–	–	–	–	?	?	?	–	D	+	–	?	?	D	–	?	–	?	?	?	?	?	?	?	–	–	–	?	?	–	–	+	–	+	–	–	N	N	–	N	N	N
19	–	–	–	–	–	–	–	?	?	?	+	+	D	?	+	?	–	–	?	?	?	?	?	?	?	?	–	?	?	V	–	+	+	–	+	–	+	–	+	–	N	N	N
20	+	–	–	–	D	D	+	?	?	?	+	+	+	–	+	?	?	–	?	–	?	?	?	?	?	?	+	?	?	–	–	+	+	–	+	–	+	–	N	N	N
21	V	V	V	V	V	V	V	?	?	?	V	V	V	–	+	?	?	–	V	V	+	+	V	V	+	+	+	+	–	V	?	?	V	+	+	–	+	–	+	–	N	N	N
22	+	+	–	–	+	+	–	?	?	?	+	+	+	–	+	?	–	–	?	?	?	?	?	?	?	?	?	?	?	+	+	–	+	–	+	–	N	N	N
23	+	+	+	–	+	+	–	?	?	?	–	+	+	–	?	+	–	?	+	+	+	+	?	?	+	?	?	–	–	+	+	–	+	–	+	–	N	N	N
24	S	+	+	–	+	+	–	?	?	?	S	+	S	–	–	?	?	–	?	S	?	?	?	?	?	?	–	?	?	–	–	+	+	–	+	–	+	–	N	N	N
25	V	+	V	–	+	V	V	?	?	?	V	V	V	–	+	?	?	+	+	+	+	+	+	+	+	+	–	–	?	?	+	–	+	+	–	+	–	N	N	N
26	+	+	+	S	+	+	+	?	?	?	+	+	+	?	+	?	?	+	?	+	+	+	+	+	+	+	+	+	?	?	–	–	+	+	–	+	–	N	N	N
27	+	–	–	–	V	V	–	?	?	?	+	+	+	–	+	+	?	–	?	–	?	?	?	?	–	–	+	?	?	–	–	+	–	–	+	+	–	N	N	N
28	+	–	+	–	+	–	–	+	+	–	+	+	+	–	+	?	?	–	–	+	+	–	+	+	+	+	+	?	–	+	?	–	–	+	–	–	+	–	–	N	N	+	–	V	V
29	+	–	–	–	+	+	–	?	?	?	–	+	+	–	S	+	?	+	?	–	?	?	?	?	?	?	–	?	?	–	–	+	–	–	+	–	–	N	N	+	+	–	–
30	+	+	+	–	+	+	–	?	?	?	V	+	+	–	+	+	?	–	?	V	?	?	?	?	?	?	+	V	V	–	–	–	–	+	–	–	+	–	–	N	N	+	+	–	–
31	+	–	–	–	–	–	–	?	?	?	+	+	–	?	+	?	?	–	?	–	?	?	?	?	?	?	–	–	–	?	?	?	–	+	–	–	+	–	+	+	+	+	–	–
32	V	–	V	–	V	+	–	V	V	–	–	–	–	–	V	+	–	–	?	–	+	+	–	+	+ +	+	V	–	–	–	+	–	–	+	–	V	V	–	+	+	–	–
33	V	–	V	–	V	V	–	+	V	–	–	–	–	S	+	–	–	–	V	+	+	S	V	+ +	+	+	+	V	–	–	–	+	–	–	+	–	V	V	–	+	+	–	–
34	+	–	V	–	+	+	–	V	V	–	+	+	V	–	+	+	–	–	–	V	+	+ +	+	V	–	V	+	+	–	+	–	–	+	–	–	N	N	+	+	–	–
35	V	–	V	–	V	+	–	V	V	–	+	+	–	–	+	+	–	+	+	–	–	+ +	+	+	–	+	+	+	+	–	+	–	–	+	–	–	+	+	–	+	+	–	–
36	V	–	–	–	V	V	–	V	V	–	–	V	–	–	S	+	–	–	–	V	+ +	V +	+ +	+	+	+	+	–	–	+	+	–	+	–	–	+	+	–	+	+	–	–
37	V	–	–	–	V	V	–	+	V	–	+	V	–	–	?	+	–	–	–	V	+ +	V +	+ +	+	+	–	–	–	–	–	–	+	–	+	–	–	N	N	+	+	–	–

N not applicable; S slow utilisation; V variable; ? result not known.

Characteristics of the species

		1 D–Glucose fermentation	2 Galactose fermentation	3 Maltose fermentation	4 Me α–D–glucoside fermn	5 Sucrose fermentation	6 Trehalose fermentation	7 Melibiose fermentation	8 Lactose fermentation	9 Cellobiose fermentn	10 Melezitose fermentn	11 Raffinose fermentation	12 D–Galactose growth	13 L–Sorbose growth	14 D–Ribose growth	15 D–Xylose growth	16 L–Arabinose growth	17 D–Arabinose growth	18 L–Rhamnose growth	19 Sucrose growth	20 Maltose growth	21 Trehalose growth	22 Me α–D–glucoside grth	23 Cellobiose growth	24 Salicin growth	25 Arbutin growth	26 Melibiose growth	27 Lactose growth	28 Raffinose growth	29 Melezitose growth	30 Inulin growth	31 Starch growth
438	Zygosaccharomyces mrakii**	+	+	–	–	+	–	+	–	–	–	+	+	–	–	V	–	–	–	+	–	–	–	–	–	–	+	–	+	–	–	–
439	Zygosaccharomyces rouxii**	+	–	V	–	V	V	–	–	–	–	–	V	V	V	–	–	–	–	V	V	V	–	–	–	–	–	–	–	–	–	–

Codes in table: + positive;   – negative;   D delayed for longer than 7 days;

Characteristics of the species

	32 Glycerol growth	33 Erythritol growth	34 Ribitol growth	35 Galactitol growth	36 D–Mannitol growth	37 D–Glucitol growth	38 myo–Inositol growth	39 Gluconolactone growth	40 2–Ketogluconate growth	41 5–Ketogluconate growth	42 D,L–Lactate growth	43 Succinate growth	44 Citrate growth	45 Methanol growth	46 Ethanol growth	47 Ethylamine growth	48 D–Glucosamine growth	49 Nitrate growth	50 Nitrite growth	51 Growth sans vitamins	52 Growth sans inositol	53 Grth sans pantothenate	54 Growth sans biotin	55 Growth sans thiamin	56 Grth sans pyridoxine	57 Growth sans niacin	58 Growth sans folic acid	59 Growth sans PABA	60 Growth in 50% glucose	61 Growth in 60% glucose	62 Growth at 37 degrees	63 0.01% cyclohex. growth	64 0.1% cyclohex. growth	65 Starch formation	66 Pink colonies	67 Budding cells	68 Splitting cells	69 Apical budding	70 Cells sph, ovl, cylind	71 Cells of other shapes	72 Filamentous	73 Pseudohyphae	74 Septate hyphae	75 Ascosporogenous	76 Ascosp sph, ovl, renfm	77 Ascosp hat or saturn	78 Ascosp of other shapes
438	S	–	–	–	+	+	–	+	+	–	–	+	–	–	V	+	–	–	–	+	+	+	+	+	+	+	+	+	+	–	–	+	+	–	–	+	–	–	+	–	+	+	–	+	+	–	–
439	V	–	V	–	V	V	–	V	V	–	–	–	–	–	V	+	–	–	–	V	+	V	V	+	+	+	+	+	+	+	V	–	–	–	–	+	–	–	+	–	V	V	–	+	+	–	–

N not applicable; S slow utilisation; V variable; ? result not known.

7
The keys

This chapter contains identification keys for all the species of yeasts accepted by the Centraalbureau voor Schimmelcultures in January 1978 except the genera *Cyniclomyces*, *Oosporidium* and *Pityrosporum*, which require special culturing. There are also keys for particular groups of yeasts, and all these keys are listed on p. 73. These are conventional dichotomous keys, arranged compactly, as Payne, Walton & Barnett (1974) proposed, for ease of use and economy of space. Further economy is achieved by reticulation (Payne, 1977), by which the user may be referred back to a previous entry, so avoiding the printing of duplicate sections.

The keys are based on the results of the tests and observations on each species tabulated in Chapter 6. There are two kinds of key: (i) those involving the results of visual examination of the yeasts and physiological tests and (ii) those involving physiological tests only. Each key is preceded by a list of yeasts and a list of tests occurring in that key. In these lists, the yeasts and tests are both numbered as in the table of results in Chapter 6. Since many tests take a number of days to complete, all the tests for a given key must usually be done before attempting to identify the yeasts with that key.

How to use the keys
Consider the following example of identifying a yeast known to utilize methanol. Suppose the tests required for the key to such yeasts (see p. 154) give the following results.

1	D-Glucose fermentation	—
12	D-Galactose growth	—
15	D-Xylose growth	+
20	Maltose growth	—
23	Cellobiose growth	+
33	Erythritol growth	+
42	D,L-Lactate growth	—
49	Nitrate growth	—
55	Growth without thiamin	+

The key (no. 12, p. 154) is a list of numbered questions in the left-hand column, with instructions in the middle and right-hand columns for negative and positive responses, respectively. The instructions give either the number of the next question or the identity of the yeast.

	Negative	Positive
1 Erythritol growth	2	15
.		
15 Maltose growth	16	26
16 Cellobiose growth	17	22
.		
22 D-Galactose growth	23	25
23 Nitrate growth	Pichia pinus	24

To question number 1, erythritol growth, the answer (see above) is positive, so the next question is '15 Maltose growth'. This test gives a negative response, so the subsequent question is '16 Cellobiose growth' which gives a positive result and leads to '22 D-Galactose growth'. Since the yeast did not grow on

How to use the keys   73

D-galactose, the question that follows is '23 Nitrate growth' and, as this is also negative, the answer is *Pichia pinus*. For any one such identification, the results of some of the tests in the key remain unused, and so can be used for confirmation, by comparing them with those given in Chapter 6 for the species obtained. In the example above, this would apply to tests 1, 15, 42 and 55. Furthermore, it is always important to check that the appearance of the yeast under the microscope is consistent with that published reliably for its putative species; appropriate descriptions may be found in *The Yeasts* (Lodder, 1970) or in Chapter 4 in this *Guide* for the new species. The yeast cells should be compared with published drawings or photographs of the supposed species.

Sometimes, the key may lead to two or more specific names between which it does not differentiate. For example, the following entry appears in the key to all yeasts that do not ferment D-glucose, with physiological tests only (key no. 2, p. 89).

	Negative	Positive
12 Maltose growth	13	Arthroascus javanensis**
		Nadsonia commutata*

showing that no distinction is made between *Arthroascus javanensis* and *Nadsonia commutata*. If, as here, there is a corresponding key with both microscopical and physiological test results, this key will often separate the two yeasts. For such cases, the tests have been chosen so that no further physiological tests have to be done for the key that includes both microscopical and physiological tests. Alternatively, reference may be made to published descriptions of the yeasts. Two asterisks after a specific name, for example *Arthroascus javanensis***, means that it is fully described in *The Yeasts* (Lodder, 1970) under a different name, found from Chapter 4 in this *Guide* to be *Endomycopis javanensis*. A single asterisk after a specific name, for example *Nadsonia commutata**, means that it is a new species, so Chapter 4 gives a brief description and a reference to the original published description (Golubev, 1973) which included photographs of the cells.

Some species are not separated by any of the keys; this is so for *Pichia terricola*, to which *Pichia membranaefaciens* always appears as an alternative. As can be seen from Lodder (1970), *Pichia terricola* is distinguished from *Pichia membranaefaciens* chiefly by having warty ascospores and, also, by fermenting D-glucose slowly, whilst *Pichia membranaefaciens* ferments D-glucose very weakly or not at all.

List of keys

		Page
1	All yeasts that do not ferment D-glucose: physiological and microscopical tests	77
2	All yeasts that do not ferment D-glucose: physiological tests only	89
3	All yeasts that ferment D-glucose: physiological and microscopical tests	102
4	All yeasts that ferment D-glucose: physiological tests only	120
5	Ascosporogenous yeasts with spherical, oval or reniform ascospores: physiological and microscopical tests	138
6	Ascosporogenous yeasts with hat- or Saturn-shaped ascospores: physiological and microscopical tests	143
7	Ascosporogenous yeasts with ascospores of other shapes: physiological and microscopical tests	145
8	Basidiomycetous yeasts: physiological and microscopical tests	146
9	Yeasts with pink colonies: physiological and microscopical tests	147
10	Yeasts that utilize hydrocarbons: physiological and microscopical tests	148
11	Yeasts that utilize hydrocarbons: physiological tests only	151
12	Yeasts that utilize methanol: physiological tests only	154
13	Yeasts most commonly isolated clinically: physiological tests only	155
14	Extended list of yeasts isolated clinically: physiological tests only	157
15	Yeasts associated with food: physiological and microscopical tests	160

16 Yeasts associated with wine and wine-making: physiological and microscopical tests 167
17 Yeasts found in brewing: physiological and microscopical tests 172
18 Yeasts found in brewing: physiological tests only 174

The keys

Keys No. 1 to No. 4 are to all the yeasts

These keys are divided into two parts according to the results of test No. 1 (D−Glucose fermentation). − If the results of test 1 are negative, use keys for yeasts that do not ferment glucose; if positive, use keys for yeasts that ferment glucose.

Tests in keys that involve physiological tests and microscopical examination (Keys No. 1 and No. 3)

1	D−Glucose fermentation	26	Melibiose growth	52	Growth without inositol
2	D−Galactose fermentation	27	Lactose growth	53	Growth without pantothenate
3	Maltose fermentation	28	Raffinose growth	54	Growth without biotin
4	Methyl α−D−glucopyranoside fermentation	29	Melezitose growth	55	Growth without thiamin
5	Sucrose fermentation	30	Inulin growth	56	Growth without pyridoxine
6	Trehalose fermentation	31	Starch growth	57	Growth without niacin
7	Melibiose fermentation	32	Glycerol growth	59	Growth without p−aminobenzoate
8	Lactose fermentation	33	Erythritol growth	60	Growth in 50% D−glucose
9	Cellobiose fermentation	35	Galactitol growth	61	Growth in 60% D−glucose
12	D−Galactose growth	37	D−Glucitol growth	62	Growth at 37 degrees
13	L−Sorbose growth	38	myo−Inositol growth	63	Growth with 0.01% cycloheximide
14	D−Ribose growth	39	D−δ−Gluconolactone growth	64	Growth with 0.1% cycloheximide
15	D−Xylose growth	40	2−Ketogluconate growth	65	Starch formation
16	L−Arabinose growth	41	5−Ketogluconate growth	66	Pink colonies
18	L−Rhamnose growth	42	D,L−Lactate growth	67	Budding cells
19	Sucrose growth	43	Succinate growth	68	Splitting cells
20	Maltose growth	44	Citrate growth	69	Apical budding
21	Trehalose growth	46	Ethanol growth	71	Cells of other shapes
22	Methyl α−D−glucopyranoside growth	47	Ethylamine growth	72	Filamentous
23	Cellobiose growth	49	Nitrate growth	73	Pseudohyphae
24	Salicin growth	51	Growth without vitamins	74	Septate hyphae

number of different tests 63

All yeasts that do not ferment D-glucose (test No. 1 negative) (Keys No. 1 and No. 2)

Yeasts in Keys No. 1 and No. 2

1	Aciculoconidium aculeatum**	66	Candida insectamans*	138	Cryptococcus luteolus
2	Aessosporon dendrophilum*	70	Candida javanica	139	Cryptococcus macerans
3	Aessosporon salmonicolor**	72	Candida krissii*	140	Cryptococcus magnus**
5	Ambrosiozyma monospora**	77	Candida marina	141	Cryptococcus melibiosum
6	Ambrosiozyma philentoma*	80	Candida melinii	142	Cryptococcus neoformans
7	Arthroascus javanensis**	82	Candida mesenterica	143	Cryptococcus skinneri
8	Botryoascus synnaedendrus*	85	Candida muscorum	144	Cryptococcus terreus
12	Brettanomyces custersianus	86	Candida naeodendra*	145	Cryptococcus uniguttulatus
15	Brettanomyces naardenensis*	87	Candida norvegensis	147	Debaryomyces coudertii
16	Bullera alba	91	Candida podzolica*	148	Debaryomyces hansenii
17	Bullera piricola*	94	Candida quercuum*	149	Debaryomyces marama
18	Bullera tsugae	95	Candida ravautii	150	Debaryomyces melissophila*
19	Candida aaseri	98	Candida rugosa	151	Debaryomyces nepalensis*
21	Candida amylolenta*	102	Candida savonica*	152	Debaryomyces phaffii
22	Candida aquatica	103	Candida silvae	154	Debaryomyces tamarii
24	Candida berthetii	109	Candida suecica*	155	Debaryomyces vanriji
25	Candida blankii	111	Candida tepae*	156	Debaryomyces yarrowii*
26	Candida bogoriensis	112	Candida terebra*	159	Filobasidium capsuligenum**
31	Candida brumptii	114	Candida tsukubaensis*	160	Filobasidium floriforme*
32	Candida buffonii	116	Candida valdiviana*	161	Geotrichum capitatum**
36	Candida catenulata	117	Candida valida	162	Geotrichum fermentans**
38	Candida chiropterorum*	120	Candida vinaria*	163	Geotrichum penicillatum**
42	Candida curvata	121	Candida vini	164	Guilliermondella selenospora**
44	Candida diddensii	123	Candida zeylanoides	172	Hansenula beckii**
45	Candida diffluens	125	Cryptococcus albidus	174	Hansenula bimundalis
47	Candida edax*	126	Cryptococcus ater	176	Hansenula canadensis
51	Candida flavificans*	127	Cryptococcus bhutanensis*	180	Hansenula dryadoides*
52	Candida fluviotilis*	128	Cryptococcus cereanus*	182	Hansenula glucozyma
53	Candida foliarum	129	Cryptococcus dimennae	183	Hansenula henricii
57	Candida glaebosa	130	Cryptococcus flavus	185	Hansenula jadinii
58	Candida graminis*	131	Cryptococcus gastricus	187	Hansenula minuta
60	Candida humicola	132	Cryptococcus heveanensis*	190	Hansenula nonfermentans
61	Candida hydrocarbofumarica*	133	Cryptococcus himalayensis*	191	Hansenula ofunaensis*
62	Candida hylophila*	134	Cryptococcus hungaricus	193	Hansenula philodendra*
63	Candida iberica*	135	Cryptococcus kuetzingii	199	Hansenula wickerhamii
64	Candida incommunis*	136	Cryptococcus lactativorus	200	Hansenula wingei
		137	Cryptococcus laurentii	201	Hormoascus platypodis**

202 Hyphopichia burtonii**	296 Rhodosporidium bisporidiis*	352 Sterigmatomyces indicus
219 Leucosporidium antarcticum	297 Rhodosporidium capitatum*	353 Sterigmatomyces nectairii*
220 Leucosporidium frigidum	298 Rhodosporidium dacryoidum*	354 Sterigmatomyces penicillatus*
221 Leucosporidium gelidum	299 Rhodosporidium infirmo−miniatum**	355 Sterigmatomyces polyborus*
222 Leucosporidium nivalis	300 Rhodosporidium malvinellum*	356 Sympodiomyces parvus*
223 Leucosporidium scottii	301 Rhodotorula acheniorum*	362 Torulopsis apis
224 Leucosporidium stokesii	302 Rhodotorula araucariae*	363 Torulopsis auriculariae*
225 Lipomyces anomalus*	303 Rhodotorula aurantiaca	364 Torulopsis austromarina*
226 Lipomyces kononenkoae	304 Rhodotorula glutinis	365 Torulopsis bacarum*
227 Lipomyces lipofer	305 Rhodotorula graminis	369 Torulopsis dendrica*
228 Lipomyces starkeyi	306 Rhodotorula lactosa	372 Torulopsis fragaria*
229 Lipomyces tetrasporus*	307 Rhodotorula marina	374 Torulopsis fujisanensis
230 Lodderomyces elongisporus	308 Rhodotorula minuta	378 Torulopsis halonitratophila
232 Metschnikowia krissii	309 Rhodotorula pallida	382 Torulopsis inconspicua
233 Metschnikowia lunata*	310 Rhodotorula pilimanae	383 Torulopsis ingeniosa
236 Metschnikowia zobellii	311 Rhodotorula rubra	384 Torulopsis insectalens*
237 Nadsonia commutata*	320 Saccharomycopsis capsularis**	390 Torulopsis maris
240 Nematospora coryli	321 Saccharomycopsis crataegensis*	391 Torulopsis methanolovescens*
243 Phaffia rhodozyma*	322 Saccharomycopsis fibuligera**	393 Torulopsis multis−gemmis*
244 Pichia abadieae*	323 Saccharomycopsis lipolytica**	397 Torulopsis nemodendra*
246 Pichia ambrosiae*	324 Saccharomycopsis malanga*	398 Torulopsis nitratophila
250 Pichia castillae*	325 Saccharomycopsis vini**	400 Torulopsis norvegica
251 Pichia chambardii	326 Sarcinosporon inkin**	402 Torulopsis philyla*
252 Pichia delftensis	327 Schizoblastosporion starkeyi−henricii	404 Torulopsis pinus
254 Pichia etchellsii	335 Sporidiobolus johnsonii	405 Torulopsis psychrophila*
257 Pichia fluxuum	336 Sporidiobolus ruinenii	406 Torulopsis pustula*
259 Pichia haplophila	337 Sporobolomyces albo−rubescens	407 Torulopsis schatavii*
261 Pichia humboldtii*	338 Sporobolomyces antarcticus*	408 Torulopsis silvatica*
264 Pichia lindnerii*	339 Sporobolomyces gracilis	410 Torulopsis sorbophila*
265 Pichia media	340 Sporobolomyces hispanicus	415 Torulopsis vanderwaltii
266 Pichia membranaefaciens	341 Sporobolomyces holsaticus	418 Torulopsis xestobii*
271 Pichia norvegensis*	342 Sporobolomyces odorus	419 Trichosporon aquatile*
277 Pichia pinus	343 Sporobolomyces pararoseus	420 Trichosporon brassicae*
278 Pichia pseudopolymorpha	344 Sporobolomyces puniceus*	421 Trichosporon cutaneum
279 Pichia quercuum	345 Sporobolomyces roseus	422 Trichosporon eriense*
283 Pichia salictaria	346 Sporobolomyces salmonicolor	425 Trichosporon pullulans
285 Pichia scolyti	347 Sporobolomyces singularis	426 Trichosporon terrestre*
286 Pichia scutulata*	348 Stephanoascus ciferrii**	427 Trigonopsis variabilis
290 Pichia terricola	349 Sterigmatomyces aphidis*	429 Wickerhamiella domercqii**
291 Pichia toletana	350 Sterigmatomyces elviae*	431 Zendera ovetensis
294 Pichia vini	351 Sterigmatomyces halophilus	

Tests in keys that involve physiological tests only (Keys No. 2 and No. 4)

1 D−Glucose fermentation	23 Cellobiose growth	45 Methanol growth
2 D−Galactose fermentation	24 Salicin growth	46 Ethanol growth
3 Maltose fermentation	26 Melibiose growth	47 Ethylamine growth
4 Methyl α−D−glucopyranoside fermentation	27 Lactose growth	49 Nitrate growth
5 Sucrose fermentation	28 Raffinose growth	51 Growth without vitamins
6 Trehalose fermentation	29 Melezitose growth	52 Growth without inositol
7 Melibiose fermentation	30 Inulin growth	53 Growth without pantothenate
8 Lactose fermentation	31 Starch growth	54 Growth without biotin
9 Cellobiose fermentation	32 Glycerol growth	55 Growth without thiamin
12 D−Galactose growth	33 Erythritol growth	56 Growth without pyridoxine
13 L−Sorbose growth	35 Galactitol growth	57 Growth without niacin
14 D−Ribose growth	36 D−Mannitol growth	59 Growth without p−aminobenzoate
15 D−Xylose growth	37 D−Glucitol growth	60 Growth in 50% D−glucose
16 L−Arabinose growth	38 myo−Inositol growth	61 Growth in 60% D−glucose
17 D−Arabinose growth	39 D−δ−Gluconolactone growth	62 Growth at 37 degrees
18 L−Rhamnose growth	40 2−Ketogluconate growth	63 Growth with 0.01% cycloheximide
19 Sucrose growth	41 5−Ketogluconate growth	64 Growth with 0.1% cycloheximide
20 Maltose growth	42 D,L−Lactate growth	65 Starch formation
21 Trehalose growth	43 Succinate growth	
22 Methyl α−D−glucopyranoside growth	44 Citrate growth	number of different tests 58

Key involving physiological tests and microscopical examination (Key No. 1)

Tests in Key No. 1

12 D−Galactose growth	22 Methyl α−D−glucopyranoside growth	32 Glycerol growth
15 D−Xylose growth	23 Cellobiose growth	33 Erythritol growth
16 L−Arabinose growth	24 Salicin growth	35 Galactitol growth
18 L−Rhamnose growth	26 Melibiose growth	37 D−Glucitol growth
19 Sucrose growth	27 Lactose growth	38 myo−Inositol growth
20 Maltose growth	29 Melezitose growth	39 D−δ−Gluconolactone growth
21 Trehalose growth	30 Inulin growth	42 D,L−Lactate growth
	31 Starch growth	44 Citrate growth

Key No. 1: non-fermenting yeasts

46 Ethanol growth
49 Nitrate growth
55 Growth without thiamin
56 Growth without pyridoxine
59 Growth without p-aminobenzoate
60 Growth in 50% D-glucose
62 Growth at 37 degrees
65 Starch formation
66 Pink colonies
67 Budding cells
68 Splitting cells
69 Apical budding
71 Cells of other shapes
72 Filamentous
73 Pseudohyphae
74 Septate hyphae

number of different tests 39

KEY No. 1

	Negative	Positive
1 Erythritol growth.	2	454
2 myo-Inositol growth.	3	365
3 D-Glucitol growth.	4	113
4 Sucrose growth.	5	71
5 D-Galactose growth.	6	33
6 Nitrate growth.	7	31
7 Ethanol growth.	8	13
8 Trehalose growth.	9	11
9 Maltose growth.	10	Nadsonia commutata*
10 D,L-Lactate growth.	Torulopsis dendrica*	Pichia norvegensis*
11 D-Xylose growth.	12	Sporobolomyces gracilis
12 Growth at 37 degrees.	Nadsonia commutata*	Nematospora coryli
13 D,L-Lactate growth.	14	24
14 Citrate growth.	15	22
15 Lactose growth.	16	Trichosporon cutaneum
16 Growth at 37 degrees.	17	20
17 Maltose growth.	18	19
18 Apical budding.	Candida valida / Pichia membranaefaciens	Arthroascus javanensis**
19 Septate hyphae.	Nadsonia commutata*	Arthroascus javanensis**
20 Trehalose growth.	21	Nematospora coryli
21 Budding cells.	Geotrichum capitatum**	Candida valida / Pichia membranaefaciens
22 Lactose growth.	23	Trichosporon cutaneum
23 Salicin growth.	Pichia membranaefaciens / Pichia terricola	Torulopsis dendrica*
24 Cellobiose growth.	25	30
25 Lactose growth.	26	Trichosporon cutaneum
26 Trehalose growth.	27	Brettanomyces custersianus
27 Citrate growth.	28	Pichia membranaefaciens / Pichia terricola
28 Growth without pyridoxine	Pichia membranaefaciens / Torulopsis inconspicua	29
29 Growth without thiamin.	21	Candida valida / Pichia membranaefaciens / Pichia scutulata*
30 Lactose growth.	Candida norvegensis / Pichia norvegensis*	Trichosporon cutaneum
31 Cellobiose growth.	32	Candida berthetii
32 Growth without thiamin.	Torulopsis halonitratophila	Leucosporidium antarcticum
33 Glycerol growth.	34	38
34 Ethanol growth.	35	36
35 L-Arabinose growth.	Leucosporidium antarcticum	Lipomyces anomalus*
36 Lactose growth.	37	Trichosporon cutaneum
37 L-Arabinose growth.	Leucosporidium antarcticum	Guilliermondella selenospora**
38 D,L-Lactate growth.	39	58
39 Nitrate growth.	40	32
40 D-Xylose growth.	41	56
41 Trehalose growth.	42	50
42 Maltose growth.	43	Nadsonia commutata*
43 Growth at 37 degrees.	44	49
44 Growth without thiamin.	45	47
45 Apical budding.	46	Schizoblastosporion starkeyi-henricii
46 Filamentous.	Torulopsis austromarina*	Candida vinaria*
47 Apical budding.	48	Schizoblastosporion starkeyi-henricii

		Negative	Positive
48	Filamentous.	Torulopsis austromarina*	Pichia humboldtii*
49	Growth without thiamin.	Geotrichum capitatum**	Pichia humboldtii*
50	Maltose growth.	51	54
51	Growth at 37 degrees.	52	Nematospora coryli
52	Pink colonies.	53	Rhodosporidium dacryoidum*
53	Apical budding.	Torulopsis austromarina*	Schizoblastosporion starkeyi–henricii
54	Growth at 37 degrees.	55	Nematospora coryli
55	Pink colonies.	Nadsonia commutata*	Rhodosporidium dacryoidum*
56	Lactose growth.	57	Trichosporon cutaneum
57	Growth without thiamin.	Candida vinaria*	Geotrichum penicillatum**
58	D–Xylose growth.	59	68
59	Trehalose growth.	60	65
60	Cellobiose growth.	61	Pichia chambardii
61	Nitrate growth.	62	Torulopsis halonitratophila
62	Growth without thiamin.	63	64
63	Growth at 37 degrees.	Zendera ovetensis	Geotrichum capitatum**
64	Septate hyphae.	Pichia humboldtii*	Zendera ovetensis
65	Citrate growth.	66	67
66	Pink colonies.	Brettanomyces custersianus	Rhodosporidium dacryoidum*
67	Growth at 37 degrees.	Rhodosporidium dacryoidum*	Trigonopsis variabilis
68	Lactose growth.	69	Trichosporon cutaneum
69	Trehalose growth.	70	Trigonopsis variabilis
70	L–Arabinose growth.	Geotrichum penicillatum**	Guilliermondella selenospora**
71	D–Xylose growth.	72	88
72	Cellobiose growth.	73	78
73	Nitrate growth.	74	76
74	Glycerol growth.	Torulopsis auriculariae*	75
75	D,L–Lactate growth.	54	66
76	Melezitose growth.	77	Rhodotorula glutinis
77	Salicin growth.	Leucosporidium antarcticum	Rhodosporidium malvinellum*
78	Nitrate growth.	79	87
79	D–Galactose growth.	80	85
80	L–Arabinose growth.	81	Phaffia rhodozyma*
81	Starch growth.	Metschnikowia krissii	82
82	Growth without thiamin.	83	84
83	Growth at 37 degrees.	Aciculoconidium aculeatum**	Saccharomycopsis fibuligera**
84	Pink colonies.	Saccharomycopsis fibuligera**	Sporobolomyces pararoseus
85	Methyl α–D–glucopyranoside growth	86	Sporobolomyces pararoseus
86	Melibiose growth.	Rhodosporidium dacryoidum*	Debaryomyces tamarii
87	Melezitose growth.	Rhodosporidium malvinellum*	Rhodotorula glutinis
88	Nitrate growth.	89	109
89	Maltose growth.	90	93
90	Lactose growth.	91	92
91	Melezitose growth.	Rhodotorula pilimanae	Rhodotorula minuta
92	Growth without p–aminobenzoate	Rhodotorula minuta	Trichosporon cutaneum
93	Starch growth.	94	100
94	Trehalose growth.	95	96
95	Ethanol growth.	Lipomyces anomalus*	Trichosporon cutaneum
96	Lactose growth.	97	Trichosporon cutaneum
97	D–Galactose growth.	98	99
98	Growth without thiamin.	Rhodotorula rubra	Phaffia rhodozyma*
99	Pink colonies.	Torulopsis xestobii*	Rhodotorula rubra Sporobolomyces albo–rubescens
100	Galactitol growth.	101	108
101	Ethanol growth.	102	104
102	L–Arabinose growth.	Sporobolomyces pararoseus	103
103	D–Galactose growth.	Phaffia rhodozyma*	Lipomyces anomalus*
104	L–Arabinose growth.	105	106
105	Lactose growth.	Sporobolomyces pararoseus	Trichosporon cutaneum

Key No. 1: non-fermenting yeasts

	Negative	Positive
106 D−Galactose growth.	107	Trichosporon aquatile* Trichosporon cutaneum
107 Lactose growth.	Phaffia rhodozyma*	Trichosporon cutaneum
108 Splitting cells.	Aessosporon dendrophilum*	Trichosporon cutaneum
109 Melezitose growth.	110	112
110 Growth without thiamin.	Rhodosporidium malvinellum*	111
111 Pink colonies.	Leucosporidium antarcticum	Rhodotorula graminis
112 Pink colonies.	Leucosporidium scottii	Rhodotorula glutinis
113 Nitrate growth.	114	286
114 Glycerol growth.	115	154
115 Sucrose growth.	116	132
116 Ethanol growth.	117	119
117 D−Galactose growth.	Candida insectamans*	118
118 L−Arabinose growth.	Torulopsis sorbophila*	Lipomyces anomalus*
119 Cellobiose growth.	120	126
120 D−Galactose growth.	121	124
121 D−Xylose growth.	122	Trichosporon cutaneum
122 Trehalose growth.	123	Torulopsis philyla*
123 D−δ−Gluconolactone growth	Candida vini Pichia fluxuum	Candida vini Pichia delftensis
124 Lactose growth.	125	Trichosporon cutaneum
125 D,L−Lactate growth.	Torulopsis sorbophila*	Candida rugosa
126 Galactitol growth.	127	131
127 D−Xylose growth.	128	129
128 D−Galactose growth.	Sporobolomyces singularis	Torulopsis insectalens*
129 Lactose growth.	130	Trichosporon cutaneum
130 L−Arabinose growth.	Brettanomyces naardenensis*	Candida bogoriensis
131 Lactose growth.	Torulopsis fujisanensis	Trichosporon cutaneum
132 L−Arabinose growth.	133	144
133 Ethanol growth.	134	137
134 Melibiose growth.	135	Lipomyces kononenkoae
135 Melezitose growth.	Candida suecica*	136
136 Starch growth.	Rhodotorula rubra	Sporobolomyces pararoseus
137 D−Xylose growth.	138	141
138 Methyl α−D−glucopyranoside growth	Torulopsis auriculariae*	139
139 Melibiose growth.	140	Lipomyces kononenkoae
140 Growth without thiamin.	Aciculoconidium aculeatum**	Sporobolomyces pararoseus
141 Lactose growth.	142	Trichosporon cutaneum
142 Trehalose growth.	Debaryomyces yarrowii*	143
143 Melibiose growth.	136	Lipomyces kononenkoae
144 Starch formation.	145	152
145 Maltose growth.	146	147
146 Lactose growth.	Rhodotorula pilimanae	Trichosporon cutaneum
147 Ethanol growth.	148	149
148 Trehalose growth.	Lipomyces anomalus*	Rhodotorula rubra Sporobolomyces albo−rubescens
149 Lactose growth.	150	Trichosporon cutaneum
150 Growth without thiamin.	Rhodotorula rubra Sporobolomyces albo−rubescens	151
151 Pink colonies.	Pichia etchellsii	Sporobolomyces albo−rubescens
152 Galactitol growth.	153	108
153 Ethanol growth.	103	107
154 Maltose growth.	155	225
155 Trehalose growth.	156	185
156 Galactitol growth.	157	184
157 D−Galactose growth.	158	172
158 D−Xylose growth.	159	169
159 Ethanol growth.	160	162
160 Sucrose growth.	161	Torulopsis apis
161 D,L−Lactate growth.	Candida krissii*	Candida hylophila*
162 Citrate growth.	163	167

	Negative	Positive
163 Lactose growth.	164	Sporobolomyces singularis
164 D,L–Lactate growth.	165	166
165 Septate hyphae.	Candida silvae / Candida vini	Saccharomycopsis vini**
166 Cells of other shapes.	Candida silvae / Candida vini	Candida hylophila*
167 Lactose growth.	168	Sporobolomyces singularis
168 Growth without thiamin.	Pichia quercuum	Candida krissii*
169 Lactose growth.	170	Trichosporon cutaneum
170 L–Rhamnose growth.	Saccharomycopsis crataegensis*	171
171 Cellobiose growth.	Torulopsis maris	Pichia salictaria
172 D,L–Lactate growth.	173	180
173 D–Xylose growth.	174	177
174 Sucrose growth.	175	Torulopsis apis
175 Growth at 37 degrees.	176	Torulopsis sorbophila*
176 Apical budding.	Candida vinaria*	Schizoblastosporion starkeyi–henricii
177 Lactose growth.	178	Trichosporon cutaneum
178 Growth without thiamin.	179	Geotrichum penicillatum**
179 D–δ–Gluconolactone growth	Torulopsis sorbophila*	Candida vinaria*
180 Lactose growth.	181	Trichosporon cutaneum
181 Cellobiose growth.	182	Geotrichum fermentans**
182 Citrate growth.	183	Candida catenulata / Candida rugosa
183 Growth at 37 degrees.	Geotrichum penicillatum**	Candida rugosa
184 Ethanol growth.	Rhodotorula marina	131
185 Citrate growth.	186	199
186 D–Xylose growth.	187	194
187 D,L–Lactate growth.	188	190
188 D–Galactose growth.	Saccharomycopsis vini**	189
189 Pink colonies.	Schizoblastosporion starkeyi–henricii	Rhodosporidium dacryoidum*
190 Growth at 37 degrees.	191	193
191 Pink colonies.	Brettanomyces custersianus	192
192 Filamentous.	Rhodotorula pallida	Rhodosporidium dacryoidum*
193 Filamentous.	Torulopsis silvatica*	Brettanomyces custersianus
194 Lactose growth.	195	92
195 Melezitose growth.	196	Rhodotorula minuta
196 D–Galactose growth.	197	198
197 Pink colonies.	Saccharomycopsis crataegensis*	Rhodotorula pallida
198 L–Arabinose growth.	Rhodotorula pallida	Rhodotorula pilimanae
199 L–Rhamnose growth.	200	217
200 D,L–Lactate growth.	201	211
201 D–Xylose growth.	202	206
202 Cellobiose growth.	203	205
203 Salicin growth.	204	Candida iberica* / Candida zeylanoides
204 Pink colonies.	Candida iberica*	Rhodosporidium dacryoidum*
205 Salicin growth.	Rhodosporidium dacryoidum*	Candida savonica* / Candida zeylanoides
206 Ethanol growth.	207	209
207 D–Galactose growth.	Sporobolomyces gracilis	208
208 L–Arabinose growth.	Candida savonica*	Rhodotorula pilimanae
209 Lactose growth.	210	Trichosporon cutaneum
210 Sucrose growth.	Candida savonica* / Candida zeylanoides	Rhodotorula pilimanae
211 Lactose growth.	212	Trichosporon cutaneum
212 D–Galactose growth.	Rhodotorula pallida / Sporobolomyces gracilis	213
213 L–Arabinose growth.	214	216
214 Growth at 37 degrees.	192	215
215 Cells of other shapes.	Candida catenulata	Trigonopsis variabilis
216 Sucrose growth.	Trigonopsis variabilis	Rhodotorula pilimanae
217 D–Galactose growth.	218	221
218 Lactose growth.	219	Trichosporon cutaneum

Key No. 1: non-fermenting yeasts

		Negative	Positive
219	D,L–Lactate growth.	Pichia lindnerii* / Torulopsis methanolovescens*	220
220	Growth without thiamin.	Rhodotorula pallida	Pichia lindnerii*
221	D–Xylose growth.	192	222
222	Ethanol growth.	223	224
223	L–Arabinose growth.	Rhodotorula pallida	Rhodotorula marina
224	Lactose growth.	Rhodotorula pallida	Trichosporon cutaneum
225	L–Arabinose growth.	226	267
226	Melezitose growth.	227	246
227	D–Xylose growth.	228	238
228	Cellobiose growth.	229	234
229	D,L–Lactate growth.	230	232
230	D–Galactose growth.	231	55
231	Apical budding.	Saccharomycopsis vini**	Nadsonia commutata*
232	Citrate growth.	66	233
233	Growth at 37 degrees.	Rhodosporidium dacryoidum*	Candida catenulata
234	Methyl α–D–glucopyranoside growth	235	237
235	D–Galactose growth.	236	Rhodosporidium dacryoidum*
236	Starch growth.	Saccharomycopsis vini**	Saccharomycopsis malanga*
237	Starch growth.	Saccharomycopsis vini**	83
238	Sucrose growth.	239	243
239	Lactose growth.	240	Trichosporon cutaneum
240	D–Galactose growth.	Saccharomycopsis malanga*	241
241	Trehalose growth.	242	Candida catenulata / Candida ravautii
242	Growth at 37 degrees.	Candida brumptii	Candida catenulata
243	Lactose growth.	244	245
244	Citrate growth.	Candida tepae*	Pichia toletana
245	Splitting cells.	Candida glaebosa	Trichosporon cutaneum
246	Lactose growth.	247	265
247	D–Galactose growth.	248	252
248	D–Xylose growth.	82	249
249	Starch growth.	250	Sporobolomyces pararoseus
250	Growth without pyridoxine	Candida quercuum*	251
251	Pink colonies.	Pichia toletana	Rhodotorula rubra
252	Trehalose growth.	Debaryomyces yarrowii*	253
253	Melibiose growth.	254	264
254	Starch growth.	255	263
255	Methyl α–D–glucopyranoside growth	256	259
256	D–Xylose growth.	257	258
257	Salicin growth.	Rhodosporidium dacryoidum*	Metschnikowia lunata*
258	Pink colonies.	Metschnikowia lunata*	Rhodotorula rubra
259	Cellobiose growth.	260	261
260	Pink colonies.	Lodderomyces elongisporus	Rhodotorula rubra
261	Growth at 37 degrees.	262	258
262	Pink colonies.	Metschnikowia zobellii	Rhodotorula rubra
263	Growth without thiamin.	Pichia vini	Sporobolomyces pararoseus
264	Inulin growth.	Pichia vini	Lipomyces kononenkoae
265	Melibiose growth.	266	245
266	Splitting cells.	Candida fluviotilis*	Trichosporon cutaneum
267	Galactitol growth.	268	279
268	Sucrose growth.	269	271
269	Lactose growth.	270	Trichosporon cutaneum
270	Melezitose growth.	Saccharomycopsis malanga*	Candida bogoriensis
271	Lactose growth.	272	277
272	D–Galactose growth.	98	273
273	Methyl α–D–glucopyranoside growth	274	275
274	Pink colonies.	Torulopsis multis–gemmis*	Rhodotorula rubra / Sporobolomyces albo–rubescens
275	Starch growth.	150	276
276	Growth without thiamin.	Pichia vini	Debaryomyces hansenii
277	Starch growth.	245	278
278	Growth in 50% D–glucose.	Trichosporon cutaneum	Debaryomyces hansenii

		Negative	Positive
279	Ethanol growth.	280	281
280	Growth without p–aminobenzoate	Rhodotorula marina	Rhodotorula rubra
281	Lactose growth.	282	285
282	D,L–Lactate growth.	283	284
283	Starch growth.	Rhodotorula rubra	Aessosporon dendrophilum*
284	Starch growth.	Rhodotorula rubra	Debaryomyces hansenii
285	D,L–Lactate growth.	108	278
286	Melezitose growth.	287	328
287	Sucrose growth.	288	314
288	Cellobiose growth.	289	302
289	Trehalose growth.	290	291
290	Citrate growth.	Torulopsis halonitratophila	Wickerhamiella domercqii**
291	L–Arabinose growth.	292	299
292	Glycerol growth.	293	295
293	Growth at 37 degrees.	294	Candida foliarum
294	Pink colonies.	Sterigmatomyces nectairii*	Sporobolomyces hispanicus
295	Growth at 37 degrees.	296	298
296	Filamentous.	Rhodotorula araucariae*	297
297	Septate hyphae.	Sporobolomyces odorus	Sporobolomyces hispanicus
298	Pink colonies.	Candida foliarum	Sporobolomyces odorus
299	Ethanol growth.	Torulopsis vanderwaltii	300
300	L–Rhamnose growth.	301	Torulopsis nitratophila
301	Growth at 37 degrees.	Rhodotorula araucariae*	Candida foliarum
302	Galactitol growth.	303	313
303	Trehalose growth.	304	305
304	D–Xylose growth.	Hansenula dryadoides*	Torulopsis norvegica
305	D–Xylose growth.	306	308
306	Growth at 37 degrees.	Rhodotorula araucariae*	307
307	Salicin growth.	Candida foliarum	Hansenula nonfermentans
308	L–Arabinose growth.	309	311
309	Salicin growth.	301	310
310	Pink colonies.	Hansenula minuta	Rhodotorula araucariae*
311	Growth at 37 degrees.	312	Candida foliarum
312	Pink colonies.	Torulopsis pustula*	Rhodotorula araucariae*
313	Maltose growth.	Rhodotorula araucariae*	Candida javanica
314	Maltose growth.	315	324
315	Cellobiose growth.	316	321
316	Trehalose growth.	Wickerhamiella domercqii**	317
317	D–Galactose growth.	318	319
318	Septate hyphae.	Sporobolomyces odorus	Aessosporon salmonicolor** Sporobolomyces salmonicolor
319	Growth without thiamin.	Torulopsis vanderwaltii	320
320	Septate hyphae.	Rhodotorula graminis	Aessosporon salmonicolor** Sporobolomyces salmonicolor
321	Ethanol growth.	322	323
322	Septate hyphae.	Rhodotorula graminis	Sporidibolus ruinenii
323	Lactose growth.	Rhodotorula graminis	Leucosporidium frigidum
324	Galactitol growth.	325	327
325	Growth without thiamin.	Rhodosporidium malvinellum*	326
326	Septate hyphae.	Rhodotorula graminis	Sporobolomyces holsaticus
327	Growth without thiamin.	Rhodosporidium malvinellum*	322
328	Glycerol growth.	329	336
329	Melibiose growth.	330	335
330	Sucrose growth.	Candida buffonii	331
331	L–Rhamnose growth.	332	334
332	Starch formation.	333	Cryptococcus bhutanensis*
333	Septate hyphae.	Rhodotorula glutinis	Sporobolomyces holsaticus
334	Starch growth.	Rhodotorula glutinis	Leucosporidium stokesii
335	L–Rhamnose growth.	Candida aquatica	Leucosporidium gelidum
336	L–Rhamnose growth.	337	360
337	Methyl α–D–glucopyranoside growth	338	348
338	Lactose growth.	339	346

Key No. 1: non-fermenting yeasts

		Negative	Positive
339	Sucrose growth.	............Candida buffonii	340
340	D–Galactose growth.	341	344
341	Starch formation.	342	............Sporobolomyces puniceus*
342	Starch growth.	333	343
343	Septate hyphae.	............Sporobolomyces roseus	............Sporobolomyces holsaticus
344	Starch growth.	345	343
345	Growth without thiamin.	............Rhodotorula aurantiaca Rhodotorula glutinis	333
346	D–Galactose growth.	............Bullera tsugae	347
347	Melibiose growth.	............Rhodotorula aurantiaca	............Candida aquatica
348	Lactose growth.	349	358
349	Inulin growth.	350	............Hansenula jadinii
350	D,L–Lactate growth.	351	353
351	Septate hyphae.	352	............Sporidibolus johnsonii Sporobolomyces holsaticus
352	Starch growth.	............Rhodotorula glutinis	............Sporobolomyces roseus
353	Starch growth.	354	355
354	Septate hyphae.	............Rhodotorula glutinis	............Sporidibolus johnsonii
355	Growth at 37 degrees.	356	357
356	Pink colonies.	............Torulopsis ingeniosa	............Sporobolomyces roseus
357	Pink colonies.	............Torulopsis ingeniosa	............Sporidibolus johnsonii
358	L–Arabinose growth.	359	............Torulopsis fragaria*
359	D,L–Lactate growth.	............Candida muscorum	............Torulopsis ingeniosa
360	Galactitol growth.	361	112
361	Melibiose growth.	362	............Rhodotorula lactosa
362	Septate hyphae.	363	364
363	Pink colonies.	............Candida melinii Hansenula canadensis	............Rhodotorula glutinis
364	Pseudohyphae.	............Hansenula beckii** Hansenula bimundalis	............Hansenula wingei
365	Glycerol growth.	366	415
366	Nitrate growth.	367	400
367	Galactitol growth.	368	382
368	Ethanol growth.	369	375
369	Lactose growth.	370	372
370	Maltose growth.	............Cryptococcus skinneri	371
371	D–δ–Gluconolactone growth	............Cryptococcus uniguttulatus	............Cryptococcus gastricus
372	Methyl α–D–glucopyranoside growth	373	374
373	Melibiose growth.	............Cryptococcus gastricus	............Cryptococcus magnus**
374	Melibiose growth.	............Bullera alba Cryptococcus ater	............Bullera alba Cryptococcus magnus**
375	Maltose growth.	376	378
376	Lactose growth.	............Cryptococcus skinneri	377
377	Splitting cells.	............Cryptococcus dimennae	............Trichosporon cutaneum
378	Melibiose growth.	379	381
379	Lactose growth.	371	380
380	Splitting cells.	............Cryptococcus gastricus	............Trichosporon cutaneum
381	Splitting cells.	............Cryptococcus magnus**	............Trichosporon cutaneum
382	Ethanol growth.	383	395
383	Lactose growth.	384	387
384	Sucrose growth.	............Cryptococcus skinneri	385
385	Growth at 37 degrees.	386	............Cryptococcus neoformans
386	Pink colonies.	............Cryptococcus luteolus	............Cryptococcus hungaricus
387	D,L–Lactate growth.	388	390
388	Starch growth.	............Bullera alba Cryptococcus luteolus Cryptococcus magnus**	389
389	Filamentous.	............Bullera alba Cryptococcus magnus**	............Candida podzolica*
390	Starch growth.	391	392
391	Pink colonies.	............Bullera alba Cryptococcus luteolus	............Cryptococcus hungaricus
392	Growth without thiamin.	393	394
393	Pink colonies.	............Bullera alba	............Cryptococcus hungaricus
394	Filamentous.	............Bullera alba	............Candida podzolica*
395	Maltose growth.	376	396
396	Lactose growth.	397	398

		Negative	Positive
397	Growth at 37 degrees.	Cryptococcus luteolus	Cryptococcus neoformans
398	Filamentous.	Cryptococcus luteolus Cryptococcus magnus**	399
399	Splitting cells.	Candida podzolica*	Trichosporon cutaneum
400	Melibiose growth.	401	409
401	Ethanol growth.	402	405
402	Sucrose growth.	403	404
403	Citrate growth.	Cryptococcus terreus	Cryptococcus himalayensis*
404	Pink colonies.	Cryptococcus albidus	Rhodosporidium infirmo−miniatum**
405	Maltose growth.	406	407
406	D−Galactose growth.	Cryptococcus kuetzingii	Leucosporidium frigidum
407	Lactose growth.	408	404
408	Pink colonies.	Leucosporidium stokesii	Rhodosporidium infirmo−miniatum**
409	Lactose growth.	410	413
410	L−Rhamnose growth.	411	412
411	Maltose growth.	Leucosporidium nivalis	Candida valdiviana*
412	Pink colonies.	Leucosporidium gelidum	Rhodosporidium bisporidiis*
413	Starch formation.	414	Cryptococcus albidus Cryptococcus heveanensis*
414	Methyl α−D−glucopyranoside growth	Bullera piricola*	Candida valdiviana*
415	Nitrate growth.	416	448
416	Galactitol growth.	417	437
417	D−Glucitol growth.	418	426
418	Ethanol growth.	419	420
419	L−Rhamnose growth.	Cryptococcus uniguttulatus	Bullera alba
420	D−Xylose growth.	421	423
421	D−Galactose growth.	Saccharomycopsis fibuligera**	422
422	Maltose growth.	Cryptococcus melibiosum	Trichosporon brassicae*
423	Lactose growth.	424	Trichosporon cutaneum
424	L−Arabinose growth.	Cryptococcus lactativorus	425
425	Melezitose growth.	Trichosporon brassicae*	Cryptococcus uniguttulatus
426	Lactose growth.	427	432
427	L−Rhamnose growth.	428	Cryptococcus skinneri
428	L−Arabinose growth.	429	430
429	Growth at 37 degrees.	Filobasidium capsuligenum**	Saccharomycopsis fibuligera**
430	Melezitose growth.	431	Cryptococcus uniguttulatus
431	Methyl α−D−glucopyranoside growth	Trichosporon brassicae*	Filobasidium capsuligenum**
432	Ethanol growth.	433	434
433	Sucrose growth.	Pichia abadieae*	Bullera alba Cryptococcus magnus**
434	Sucrose growth.	435	436
435	Splitting cells.	Pichia abadieae*	Trichosporon cutaneum
436	Maltose growth.	377	381
437	Melezitose growth.	438	443
438	Lactose growth.	439	440
439	L−Rhamnose growth.	Filobasidium capsuligenum**	Cryptococcus skinneri
440	Sucrose growth.	441	377
441	D,L−Lactate growth.	442	435
442	Splitting cells.	Hansenula ofunaensis*	Trichosporon cutaneum
443	Ethanol growth.	444	446
444	Lactose growth.	385	445
445	D,L−Lactate growth.	Bullera alba Cryptococcus luteolus Cryptococcus magnus**	391
446	Lactose growth.	397	447
447	Splitting cells.	Cryptococcus luteolus Cryptococcus magnus**	Trichosporon cutaneum
448	Melibiose growth.	449	452
449	Maltose growth.	Cryptococcus kuetzingii	450

Key No. 1: non-fermenting yeasts

	Negative	Positive
450 Methyl α–D–glucopyranoside growth	451	Cryptococcus albidus Filobasidium floriforme*
451 Pink colonies	Cryptococcus albidus	Rhodosporidium capitatum* Rhodosporidium infirmo–miniatum**
452 Lactose growth	453	413
453 L–Rhamnose growth	411	Rhodosporidium bisporidiis*
454 myo–Inositol growth	455	589
455 Galactitol growth	456	551
456 Sucrose growth	457	488
457 D–Galactose growth	458	471
458 Nitrate growth	459	468
459 D,L–Lactate growth	460	465
460 Maltose growth	461	463
461 Lactose growth	462	Trichosporon cutaneum
462 Cellobiose growth	Torulopsis pinus	Pichia pinus
463 Lactose growth	464	Trichosporon cutaneum
464 Methyl α–D–glucopyranoside growth	Saccharomycopsis malanga*	Saccharomycopsis capsularis**
465 Lactose growth	466	Trichosporon cutaneum
466 Trehalose growth	467	Rhodotorula pallida
467 D–Xylose growth	Saccharomycopsis lipolytica**	Trichosporon eriense*
468 Cellobiose growth	469	470
469 L–Rhamnose growth	Hansenula philodendra*	Hansenula wickerhamii
470 Growth without thiamin	Hansenula glucozyma	Hansenula henricii
471 Ethanol growth	472	477
472 Cellobiose growth	473	475
473 D,L–Lactate growth	474	Rhodotorula pallida
474 Citrate growth	Torulopsis psychrophila*	Torulopsis schatavii*
475 Methyl α–D–glucopyranoside growth	476	Sympodiomyces parvus*
476 D,L–Lactate growth	Candida flavificans*	Rhodotorula pallida
477 Maltose growth	478	485
478 D,L–Lactate growth	479	465
479 Nitrate growth	480	Sterigmatomyces halophilus
480 Lactose growth	481	Trichosporon cutaneum
481 Cellobiose growth	482	Candida flavificans*
482 Citrate growth	483	484
483 Salicin growth	Torulopsis psychrophila*	Sterigmatomyces indicus
484 Salicin growth	Torulopsis schatavii*	Sterigmatomyces indicus
485 Lactose growth	486	Trichosporon cutaneum
486 Melibiose growth	487	Pichia castillae*
487 Citrate growth	Debaryomyces coudertii	Pichia media
488 Nitrate growth	489	542
489 D–Xylose growth	490	494
490 Maltose growth	Rhodotorula pallida	491
491 D–Galactose growth	492	Debaryomyces melissophila*
492 Starch growth	493	Saccharomycopsis fibuligera**
493 Growth without thiamin	Pichia ambrosiae*	Candida mesenterica
494 Melibiose growth	495	527
495 Melezitose growth	496	503
496 Ethanol growth	497	499
497 Methyl α–D–glucopyranoside growth	498	Sympodiomyces parvus*
498 L–Arabinose growth	Rhodotorula pallida	Sterigmatomyces elviae*
499 Lactose growth	500	502
500 Maltose growth	Rhodotorula pallida	501
501 Starch growth	Candida diddensii	Hyphopichia burtonii**
502 Budding cells	Sterigmatomyces elviae*	Trichosporon cutaneum
503 D–Galactose growth	504	506
504 Lactose growth	505	Trichosporon cutaneum
505 L–Arabinose growth	Ambrosiozyma philentoma*	Ambrosiozyma monospora**
506 L–Rhamnose growth	507	519
507 D,L–Lactate growth	508	517
508 Lactose growth	509	513

		Negative	Positive
509	Starch growth.	510	511
510	Cells of other shapes.	Candida aaseri	Candida diddensii
511	Inulin growth.	512	Debaryomyces marama / Lipomyces tetrasporus*
512	Filamentous.	Debaryomyces marama	Hyphopichia burtonii**
513	Inulin growth.	514	Debaryomyces marama / Lipomyces tetrasporus*
514	Starch growth.	515	516
515	Splitting cells.	Candida aaseri	Trichosporon cutaneum
516	Splitting cells.	Debaryomyces marama	Trichosporon cutaneum
517	Growth in 50% D–glucose.	518	Debaryomyces hansenii / Debaryomyces marama
518	Inulin growth.	516	Debaryomyces marama / Lipomyces tetrasporus*
519	Lactose growth.	520	525
520	D,L–Lactate growth.	521	524
521	Starch growth.	522	523
522	Cells of other shapes.	Candida terebra*	Candida diddensii
523	Inulin growth.	Candida naeodendra*	Lipomyces tetrasporus*
524	Growth in 50% D–glucose.	Lipomyces tetrasporus*	Debaryomyces hansenii
525	Growth in 50% D–glucose.	526	Debaryomyces hansenii
526	Inulin growth.	Trichosporon cutaneum	Lipomyces tetrasporus*
527	Glycerol growth.	528	531
528	Inulin growth.	529	Lipomyces lipofer / Lipomyces tetrasporus*
529	Splitting cells.	530	Trichosporon cutaneum
530	Filamentous.	Lipomyces lipofer	Candida humicola
531	Inulin growth.	532	541
532	L–Rhamnose growth.	533	538
533	Lactose growth.	534	535
534	Starch formation.	Debaryomyces hansenii / Debaryomyces marama / Debaryomyces nepalensis*	Candida humicola
535	Starch formation.	536	537
536	Growth in 50% D–glucose.	516	Debaryomyces hansenii / Debaryomyces marama
537	Splitting cells.	Candida humicola	Trichosporon cutaneum
538	Starch formation.	539	537
539	D,L–Lactate growth.	540	278
540	Splitting cells.	Pichia scolyti	Trichosporon cutaneum
541	Growth in 50% D–glucose.	Debaryomyces marama / Lipomyces tetrasporus*	Debaryomyces hansenii / Debaryomyces marama
542	Ethanol growth.	543	545
543	L–Arabinose growth.	Candida graminis*	544
544	D–Galactose growth.	Torulopsis bacarum*	Rhodotorula acheniorum*
545	D–Galactose growth.	546	548
546	D–Xylose growth.	Candida diffluens	547
547	L–Arabinose growth.	Hormoascus platypodis**	Torulopsis bacarum*
548	D–Xylose growth.	549	550
549	Cellobiose growth.	Candida diffluens	Trichosporon pullulans
550	Salicin growth.	Rhodotorula acheniorum*	Trichosporon pullulans
551	Melezitose growth.	552	559
552	Maltose growth.	553	556
553	Lactose growth.	554	Trichosporon cutaneum
554	L–Arabinose growth.	Rhodotorula pallida	555
555	Trehalose growth.	Pichia haplophila	Torulopsis nemodendra*
556	Sucrose growth.	557	537
557	Lactose growth.	558	Trichosporon cutaneum
558	Melibiose growth.	Pichia media	Pichia castillae*
559	Ethanol growth.	560	566
560	Melibiose growth.	561	562
561	Growth at 37 degrees.	Sterigmatomyces polyborus*	Candida terebra*
562	L–Rhamnose growth.	563	564
563	Inulin growth.	530	Lipomyces lipofer / Lipomyces starkeyi
564	Inulin growth.	565	Lipomyces starkeyi
565	Starch formation.	Sterigmatomyces polyborus*	Candida humicola
566	Glycerol growth.	567	568

Key No. 1: non-fermenting yeasts

		Negative	Positive
567	Inulin growth.	529	Lipomyces lipofer Lipomyces starkeyi Lipomyces tetrasporus*
568	Melibiose growth.	569	575
569	Lactose growth.	570	572
570	D,L–Lactate growth.	571	524
571	Inulin growth.	Candida terebra*	Lipomyces tetrasporus*
572	D,L–Lactate growth.	573	525
573	Inulin growth.	574	Lipomyces tetrasporus*
574	Splitting cells.	Candida blankii	Trichosporon cutaneum
575	D,L–Lactate growth.	576	582
576	Lactose growth.	577	580
577	Starch formation.	578	579
578	Growth in 50% D–glucose.	Debaryomyces vanriji Lipomyces tetrasporus*	Debaryomyces phaffii Debaryomyces vanriji
579	Inulin growth.	Candida humicola	Lipomyces starkeyi Lipomyces tetrasporus*
580	Inulin growth.	537	581
581	Growth in 50% D–glucose.	Lipomyces starkeyi Lipomyces tetrasporus*	Debaryomyces phaffii
582	Lactose growth.	583	586
583	Starch formation.	584	585
584	Growth in 50% D–glucose.	Debaryomyces vanriji Lipomyces tetrasporus*	Debaryomyces hansenii Debaryomyces vanriji
585	Inulin growth.	Candida humicola	Lipomyces tetrasporus*
586	Growth in 50% D–glucose.	587	588
587	Inulin growth.	537	Lipomyces tetrasporus*
588	Starch formation.	Debaryomyces hansenii Pichia pseudopolymorpha	Candida humicola
589	Galactitol growth.	590	627
590	Ethanol growth.	591	601
591	Nitrate growth.	592	598
592	Glycerol growth.	593	596
593	Starch formation.	Bullera alba Cryptococcus flavus	594
594	Budding cells.	Sterigmatomyces penicillatus*	595
595	Filamentous.	Bullera alba Cryptococcus laurentii	Candida humicola Cryptococcus laurentii
596	Melibiose growth.	597	595
597	Lactose growth.	Sympodiomyces parvus*	Bullera alba
598	L–Arabinose growth.	Candida incommunis*	599
599	Cellobiose growth.	Sporobolomyces antarcticus*	600
600	Pink colonies.	Cryptococcus albidus	Cryptococcus macerans
601	Nitrate growth.	602	617
602	D–Galactose growth.	603	608
603	Sucrose growth.	604	607
604	Lactose growth.	605	Trichosporon cutaneum
605	L–Arabinose growth.	606	Candida marina
606	L–Rhamnose growth.	Cryptococcus cereanus*	Botryoascus synnaedendrus*
607	D–Xylose growth.	Saccharomycopsis fibuligera**	537
608	Melibiose growth.	609	612
609	Inulin growth.	610	Lipomyces tetrasporus*
610	L–Rhamnose growth.	Sarcinosporon inkin** Trichosporon cutaneum	611
611	Splitting cells.	Candida curvata	Trichosporon cutaneum
612	Lactose growth.	613	615
613	Inulin growth.	614	Lipomyces tetrasporus*
614	Septate hyphae.	Candida amylolenta*	Candida humicola
615	Inulin growth.	616	Lipomyces tetrasporus*
616	Splitting cells.	Candida humicola Cryptococcus laurentii	Trichosporon cutaneum
617	Melibiose growth.	618	623
618	D–Glucitol growth.	619	620
619	Salicin growth.	Candida tsukubaensis*	Cryptococcus albidus
620	L–Arabinose growth.	Candida incommunis*	621

	Negative	Positive
621 Lactose growth.	622	600
622 Growth without thiamin.	...Cryptococcus macerans	...Trichosporon terrestre*
623 Cellobiose growth.	...Sporobolomyces antarcticus*	624
624 Growth at 37 degrees.	625	626
625 Splitting cells.	...Cryptococcus albidus	...Trichosporon pullulans
626 Septate hyphae.	...Sterigmatomyces aphidis*	...Candida edax*
627 Ethanol growth.	628	643
628 Melibiose growth.	629	637
629 Lactose growth.	630	632
630 Growth at 37 degrees.	631	...Candida chiropterorum* Cryptococcus neoformans
631 Starch formation.	...Sterigmatomyces polyborus*	...Cryptococcus luteolus
632 Nitrate growth.	633	...Cryptococcus albidus
633 Starch formation.	634	636
634 Growth at 37 degrees.	635	...Candida hydrocarbofumarica*
635 Budding cells.	...Sterigmatomyces polyborus*	...Bullera alba
636 Budding cells.	...Sterigmatomyces penicillatus*	...Bullera alba Cryptococcus luteolus
637 Nitrate growth.	638	...Cryptococcus albidus
638 Starch formation.	639	640
639 Budding cells.	...Sterigmatomyces polyborus*	...Bullera alba Cryptococcus flavus
640 Inulin growth.	641	...Lipomyces starkeyi
641 Budding cells.	...Sterigmatomyces penicillatus* Bullera alba Cryptococcus laurentii Cryptococcus luteolus	642
642 Filamentous.		...Candida humicola Cryptococcus laurentii
643 Nitrate growth.	644	661
644 Melibiose growth.	645	652
645 Lactose growth.	646	648
646 Inulin growth.	647	...Lipomyces tetrasporus*
647 Growth at 37 degrees.	...Cryptococcus luteolus	...Candida chiropterorum* Cryptococcus neoformans
648 Inulin growth.	649	...Lipomyces tetrasporus*
649 Starch growth.	650	651
650 Splitting cells.	...Cryptococcus luteolus	...Trichosporon cutaneum
651 Splitting cells.	...Candida blankii Candida hydrocarbofumarica*	...Trichosporon cutaneum
652 Lactose growth.	653	658
653 Melezitose growth.	654	655
654 Starch formation.	...Stephanoascus ciferrii**	...Candida humicola
655 Inulin growth.	656	...Lipomyces starkeyi Lipomyces tetrasporus*
656 Starch growth.	657	614
657 Filamentous.	...Cryptococcus luteolus	...Candida humicola
658 Inulin growth.	659	...Lipomyces starkeyi Lipomyces tetrasporus*
659 Splitting cells.	660	...Trichosporon cutaneum
660 Filamentous.	...Cryptococcus laurentii Cryptococcus luteolus	...Candida humicola Cryptococcus laurentii
661 Melezitose growth.	...Trichosporon terrestre*	662
662 Growth at 37 degrees.	...Cryptococcus albidus	626

Key involving physiological tests only (Key No. 2)

Tests in Key No. 2
- 12 D–Galactose growth
- 13 L–Sorbose growth
- 14 D–Ribose growth
- 15 D–Xylose growth
- 16 L–Arabinose growth
- 17 D–Arabinose growth
- 18 L–Rhamnose growth
- 19 Sucrose growth
- 20 Maltose growth
- 21 Trehalose growth
- 22 Methyl α–D–glucopyranoside growth
- 23 Cellobiose growth
- 24 Salicin growth
- 26 Melibiose growth
- 27 Lactose growth
- 28 Raffinose growth
- 29 Melezitose growth
- 30 Inulin growth
- 31 Starch growth
- 32 Glycerol growth
- 33 Erythritol growth
- 35 Galactitol growth
- 36 D–Mannitol growth

Key No. 2: non-fermenting yeasts 89

37 D–Glucitol growth
38 myo–Inositol growth
39 D–δ–Gluconolactone growth
40 2–Ketogluconate growth
42 D,L–Lactate growth
43 Succinate growth
44 Citrate growth

45 Methanol growth
46 Ethanol growth
49 Nitrate growth
51 Growth without vitamins
52 Growth without inositol
54 Growth without biotin
55 Growth without thiamin

56 Growth without pyridoxine
59 Growth without p–aminobenzoate
60 Growth in 50% D–glucose
62 Growth at 37 degrees
65 Starch formation

number of different tests 42

KEY No. 2

		Negative	Positive
1	Erythritol growth.	2	394
2	myo–Inositol growth.	3	319
3	D–Mannitol growth.	4	99
4	Trehalose growth.	5	47
5	D–Galactose growth.	6	27
6	D–Glucitol growth.	7	23
7	Cellobiose growth.	8	18
8	Citrate growth.	9	16
9	Lactose growth.	10	Trichosporon cutaneum
10	Nitrate growth.	11	Leucosporidium antarcticum
11	D,L–Lactate growth.	12	14
12	Maltose growth.	13	Arthroascus javanensis** / Nadsonia commutata*
13	Growth at 37 degrees.	Arthroascus javanensis** / Candida valida / Pichia membranaefaciens	Candida valida / Geotrichum capitatum** / Pichia membranaefaciens
14	Growth without pyridoxine	Pichia membranaefaciens / Torulopsis inconspicua	15
15	Growth without vitamins.	Candida valida / Geotrichum capitatum** / Pichia membranaefaciens	Candida valida / Pichia membranaefaciens / Pichia scutulata*
16	Lactose growth.	17	Trichosporon cutaneum
17	Salicin growth.	Pichia membranaefaciens / Pichia terricola	Torulopsis dendrica*
18	Lactose growth.	19	Trichosporon cutaneum
19	D,L–Lactate growth.	20	22
20	Sucrose growth.	21	Saccharomycopsis fibuligera**
21	Nitrate growth.	Torulopsis dendrica*	Candida berthetii
22	Sucrose growth.	Candida norvegensis / Pichia norvegensis*	Saccharomycopsis fibuligera**
23	D–Xylose growth.	24	26
24	Methyl α–D–glucopyranoside growth	Nadsonia commutata* / Saccharomycopsis vini**	25
25	Starch growth.	Saccharomycopsis vini**	Saccharomycopsis fibuligera**
26	Lactose growth.	Hansenula canadensis	Trichosporon cutaneum
27	Glycerol growth.	28	32
28	Ethanol growth.	29	30
29	L–Arabinose growth.	Leucosporidium antarcticum	Lipomyces anomalus*
30	Lactose growth.	31	Trichosporon cutaneum
31	L–Arabinose growth.	Leucosporidium antarcticum	Guilliermondella selenospora**
32	D–Xylose growth.	33	44
33	D,L–Lactate growth.	34	41
34	Succinate growth.	35	38
35	Nitrate growth.	36	Leucosporidium antarcticum
36	Growth without vitamins.	37	Pichia humboldtii*
37	Growth at 37 degrees.	Candida vinaria*	Geotrichum capitatum**
38	Maltose growth.	39	Nadsonia commutata*
39	Growth without vitamins.	40	Pichia humboldtii*
40	Growth at 37 degrees.	Schizoblastosporion starkeyi–henricii / Torulopsis austromarina*	Geotrichum capitatum**
41	Cellobiose growth.	42	Pichia chambardii
42	Growth without vitamins.	43	Pichia humboldtii*
43	Growth at 37 degrees.	Zendera ovetensis	Geotrichum capitatum**
44	L–Sorbose growth.	30	45

	Negative	Positive
45 Lactose growth.	46	Trichosporon cutaneum
46 Growth without vitamins.	Candida vinaria*	Geotrichum penicillatum**
47 D−Glucitol growth.	48	85
48 D−Xylose growth.	49	65
49 Cellobiose growth.	50	61
50 Nitrate growth.	51	60
51 D,L−Lactate growth.	52	58
52 D−Galactose growth.	53	54
53 Growth at 37 degrees.	Arthroascus javanensis** Nadsonia commutata*	Nematospora coryli
54 Maltose growth.	55	57
55 Growth at 37 degrees.	56	Nematospora coryli
56 Growth without biotin.	Schizoblastosporion starkeyi−henricii Torulopsis austromarina*	Rhodosporidium dacryoidum* Schizoblastosporion starkeyi−henricii
57 Growth at 37 degrees.	Nadsonia commutata* Rhodosporidium dacryoidum*	Nematospora coryli
58 L−Sorbose growth.	Brettanomyces custersianus Rhodosporidium dacryoidum*	59
59 Growth at 37 degrees.	Rhodosporidium dacryoidum*	Trigonopsis variabilis
60 Melezitose growth.	Leucosporidium antarcticum	Rhodotorula glutinis
61 Nitrate growth.	62	Rhodotorula glutinis
62 L−Arabinose growth.	63	Phaffia rhodozyma*
63 Methyl α−D−glucopyranoside growth	Rhodosporidium dacryoidum*	64
64 Growth without vitamins.	Saccharomycopsis fibuligera**	Sporobolomyces pararoseus
65 Nitrate growth.	66	83
66 Glycerol growth.	67	73
67 Lactose growth.	68	Trichosporon aquatile* Trichosporon cutaneum
68 Raffinose growth.	Trichosporon aquatile*	69
69 L−Arabinose growth.	70	71
70 Starch growth.	Rhodotorula rubra	Sporobolomyces pararoseus
71 D−Galactose growth.	72	Rhodotorula rubra Sporobolomyces albo−rubescens
72 2−Ketogluconate growth.	Rhodotorula rubra	Phaffia rhodozyma*
73 Maltose growth.	74	77
74 D−Arabinose growth.	75	76
75 Lactose growth.	Trigonopsis variabilis	Trichosporon cutaneum
76 Growth without p−aminobenzoate	Rhodotorula minuta	Trichosporon cutaneum
77 Lactose growth.	78	Trichosporon cutaneum
78 D−Galactose growth.	79	80
79 L−Arabinose growth.	70	72
80 L−Arabinose growth.	81	82
81 Starch growth.	Rhodotorula rubra Torulopsis xestobii*	Sporobolomyces pararoseus
82 D−Arabinose growth.	Rhodotorula rubra Torulopsis xestobii*	Rhodotorula rubra Sporobolomyces albo−rubescens
83 Melezitose growth.	84	Rhodotorula glutinis
84 Succinate growth.	Leucosporidium antarcticum	Rhodotorula graminis
85 D−Xylose growth.	86	92
86 Melezitose growth.	87	61
87 D−Galactose growth.	88	90
88 Methyl α−D−glucopyranoside growth	89	25
89 D,L−Lactate growth.	Nadsonia commutata* Saccharomycopsis vini**	Brettanomyces custersianus
90 D,L−Lactate growth.	91	58
91 Maltose growth.	Rhodosporidium dacryoidum* Schizoblastosporion starkeyi−henricii	Nadsonia commutata* Rhodosporidium dacryoidum*
92 Nitrate growth.	93	98
93 Raffinose growth.	94	97
94 D−Arabinose growth.	95	76
95 Lactose growth.	96	Trichosporon cutaneum

Key No. 2: non-fermenting yeasts

		Negative	Positive
96	Sucrose growth.	......Trigonopsis variabilis	......Torulopsis multis−gemmis*
97	Lactose growth.	69	......Trichosporon cutaneum
98	Melezitose growth.	......Rhodotorula graminis	......Hansenula canadensis Rhodotorula glutinis
99	Nitrate growth.	100	241
100	Sucrose growth.	101	172
101	Glycerol growth.	102	116
102	Cellobiose growth.	103	109
103	D−Galactose growth.	104	107
104	D−Xylose growth.	105	......Trichosporon cutaneum
105	Trehalose growth.	106	......Torulopsis philyla*
106	D−δ−Gluconolactone growth	......Candida vini Pichia fluxuum	......Candida vini Pichia delftensis
107	Lactose growth.	108	......Trichosporon cutaneum
108	D,L−Lactate growth.	......Torulopsis sorbophila*	......Candida rugosa
109	Ethanol growth.	......Candida insectamans*	110
110	Galactitol growth.	111	115
111	D−Xylose growth.	112	113
112	D−Galactose growth.	......Sporobolomyces singularis	......Torulopsis insectalens*
113	Lactose growth.	114	......Trichosporon cutaneum
114	L−Arabinose growth.	......Brettanomyces naardenensis*	......Candida bogoriensis
115	Lactose growth.	......Torulopsis fujisanensis	......Trichosporon cutaneum
116	Methanol growth.	117	171
117	D−Galactose growth.	118	145
118	D−Xylose growth.	119	131
119	Citrate growth.	120	127
120	D,L−Lactate growth.	121	124
121	Maltose growth.	122	123
122	Lactose growth.	......Candida silvae Candida vini Saccharomycopsis vini**	......Sporobolomyces singularis
123	D−Ribose growth.	......Nadsonia commutata* Saccharomycopsis vini**	......Saccharomycopsis malanga*
124	Trehalose growth.	125	126
125	L−Sorbose growth.	......Candida silvae Candida vini	......Candida hylophila*
126	Growth at 37 degrees.	......Rhodotorula pallida	......Torulopsis silvatica*
127	Trehalose growth.	128	130
128	Lactose growth.	129	......Sporobolomyces singularis
129	2−Ketogluconate growth.	......Pichia quercuum	......Candida krissii*
130	D,L−Lactate growth.	......Candida iberica* Candida zeylanoides	......Rhodotorula pallida
131	Ethanol growth.	......Rhodotorula pallida Sporobolomyces gracilis	132
132	Galactitol growth.	133	143
133	Citrate growth.	134	138
134	Lactose growth.	135	......Trichosporon cutaneum
135	Maltose growth.	136	......Saccharomycopsis malanga*
136	L−Sorbose growth.	137	......Rhodotorula pallida Saccharomycopsis crataegensis*
137	Trehalose growth.	......Torulopsis maris	......Rhodotorula pallida
138	D,L−Lactate growth.	139	141
139	Lactose growth.	140	......Trichosporon cutaneum
140	Maltose growth.	......Candida zeylanoides	......Candida bogoriensis
141	Lactose growth.	142	......Trichosporon cutaneum
142	Trehalose growth.	......Pichia salictaria	......Rhodotorula pallida
143	Lactose growth.	144	......Trichosporon cutaneum
144	L−Arabinose growth.	......Rhodotorula pallida	......Torulopsis fujisanensis
145	D,L−Lactate growth.	146	159
146	Galactitol growth.	147	115
147	Citrate growth.	148	153
148	Lactose growth.	149	......Trichosporon cutaneum
149	Maltose growth.	150	......Nadsonia commutata* Rhodosporidium dacryoidum*

		Negative	Positive
150	Trehalose growth.	151	........Rhodosporidium dacryoidum* Schizoblastosporion starkeyi–henricii
151	Succinate growth.	...............Candida vinaria*	152
152	Growth at 37 degrees.	...............Schizoblastosporion starkeyi–henricii	...............Torulopsis sorbophila*
153	D–Xylose growth.	154	157
154	L–Sorbose growth.	155	156
155	Salicin growth.	...............Rhodosporidium dacryoidum*	...............Candida savonica* Candida zeylanoides
156	Salicin growth.	...............Candida iberica* Rhodosporidium dacryoidum*	...............Candida iberica* Candida zeylanoides
157	Lactose growth.	158	...............Trichosporon cutaneum
158	Maltose growth.	...............Candida savonica* Candida zeylanoides	...............Candida bogoriensis
159	Trehalose growth.	160	164
160	Lactose growth.	161	...............Trichosporon cutaneum
161	Cellobiose growth.	162	...............Geotrichum fermentans**
162	Maltose growth.	...............Candida catenulata Candida rugosa	163
163	Growth at 37 degrees.	...............Candida brumptii	...............Candida catenulata
164	Lactose growth.	165	...............Trichosporon cutaneum
165	L–Sorbose growth.	166	170
166	D–Xylose growth.	167	168
167	Growth at 37 degrees.	...............Rhodosporidium dacryoidum* Rhodotorula pallida	...............Candida catenulata
168	Maltose growth.	169	...............Candida catenulata Candida ravautii
169	Growth without p–aminobenzoate.	...............Rhodotorula pallida	...............Candida catenulata
170	Growth at 37 degrees.	...............Rhodosporidium dacryoidum* Rhodotorula pallida	...............Trigonopsis variabilis
171	L–Rhamnose growth.	...............Nadsonia commutata*	...............Pichia lindnerii* Torulopsis methanolovescens*
172	Ethanol growth.	173	192
173	Raffinose growth.	174	184
174	Glycerol growth.	...............Candida suecica*	175
175	Galactitol growth.	176	183
176	D–Glucitol growth.	177	179
177	D–Xylose growth.	178	...............Rhodotorula minuta
178	Methyl α–D–glucopyranoside growth	...............Nadsonia commutata* Rhodosporidium dacryoidum*	...............Metschnikowia krissii
179	Maltose growth.	180	181
180	L–Arabinose growth.	...............Rhodosporidium dacryoidum* Rhodotorula pallida	...............Rhodotorula minuta
181	Cellobiose growth.	...............Nadsonia commutata* Rhodosporidium dacryoidum*	182
182	Salicin growth.	...............Rhodosporidium dacryoidum*	...............Metschnikowia lunata*
183	L–Arabinose growth.	...............Rhodotorula pallida	...............Rhodotorula marina
184	L–Arabinose growth.	185	188
185	Melezitose growth.	186	187
186	Maltose growth.	...............Torulopsis apis	...............Nadsonia commutata*
187	Melibiose growth.	70	...............Lipomyces kononenkoae
188	Galactitol growth.	189	190
189	Maltose growth.	...............Rhodotorula pilimanae	71
190	Melezitose growth.	...............Rhodotorula pilimanae	191
191	Growth without p–aminobenzoate.	...............Rhodotorula marina	...............Rhodotorula rubra
192	Raffinose growth.	193	222
193	D–Xylose growth.	194	206
194	Glycerol growth.	195	196
195	Methyl α–D–glucopyranoside growth	...............Torulopsis auriculariae*	...............Aciculoconidium aculeatum**
196	D–Galactose growth.	197	202
197	Methyl α–D–glucopyranoside growth	198	199
198	D,L–Lactate growth.	...............Nadsonia commutata* Saccharomycopsis vini**	...............Rhodotorula pallida
199	Starch growth.	200	201

Key No. 2: non-fermenting yeasts

	Negative	Positive
200 D–Glucitol growth.	Metschnikowia krissii	Saccharomycopsis vini**
201 Growth at 37 degrees.	Aciculoconidium aculeatum**	Saccharomycopsis fibuligera**
202 Methyl α–D–glucopyranoside growth	203	204
203 Maltose growth.	Rhodosporidium dacryoidum* Rhodotorula pallida	181
204 Cellobiose growth.	Lodderomyces elongisporus	205
205 Growth at 37 degrees.	Metschnikowia zobellii	Metschnikowia lunata*
206 Galactitol growth.	207	221
207 Maltose growth.	208	210
208 L–Arabinose growth.	209	76
209 Lactose growth.	Rhodotorula pallida	Trichosporon cutaneum
210 Lactose growth.	211	Candida fluviotilis* Trichosporon cutaneum
211 D–Galactose growth.	212	214
212 Citrate growth.	Candida tepae*	213
213 Growth without pyridoxine	Candida quercuum*	Pichia toletana
214 Melezitose growth.	Candida tepae*	215
215 L–Arabinose growth.	216	220
216 Cellobiose growth.	217	218
217 Salicin growth.	Lodderomyces elongisporus	Pichia vini
218 Citrate growth.	205	219
219 Starch growth.	Metschnikowia lunata*	Pichia vini
220 Starch growth.	Pichia etchellsii	Pichia vini
221 L–Arabinose growth.	209	Aessosporon dendrophilum* Trichosporon cutaneum
222 D–Xylose growth.	223	229
223 Melezitose growth.	224	227
224 D–Glucitol growth.	225	24
225 Maltose growth.	Debaryomyces tamarii	226
226 Methyl α–D–glucopyranoside growth	Nadsonia commutata*	Saccharomycopsis fibuligera**
227 L–Arabinose growth.	228	Phaffia rhodozyma*
228 Melibiose growth.	64	Lipomyces kononenkoae
229 Lactose growth.	230	239
230 L–Arabinose growth.	231	235
231 Trehalose growth.	Debaryomyces yarrowii*	232
232 Succinate growth.	Lipomyces kononenkoae	233
233 Starch growth.	Rhodotorula rubra	234
234 Growth without vitamins.	Pichia vini	Sporobolomyces pararoseus
235 Maltose growth.	Rhodotorula pilimanae	236
236 D–Galactose growth.	72	237
237 Starch growth.	Rhodotorula rubra Sporobolomyces albo–rubescens	238
238 Growth without thiamin.	Pichia vini	Debaryomyces hansenii
239 Growth without inositol.	Candida glaebosa	240
240 Growth in 50% D–glucose.	Trichosporon cutaneum	Debaryomyces hansenii
241 Melezitose growth.	242	287
242 Methanol growth.	243	285
243 Galactitol growth.	244	277
244 Sucrose growth.	245	264
245 D–Glucitol growth.	246	247
246 Growth without vitamins.	Torulopsis halonitratophila	Leucosporidium antarcticum
247 Trehalose growth.	248	251
248 Cellobiose growth.	249	250
249 L–Sorbose growth.	Torulopsis halonitratophila	Wickerhamiella domercqii**
250 D–Xylose growth.	Hansenula dryadoides*	Torulopsis norvegica
251 Glycerol growth.	252	256
252 L–Sorbose growth.	253	255
253 Growth at 37 degrees.	254	Candida foliarum
254 Growth without vitamins.	Sterigmatomyces nectairii*	Sporobolomyces hispanicus
255 L–Arabinose growth.	Sporobolomyces hispanicus	Torulopsis pustula*
256 Ethanol growth.	257	259

		Negative	Positive
257	L–Arabinose growth.	Sporobolomyces hispanicus / Sporobolomyces odorus	258
258	Cellobiose growth.	Torulopsis vanderwaltii	Torulopsis pustula*
259	L–Arabinose growth.	260	261
260	Growth at 37 degrees.	Rhodotorula araucariae* / Sporobolomyces hispanicus / Sporobolomyces odorus	Candida foliarum / Sporobolomyces odorus
261	L–Sorbose growth.	262	263
262	Growth at 37 degrees.	Rhodotorula araucariae*	Candida foliarum
263	D–Ribose growth.	Rhodotorula araucariae*	Torulopsis pustula*
264	Maltose growth.	265	271
265	D–Glucitol growth.	84	266
266	Lactose growth.	267	Leucosporidium frigidum
267	Trehalose growth.	Wickerhamiella domercqii**	268
268	D–Galactose growth.	Aessosporon salmonicolor** / Sporobolomyces odorus / Sporobolomyces salmonicolor	269
269	2–Ketogluconate growth.	Aessosporon salmonicolor** / Sporobolomyces salmonicolor	270
270	Raffinose growth.	Torulopsis vanderwaltii	Rhodotorula graminis
271	D–Glucitol growth.	272	275
272	Succinate growth.	273	274
273	Salicin growth.	Leucosporidium antarcticum	Rhodosporidium malvinellum*
274	Growth without vitamins.	Rhodosporidium malvinellum*	Rhodotorula graminis
275	Growth without vitamins.	Rhodosporidium malvinellum*	276
276	2–Ketogluconate growth.	Sporobolomyces holsaticus	Rhodotorula graminis
277	Ethanol growth.	278	281
278	Raffinose growth.	Torulopsis vanderwaltii	279
279	Growth without vitamins.	Rhodosporidium malvinellum*	280
280	2–Ketogluconate growth.	Sporidibolus ruinenii	Rhodotorula graminis
281	Glycerol growth.	282	283
282	Lactose growth.	Rhodotorula graminis	Leucosporidium frigidum
283	Sucrose growth.	284	274
284	D–Ribose growth.	Rhodotorula araucariae*	Candida javanica
285	D–Galactose growth.	286	Torulopsis nitratophila
286	D–Xylose growth.	Hansenula nonfermentans	Hansenula minuta
287	Glycerol growth.	288	293
288	Melibiose growth.	289	292
289	Succinate growth.	290	291
290	D–Galactose growth.	Candida buffonii	Leucosporidium stokesii
291	Starch formation.	Rhodotorula glutinis / Sporobolomyces holsaticus	Cryptococcus bhutanensis*
292	L–Rhamnose growth.	Candida aquatica	Leucosporidium gelidum
293	Raffinose growth.	294	300
294	L–Rhamnose growth.	295	Candida melinii / Hansenula beckii** / Hansenula bimundalis / Hansenula canadensis / Hansenula wingei / Rhodotorula glutinis
295	Sucrose growth.	Candida buffonii	296
296	Lactose growth.	297	299
297	Methyl α–D–glucopyranoside growth	298	Rhodotorula glutinis / Sporidibolus johnsonii / Sporobolomyces holsaticus
298	2–Ketogluconate growth.	Rhodotorula glutinis / Sporobolomyces holsaticus	Rhodotorula aurantiaca / Rhodotorula glutinis
299	D–Galactose growth.	Bullera tsugae	Rhodotorula aurantiaca
300	Ethanol growth.	301	311
301	Lactose growth.	302	308
302	Melibiose growth.	303	Rhodotorula lactosa
303	D–Arabinose growth.	304	307

Key No. 2: non-fermenting yeasts

		Negative	Positive
304	D,L−Lactate growth.	.305	.306
305	Growth without biotin.	Sporobolomyces puniceus*	Rhodotorula glutinis Sporobolomyces holsaticus
306	Starch growth.	Rhodotorula glutinis	Torulopsis ingeniosa
307	Starch growth.	Rhodotorula glutinis Sporobolomyces holsaticus	Sporobolomyces holsaticus Sporobolomyces roseus
308	L−Arabinose growth.	.309	.310
309	D,L−Lactate growth.	Candida muscorum	Torulopsis ingeniosa
310	L−Sorbose growth.	Rhodotorula lactosa	Torulopsis fragaria*
311	Galactitol growth.	.312	.318
312	Melibiose growth.	.313	.317
313	Lactose growth.	.314	Torulopsis fragaria*
314	D−Arabinose growth.	.315	.307
315	D,L−Lactate growth.	Rhodotorula glutinis Sporobolomyces holsaticus	.316
316	Inulin growth.	Rhodotorula glutinis	Hansenula jadinii
317	D−Galactose growth.	Rhodotorula lactosa	Candida aquatica
318	Succinate growth.	Leucosporidium scottii	Rhodotorula glutinis
319	Glycerol growth.	.320	.356
320	Nitrate growth.	.321	.344
321	Galactitol growth.	.322	.333
322	Ethanol growth.	.323	.329
323	Lactose growth.	.324	.326
324	Maltose growth.	Cryptococcus skinneri	.325
325	D−δ−Gluconolactone growth	Cryptococcus uniguttulatus	Cryptococcus gastricus
326	Methyl α−D−glucopyranoside growth	.327	.328
327	Melibiose growth.	Cryptococcus gastricus	Cryptococcus magnus**
328	Melibiose growth.	Bullera alba Cryptococcus ater	Bullera alba Cryptococcus magnus**
329	Maltose growth.	.330	.331
330	Lactose growth.	Cryptococcus skinneri	Cryptococcus dimennae Trichosporon cutaneum
331	Melibiose growth.	.332	Cryptococcus magnus** Trichosporon cutaneum
332	Lactose growth.	.325	Cryptococcus gastricus Trichosporon cutaneum
333	Ethanol growth.	.334	.340
334	D−Mannitol growth.	.335	.336
335	Growth without vitamins.	Bullera alba Cryptococcus magnus**	Candida podzolica* Cryptococcus magnus**
336	Sucrose growth.	Cryptococcus skinneri	.337
337	Lactose growth.	.338	.339
338	Growth at 37 degrees.	Cryptococcus hungaricus Cryptococcus luteolus	Cryptococcus neoformans
339	D,L−Lactate growth.	Bullera alba Cryptococcus luteolus Cryptococcus magnus**	Bullera alba Cryptococcus hungaricus Cryptococcus luteolus
340	D−Mannitol growth.	Candida podzolica* Cryptococcus magnus** Trichosporon cutaneum	.341
341	Maltose growth.	.330	.342
342	Lactose growth.	.343	Cryptococcus luteolus Cryptococcus magnus** Trichosporon cutaneum
343	Growth at 37 degrees.	Cryptococcus luteolus	Cryptococcus neoformans
344	Melibiose growth.	.345	.351
345	Ethanol growth.	.346	.348
346	Sucrose growth.	.347	Cryptococcus albidus Rhodosporidium infirmo−miniatum**
347	Citrate growth.	Cryptococcus terreus	Cryptococcus himalayensis*
348	Maltose growth.	.349	.350
349	D−Galactose growth.	Cryptococcus kuetzingii	Leucosporidium frigidum
350	Lactose growth.	Leucosporidium stokesii Rhodosporidium infirmo−miniatum**	Cryptococcus albidus Rhodosporidium infirmo−miniatum**
351	Lactose growth.	.352	.354

		Negative	Positive
352	L–Rhamnose growth.	353	Leucosporidium gelidum Rhodosporidium bisporidiis*
353	Maltose growth.	Leucosporidium nivalis	Candida valdiviana*
354	Starch formation.	355	Cryptococcus albidus Cryptococcus heveanensis*
355	Methyl α–D–glucopyranoside growth	Bullera piricola*	Candida valdiviana*
356	Nitrate growth.	357	386
357	Galactitol growth.	358	377
358	D–Glucitol growth.	359	367
359	Ethanol growth.	360	361
360	L–Rhamnose growth.	Cryptococcus uniguttulatus	Bullera alba
361	D–Xylose growth.	362	364
362	D–Galactose growth.	Saccharomycopsis fibuligera**	363
363	Maltose growth.	Cryptococcus melibiosum	Trichosporon brassicae*
364	Lactose growth.	365	Trichosporon cutaneum
365	L–Sorbose growth.	Cryptococcus uniguttulatus	366
366	D–Galactose growth.	Cryptococcus lactativorus	Trichosporon brassicae*
367	Lactose growth.	368	373
368	L–Rhamnose growth.	369	Cryptococcus skinneri
369	L–Sorbose growth.	370	Trichosporon brassicae*
370	D–Xylose growth.	371	372
371	Growth at 37 degrees.	Filobasidium capsuligenum**	Saccharomycopsis fibuligera**
372	Melezitose growth.	Filobasidium capsuligenum**	Cryptococcus uniguttulatus
373	Ethanol growth.	374	375
374	Sucrose growth.	Pichia abadieae*	Bullera alba Cryptococcus magnus**
375	Sucrose growth.	Pichia abadieae* Trichosporon cutaneum	376
376	Maltose growth.	Cryptococcus dimennae Trichosporon cutaneum	Cryptococcus magnus** Trichosporon cutaneum
377	Methanol growth.	378	Hansenula ofunaensis*
378	Ethanol growth.	379	381
379	Sucrose growth.	380	337
380	Cellobiose growth.	Pichia abadieae*	Cryptococcus skinneri
381	Maltose growth.	382	384
382	Lactose growth.	Cryptococcus skinneri	383
383	D–Ribose growth.	Pichia abadieae* Trichosporon cutaneum	Cryptococcus dimennae Trichosporon cutaneum
384	Lactose growth.	385	Cryptococcus luteolus Cryptococcus magnus** Trichosporon cutaneum
385	L–Rhamnose growth.	Filobasidium capsuligenum**	343
386	Melibiose growth.	387	392
387	Galactitol growth.	388	390
388	Maltose growth.	Cryptococcus kuetzingii	389
389	Methyl α–D–glucopyranoside growth	Cryptococcus albidus Rhodosporidium infirmo–miniatum**	Cryptococcus albidus Filobasidium floriforme*
390	Methyl α–D–glucopyranoside growth	391	Cryptococcus albidus Filobasidium floriforme*
391	Growth without biotin.	Rhodosporidium capitatum* Rhodosporidium infirmo–miniatum**	Cryptococcus albidus Rhodosporidium infirmo–miniatum**
392	Lactose growth.	393	354
393	L–Rhamnose growth.	353	Rhodosporidium bisporidiis*
394	myo–Inositol growth.	395	523
395	Methanol growth.	396	517
396	Galactitol growth.	397	480
397	Melezitose growth.	398	436
398	Maltose growth.	399	419
399	D–Mannitol growth.	400	404

Key No. 2: non-fermenting yeasts

	Negative	Positive
400 Ethanol growth.	401	402
401 Methyl α–D–glucopyranoside growth	Candida flavificans*	Sympodiomyces parvus*
402 Lactose growth.	403	Trichosporon cutaneum
403 D,L–Lactate growth.	Candida flavificans*	Saccharomycopsis lipolytica**
404 Ethanol growth.	405	409
405 Succinate growth.	406	407
406 Methyl α–D–glucopyranoside growth	Torulopsis psychrophila*	Sympodiomyces parvus*
407 Lactose growth.	408	Sterigmatomyces elviae*
408 D,L–Lactate growth.	Torulopsis schatavii*	Rhodotorula pallida
409 L–Arabinose growth.	410	414
410 Trehalose growth.	411	413
411 D–Xylose growth.	Saccharomycopsis lipolytica**	412
412 Lactose growth.	Trichosporon eriense*	Trichosporon cutaneum
413 Lactose growth.	408	Trichosporon cutaneum
414 Lactose growth.	415	418
415 D–Ribose growth.	Torulopsis psychrophila*	416
416 Nitrate growth.	417	Sterigmatomyces halophilus
417 Salicin growth.	Torulopsis schatavii*	Sterigmatomyces indicus
418 Nitrate growth.	Sterigmatomyces elviae* Trichosporon cutaneum	Sterigmatomyces halophilus
419 Ethanol growth.	420	422
420 D–Xylose growth.	Pichia ambrosiae*	421
421 Melibiose growth.	Sympodiomyces parvus*	Candida humicola
422 Sucrose growth.	423	429
423 D–Galactose growth.	424	426
424 Lactose growth.	425	Trichosporon cutaneum
425 Methyl α–D–glucopyranoside growth	Saccharomycopsis malanga*	Saccharomycopsis capsularis**
426 Lactose growth.	427	Trichosporon cutaneum
427 D–Ribose growth.	Pichia media	428
428 D–Arabinose growth.	Debaryomyces coudertii	Pichia castillae*
429 D–Xylose growth.	430	433
430 D–Ribose growth.	431	432
431 Starch growth.	Candida mesenterica	Saccharomycopsis fibuligera**
432 Growth without thiamin.	Pichia ambrosiae*	Candida mesenterica
433 Melibiose growth.	434	Candida humicola Trichosporon cutaneum
434 Lactose growth.	435	Trichosporon cutaneum
435 Raffinose growth.	Candida diddensii	Hyphopichia burtonii**
436 Raffinose growth.	437	457
437 L–Arabinose growth.	438	449
438 D–Xylose growth.	439	443
439 Ethanol growth.	440	441
440 L–Sorbose growth.	Debaryomyces melissophila*	Candida graminis*
441 Nitrate growth.	442	Candida diffluens
442 D–Galactose growth.	431	Debaryomyces melissophila*
443 Ethanol growth.	444	446
444 Melibiose growth.	445	Candida humicola
445 L–Sorbose growth.	Ambrosiozyma philentoma*	Candida graminis*
446 Nitrate growth.	447	Hormoascus platypodis**
447 Melibiose growth.	448	Candida humicola Trichosporon cutaneum
448 Lactose growth.	Ambrosiozyma philentoma*	Trichosporon cutaneum
449 L–Rhamnose growth.	450	454
450 Melibiose growth.	451	Candida humicola Trichosporon cutaneum
451 D–Galactose growth.	452	453
452 Lactose growth.	Ambrosiozyma monospora**	Trichosporon cutaneum
453 Lactose growth.	Candida aaseri Candida diddensii	Candida aaseri Trichosporon cutaneum
454 Melibiose growth.	455	Candida humicola Trichosporon cutaneum
455 Lactose growth.	456	Trichosporon cutaneum

		Negative	Positive
456	Starch growth.	Candida diddensii Candida terebra*	Candida naeodendra*
457	Nitrate growth.	458	478
458	Glycerol growth.	459	462
459	D–Ribose growth.	460	461
460	Inulin growth.	Lipomyces lipofer Trichosporon cutaneum	Lipomyces lipofer Lipomyces tetrasporus*
461	Inulin growth.	Candida humicola Trichosporon cutaneum	Lipomyces tetrasporus*
462	D–Xylose growth.	Saccharomycopsis fibuligera**	463
463	Lactose growth.	464	472
464	Melibiose growth.	465	469
465	D,L–Lactate growth.	466	468
466	Inulin growth.	467	Debaryomyces marama Lipomyces tetrasporus*
467	Growth without vitamins.	Debaryomyces marama	Hyphopichia burtonii**
468	Growth in 50% D–glucose.	Debaryomyces marama Lipomyces tetrasporus*	Debaryomyces hansenii Debaryomyces marama
469	Starch formation.	470	471
470	Inulin growth.	Debaryomyces hansenii Debaryomyces marama Debaryomyces nepalensis*	468
471	Inulin growth.	Candida humicola	Lipomyces tetrasporus*
472	Succinate growth.	473	475
473	L–Sorbose growth.	474	461
474	Growth without biotin.	Pichia scolyti Trichosporon cutaneum	Candida humicola Trichosporon cutaneum
475	Inulin growth.	476	468
476	Starch formation.	477	Candida humicola Trichosporon cutaneum
477	Growth in 50% D–glucose.	Debaryomyces marama Trichosporon cutaneum	Debaryomyces hansenii Debaryomyces marama
478	D–Galactose growth.	Torulopsis bacarum*	479
479	Salicin growth.	Rhodotorula acheniorum*	Trichosporon pullulans
480	Glycerol growth.	481	488
481	Ethanol growth.	482	485
482	D–Ribose growth.	Lipomyces lipofer Lipomyces starkeyi	483
483	Inulin growth.	484	Lipomyces starkeyi
484	Starch formation.	Sterigmatomyces polyborus*	Candida humicola
485	D–Ribose growth.	486	487
486	Inulin growth.	Lipomyces lipofer Trichosporon cutaneum	Lipomyces lipofer Lipomyces starkeyi Lipomyces tetrasporus*
487	Inulin growth.	Candida humicola Trichosporon cutaneum	Lipomyces starkeyi Lipomyces tetrasporus*
488	Ethanol growth.	489	492
489	Melibiose growth.	490	483
490	Maltose growth.	Rhodotorula pallida	491
491	Raffinose growth.	Candida terebra*	Sterigmatomyces polyborus*
492	Methyl α–D–glucopyranoside growth	493	498
493	Maltose growth.	494	496
494	Lactose growth.	495	Trichosporon cutaneum
495	L–Arabinose growth.	Rhodotorula pallida	Pichia haplophila
496	Lactose growth.	497	Trichosporon cutaneum
497	D–Ribose growth.	Pichia media	Pichia castillae*
498	Melibiose growth.	499	505
499	Raffinose growth.	500	501
500	Lactose growth.	Candida terebra*	Trichosporon cutaneum
501	D,L–Lactate growth.	502	503
502	Inulin growth.	Candida blankii Trichosporon cutaneum	Lipomyces tetrasporus*
503	Growth in 50% D–glucose.	504	Debaryomyces hansenii
504	Inulin growth.	Trichosporon cutaneum	Lipomyces tetrasporus*
505	D,L–Lactate growth.	506	512
506	Lactose growth.	507	510

Key No. 2: non-fermenting yeasts

	Negative	Positive
507 Starch formation.	508	509
508 Growth in 50% D−glucose.	......Debaryomyces vanriji Lipomyces tetrasporus*	......Debaryomyces phaffii Debaryomyces vanriji
509 Inulin growth.	......Candida humicola	......Lipomyces starkeyi Lipomyces tetrasporus*
510 Inulin growth.	......Candida humicola Trichosporon cutaneum	511
511 Growth in 50% D−glucose.	......Lipomyces starkeyi Lipomyces tetrasporus*	......Debaryomyces phaffii
512 Lactose growth.	513	515
513 Starch formation.	514	471
514 Growth in 50% D−glucose.	......Debaryomyces vanriji Lipomyces tetrasporus*	......Debaryomyces hansenii Debaryomyces vanriji
515 Growth in 50% D−glucose.	461	516
516 Starch formation.	......Debaryomyces hansenii Pichia pseudopolymorpha	......Candida humicola
517 Galactitol growth.	518	......Torulopsis nemodendra*
518 Cellobiose growth.	519	521
519 D−Arabinose growth.	520	......Hansenula wickerhamii
520 Succinate growth.	......Torulopsis pinus	......Hansenula philodendra*
521 Nitrate growth.	......Pichia pinus	522
522 Growth without thiamin.	......Hansenula glucozyma	......Hansenula henricii
523 Galactitol growth.	524	555
524 Ethanol growth.	525	534
525 Nitrate growth.	526	531
526 Glycerol growth.	527	529
527 Raffinose growth.	......Candida humicola Sterigmatomyces penicillatus*	528
528 Starch formation.	......Bullera alba Cryptococcus flavus	......Bullera alba Candida humicola Cryptococcus laurentii
529 Melibiose growth.	530	......Bullera alba Candida humicola Cryptococcus laurentii
530 Lactose growth.	......Sympodiomyces parvus*	......Bullera alba
531 L−Arabinose growth.	......Candida incommunis*	532
532 Cellobiose growth.	......Sporobolomyces antarcticus*	533
533 Growth without biotin.	......Cryptococcus macerans	......Cryptococcus albidus
534 Nitrate growth.	535	547
535 D−Galactose growth.	536	541
536 Sucrose growth.	537	540
537 Lactose growth.	538	......Trichosporon cutaneum
538 L−Sorbose growth.	......Botryoascus synnaedendrus*	539
539 L−Arabinose growth.	......Cryptococcus cereanus*	......Candida marina
540 D−Xylose growth.	......Saccharomycopsis fibuligera**	......Candida humicola Trichosporon cutaneum
541 Melibiose growth.	542	544
542 Raffinose growth.	......Sarcinosporon inkin** Trichosporon cutaneum	543
543 Inulin growth.	......Candida curvata Trichosporon cutaneum	......Lipomyces tetrasporus*
544 Lactose growth.	545	546
545 Inulin growth.	......Candida amylolenta* Candida humicola	......Lipomyces tetrasporus*
546 Inulin growth.	......Candida humicola Cryptococcus laurentii Trichosporon cutaneum	......Lipomyces tetrasporus*
547 Melibiose growth.	548	553
548 D−Mannitol growth.	549	550
549 Salicin growth.	......Candida tsukubaensis*	......Cryptococcus albidus
550 L−Arabinose growth.	......Candida incommunis*	551
551 Lactose growth.	552	533
552 Growth without vitamins.	......Cryptococcus macerans	......Trichosporon terrestre*
553 Cellobiose growth.	......Sporobolomyces antarcticus*	554

		Negative	Positive
554	Growth at 37 degrees.	Cryptococcus albidus / Trichosporon pullulans	Candida edax* / Sterigmatomyces aphidis*
555	Ethanol growth.	556	572
556	Melibiose growth.	557	566
557	Lactose growth.	558	561
558	Succinate growth.	Candida chiropterorum* / Cryptococcus neoformans	559
559	Growth at 37 degrees.	560	Cryptococcus neoformans
560	Starch formation.	Sterigmatomyces polyborus*	Cryptococcus luteolus
561	Nitrate growth.	562	Cryptococcus albidus
562	Raffinose growth.	563	564
563	Starch growth.	Sterigmatomyces penicillatus*	Candida hydrocarbofumarica*
564	Growth at 37 degrees.	565	Candida hydrocarbofumarica*
565	Starch formation.	Bullera alba / Sterigmatomyces polyborus*	Bullera alba / Cryptococcus luteolus
566	Nitrate growth.	567	Cryptococcus albidus
567	Raffinose growth.	Candida humicola / Sterigmatomyces penicillatus*	568
568	L–Sorbose growth.	569	570
569	Starch formation.	Bullera alba / Cryptococcus flavus / Sterigmatomyces polyborus*	Bullera alba / Candida humicola / Cryptococcus laurentii / Cryptococcus luteolus
570	Inulin growth.	571	Lipomyces starkeyi
571	Starch formation.	Bullera alba / Sterigmatomyces polyborus*	Bullera alba / Candida humicola / Cryptococcus laurentii / Cryptococcus luteolus
572	Nitrate growth.	573	586
573	Melibiose growth.	574	580
574	Lactose growth.	575	578
575	Succinate growth.	576	577
576	Inulin growth.	Candida chiropterorum* / Cryptococcus neoformans	Lipomyces tetrasporus*
577	Inulin growth.	343	Lipomyces tetrasporus*
578	Inulin growth.	579	Lipomyces tetrasporus*
579	Starch growth.	Cryptococcus luteolus / Trichosporon cutaneum	Candida blankii / Candida hydrocarbofumarica* / Trichosporon cutaneum
580	Lactose growth.	581	585
581	Melezitose growth.	582	583
582	Starch formation.	Stephanoascus ciferrii**	Candida humicola
583	Inulin growth.	584	Lipomyces starkeyi / Lipomyces tetrasporus*
584	Starch growth.	Candida humicola / Cryptococcus luteolus	Candida amylolenta* / Candida humicola
585	Inulin growth.	Candida humicola / Cryptococcus laurentii / Cryptococcus luteolus / Trichosporon cutaneum	Lipomyces starkeyi / Lipomyces tetrasporus*
586	Melezitose growth.	Trichosporon terrestre*	587
587	Growth at 37 degrees.	Cryptococcus albidus	Candida edax* / Sterigmatomyces aphidis*

All yeasts that ferment D-glucose (test No. 1 positive) (Keys No. 3 and No. 4)

Yeasts in Keys No. 3 and No. 4

- 1 Aciculoconidium aculeatum**
- 4 Ambrosiozyma cicatricosa*
- 5 Ambrosiozyma monospora**
- 6 Ambrosiozyma philentoma*
- 9 Brettanomyces abstinens*
- 10 Brettanomyces anomalus
- 11 Brettanomyces claussenii
- 12 Brettanomyces custersianus
- 13 Brettanomyces custersii
- 14 Brettanomyces lambicus
- 15 Brettanomyces naardenensis*
- 20 Candida albicans
- 22 Candida aquatica
- 23 Candida beechii
- 24 Candida berthetii
- 25 Candida blankii
- 27 Candida boidinii
- 28 Candida boleticola*
- 29 Candida bombi*
- 30 Candida brassicae*

31 Candida brumptii	157 Dekkera bruxellensis	255 Pichia farinosa
33 Candida buinensis*	158 Dekkera intermedia	256 Pichia fermentans
34 Candida butyri*	159 Filobasidium capsuligenum**	257 Pichia fluxuum
35 Candida cacaoi	162 Geotrichum fermentans**	258 Pichia guilliermondii
36 Candida catenulata	163 Geotrichum penicillatum**	259 Pichia haplophila
37 Candida chilensis*	164 Guilliermondella selenospora**	260 Pichia heimii*
39 Candida citrea*	165 Hanseniaspora guilliermondii*	262 Pichia kluyveri
40 Candida conglobata	166 Hanseniaspora occidentalis*	263 Pichia kudriavzevii
41 Candida curiosa	167 Hanseniaspora osmophila	264 Pichia lindnerii*
43 Candida dendronema*	168 Hanseniaspora uvarum	266 Pichia membranaefaciens
44 Candida diddensii	169 Hanseniaspora valbyensis	267 Pichia methanolica*
46 Candida diversa	170 Hanseniaspora vineae*	268 Pichia mucosa*
48 Candida entomaea*	171 Hansenula anomala	269 Pichia naganishii*
49 Candida entomophila*	172 Hansenula beckii**	270 Pichia nakazawae*
50 Candida ergatensis*	173 Hansenula beijerinckii	271 Pichia norvegensis*
51 Candida flavificans*	174 Hansenula bimundalis	272 Pichia ohmeri
52 Candida fluviotilis*	175 Hansenula californica	273 Pichia onychis
54 Candida fragicola*	177 Hansenula capsulata	274 Pichia pastoris
55 Candida freyschussii	178 Hansenula ciferrii	275 Pichia philogaea*
56 Candida friedrichii	179 Hansenula dimennae	276 Pichia pijperi
59 Candida homilentoma*	181 Hansenula fabianii	277 Pichia pinus
63 Candida iberica*	182 Hansenula glucozyma	278 Pichia pseudopolymorpha
64 Candida incommunis*	183 Hansenula henricii	279 Pichia quercuum
65 Candida inositophila*	184 Hansenula holstii	280 Pichia rabaulensis*
66 Candida insectamans*	185 Hansenula jadinii	281 Pichia rhodanensis
67 Candida insectorum*	186 Hansenula lynferdii*	282 Pichia saitoi
68 Candida intermedia	187 Hansenula minuta	284 Pichia sargentensis*
69 Candida ishiwadae*	188 Hansenula mrakii	285 Pichia scolyti
71 Candida kefyr	189 Hansenula muscicola*	286 Pichia scutulata*
72 Candida krissii*	192 Hansenula petersonii	287 Pichia spartinae*
73 Candida lambica	193 Hansenula philodendra*	288 Pichia stipitis
74 Candida lusitaniae	194 Hansenula polymorpha	289 Pichia strasburgensis
75 Candida macedoniensis	195 Hansenula saturnus	290 Pichia terricola
76 Candida maltosa*	196 Hansenula silvicola	291 Pichia toletana
78 Candida maritima	197 Hansenula subpelliculosa	292 Pichia trehalophila
79 Candida melibiosica	198 Hansenula sydowiorum*	293 Pichia veronae*
81 Candida membranaefaciens	201 Hormoascus platypodis**	295 Pichia wickerhamii
83 Candida milleri*	202 Hyphopichia burtonii**	312 Saccharomyces cerevisiae
84 Candida mogii	203 Kluyveromyces aestuarii	313 Saccharomyces dairensis
86 Candida naeodendra*	204 Kluyveromyces africanus	314 Saccharomyces exiguus
87 Candida norvegensis	205 Kluyveromyces blattae*	315 Saccharomyces kluyveri
88 Candida oleophila*	206 Kluyveromyces bulgaricus	316 Saccharomyces servazzii*
89 Candida oregonensis	207 Kluyveromyces delphensis	317 Saccharomyces telluris
90 Candida parapsilosis	208 Kluyveromyces dobzhanskii	318 Saccharomyces unisporus
92 Candida pseudointermedia*	209 Kluyveromyces drosophilarum	319 Saccharomycodes ludwigii
93 Candida pseudotropicalis	210 Kluyveromyces lactis	320 Saccharomycopsis capsularis**
94 Candida quercuum*	211 Kluyveromyces lodderi	321 Saccharomycopsis crataegensis*
95 Candida ravautii	212 Kluyveromyces marxianus	322 Saccharomycopsis fibuligera**
96 Candida rhagii	213 Kluyveromyces phaffii	324 Saccharomycopsis malanga*
97 Candida rugopelliculosa*	214 Kluyveromyces phaseolosporus	325 Saccharomycopsis vini**
99 Candida sake	215 Kluyveromyces polysporus	328 Schizosaccharomyces japonicus
100 Candida salmanticensis	216 Kluyveromyces thermotolerans**	329 Schizosaccharomyces malidevorans
101 Candida santamariae	217 Kluyveromyces waltii*	330 Schizosaccharomyces octosporus
102 Candida savonica*	218 Kluyveromyces wickerhamii	331 Schizosaccharomyces pombe
104 Candida silvanorum*	220 Leucosporidium frigidum	332 Schizosaccharomyces slooffiae*
105 Candida silvicultrix*	221 Leucosporidium gelidum	333 Schwanniomyces occidentalis
106 Candida solani	222 Leucosporidium nivalis	334 Selenozyma peltata*
107 Candida sorboxylosa*	224 Leucosporidium stokesii	357 Torulaspora delbrueckii**
108 Candida steatolytica*	230 Lodderomyces elongisporus	358 Torulaspora globosa**
109 Candida suecica*	231 Metschnikowia bicuspidata	359 Torulaspora pretoriensis**
110 Candida tenuis	233 Metschnikowia lunata*	360 Torulopsis anatomiae
112 Candida terebra*	234 Metschnikowia pulcherrima	361 Torulopsis apicola
113 Candida tropicalis	235 Metschnikowia reukaufii	366 Torulopsis bombicola*
115 Candida utilis	236 Metschnikowia zobellii	367 Torulopsis cantarelii
116 Candida valdiviana*	238 Nadsonia elongata	368 Torulopsis castellii
117 Candida valida	239 Nadsonia fulvescens	369 Torulopsis dendrica*
118 Candida vartiovaarai	240 Nematospora coryli	370 Torulopsis ernobii
119 Candida veronae	241 Pachysolen tannophilus	371 Torulopsis etchellsii
122 Candida viswanathii	242 Pachytichospora transvaalensis	373 Torulopsis fructus*
123 Candida zeylanoides	243 Phaffia rhodozyma*	375 Torulopsis glabrata
124 Citeromyces matritensis	244 Pichia abadieae*	376 Torulopsis gropengiesseri
146 Debaryomyces castellii	245 Pichia acaciae	377 Torulopsis haemulonii
148 Debaryomyces hansenii	246 Pichia ambrosiae*	378 Torulopsis halonitratophila
149 Debaryomyces marama	247 Pichia angophorae	379 Torulopsis halophilus*
151 Debaryomyces nepalensis*	248 Pichia besseyi*	380 Torulopsis holmii
152 Debaryomyces phaffii	249 Pichia bovis	381 Torulopsis humilis*
153 Debaryomyces polymorpha**	252 Pichia delftensis	385 Torulopsis karawaiewi*
154 Debaryomyces tamarii	253 Pichia dispora	386 Torulopsis kruisii*
155 Debaryomyces vanriji	254 Pichia etchellsii	387 Torulopsis lactis−condensi

388 Torulopsis magnoliae
389 Torulopsis mannitofaciens*
391 Torulopsis methanolovescens*
392 Torulopsis molischiana
393 Torulopsis multis−gemmis*
394 Torulopsis musae*
395 Torulopsis nagoyaensis*
396 Torulopsis navarrensis*
398 Torulopsis nitratophila
399 Torulopsis nodaensis*
400 Torulopsis norvegica
401 Torulopsis pampelonensis*
403 Torulopsis pignaliae*
407 Torulopsis schatavii*
409 Torulopsis sonorensis*
411 Torulopsis spandovensis*
412 Torulopsis stellata
413 Torulopsis tannotolerans*
414 Torulopsis torresii
416 Torulopsis versatilis
417 Torulopsis wickerhamii
418 Torulopsis xestobii*
423 Trichosporon fennicum*
424 Trichosporon melibiosaceum*
428 Wickerhamia fluorescens
430 Wingea robertsii
432 Zygosaccharomyces bailii**
433 Zygosaccharomyces bisporus**
434 Zygosaccharomyces cidri**
435 Zygosaccharomyces fermentati**
436 Zygosaccharomyces florentinus**
437 Zygosaccharomyces microellipsodes**
438 Zygosaccharomyces mrakii**
439 Zygosaccharomyces rouxii**

Key involving physiological tests and microscopical examination (Key No. 3)

Tests in Key No. 3
2 D−Galactose fermentation
3 Maltose fermentation
4 Methyl δ−D−glucopyranoside fermentation
5 Sucrose fermentation
6 Trehalose fermentation
7 Melibiose fermentation
8 Lactose fermentation
9 Cellobiose fermentation
12 D−Galactose growth
13 L−Sorbose growth
14 D−Ribose growth
15 D−Xylose growth
16 L−Arabinose growth
18 L−Rhamnose growth
19 Sucrose growth
20 Maltose growth
21 Trehalose growth
22 Methyl α−D−glucopyranoside growth
23 Cellobiose growth
24 Salicin growth
26 Melibiose growth
27 Lactose growth
28 Raffinose growth
30 Inulin growth
31 Starch growth
32 Glycerol growth
33 Erythritol growth
35 Galactitol growth
37 D−Glucitol growth
39 D−δ−Gluconolactone growth
40 2−Ketogluconate growth
41 5−Ketogluconate growth
42 D,L−Lactate growth
43 Succinate growth
44 Citrate growth
46 Ethanol growth
47 Ethylamine growth
49 Nitrate growth
51 Growth without vitamins
52 Growth without inositol
53 Growth without pantothenate
54 Growth without biotin
55 Growth without thiamin
56 Growth without pyridoxine
57 Growth without niacin
61 Growth in 60% D−glucose
62 Growth at 37 degrees
63 Growth with 0.01% cycloheximide
64 Growth with 0.1% cycloheximide
66 Pink colonies
67 Budding cells
69 Apical budding
71 Cells of other shapes
72 Filamentous
74 Septate hyphae

number of different tests 55

KEY No. 3

		Negative	Positive
1	Erythritol growth.	2	786
2	D−Glucitol growth.	3	338
3	Ethanol growth.	4	142
4	Glycerol growth.	5	76
5	D−Galactose growth.	6	47
6	Cellobiose growth.	7	40
7	Raffinose growth.	8	29
8	Nitrate growth.	9	28
9	Succinate growth.	10	21
10	Maltose growth.	11	17
11	Trehalose growth.	12	15
12	Growth in 60% D−glucose.	13	14
13	Growth at 37 degrees.	Saccharomyces cerevisiae / Torulopsis karawaiewi*	Saccharomyces cerevisiae / Saccharomyces telluris
14	Growth without inositol.	Schizosaccharomyces slooffiae*	Torulaspora delbrueckii** / Zygosaccharomyces bisporus** / Zygosaccharomyces rouxii**
15	Growth without pyridoxine	Saccharomyces cerevisiae / Torulopsis glabrata	16
16	Growth in 60% D−glucose.	Saccharomyces cerevisiae	Torulaspora delbrueckii** / Zygosaccharomyces rouxii**
17	Growth without inositol.	18	16
18	Maltose fermentation.	19	20
19	Growth in 60% D−glucose.	Saccharomyces cerevisiae	Schizosaccharomyces slooffiae*
20	Budding cells.	Schizosaccharomyces octosporus	Saccharomyces cerevisiae
21	D,L−Lactate growth.	22	24
22	Citrate growth.	23	Torulopsis dendrica* / Torulopsis karawaiewi*

Key No. 3: fermenting yeasts

	Negative	Positive
23 Growth in 60% D-glucose.	Saccharomyces cerevisiae Torulopsis karawaiewi*	Torulaspora delbrueckii**
24 D-δ-Gluconolactone growth	25	27
25 Growth without thiamin.	Candida sorboxylosa* Saccharomyces cerevisiae	26
26 Growth in 60% D-glucose.	Saccharomyces cerevisiae	Torulaspora delbrueckii**
27 Filamentous.	Torulaspora delbrueckii**	Candida rugopelliculosa*
28 D-Galactose fermentation.	Torulopsis lactis-condensi	Torulopsis halophilus*
29 Nitrate growth.	30	Torulopsis lactis-condensi
30 Maltose growth.	31	35
31 L-Sorbose growth.	32	34
32 Growth without inositol.	33	26
33 Budding cells.	Schizosaccharomyces malidevorans	Saccharomyces cerevisiae Torulopsis stellata
34 Growth in 60% D-glucose.	Zygosaccharomyces florentinus**	Torulaspora delbrueckii**
35 L-Sorbose growth.	36	34
36 Growth without inositol.	37	26
37 Melibiose fermentation.	38	39
38 Budding cells.	Schizosaccharomyces pombe	Saccharomyces cerevisiae
39 Budding cells.	Schizosaccharomyces japonicus	Saccharomyces cerevisiae
40 Maltose growth.	41	46
41 L-Rhamnose growth.	42	Torulopsis anatomiae
42 Nitrate growth.	43	Torulopsis halophilus*
43 Succinate growth.	44	Torulopsis dendrica*
44 2-Ketogluconate growth.	Hanseniaspora valbyensis	45
45 Growth at 37 degrees.	Hanseniaspora uvarum	Hanseniaspora guilliermondii*
46 L-Arabinose growth.	Hanseniaspora osmophila Hanseniaspora vineae*	Phaffia rhodozyma*
47 Raffinose growth.	48	64
48 D-Xylose growth.	49	61
49 Nitrate growth.	50	Torulopsis halophilus*
50 D-Galactose fermentation.	51	52
51 Growth in 60% D-glucose.	Saccharomyces cerevisiae	Torulaspora delbrueckii** Zygosaccharomyces bisporus** Zygosaccharomyces rouxii**
52 Sucrose growth.	53	58
53 Growth without niacin.	54	56
54 Growth with 0.1% cycloheximide	Saccharomyces cerevisiae Saccharomyces dairensis	55
55 Ethylamine growth.	Saccharomyces servazzii*	Saccharomyces unisporus
56 Growth in 60% D-glucose.	57	Torulaspora delbrueckii**
57 Growth without pantothenate	Pachytichospora transvaalensis Saccharomyces cerevisiae	Kluyveromyces africanus Saccharomyces cerevisiae
58 Sucrose fermentation.	59	60
59 Growth in 60% D-glucose.	Kluyveromyces africanus Saccharomyces cerevisiae	Torulaspora delbrueckii**
60 Growth in 60% D-glucose.	Saccharomyces cerevisiae Saccharomyces exiguus	Torulaspora delbrueckii**
61 Cellobiose growth.	62	63
62 Starch growth.	Torulaspora delbrueckii**	Candida albicans
63 Starch growth.	Candida buinensis*	Candida albicans
64 Melibiose growth.	65	71
65 Lactose growth.	66	Candida kefyr
66 Maltose growth.	67	69
67 Growth in 60% D-glucose.	68	Torulaspora delbrueckii**
68 D-δ-Gluconolactone growth	Candida milleri* Saccharomyces cerevisiae Saccharomyces exiguus	Candida milleri* Saccharomyces exiguus Torulopsis holmii
69 Ethylamine growth.	26	70
70 Growth without niacin.	Kluyveromyces thermotolerans**	Torulaspora delbrueckii**
71 L-Sorbose growth.	72	74
72 Growth in 60% D-glucose.	73	Torulaspora delbrueckii**

		Negative	Positive
73	D–δ–Gluconolactone growth	Saccharomyces cerevisiae	Zygosaccharomyces microellipsodes**
74	D,L–Lactate growth.	34	75
75	Growth in 60% D–glucose.	Zygosaccharomyces microellipsodes**	Torulaspora delbrueckii**
76	Sucrose growth.	77	105
77	Succinate growth.	78	93
78	Nitrate growth.	79	92
79	D–Galactose growth.	80	85
80	Trehalose growth.	81	82
81	Growth in 60% D→glucose.	Saccharomyces cerevisiae Torulopsis karawaiewi*	Torulaspora delbrueckii** Zygosaccharomyces bisporus** Zygosaccharomyces rouxii**
82	Growth without niacin.	83	16
83	2–Ketogluconate growth.	Saccharomyces cerevisiae Torulopsis glabrata	84
84	Growth without inositol.	Torulopsis castellii	Torulopsis glabrata
85	D–Galactose fermentation.	51	86
86	Growth without niacin.	87	88
87	Growth with 0.1% cycloheximide	Saccharomyces cerevisiae Saccharomyces dairensis	Saccharomyces servazzii*
88	Growth without inositol.	Kluyveromyces africanus Saccharomyces cerevisiae	89
89	D–δ–Gluconolactone growth	26	90
90	Growth without thiamin.	91	Torulaspora delbrueckii**
91	Growth without biotin.	Kluyveromyces phaffii	Torulopsis tannotolerans*
92	D–Galactose fermentation.	Torulopsis halonitratophila	Torulopsis halophilus*
93	Nitrate growth.	94	92
94	D,L–Lactate growth.	95	104
95	Trehalose growth.	96	102
96	Citrate growth.	97	101
97	Growth in 60% D–glucose.	98	100
98	D–δ–Gluconolactone growth	Saccharomyces cerevisiae Torulopsis karawaiewi*	99
99	Filamentous.	Torulopsis karawaiewi*	Candida citrea*
100	Growth without thiamin.	Candida citrea*	Torulaspora delbrueckii**
101	Salicin growth.	99	Torulopsis dendrica* Torulopsis karawaiewi*
102	Growth in 60% D–glucose.	103	Torulaspora delbrueckii**
103	Septate hyphae.	Saccharomyces cerevisiae	Nematospora coryli
104	Cellobiose growth.	25	Pichia norvegensis*
105	Maltose growth.	106	124
106	Raffinose growth.	107	114
107	Nitrate growth.	108	Torulopsis halophilus*
108	Cellobiose growth.	109	Hanseniaspora occidentalis*
109	Succinate growth.	110	102
110	D–Galactose fermentation.	16	111
111	Growth in 60% D–glucose.	112	Torulaspora delbrueckii**
112	D–δ–Gluconolactone growth	Kluyveromyces africanus Saccharomyces cerevisiae	113
113	Growth without inositol.	Kluyveromyces africanus	Kluyveromyces phaffii
114	Melibiose growth.	115	71
115	Trehalose growth.	116	118
116	Lactose growth.	117	Candida kefyr
117	Cellobiose growth.	26	Saccharomycodes ludwigii
118	Succinate growth.	119	121
119	Growth in 60% D–glucose.	120	Torulaspora delbrueckii**
120	D–δ–Gluconolactone growth	Candida milleri* Saccharomyces cerevisiae	Candida milleri* Kluyveromyces polysporus Torulopsis holmii
121	D–Galactose fermentation.	102	122
122	Growth in 60% D–glucose.	123	Torulaspora delbrueckii**
123	D–δ–Gluconolactone growth	Candida milleri* Saccharomyces cerevisiae	Candida milleri* Kluyveromyces polysporus
124	Cellobiose growth.	125	138

Key No. 3: fermenting yeasts

	Negative	Positive
125 D–Xylose growth.	126	133
126 Melibiose growth.	127	132
127 Trehalose growth.	128	130
128 Raffinose growth.	16	129
129 Growth without inositol.	38	26
130 Methyl α–D–glucopyranoside growth	131	69
131 Growth in 60% D–glucose.	103	Torulaspora delbrueckii** Zygosaccharomyces rouxii**
132 L–Sorbose growth.	26	34
133 Citrate growth.	134	135
134 Melibiose growth.	62	34
135 Methyl α–D–glucopyranoside growth	136	137
136 Septate hyphae.	Candida mogii	Candida albicans
137 Maltose fermentation.	Torulopsis xestobii*	Candida albicans
138 D–Galactose growth.	139	140
139 L–Arabinose growth.	Candida utilis	Phaffia rhodozyma*
140 Nitrate growth.	141	Torulopsis mannitofaciens*
141 D–Galactose fermentation.	137	63
142 Sucrose growth.	143	248
143 Glycerol growth.	144	171
144 D–Galactose growth.	145	164
145 D,L–Lactate growth.	146	155
146 L–Rhamnose growth.	147	Torulopsis anatomiae
147 Nitrate growth.	148	Torulopsis halophilus*
148 Citrate growth.	149	154
149 Trehalose growth.	150	15
150 Growth in 60% D–glucose.	151	Pichia membranaefaciens Torulaspora delbrueckii** Zygosaccharomyces bisporus** Zygosaccharomyces rouxii**
151 L–Sorbose growth.	152	153
152 Growth at 37 degrees.	Pichia membranaefaciens Saccharomyces cerevisiae Torulopsis karawaiewi*	Pichia membranaefaciens Saccharomyces cerevisiae Saccharomyces telluris
153 Apical budding.	Pichia membranaefaciens	Nadsonia elongata
154 Salicin growth.	Pichia membranaefaciens Torulopsis karawaiewi*	Torulopsis dendrica* Torulopsis karawaiewi*
155 D–Xylose growth.	156	162
156 Succinate growth.	157	158
157 Growth in 60% D–glucose.	Pichia membranaefaciens Saccharomyces cerevisiae Saccharomyces telluris	Pichia membranaefaciens Torulaspora delbrueckii**
158 D–δ–Gluconolactone growth	159	161
159 Growth without thiamin.	Candida sorboxylosa* Pichia membranaefaciens Saccharomyces cerevisiae	160
160 Growth in 60% D–glucose.	Pichia membranaefaciens Saccharomyces cerevisiae	Pichia membranaefaciens Torulaspora delbrueckii**
161 Filamentous.	Pichia membranaefaciens Torulaspora delbrueckii**	Candida rugopelliculosa* Pichia membranaefaciens
162 Growth without pyridoxine	Candida sorboxylosa* Pichia membranaefaciens	163
163 Filamentous.	Pichia membranaefaciens Torulaspora delbrueckii**	Candida lambica Pichia membranaefaciens
164 Nitrate growth.	165	170
165 L–Arabinose growth.	166	Guilliermondella selenospora**
166 D–Galactose fermentation.	51	167
167 Growth with 0.1% cycloheximide	168	55
168 Growth in 60% D–glucose.	169	Torulaspora delbrueckii**
169 Growth without inositol.	Pachytichospora transvaalensis Saccharomyces cerevisiae	Saccharomyces cerevisiae Saccharomyces dairensis
170 Filamentous.	Torulopsis halophilus*	Brettanomyces abstinens*
171 D,L–Lactate growth.	172	216

		Negative	Positive
172	Nitrate growth.	...173	...211
173	D–Galactose growth.	...174	...200
174	Succinate growth.	...175	...188
175	L–Sorbose growth.	...176	...187
176	Trehalose growth.	...177	...184
177	Growth in 60% D–glucose.	...178	...182
178	Growth without pyridoxine	...179	...181
179	Growth at 37 degrees.	Pichia membranaefaciens Saccharomyces cerevisiae Torulopsis karawaiewi*	...180
180	Growth with 0.01% cycloheximide	Pichia membranaefaciens Saccharomyces cerevisiae	Kluyveromyces delphensis Pichia membranaefaciens
181	Filamentous.	Pichia membranaefaciens Saccharomyces cerevisiae Torulopsis karawaiewi*	Candida valida Pichia membranaefaciens Saccharomyces cerevisiae
182	Growth without pyridoxine	Kluyveromyces delphensis Pichia membranaefaciens	...183
183	Filamentous.	Pichia membranaefaciens Torulaspora delbrueckii** Zygosaccharomyces bisporus** Zygosaccharomyces rouxii**	Candida valida Pichia membranaefaciens Zygosaccharomyces bisporus** Zygosaccharomyces rouxii**
184	Growth without pyridoxine	...185	...16
185	Growth without niacin.	Saccharomyces cerevisiae Torulopsis glabrata	...186
186	Growth with 0.01% cycloheximide	Saccharomyces cerevisiae	Kluyveromyces delphensis
187	Growth in 60% D–glucose.	...153	Pichia membranaefaciens Torulaspora delbrueckii** Zygosaccharomyces bisporus** Zygosaccharomyces rouxii**
188	Citrate growth.	...189	...197
189	L–Sorbose growth.	...190	...196
190	Trehalose growth.	...191	...102
191	Growth in 60% D–glucose.	...192	...194
192	D–δ–Gluconolactone growth	...181	...193
193	Filamentous.	Pichia membranaefaciens Torulopsis karawaiewi*	Candida citrea* Candida valida Pichia membranaefaciens
194	Growth without thiamin.	Candida citrea* Candida valida Pichia membranaefaciens	...195
195	Filamentous.	Pichia membranaefaciens Torulaspora delbrueckii**	Candida valida Pichia membranaefaciens
196	Growth in 60% D–glucose.	...153	Pichia membranaefaciens Torulaspora delbrueckii**
197	Salicin growth.	...198	Torulopsis dendrica* Torulopsis karawaiewi*
198	Growth at 37 degrees.	...199	Pichia membranaefaciens Pichia terricola
199	Filamentous.	Pichia membranaefaciens Torulopsis karawaiewi*	Candida citrea* Pichia membranaefaciens
200	D–Xylose growth.	...201	...209
201	Succinate growth.	...202	...102
202	D–Ribose growth.	...203	...207
203	D–Galactose fermentation.	...51	...204
204	Growth without niacin.	...87	...205
205	D–α–Gluconolactone growth	...26	...206
206	Growth without thiamin.	Torulopsis tannotolerans*	Torulaspora delbrueckii**
207	D–Galactose fermentation.	Zygosaccharomyces rouxii**	...208
208	Growth with 0.1% cycloheximide	Kluyveromyces blattae* Saccharomyces dairensis	Saccharomyces servazzii*
209	L–Sorbose growth.	...206	...210
210	Growth in 60% D–glucose.	Geotrichum penicillatum**	Torulaspora delbrueckii**
211	L–Sorbose growth.	...212	Torulopsis nodaensis*
212	D–Galactose fermentation.	...213	...170
213	D–Galactose growth.	...214	...215

Key No. 3: fermenting yeasts

	Negative	Positive
214 Cellobiose growth.	Torulopsis halonitratophila	Candida berthetii
215 Growth at 37 degrees.	Torulopsis halonitratophila	Brettanomyces abstinens*
216 Cellobiose growth.	217	245
217 D–Galactose growth.	218	235
218 Citrate growth.	219	230
219 Nitrate growth.	220	Torulopsis halonitratophila
220 Trehalose growth.	221	228
221 D–Xylose growth.	222	226
222 Growth without pyridoxine	Candida sorboxylosa* Pichia membranaefaciens Saccharomyces cerevisiae	223
223 Growth in 60% D–glucose.	224	225
224 Ethylamine growth.	Candida valida Pichia kudriavzevii Pichia membranaefaciens Saccharomyces cerevisiae	Candida valida Pichia kudriavzevii Pichia membranaefaciens Pichia scutulata*
225 Filamentous.	Pichia membranaefaciens Torulaspora delbrueckii**	Candida valida Pichia kudriavzevii Pichia membranaefaciens
226 Growth without pyridoxine	Candida sorboxylosa* Pichia membranaefaciens	227
227 Filamentous.	Pichia membranaefaciens Torulaspora delbrueckii**	Candida lambica Candida valida Pichia membranaefaciens Torulaspora delbrueckii**
228 Growth in 60% D–glucose.	229	
229 Growth with 0.01% cycloheximide	Saccharomyces cerevisiae	Brettanomyces custersianus
230 D–Xylose growth.	231	234
231 Growth without thiamin.	232	233
232 Growth without pyridoxine	Candida sorboxylosa* Pichia membranaefaciens	Pichia membranaefaciens Pichia terricola
233 Growth without vitamins.	Pichia kluyveri Pichia membranaefaciens	Pichia kudriavzevii Pichia membranaefaciens
234 Growth without pyridoxine	Candida sorboxylosa* Pichia membranaefaciens	Candida lambica Pichia fermentans Pichia membranaefaciens
235 D–Xylose growth.	236	241
236 Trehalose growth.	237	239
237 Nitrate growth.	238	Torulopsis halonitratophila
238 5–Ketogluconate growth.	26	Candida fragicola*
239 Growth with 0.01% cycloheximide	26	240
240 Succinate growth.	Torulopsis humilis*	Brettanomyces custersianus
241 L–Sorbose growth.	242	210
242 L–Arabinose growth.	243	244
243 5–Ketogluconate growth.	Torulaspora delbrueckii**	Candida fragicola*
244 D–Galactose fermentation.	Guilliermondella selenospora**	Candida fragicola*
245 Nitrate growth.	246	247
246 D–Galactose growth.	Candida norvegensis Pichia norvegensis*	Candida fragicola*
247 L–Sorbose growth.	Hansenula mrakii	Hansenula dimennae
248 Cellobiose growth.	249	308
249 Maltose growth.	250	284
250 D,L–Lactate growth.	251	269
251 Glycerol growth.	252	261
252 Melibiose growth.	253	256
253 Nitrate growth.	254	Torulopsis halophilus*
254 D–Galactose fermentation.	16	255
255 Growth in 60% D–glucose.	Candida milleri* Saccharomyces cerevisiae Saccharomyces exiguus	Torulaspora delbrueckii**
256 L–Sorbose growth.	257	259
257 Growth in 60% D–glucose.	258	Torulaspora delbrueckii**
258 Ethylamine growth.	Saccharomyces cerevisiae	Saccharomyces kluyveri
259 Growth in 60% D–glucose.	260	Torulaspora delbrueckii**

		Negative	Positive
260	Growth at 37 degrees.	Zygosaccharomyces florentinus**	Saccharomyces kluyveri
261	Melibiose growth.	262	256
262	Raffinose growth.	263	264
263	Nitrate growth.	131	Torulopsis halophilus*
264	Trehalose growth.	265	266
265	Growth in 60% D–glucose.	Saccharomyces cerevisiae Torulopsis bombicola*	Torulaspora delbrueckii** Torulopsis bombicola*
266	D–Galactose fermentation.	102	267
267	Growth in 60% D–glucose.	268	Torulaspora delbrueckii**
268	D–δ–Gluconolactone growth	Candida milleri* Saccharomyces cerevisiae	Candida milleri* Kluyveromyces lodderi Kluyveromyces polysporus
269	Melibiose growth.	270	281
270	Glycerol growth.	271	275
271	Lactose growth.	272	Candida kefyr
272	D–Galactose growth.	273	255
273	Growth in 60% D–glucose.	274	Torulaspora delbrueckii**
274	Growth with 0.01% cycloheximide	Saccharomyces cerevisiae	Torulaspora globosa**
275	Lactose growth.	276	Candida kefyr
276	Citrate growth.	277	Wickerhamia fluorescens
277	Raffinose growth.	228	278
278	D–Galactose growth.	273	279
279	Growth in 60% D–glucose.	280	Torulaspora delbrueckii**
280	D–δ–Gluconolactone growth	Candida milleri* Saccharomyces cerevisiae	Candida milleri* Kluyveromyces lodderi
281	Growth in 60% D–glucose.	282	Torulaspora delbrueckii**
282	Ethylamine growth.	Saccharomyces cerevisiae	283
283	Growth at 37 degrees.	Zygosaccharomyces microellipsodes**	Saccharomyces kluyveri
284	Methyl α–D–glucopyranoside growth	285	296
285	Melibiose growth.	286	292
286	D–Xylose growth.	287	290
287	D,L–Lactate growth.	288	228
288	Growth in 60% D–glucose.	289	Torulaspora delbrueckii** Zygosaccharomyces rouxii**
289	Growth without inositol.	103	Saccharomyces cerevisiae Schwanniomyces occidentalis
290	Starch growth.	Torulaspora delbrueckii**	291
291	Septate hyphae.	Schwanniomyces occidentalis	Candida albicans
292	L–Sorbose growth.	293	295
293	Growth in 60% D–glucose.	294	Torulaspora delbrueckii**
294	Starch growth.	258	Saccharomyces cerevisiae Schwanniomyces occidentalis
295	Starch growth.	259	Schwanniomyces occidentalis
296	D–Xylose growth.	297	304
297	Melibiose growth.	298	292
298	Nitrate growth.	299	Brettanomyces lambicus Dekkera bruxellensis
299	D–Galactose growth.	300	303
300	Growth in 60% D–glucose.	301	Torulaspora delbrueckii**
301	Starch growth.	302	Saccharomyces cerevisiae Schwanniomyces occidentalis
302	Ethylamine growth.	Saccharomyces cerevisiae	Dekkera bruxellensis
303	Starch growth.	69	Saccharomyces cerevisiae Schwanniomyces occidentalis
304	Citrate growth.	305	306
305	Melibiose growth.	290	295
306	Maltose fermentation.	307	291
307	Sucrose fermentation.	Torulopsis xestobii*	Schwanniomyces occidentalis

Key No. 3: fermenting yeasts

		Negative	Positive
308	Lactose growth.	309	330
309	D-Galactose growth.	310	321
310	Nitrate growth.	311	318
311	L-Sorbose growth.	312	317
312	L-Arabinose growth.	313	316
313	Sucrose fermentation.	314	315
314	Growth at 37 degrees.	Aciculoconidium aculeatum**	Saccharomycopsis fibuligera**
315	Septate hyphae.	Schwanniomyces occidentalis	Saccharomycopsis fibuligera**
316	Pink colonies.	Schwanniomyces occidentalis	Phaffia rhodozyma*
317	D,L-Lactate growth.	Schwanniomyces occidentalis	Candida solani
318	Raffinose growth.	319	320
319	L-Sorbose growth.	Torulopsis halophilus*	Hansenula californica
320	Maltose growth.	Hansenula saturnus	Candida utilis
321	Nitrate growth.	322	328
322	Citrate growth.	323	327
323	Melibiose growth.	324	326
324	D-Xylose growth.	325	291
325	Starch growth.	Dekkera intermedia	Schwanniomyces occidentalis
326	Starch growth.	Saccharomyces kluyveri	Schwanniomyces occidentalis
327	Sucrose fermentation.	141	291
328	Methyl α-D-glucopyranoside growth	329	Brettanomyces custersii Dekkera intermedia
329	Maltose growth.	Torulopsis halophilus*	Torulopsis mannitofaciens*
330	D-Xylose growth.	331	336
331	Maltose growth.	332	334
332	Trehalose growth.	Candida kefyr	333
333	Melibiose growth.	Brettanomyces anomalus	Debaryomyces tamarii
334	Lactose fermentation.	Brettanomyces custersii	335
335	Septate hyphae.	Brettanomyces claussenii	Brettanomyces anomalus
336	L-Arabinose growth.	Kluyveromyces wickerhamii	337
337	Lactose fermentation.	Candida macedoniensis	Candida pseudotropicalis
338	Glycerol growth.	339	435
339	Cellobiose growth.	340	375
340	Raffinose growth.	341	362
341	Maltose growth.	342	360
342	Succinate growth.	343	345
343	Growth in 60% D-glucose.	344	Torulaspora delbrueckii** Zygosaccharomyces bailii** Zygosaccharomyces bisporus** Zygosaccharomyces rouxii**
344	L-Sorbose growth.	Saccharomyces cerevisiae Torulopsis karawaiewi*	Nadsonia elongata
345	D,L-Lactate growth.	346	354
346	L-Sorbose growth.	347	353
347	Citrate growth.	348	Candida diversa Torulopsis karawaiewi*
348	Trehalose growth.	349	351
349	Growth in 60% D-glucose.	350	Pichia fluxuum Torulaspora delbrueckii**
350	D-δ-Gluconolactone growth	Pichia fluxuum Saccharomyces cerevisiae Torulopsis karawaiewi*	Pichia delftensis Torulopsis karawaiewi*
351	Growth in 60% D-glucose.	352	Torulaspora delbrueckii**
352	2-Ketogluconate growth.	Saccharomyces cerevisiae	Pichia dispora
353	Growth in 60% D-glucose.	Nadsonia elongata	Torulaspora delbrueckii**
354	Trehalose growth.	355	358
355	Growth without thiamin.	356	357
356	2-Ketogluconate growth.	Pichia saitoi Saccharomyces cerevisiae	Pichia besseyi*
357	Growth in 60% D-glucose.	Pichia fluxuum Saccharomyces cerevisiae	Pichia fluxuum Torulaspora delbrueckii**

		Negative	Positive
358	Growth in 60% D–glucose.	359	Torulaspora delbrueckii**
359	2–Ketogluconate growth.	Pichia saitoi / Saccharomyces cerevisiae	Pichia dispora
360	D–Xylose growth.	361	290
361	Growth in 60% D–glucose.	301	Torulaspora delbrueckii** / Zygosaccharomyces rouxii**
362	Maltose growth.	363	369
363	Melibiose growth.	364	367
364	L–Sorbose growth.	365	366
365	Lactose growth.	26	Candida kefyr
366	Filamentous.	Torulaspora delbrueckii** / Torulopsis apicola	Kluyveromyces waltii*
367	Growth with 0.01% cycloheximide.	281	368
368	L–Sorbose growth.	Zygosaccharomyces mrakii**	Zygosaccharomyces florentinus**
369	Melibiose growth.	370	292
370	D–Galactose growth.	371	373
371	Nitrate growth.	300	372
372	Maltose fermentation.	Citeromyces matritensis	Dekkera bruxellensis
373	D,L–Lactate growth.	303	374
374	2–Ketogluconate growth.	26	Torulaspora delbrueckii** / Torulaspora pretoriensis**
375	Raffinose growth.	376	411
376	Ethanol growth.	377	383
377	D–Galactose growth.	378	380
378	Sucrose growth.	Candida insectamans*	379
379	Maltose growth.	Candida curiosa	Candida suecica*
380	Maltose growth.	Candida curiosa	381
381	Citrate growth.	382	63
382	Maltose fermentation.	Metschnikowia bicuspidata	Candida albicans
383	D–Galactose growth.	384	395
384	Sucrose growth.	385	386
385	L–Arabinose growth.	Brettanomyces naardenensis*	Torulopsis sonorensis*
386	L–Rhamnose growth.	387	393
387	Maltose growth.	Candida curiosa	388
388	L–Sorbose growth.	389	392
389	D–Xylose growth.	390	391
390	Sucrose fermentation.	Aciculoconidium aculeatum**	Schwanniomyces occidentalis
391	D,L–Lactate growth.	Schwanniomyces occidentalis	Pichia angophorae
392	Starch growth.	Pichia spartinae*	Schwanniomyces occidentalis
393	Maltose growth.	Candida curiosa	394
394	L–Sorbose growth.	Candida oregonensis	Candida lusitaniae
395	Nitrate growth.	396	410
396	Citrate growth.	397	406
397	L–Rhamnose growth.	398	405
398	Sucrose growth.	399	400
399	Maltose fermentation.	Brettanomyces naardenensis*	Candida tropicalis
400	D–Xylose growth.	401	403
401	L–Sorbose growth.	325	402
402	Sucrose fermentation.	Metschnikowia bicuspidata	Schwanniomyces occidentalis
403	Maltose fermentation.	402	404
404	5–Ketogluconate growth.	291	Candida albicans / Candida tropicalis
405	Sucrose growth.	Brettanomyces naardenensis*	Candida lusitaniae
406	L–Rhamnose growth.	407	Candida lusitaniae
407	D–Galactose fermentation.	408	409
408	Starch growth.	Pichia etchellsii	291
409	Starch growth.	Candida buinensis*	404
410	Maltose growth.	Candida curiosa	Brettanomyces custersii / Dekkera intermedia
411	Nitrate growth.	412	427

Key No. 3: fermenting yeasts

	Negative	Positive
412 Galactitol growth.	413	426
413 Melibiose growth.	414	422
414 Lactose growth.	415	420
415 D–Galactose growth.	316	416
416 D–Ribose growth.	417	419
417 D,L–Lactate growth.	325	418
418 Cellobiose fermentation.	Zygosaccharomyces fermentati**	Dekkera intermedia
419 L–Sorbose growth.	Dekkera intermedia	Candida pseudointermedia*
420 L–Sorbose growth.	421	Candida intermedia
421 D–Xylose growth.	Candida kefyr	Candida pseudotropicalis
422 D,L–Lactate growth.	423	425
423 Citrate growth.	326	424
424 Sucrose fermentation.	Candida melibiosica	Schwanniomyces occidentalis
425 5–Ketogluconate growth.	Saccharomyces kluyveri	Candida salmanticensis
426 D–Ribose growth.	Debaryomyces castellii	Candida pseudointermedia*
427 Maltose growth.	428	431
428 Melibiose growth.	429	430
429 Apical budding.	Candida curiosa	Leucosporidium frigidum
430 Apical budding.	Candida curiosa	Leucosporidium nivalis
431 Melibiose growth.	432	433
432 D–Xylose growth.	Brettanomyces custersii Dekkera intermedia	Leucosporidium stokesii
433 L–Rhamnose growth.	434	Leucosporidium gelidum
434 Methyl α–D–glucopyranoside growth	Candida aquatica	Candida valdiviana*
435 Maltose growth.	436	540
436 Cellobiose growth.	437	497
437 Raffinose growth.	438	467
438 Nitrate growth.	439	462
439 Citrate growth.	440	456
440 L–Sorbose growth.	441	452
441 Succinate growth.	442	447
442 Growth in 60% D–glucose.	443	445
443 Filamentous.	Saccharomyces cerevisiae Torulopsis karawaiewi*	444
444 Septate hyphae.	Saccharomyces cerevisiae	Saccharomycopsis vini**
445 Filamentous.	Torulaspora delbrueckii** Zygosaccharomyces bailii** Zygosaccharomyces bisporus** Zygosaccharomyces rouxii**	446
446 Septate hyphae.	Zygosaccharomyces bailii** Zygosaccharomyces bisporus** Zygosaccharomyces rouxii**	Saccharomycopsis vini**
447 L–Rhamnose growth.	448	Pichia pastoris
448 D,L–Lactate growth.	449	451
449 Growth in 60% D–glucose.	443	450
450 Filamentous.	Torulaspora delbrueckii**	Saccharomycopsis vini**
451 Trehalose growth.	238	228
452 D–Xylose growth.	453	455
453 Growth in 60% D–glucose.	454	445
454 Growth without vitamins.	Saccharomycopsis vini**	Nadsonia elongata
455 D–Ribose growth.	210	Saccharomycopsis crataegensis*
456 D,L–Lactate growth.	457	461
457 Trehalose growth.	458	459
458 Filamentous.	Torulopsis karawaiewi*	Candida krissii*
459 D–Xylose growth.	Candida iberica* Candida zeylanoides	460
460 Salicin growth.	Torulopsis fructus*	Candida zeylanoides
461 L–Arabinose growth.	Candida catenulata	Pichia abadieae*
462 D–Xylose growth.	463	465
463 Sucrose growth.	464	Torulopsis magnoliae
464 L–Sorbose growth.	Torulopsis halonitratophila	Torulopsis nodaensis*

		Negative	Positive
465	Ethanol growth.	Torulopsis pignaliae*	466
466	L–Rhamnose growth.	Pachysolen tannophilus	Torulopsis nitratophila
467	Galactitol growth.	468	Torulopsis spandovensis*
468	Ethanol growth.	469	480
469	Melibiose growth.	470	476
470	L–Sorbose growth.	471	473
471	Lactose growth.	472	Candida kefyr
472	Nitrate growth.	26	Torulopsis magnoliae
473	Nitrate growth.	474	Torulopsis magnoliae
474	Succinate growth.	Torulaspora delbrueckii** Torulopsis apicola	475
475	Filamentous.	Torulaspora delbrueckii**	Candida bombi*
476	L–Sorbose growth.	477	74
477	D,L–Lactate growth.	478	72
478	Growth in 60% D–glucose.	479	Torulaspora delbrueckii**
479	D–δ–Gluconolactone growth	Saccharomyces cerevisiae	Zygosaccharomyces mrakii**
480	Melibiose growth.	481	489
481	D,L–Lactate growth.	482	486
482	Nitrate growth.	483	Torulopsis magnoliae
483	Growth in 60% D–glucose.	484	485
484	Growth at 37 degrees.	444	Saccharomyces cerevisiae Torulopsis bombicola*
485	Growth at 37 degrees.	450	Torulaspora delbrueckii** Torulopsis bombicola*
486	Lactose growth.	487	488
487	Growth with 0.01% cycloheximide	26	Kluyveromyces bulgaricus Kluyveromyces phaseolosporus
488	Growth without pantothenate	Candida kefyr	Kluyveromyces bulgaricus
489	L–Sorbose growth.	490	494
490	D,L–Lactate growth.	491	281
491	Growth in 60% D–glucose.	492	Torulaspora delbrueckii**
492	Ethylamine growth.	Saccharomyces cerevisiae	493
493	Growth at 37 degrees.	Zygosaccharomyces mrakii**	Saccharomyces kluyveri
494	D,L–Lactate growth.	259	495
495	Growth in 60% D–glucose.	496	Torulaspora delbrueckii**
496	Growth with 0.01% cycloheximide	283	Zygosaccharomyces cidri**
497	Sucrose growth.	498	520
498	Nitrate growth.	499	514
499	Trehalose growth.	500	508
500	D–Xylose growth.	501	504
501	Citrate growth.	502	503
502	D–Galactose growth.	Saccharomycopsis vini**	Candida fragicola*
503	2–Ketogluconate growth.	Pichia quercuum	Candida krissii*
504	D–Galactose growth.	505	507
505	L–Sorbose growth.	506	Pichia pijperi
506	L–Arabinose growth.	Pichia sargentensis*	Torulopsis sonorensis*
507	L–Sorbose growth.	Candida fragicola*	Geotrichum fermentans**
508	L–Rhamnose growth.	509	Pichia lindnerii* Torulopsis methanolovescens*
509	Citrate growth.	510	511
510	D–Galactose growth.	Saccharomycopsis vini**	Torulopsis torresii
511	L–Sorbose growth.	Candida savonica* Candida zeylanoides	512
512	D,L–Lactate growth.	513	Candida beechii Candida santamariae
513	Trehalose fermentation.	Candida zeylanoides	Candida beechii
514	D–Galactose growth.	515	517
515	L–Sorbose growth.	516	Hansenula dimennae
516	D–Ribose growth.	Hansenula mrakii Torulopsis norvegica	Hansenula minuta
517	Citrate growth.	518	519
518	L–Rhamnose growth.	Pachysolen tannophilus	Torulopsis wickerhamii
519	L–Sorbose growth.	Hansenula muscicola*	Torulopsis nagoyaensis*
520	Nitrate growth.	521	535
521	Ethanol growth.	522	523
522	Lactose growth.	Torulopsis gropengiesseri	Candida kefyr

Key No. 3: fermenting yeasts

	Negative	Positive
523 Lactose growth.	524	526
524 D–Galactose growth.	Saccharomycopsis vini**	525
525 Melibiose growth.	Kluyveromyces bulgaricus Kluyveromyces lactis Kluyveromyces phaseolosporus	Saccharomyces kluyveri
526 L–Arabinose growth.	527	531
527 Raffinose growth.	Kluyveromyces lactis Kluyveromyces wickerhamii	528
528 L–Sorbose growth.	529	530
529 Growth without pantothenate	Candida kefyr	Kluyveromyces bulgaricus Kluyveromyces lactis
530 D–Galactose fermentation.	Kluyveromyces aestuarii	Kluyveromyces bulgaricus Kluyveromyces lactis
531 D–Xylose growth.	529	532
532 Lactose fermentation.	533	534
533 Growth without biotin.	Kluyveromyces bulgaricus Kluyveromyces lactis Kluyveromyces marxianus	Candida macedoniensis
534 Growth without pantothenate	Candida pseudotropicalis Kluyveromyces marxianus	Kluyveromyces bulgaricus Kluyveromyces lactis Kluyveromyces marxianus
535 D–Galactose growth.	536	538
536 Methyl α–D–glucopyranoside growth	537	Hansenula californica
537 Trehalose growth.	Hansenula saturnus	Candida curiosa
538 D–Xylose growth.	539	430
539 Lactose fermentation.	Candida curiosa	Brettanomyces anomalus
540 Raffinose growth.	541	689
541 D–Galactose growth.	542	591
542 L–Rhamnose growth.	543	583
543 Cellobiose growth.	544	562
544 Methyl α–D–glucopyranoside growth	545	552
545 Citrate growth.	546	551
546 D,L–Lactate growth.	547	228
547 Growth in 60% D–glucose.	548	549
548 Starch growth.	444	Saccharomyces cerevisiae Schwanniomyces occidentalis
549 Filamentous.	Torulaspora delbrueckii** Zygosaccharomyces rouxii**	550
550 Septate hyphae.	Zygosaccharomyces rouxii**	Saccharomycopsis vini**
551 Sucrose fermentation.	Torulopsis musae*	Schwanniomyces occidentalis
552 Nitrate growth.	553	Brettanomyces lambicus Dekkera bruxellensis
553 Citrate growth.	554	560
554 Sucrose fermentation.	555	558
555 Starch growth.	556	557
556 Growth in 60% D–glucose.	444	450
557 2–Ketogluconate growth.	Saccharomyces cerevisiae	Filobasidium capsuligenum**
558 Maltose fermentation.	559	300
559 Starch growth.	556	Saccharomyces cerevisiae Schwanniomyces occidentalis
560 L–Sorbose growth.	561	551
561 Sucrose fermentation.	Filobasidium capsuligenum**	Schwanniomyces occidentalis
562 D,L–Lactate growth.	563	577
563 L–Sorbose growth.	564	572
564 Starch growth.	565	567
565 Citrate growth.	566	Pichia toletana
566 Septate hyphae.	Metschnikowia reukaufii	Saccharomycopsis vini**
567 D–Ribose growth.	568	571
568 Sucrose fermentation.	569	315

		Negative	Positive
569	Growth at 37 degrees.	570	Saccharomycopsis fibuligera**
570	Growth without biotin.	Aciculoconidium aculeatum**	Filobasidium capsuligenum**
571	Methyl α–D–glucopyranoside growth	Saccharomycopsis malanga*	Filobasidium capsuligenum**
572	Citrate growth.	573	392
573	Sucrose fermentation.	574	576
574	Growth without thiamin.	575	566
575	Filamentous.	Pichia mucosa*	Saccharomycopsis vini**
576	Starch growth.	Saccharomycopsis vini**	Schwanniomyces occidentalis
577	D–Xylose growth.	578	579
578	L–Sorbose growth.	314	Pichia spartinae*
579	Nitrate growth.	580	582
580	L–Arabinose growth.	581	Pichia bovis
581	Growth without pyridoxine	Candida quercuum*	Pichia toletana
582	L–Sorbose growth.	Candida vartiovaarai	Hansenula californica
583	Nitrate growth.	584	588
584	L–Sorbose growth.	585	Candida lusitaniae
585	Growth without pyridoxine	586	587
586	Growth without thiamin.	Pichia rhodanensis Pichia veronae*	Pichia wickerhamii
587	Growth without biotin.	Pichia toletana	Candida freyschussii
588	L–Sorbose growth.	589	590
589	Trehalose growth.	Torulopsis ernobii	Hansenula beckii** Hansenula bimundalis
590	D,L–Lactate growth.	Torulopsis ernobii	Hansenula californica
591	Cellobiose growth.	592	640
592	Ethanol growth.	593	602
593	D–Xylose growth.	594	596
594	Nitrate growth.	16	595
595	Trehalose growth.	Torulopsis etchellsii	Torulopsis versatilis
596	L–Arabinose growth.	597	601
597	Sucrose growth.	598	599
598	Starch growth.	Torulaspora delbrueckii**	Candida ravautii
599	Starch growth.	600	Candida albicans
600	Filamentous.	Torulaspora delbrueckii**	Candida sake
601	Maltose fermentation.	Torulopsis multis–gemmis*	136
602	D–Xylose growth.	603	620
603	Nitrate growth.	604	618
604	Methyl α–D–glucopyranoside growth	605	608
605	D,L–Lactate growth.	606	607
606	Growth in 60% D–glucose.	Saccharomyces cerevisiae Schwanniomyces occidentalis	Torulaspora delbrueckii** Zygosaccharomyces rouxii**
607	Citrate growth.	228	Candida catenulata
608	L–Sorbose growth.	609	612
609	Sucrose fermentation.	610	611
610	Growth in 60% D–glucose.	557	Torulaspora delbrueckii**
611	Growth in 60% D–glucose.	Saccharomyces cerevisiae Schwanniomyces occidentalis	Torulaspora delbrueckii**
612	Trehalose growth.	613	616
613	Citrate growth.	614	615
614	Starch growth.	Torulaspora delbrueckii**	Schwanniomyces occidentalis
615	Starch growth.	Nadsonia fulvescens	Schwanniomyces occidentalis
616	Starch growth.	617	Schwanniomyces occidentalis
617	Growth in 60% D–glucose.	Lodderomyces elongisporus	Torulaspora delbrueckii**
618	Sucrose growth.	619	Brettanomyces lambicus
619	Maltose fermentation.	Torulopsis nodaensis*	Torulopsis etchellsii
620	Methyl α–D–glucopyranoside growth	621	632
621	Sucrose growth.	622	627

Key No. 3: fermenting yeasts

		Negative	Positive
622	Trehalose growth.	623	626
623	Citrate growth.	624	625
624	Starch growth.	Torulaspora delbrueckii**	Candida brumptii
625	Growth at 37 degrees.	Candida brumptii	Candida catenulata
626	Citrate growth.	598	Candida catenulata / Candida ravautii
627	L–Arabinose growth.	628	629
628	Starch growth.	600	291
629	L–Rhamnose growth.	630	Torulopsis haemulonii
630	Maltose fermentation.	631	291
631	Sucrose fermentation.	Torulopsis multis–gemmis*	Schwanniomyces occidentalis
632	Starch growth.	633	638
633	L–Arabinose growth.	634	Candida parapsilosis
634	Salicin growth.	635	637
635	Growth in 60% D–glucose.	636	600
636	Growth at 37 degrees.	Candida sake	Lodderomyces elongisporus
637	Growth at 37 degrees.	Candida sake	Candida maltosa*
638	Sucrose fermentation.	639	291
639	Growth at 37 degrees.	Filobasidium capsuligenum**	Candida albicans
640	Nitrate growth.	641	683
641	Citrate growth.	642	664
642	L–Rhamnose growth.	643	Candida lusitaniae
643	Sucrose fermentation.	644	660
644	Maltose fermentation.	645	655
645	L–Sorbose growth.	646	650
646	D–Galactose fermentation.	647	649
647	Starch growth.	648	Filobasidium capsuligenum**
648	Growth at 37 degrees.	Metschnikowia reukaufii	Metschnikowia lunata*
649	Growth without niacin.	Kluyveromyces lactis	Metschnikowia reukaufii
650	D–Galactose fermentation.	651	654
651	Growth at 37 degrees.	652	653
652	Growth without thiamin.	Metschnikowia bicuspidata / Metschnikowia zobellii	Candida sake / Metschnikowia bicuspidata / Metschnikowia pulcherrima / Metschnikowia reukaufii
653	Cells of other shapes.	Metschnikowia pulcherrima	Metschnikowia lunata*
654	Growth without niacin.	Kluyveromyces lactis	Candida sake / Metschnikowia pulcherrima / Metschnikowia reukaufii
655	D–Galactose fermentation.	656	658
656	L–Sorbose growth.	639	657
657	Starch growth.	Candida sake	Candida albicans
658	Starch growth.	659	Candida albicans / Candida tropicalis
659	Growth without niacin.	Kluyveromyces lactis	Candida sake
660	Starch growth.	661	404
661	D–Xylose growth.	662	663
662	Cellobiose fermentation.	Kluyveromyces lactis	Dekkera intermedia
663	Growth without niacin.	Candida maltosa* / Kluyveromyces lactis	637
664	Starch growth.	665	674
665	L–Rhamnose growth.	666	Candida lusitaniae
666	Melibiose growth.	667	Torulopsis navarrensis*
667	L–Arabinose growth.	668	673
668	Sucrose fermentation.	669	672
669	D–Galactose fermentation.	670	671
670	Growth at 37 degrees.	Candida sake / Metschnikowia pulcherrima	653
671	2–Ketogluconate growth.	Candida buinensis*	Candida oleophila* / Candida sake / Metschnikowia pulcherrima
672	Trehalose fermentation.	Candida oleophila* / Candida sake	637
673	D–Galactose fermentation.	Pichia etchellsii	Candida buinensis*
674	Lactose growth.	675	Candida fluviotilis*

		Negative	Positive
675	Maltose fermentation.	676	678
676	Sucrose fermentation.	677	Schwanniomyces occidentalis
677	Growth without vitamins.	Filobasidium capsuligenum**	Torulopsis kruisii*
678	Sucrose fermentation.	679	404
679	D–Galactose fermentation.	680	682
680	Growth at 37 degrees.	Filobasidium capsuligenum**	681
681	Septate hyphae.	Candida viswanathii	Candida albicans
682	Septate hyphae.	Candida viswanathii	Candida albicans Candida tropicalis
683	Ethanol growth.	Torulopsis versatilis	684
684	D–Xylose growth.	685	687
685	Lactose growth.	Brettanomyces custersii Dekkera intermedia	686
686	Lactose fermentation.	Brettanomyces custersii	Brettanomyces anomalus
687	L–Sorbose growth.	Hansenula muscicola*	688
688	Starch growth.	Hansenula silvicola	Hansenula holstii
689	Citrate growth.	690	739
690	Cellobiose growth.	691	707
691	Ethanol growth.	692	697
692	Nitrate growth.	693	696
693	L–Sorbose growth.	694	695
694	D,L–Lactate growth.	69	374
695	Melibiose growth.	70	34
696	D–Galactose growth.	Citeromyces matritensis	Torulopsis versatilis
697	Melibiose growth.	698	703
698	D–Galactose growth.	699	700
699	Nitrate growth.	558	372
700	D,L–Lactate growth.	303	701
701	L–Sorbose growth.	374	702
702	Growth without niacin.	Kluyveromyces drosophilarum	Torulaspora delbrueckii**
703	L–Sorbose growth.	293	704
704	D,L–Lactate growth.	295	705
705	Growth in 60% D–glucose.	706	Torulaspora delbrueckii**
706	Growth with 0.01% cycloheximide.	Saccharomyces kluyveri	Zygosaccharomyces cidri**
707	D–Galactose growth.	708	712
708	D–Ribose growth.	709	Candida maritima
709	Nitrate growth.	710	Hansenula beijerinckii
710	L–Arabinose growth.	711	316
711	Starch growth.	Saccharomycopsis vini**	315
712	Ethanol growth.	713	714
713	L–Sorbose growth.	Torulopsis versatilis	Candida inositophila*
714	Lactose growth.	715	732
715	L–Arabinose growth.	716	727
716	Melibiose growth.	717	326
717	Nitrate growth.	718	Brettanomyces custersii Dekkera intermedia
718	D,L–Lactate growth.	719	721
719	Cellobiose fermentation.	720	325
720	Starch growth.	Kluyveromyces lactis	Schwanniomyces occidentalis
721	L–Sorbose growth.	722	726
722	Cellobiose fermentation.	723	Dekkera intermedia
723	Methyl α–D–glucopyranoside fermentation	724	725
724	Growth without inositol.	Zygosaccharomyces fermentati**	Kluyveromyces lactis
725	Growth without inositol.	Zygosaccharomyces fermentati**	Kluyveromyces dobzhanskii
726	Maltose fermentation.	Kluyveromyces drosophilarum Kluyveromyces lactis	723
727	D,L–Lactate growth.	728	729
728	L–Rhamnose growth.	720	Candida inositophila*

Key No. 3: fermenting yeasts

	Negative	Positive
729 Methyl α−D−glucopyranoside fermentation	730	731
730 Starch growth.	......Kluyveromyces drosophilarum Kluyveromyces lactis	......Debaryomyces hansenii
731 Trehalose fermentation.	......Debaryomyces hansenii	......Kluyveromyces dobzhanskii
732 Nitrate growth.	733	738
733 L−Arabinose growth.	734	737
734 L−Sorbose growth.	735	736
735 Methyl α−D−glucopyranoside fermentation	......Kluyveromyces lactis	......Brettanomyces claussenii
736 D−Galactose fermentation.	......Kluyveromyces aestuarii	......Kluyveromyces lactis
737 Starch growth.	......Kluyveromyces lactis	......Debaryomyces hansenii
738 L−Sorbose growth.	334	......Candida aquatica
739 Galactitol growth.	740	775
740 D−Galactose growth.	741	750
741 Nitrate growth.	742	747
742 L−Rhamnose growth.	743	746
743 L−Arabinose growth.	315	744
744 D,L−Lactate growth.	316	745
745 Growth without thiamin.	......Pichia onychis	......Phaffia rhodozyma*
746 D−Ribose growth.	......Pichia rabaulensis*	......Candida maritima
747 D−Xylose growth.	......Citeromyces matritensis	748
748 L−Rhamnose growth.	749	......Hansenula petersonii
749 Inulin growth.	......Hansenula fabianii	......Hansenula jadinii
750 Melibiose growth.	751	764
751 D,L−Lactate growth.	752	757
752 Ethanol growth.	753	754
753 Methyl α−D−glucopyranoside growth	......Candida mogii	......Candida pseudointermedia*
754 D−Ribose growth.	755	......Candida pseudointermedia*
755 L−Rhamnose growth.	756	......Torulopsis haemulonii
756 Starch growth.	......Pichia ohmeri	......Schwanniomyces occidentalis
757 L−Rhamnose growth.	758	763
758 Trehalose fermentation.	759	762
759 D−Xylose growth.	760	761
760 Growth without niacin.	......Kluyveromyces drosophilarum	......Pichia ohmeri
761 Starch growth.	......Kluyveromyces drosophilarum	......Debaryomyces hansenii
762 Growth without niacin.	......Kluyveromyces dobzhanskii	......Pichia ohmeri
763 Starch growth.	......Pichia strasburgensis	......Debaryomyces hansenii
764 Nitrate growth.	765	434
765 L−Arabinose growth.	766	772
766 D,L−Lactate growth.	767	......Zygosaccharomyces cidri**
767 L−Sorbose growth.	768	769
768 Starch growth.	......Torulopsis navarrensis*	......Schwanniomyces occidentalis
769 Sucrose fermentation.	770	771
770 Filamentous.	......Torulopsis pampelonensis*	......Candida melibiosica
771 Starch growth.	......Torulopsis pampelonensis*	......Schwanniomyces occidentalis
772 D−Ribose growth.	773	774
773 D,L−Lactate growth.	......Schwanniomyces occidentalis	......Debaryomyces hansenii
774 Starch growth.	......Pichia guilliermondii	......Debaryomyces hansenii
775 Melibiose growth.	776	781
776 Ethanol growth.	753	777
777 Cellobiose growth.	......Torulopsis haemulonii	778
778 D,L−Lactate growth.	779	780
779 2−Ketogluconate growth.	......Candida steatolytica*	......Candida pseudointermedia*
780 Growth without thiamin.	......Candida steatolytica*	......Debaryomyces hansenii
781 Nitrate growth.	782	......Candida valdiviana*
782 Lactose growth.	783	785
783 Sucrose fermentation.	784	774
784 Starch growth.	......Candida brassicae*	......Debaryomyces hansenii
785 Melibiose fermentation.	......Debaryomyces hansenii	......Debaryomyces castellii

		Negative	Positive
786	Galactitol growth.	787	871
787	Sucrose growth.	788	809
788	D–Galactose growth.	789	798
789	Nitrate growth.	790	794
790	Maltose growth.	791	793
791	D–Ribose growth.	Torulopsis cantarelii	792
792	Cellobiose growth.	Pichia trehalophila	Pichia pinus
793	Methyl α–D–glucopyranoside growth	Saccharomycopsis malanga*	Saccharomycopsis capsularis**
794	Maltose growth.	795	Hansenula capsulata Torulopsis molischiana
795	Cellobiose growth.	796	797
796	Trehalose growth.	Candida boidinii	Hansenula philodendra*
797	Growth without thiamin.	Hansenula glucozyma	Hansenula henricii
798	Citrate growth.	799	804
799	L–Sorbose growth.	800	803
800	L–Arabinose growth.	801	802
801	Growth without vitamins.	Candida flavificans*	Candida cacaoi
802	Filamentous.	Pichia methanolica*	Candida cacaoi
803	Growth at 37 degrees.	Candida conglobata	Pichia methanolica*
804	Methyl α–D–glucopyranoside growth	805	808
805	Growth at 37 degrees.	806	807
806	Growth without vitamins.	Candida boleticola*	Torulopsis schatavii*
807	Filamentous.	Pichia methanolica*	Pichia farinosa
808	L–Sorbose growth.	Hansenula muscicola*	Pichia acaciae
809	Raffinose growth.	810	839
810	L–Rhamnose growth.	811	824
811	D–Galactose growth.	812	816
812	L–Arabinose growth.	813	Ambrosiozyma cicatricosa* Ambrosiozyma monospora**
813	D–Ribose growth.	814	815
814	L–Sorbose growth.	Saccharomycopsis fibuligera**	Candida incommunis*
815	Nitrate growth.	Pichia ambrosiae*	Hansenula polymorpha
816	Lactose growth.	817	821
817	Nitrate growth.	818	Hansenula polymorpha
818	Starch growth.	819	Pichia nakazawae*
819	2–Ketogluconate growth.	820	Candida butyri* Pichia philogaea*
820	Cells of other shapes.	Candida butyri*	Candida diddensii
821	D,L–Lactate growth.	822	Candida chilensis*
822	Maltose fermentation.	823	Trichosporon fennicum*
823	Growth at 37 degrees.	Candida ergatensis*	Candida butyri*
824	Nitrate growth.	825	836
825	D,L–Lactate growth.	826	834
826	Starch growth.	827	830
827	L–Sorbose growth.	828	829
828	Cells of other shapes.	Candida insectorum* Candida tenuis Candida veronae	Candida diddensii
829	Cells of other shapes.	Candida insectorum* Candida tenuis Candida terebra*	Candida diddensii
830	Growth without vitamins.	831	833
831	Cells of other shapes.	832	Candida naeodendra*
832	Septate hyphae.	Candida insectorum* Candida tenuis	Pichia nakazawae*
833	Cells of other shapes.	Candida tenuis	Candida homilentoma*
834	D–Galactose growth.	Ambrosiozyma philentoma*	835
835	Maltose fermentation.	Candida tenuis	Pichia stipitis
836	D,L–Lactate growth.	837	838
837	L–Arabinose growth.	Hansenula polymorpha	Hansenula holstii
838	L–Sorbose growth.	Hormoascus platypodis**	Candida ishiwadae*
839	Nitrate growth.	840	868
840	Ethanol growth.	841	845
841	D–Ribose growth.	Pichia naganishii*	842
842	Melibiose growth.	843	844

Key No. 3: fermenting yeasts

		Negative	Positive
843	Growth at 37 degrees.	Trichosporon melibiosaceum*	Trichosporon fennicum*
844	Maltose fermentation.	Debaryomyces nepalensis*	Trichosporon melibiosaceum*
845	L–Rhamnose growth.	846	862
846	Melibiose growth.	847	856
847	D–Ribose growth.	848	850
848	D–Xylose growth.	Saccharomycopsis fibuligera**	849
849	Methyl α–D–glucopyranoside growth	Pichia naganishii*	Debaryomyces hansenii Debaryomyces marama
850	Citrate growth.	851	852
851	Starch growth.	Candida rhagii	Debaryomyces hansenii
852	D,L–Lactate growth.	853	Debaryomyces hansenii Debaryomyces marama
853	Lactose growth.	854	855
854	Growth without vitamins.	Debaryomyces marama	Hyphopichia burtonii**
855	Growth without vitamins.	Debaryomyces marama	Trichosporon fennicum*
856	Lactose growth.	857	860
857	L–Sorbose growth.	858	Debaryomyces hansenii Debaryomyces marama Debaryomyces nepalensis*
858	D–Galactose fermentation.	Debaryomyces hansenii Debaryomyces marama	859
859	Melibiose fermentation.	Debaryomyces hansenii	Candida silvicultrix*
860	D,L–Lactate growth.	861	Debaryomyces hansenii Debaryomyces marama
861	Starch growth.	Candida entomophila*	Debaryomyces marama
862	Succinate growth.	Pichia scolyti	863
863	D,L–Lactate growth.	864	867
864	Melibiose growth.	865	866
865	Maltose fermentation.	Candida insectorum*	Wingea robertsii
866	Cells of other shapes.	Candida insectorum*	Candida silvanorum*
867	Trehalose fermentation.	Debaryomyces hansenii	Wingea robertsii
868	Melibiose growth.	869	Hansenula sydowiorum*
869	Inulin growth.	870	Hansenula lynferdii*
870	Growth without vitamins.	Hansenula subpelliculosa	Hansenula anomala Hansenula ciferrii
871	Raffinose growth.	872	881
872	Nitrate growth.	873	880
873	Ethanol growth.	874	876
874	L–Rhamnose growth.	Pichia methanolica*	875
875	D,L–Lactate growth.	Candida terebra*	Selenozyma peltata*
876	L–Rhamnose growth.	877	878
877	Trehalose growth.	Pichia haplophila	Pichia methanolica*
878	L–Sorbose growth.	Candida entomaea*	879
879	Growth without thiamin.	Candida terebra*	Candida dendronema*
880	D,L–Lactate growth.	837	Candida ishiwadae*
881	L–Rhamnose growth.	882	887
882	Methyl α–D–glucopyranoside growth	Debaryomyces polymorpha**	883
883	D,L–Lactate growth.	884	886
884	Starch growth.	885	Debaryomyces phaffii Debaryomyces vanriji
885	Inulin growth.	Candida friedrichii	Candida membranaefaciens
886	Starch growth.	Candida membranaefaciens	Debaryomyces hansenii Debaryomyces vanriji
887	Melibiose growth.	888	890
888	D,L–Lactate growth.	889	Debaryomyces hansenii
889	L–Sorbose growth.	Pichia heimii*	Candida blankii
890	Lactose growth.	Debaryomyces hansenii Debaryomyces vanriji	Debaryomyces hansenii Pichia pseudopolymorpha

Key involving physiological tests only (Key No. 4)

Tests in Key No. 4
- 2 D−Galactose fermentation
- 3 Maltose fermentation
- 4 Methyl α−D−glucopyranoside fermentation
- 5 Sucrose fermentation
- 6 Trehalose fermentation
- 7 Melibiose fermentation
- 8 Lactose fermentation
- 9 Cellobiose fermentation
- 12 D−Galactose growth
- 13 L−Sorbose growth
- 14 D−Ribose growth
- 15 D−Xylose growth
- 16 L−Arabinose growth
- 18 L−Rhamnose growth
- 19 Sucrose growth
- 20 Maltose growth
- 21 Trehalose growth
- 22 Methyl α−D−glucopyranoside growth
- 23 Cellobiose growth
- 24 Salicin growth
- 26 Melibiose growth
- 27 Lactose growth
- 28 Raffinose growth
- 29 Melezitose growth
- 30 Inulin growth
- 31 Starch growth
- 32 Glycerol growth
- 33 Erythritol growth
- 35 Galactitol growth
- 36 D−Mannitol growth
- 37 D−Glucitol growth
- 39 D−δ−Gluconolactone growth
- 40 2−Ketogluconate growth
- 41 5−Ketogluconate growth
- 42 D,L−Lactate growth
- 43 Succinate growth
- 44 Citrate growth
- 45 Methanol growth
- 46 Ethanol growth
- 47 Ethylamine growth
- 49 Nitrate growth
- 51 Growth without vitamins
- 52 Growth without inositol
- 53 Growth without pantothenate
- 54 Growth without biotin
- 55 Growth without thiamin
- 56 Growth without pyridoxine
- 57 Growth without niacin
- 61 Growth in 60% D−glucose
- 62 Growth at 37 degrees
- 63 Growth with 0.01% cycloheximide
- 64 Growth with 0.1% cycloheximide
- 65 Starch formation

number of different tests 53

KEY No. 4

	Negative	Positive
1 Erythritol growth.	2	.726
2 D−Mannitol growth.	3	.310
3 Ethanol growth.	4	.121
4 Glycerol growth.	5	.62
5 Raffinose growth.	6	.42
6 Cellobiose growth.	7	.38
7 D−Galactose growth.	8	.27
8 D−Glucitol growth.	9	.26
9 Succinate growth.	10	.20
10 Nitrate growth.	11	Torulopsis lactis−condensi
11 Maltose growth.	12	.18
12 Trehalose growth.	13	.16
13 Growth in 60% D−glucose.	14	.15
14 Growth at 37 degrees.	Saccharomyces cerevisiae Torulopsis karawaiewi*	Saccharomyces cerevisiae Saccharomyces telluris
15 Growth without inositol.	Schizosaccharomyces slooffiae*	Torulaspora delbrueckii** Zygosaccharomyces bisporus** Zygosaccharomyces rouxii**
16 Growth without pyridoxine	Saccharomyces cerevisiae Torulopsis glabrata	.17
17 Growth in 60% D−glucose.	Saccharomyces cerevisiae	Torulaspora delbrueckii** Zygosaccharomyces rouxii**
18 Starch formation.	17	.19
19 Maltose fermentation.	Schizosaccharomyces slooffiae*	Schizosaccharomyces octosporus
20 D,L−Lactate growth.	21	.23
21 Citrate growth.	22	Torulopsis dendrica* Torulopsis karawaiewi*
22 Growth in 60% D−glucose.	Saccharomyces cerevisiae Torulopsis karawaiewi*	Torulaspora delbrueckii**
23 D−δ−Gluconolactone growth	24	Candida rugopelliculosa* Torulaspora delbrueckii**
24 Growth without thiamin.	Candida sorboxylosa* Saccharomyces cerevisiae	.25
25 Growth in 60% D−glucose.	Saccharomyces cerevisiae	Torulaspora delbrueckii**
26 Growth in 60% D−glucose.	Saccharomyces cerevisiae Torulopsis karawaiewi*	Torulaspora delbrueckii** Zygosaccharomyces bailii** Zygosaccharomyces bisporus** Zygosaccharomyces rouxii**
27 D−Galactose fermentation.	28	.29

Key No. 4: fermenting yeasts

	Negative	Positive
28 Growth in 60% D-glucose.	Saccharomyces cerevisiae	Torulaspora delbrueckii** Zygosaccharomyces bailii** Zygosaccharomyces bisporus** Zygosaccharomyces rouxii**
29 Sucrose growth.	30	35
30 Growth without niacin.	31	33
31 Growth with 0.1% cycloheximide	Saccharomyces cerevisiae Saccharomyces dairensis	32
32 Ethylamine growth.	Saccharomyces servazzii*	Saccharomyces unisporus
33 Growth in 60% D-glucose.	34	Torulaspora delbrueckii**
34 Growth without pantothenate	Pachytichospora transvaalensis Saccharomyces cerevisiae	Kluyveromyces africanus Saccharomyces cerevisiae
35 Sucrose fermentation.	36	37
36 Growth in 60% D-glucose.	Kluyveromyces africanus Saccharomyces cerevisiae	Torulaspora delbrueckii**
37 Growth in 60% D-glucose.	Saccharomyces cerevisiae Saccharomyces exiguus	Torulaspora delbrueckii**
38 Maltose growth.	39	Hanseniaspora osmophila Hanseniaspora vineae*
39 Succinate growth.	40	Torulopsis dendrica*
40 2-Ketogluconate growth.	Hanseniaspora valbyensis	41
41 Growth at 37 degrees.	Hanseniaspora uvarum	Hanseniaspora guilliermondii*
42 D-Galactose growth.	43	53
43 Maltose growth.	44	49
44 Nitrate growth.	45	Torulopsis lactis-condensi
45 L-Sorbose growth.	46	48
46 Starch formation.	47	Schizosaccharomyces malidevorans
47 Growth without inositol.	Saccharomyces cerevisiae Torulopsis stellata	25
48 Growth in 60% D-glucose.	Zygosaccharomyces florentinus**	Torulaspora delbrueckii**
49 L-Arabinose growth.	50	Phaffia rhodozyma*
50 L-Sorbose growth.	51	48
51 Starch formation.	25	52
52 Melibiose fermentation.	Schizosaccharomyces pombe	Schizosaccharomyces japonicus
53 Melibiose growth.	54	57
54 Lactose growth.	55	Candida kefyr
55 Growth in 60% D-glucose.	56	Torulaspora delbrueckii**
56 D-δ-Gluconolactone growth	Candida milleri* Saccharomyces cerevisiae Saccharomyces exiguus	Candida milleri* Saccharomyces exiguus Torulopsis holmii
57 L-Sorbose growth.	58	60
58 Growth in 60% D-glucose.	59	Torulaspora delbrueckii**
59 D-δ-Gluconolactone growth	Saccharomyces cerevisiae	Zygosaccharomyces microellipsodes**
60 D,L-Lactate growth.	48	61
61 Growth in 60% D-glucose.	Zygosaccharomyces microellipsodes**	Torulaspora delbrueckii**
62 D-Glucitol growth.	63	109
63 Raffinose growth.	64	91
64 D-Galactose growth.	65	81
65 Succinate growth.	66	72
66 Cellobiose growth.	67	Hanseniaspora occidentalis*
67 Trehalose growth.	68	69
68 Growth in 60% D-glucose.	Saccharomyces cerevisiae Torulopsis karawaiewi*	Torulaspora delbrueckii** Zygosaccharomyces bisporus** Zygosaccharomyces rouxii**
69 Growth without niacin.	70	17
70 2-Ketogluconate growth.	Saccharomyces cerevisiae Torulopsis glabrata	71
71 Growth without inositol.	Torulopsis castellii	Torulopsis glabrata

		Negative	Positive
72	D,L–Lactate growth.	73	80
73	Trehalose growth.	74	79
74	Citrate growth.	75	78
75	Growth in 60% D–glucose.	76	77
76	D–δ–Gluconolactone growth	Saccharomyces cerevisiae Torulopsis karawaiewi*	Candida citrea* Torulopsis karawaiewi*
77	Growth without thiamin.	Candida citrea*	Torulaspora delbrueckii**
78	Salicin growth.	Candida citrea* Torulopsis karawaiewi*	Torulopsis dendrica* Torulopsis karawaiewi*
79	Growth in 60% D–glucose.	Nematospora coryli Saccharomyces cerevisiae	Torulaspora delbrueckii**
80	Cellobiose growth.	24	Pichia norvegensis*
81	Citrate growth.	82	Torulopsis xestobii*
82	Succinate growth.	83	79
83	D–Galactose fermentation.	84	85
84	Growth in 60% D–glucose.	Saccharomyces cerevisiae	Torulaspora delbrueckii** Zygosaccharomyces bisporus** Zygosaccharomyces rouxii**
85	Growth without niacin.	86	87
86	Growth with 0.1% cycloheximide	Saccharomyces cerevisiae Saccharomyces dairensis	Saccharomyces servazzii*
87	Growth without inositol.	Kluyveromyces africanus Saccharomyces cerevisiae	88
88	D–δ–Gluconolactone growth	25	89
89	Growth without thiamin.	90	Torulaspora delbrueckii**
90	Growth without biotin.	Kluyveromyces phaffii	Torulopsis tannotolerans*
91	D–Galactose growth.	92	98
92	Cellobiose growth.	93	96
93	L–Sorbose growth.	94	48
94	Trehalose growth.	95	79
95	Starch formation.	25	Schizosaccharomyces pombe
96	L–Arabinose growth.	97	Phaffia rhodozyma*
97	Maltose growth.	Saccharomycodes ludwigii	Candida utilis
98	Melibiose growth.	99	108
99	Lactose growth.	100	Candida kefyr
100	D–Xylose growth.	101	107
101	Succinate growth.	102	104
102	Growth in 60% D–glucose.	103	Torulaspora delbrueckii**
103	D–δ–Gluconolactone growth	Candida milleri* Saccharomyces cerevisiae	Candida milleri* Kluyveromyces polysporus Torulopsis holmii
104	D–Galactose fermentation.	79	105
105	Growth in 60% D–glucose.	106	Torulaspora delbrueckii**
106	D–δ–Gluconolactone growth	Candida milleri* Saccharomyces cerevisiae	Candida milleri* Kluyveromyces polysporus
107	Citrate growth.	Torulaspora delbrueckii**	Torulopsis xestobii*
108	Citrate growth.	57	Torulopsis xestobii*
109	Raffinose growth.	110	115
110	Nitrate growth.	111	114
111	Melezitose growth.	26	112
112	L–Arabinose growth.	113	Torulopsis multis–gemmis*
113	Cellobiose growth.	25	Pichia mucosa*
114	Trehalose growth.	Torulopsis etchellsii	Torulopsis versatilis
115	L–Sorbose growth.	116	120
116	Maltose growth.	117	118
117	Lactose growth.	58	Candida kefyr
118	Nitrate growth.	119	Torulopsis versatilis
119	L–Arabinose growth.	25	Phaffia rhodozyma*
120	Melibiose growth.	Candida bombi* Torulaspora delbrueckii**	60
121	D–Glucitol growth.	122	258
122	Sucrose growth.	123	209
123	Glycerol growth.	124	147
124	D–Galactose growth.	125	141
125	Succinate growth.	126	130
126	Trehalose growth.	127	16

Key No. 4: fermenting yeasts

		Negative	Positive
127	Growth in 60% D−glucose.	128	Pichia membranaefaciens Torulaspora delbrueckii** Zygosaccharomyces bisporus** Zygosaccharomyces rouxii**
128	L−Sorbose growth.	129	Nadsonia elongata Pichia membranaefaciens
129	Growth at 37 degrees.	Pichia membranaefaciens Saccharomyces cerevisiae Torulopsis karawaiewi*	Pichia membranaefaciens Saccharomyces cerevisiae Saccharomyces telluris
130	D,L−Lactate growth.	131	136
131	L−Sorbose growth.	132	135
132	Citrate growth.	133	134
133	Growth in 60% D−glucose.	Pichia membranaefaciens Saccharomyces cerevisiae Torulopsis karawaiewi*	Pichia membranaefaciens Torulaspora delbrueckii**
134	Salicin growth.	Pichia membranaefaciens Torulopsis karawaiewi*	Torulopsis dendrica* Torulopsis karawaiewi*
135	Growth in 60% D−glucose.	Nadsonia elongata Pichia membranaefaciens	Pichia membranaefaciens Torulaspora delbrueckii**
136	D−Xylose growth.	137	140
137	D−δ−Gluconolactone growth	138	Candida rugopelliculosa* Pichia membranaefaciens Torulaspora delbrueckii**
138	Growth without thiamin.	Candida sorboxylosa* Pichia membranaefaciens Saccharomyces cerevisiae	139
139	Growth in 60% D−glucose.	Pichia membranaefaciens Saccharomyces cerevisiae	Pichia membranaefaciens Torulaspora delbrueckii**
140	Growth without pyridoxine	Candida sorboxylosa* Pichia membranaefaciens	Candida lambica Pichia membranaefaciens Torulaspora delbrueckii**
141	L−Arabinose growth.	142	Guilliermondella selenospora**
142	Nitrate growth.	143	Brettanomyces abstinens*
143	D−Galactose fermentation.	84	144
144	Growth with 0.1% cycloheximide	145	32
145	Growth in 60% D−glucose.	146	Torulaspora delbrueckii**
146	Growth without inositol.	Pachytichospora transvaalensis Saccharomyces cerevisiae	Saccharomyces cerevisiae Saccharomyces dairensis
147	D,L−Lactate growth.	148	183
148	D−Galactose growth.	149	171
149	Succinate growth.	150	162
150	Cellobiose growth.	151	Candida berthetii
151	L−Sorbose growth.	152	161
152	Trehalose growth.	153	158
153	Growth in 60% D−glucose.	154	157
154	Growth without pyridoxine	155	Candida valida Pichia membranaefaciens Saccharomyces cerevisiae Torulopsis karawaiewi*
155	Growth at 37 degrees.	Pichia membranaefaciens Saccharomyces cerevisiae Torulopsis karawaiewi*	156
156	Growth with 0.01% cycloheximide	Pichia membranaefaciens Saccharomyces cerevisiae	Kluyveromyces delphensis Pichia membranaefaciens
157	Growth without pyridoxine	Kluyveromyces delphensis Pichia membranaefaciens	Candida valida Pichia membranaefaciens Torulaspora delbrueckii** Zygosaccharomyces bisporus** Zygosaccharomyces rouxii**
158	Growth without pyridoxine	159	17
159	Growth without niacin.	Saccharomyces cerevisiae Torulopsis glabrata	160

		Negative	Positive
160	Growth with 0.01% cycloheximide	Saccharomyces cerevisiae	Kluyveromyces delphensis
161	Growth in 60% D–glucose.	Nadsonia elongata Pichia membranaefaciens	Pichia membranaefaciens Torulaspora delbrueckii** Zygosaccharomyces bisporus** Zygosaccharomyces rouxii**
162	Nitrate growth.	163	Candida berthetii
163	Citrate growth.	164	169
164	L–Sorbose growth.	165	135
165	Trehalose growth.	166	79
166	Growth in 60% D–glucose.	167	168
167	D–δ–Gluconolactone growth	Candida valida Pichia membranaefaciens Saccharomyces cerevisiae Torulopsis karawaiewi*	Candida citrea* Candida valida Pichia membranaefaciens Torulopsis karawaiewi*
168	Growth without thiamin.	Candida citrea* Candida valida Pichia membranaefaciens	Candida valida Pichia membranaefaciens Torulaspora delbrueckii**
169	Salicin growth.	170	Torulopsis dendrica* Torulopsis karawaiewi*
170	Growth at 37 degrees.	Candida citrea* Pichia membranaefaciens Torulopsis karawaiewi*	Pichia membranaefaciens Pichia terricola
171	Nitrate growth.	172	Brettanomyces abstinens*
172	D–Xylose growth.	173	181
173	Succinate growth.	174	79
174	D–Ribose growth.	175	179
175	D–Galactose fermentation.	84	176
176	Growth without niacin.	86	177
177	D–δ–Gluconolactone growth	25	178
178	Growth without thiamin.	Torulopsis tannotolerans*	Torulaspora delbrueckii**
179	D–Galactose fermentation.	Zygosaccharomyces rouxii**	180
180	Growth with 0.1% cycloheximide	Kluyveromyces blattae* Saccharomyces dairensis	Saccharomyces servazzii*
181	L–Sorbose growth.	178	182
182	Growth in 60% D–glucose.	Geotrichum penicillatum**	Torulaspora delbrueckii**
183	D–Galactose growth.	184	200
184	Cellobiose growth.	185	199
185	Citrate growth.	186	194
186	Trehalose growth.	187	192
187	D–Xylose growth.	188	191
188	Growth without pyridoxine	Candida sorboxylosa* Pichia membranaefaciens Saccharomyces cerevisiae	189
189	Growth in 60% D–glucose.	190	Candida valida Pichia kudriavzevii Pichia membranaefaciens Torulaspora delbrueckii**
190	Ethylamine growth.	Candida valida Pichia kudriavzevii Pichia membranaefaciens Saccharomyces cerevisiae	Candida valida Pichia kudriavzevii Pichia membranaefaciens Pichia scutulata*
191	Growth without pyridoxine	Candida sorboxylosa* Pichia membranaefaciens	Candida lambica Candida valida Pichia membranaefaciens Torulaspora delbrueckii**
192	Growth in 60% D–glucose.	193	Torulaspora delbrueckii**
193	Growth with 0.01% cycloheximide	Saccharomyces cerevisiae	Brettanomyces custersianus
194	D–Xylose growth.	195	198
195	Growth without thiamin.	196	197
196	Growth without pyridoxine	Candida sorboxylosa* Pichia membranaefaciens	Pichia membranaefaciens Pichia terricola
197	Growth without vitamins.	Pichia kluyveri Pichia membranaefaciens	Pichia kudriavzevii Pichia membranaefaciens
198	Growth without pyridoxine	Candida sorboxylosa* Pichia membranaefaciens	Candida lambica Pichia fermentans Pichia membranaefaciens

Key No. 4: fermenting yeasts

	Negative	Positive
199 Nitrate growth.	Candida norvegensis / Pichia norvegensis*	Hansenula mrakii
200 D–Xylose growth.	201	205
201 Trehalose growth.	202	203
202 5–Ketogluconate growth.	25	Candida fragicola*
203 Growth with 0.01% cycloheximide	25	204
204 Succinate growth.	Torulopsis humilis*	Brettanomyces custersianus
205 L–Sorbose growth.	206	182
206 L–Arabinose growth.	207	208
207 5–Ketogluconate growth.	Torulaspora delbrueckii**	Candida fragicola*
208 D–Galactose fermentation.	Guilliermondella selenospora**	Candida fragicola*
209 Cellobiose growth.	210	244
210 Melibiose growth.	211	235
211 Methyl α–D–glucopyranoside growth	212	231
212 D,L–Lactate growth.	213	221
213 Glycerol growth.	214	216
214 D–Galactose fermentation.	17	215
215 Growth in 60% D–glucose.	Candida milleri* / Saccharomyces cerevisiae / Saccharomyces exiguus	Torulaspora delbrueckii**
216 Raffinose growth.	217	218
217 Growth in 60% D–glucose.	Nematospora coryli / Saccharomyces cerevisiae	Torulaspora delbrueckii** / Zygosaccharomyces rouxii**
218 D–Galactose fermentation.	79	219
219 Growth in 60% D–glucose.	220	Torulaspora delbrueckii**
220 D–δ–Gluconolactone growth	Candida milleri* / Saccharomyces cerevisiae	Candida milleri* / Kluyveromyces lodderi / Kluyveromyces polysporus
221 Glycerol growth.	222	226
222 Lactose growth.	223	Candida kefyr
223 D–Galactose growth.	224	215
224 Growth in 60% D–glucose.	225	Torulaspora delbrueckii**
225 Growth with 0.01% cycloheximide	Saccharomyces cerevisiae	Torulaspora globosa**
226 Lactose growth.	227	Candida kefyr
227 Raffinose growth.	192	228
228 D–Galactose growth.	224	229
229 Growth in 60% D–glucose.	230	Torulaspora delbrueckii**
230 D–δ–Gluconolactone growth	Candida milleri* / Saccharomyces cerevisiae	Candida milleri* / Kluyveromyces lodderi
231 Citrate growth.	232	Torulopsis xestobii*
232 Nitrate growth.	233	Brettanomyces lambicus / Dekkera bruxellensis
233 Growth in 60% D–glucose.	234	Torulaspora delbrueckii**
234 Ethylamine growth.	Saccharomyces cerevisiae	Dekkera bruxellensis
235 Citrate growth.	236	Torulopsis xestobii*
236 L–Sorbose growth.	237	240
237 Growth in 60% D–glucose.	238	Torulaspora delbrueckii**
238 Ethylamine growth.	Saccharomyces cerevisiae	239
239 Growth at 37 degrees.	Zygosaccharomyces microellipsodes**	Saccharomyces kluyveri
240 D,L–Lactate growth.	241	243
241 Growth in 60% D–glucose.	242	Torulaspora delbrueckii**
242 Growth at 37 degrees.	Zygosaccharomyces florentinus**	Saccharomyces kluyveri
243 Growth in 60% D–glucose.	239	Torulaspora delbrueckii**
244 D–Galactose growth.	245	249
245 Nitrate growth.	246	248
246 L–Sorbose growth.	247	Candida solani
247 L–Arabinose growth.	Saccharomycopsis fibuligera**	Phaffia rhodozyma*
248 Maltose growth.	Hansenula saturnus	Candida utilis
249 D–Xylose growth.	250	254
250 Lactose growth.	251	252
251 Melibiose growth.	Brettanomyces custersii / Dekkera intermedia	Saccharomyces kluyveri
252 Trehalose growth.	Candida kefyr	253

		Negative	Positive
253	Lactose fermentation.	Brettanomyces custersii	Brettanomyces anomalus Brettanomyces claussenii
254	Lactose growth.	255	256
255	Citrate growth.	Saccharomyces kluyveri	Torulopsis xestobii*
256	L–Arabinose growth.	Kluyveromyces wickerhamii	257
257	Lactose fermentation.	Candida macedoniensis	Candida pseudotropicalis
258	Cellobiose growth.	259	283
259	Sucrose growth.	260	270
260	D–Xylose growth.	261	268
261	D,L–Lactate growth.	262	266
262	Nitrate growth.	263	Torulopsis etchellsii
263	Growth in 60% D–glucose.	264	Saccharomycopsis vini** Torulaspora delbrueckii** Zygosaccharomyces bailii** Zygosaccharomyces bisporus** Zygosaccharomyces rouxii**
264	L–Sorbose growth.	Saccharomyces cerevisiae Saccharomycopsis vini** Torulopsis karawaiewi*	265
265	Growth without vitamins.	Saccharomycopsis vini**	Nadsonia elongata
266	Nitrate growth.	267	Torulopsis etchellsii
267	Trehalose growth.	202	192
268	Citrate growth.	269	Torulopsis fructus*
269	L–Sorbose growth.	207	182
270	Raffinose growth.	271	275
271	Melezitose growth.	272	274
272	D,L–Lactate growth.	273	192
273	Growth in 60% D–glucose.	Saccharomyces cerevisiae Saccharomycopsis vini**	Saccharomycopsis vini** Torulaspora delbrueckii** Zygosaccharomyces bailii** Zygosaccharomyces rouxii**
274	L–Arabinose growth.	232	Torulopsis multis–gemmis*
275	Melibiose growth.	276	236
276	D–Galactose growth.	277	279
277	Melezitose growth.	278	233
278	Growth in 60% D–glucose.	Saccharomyces cerevisiae Saccharomycopsis vini**	Saccharomycopsis vini** Torulaspora delbrueckii**
279	Lactose growth.	280	282
280	Growth in 60% D–glucose.	281	Torulaspora delbrueckii**
281	Inulin growth.	Saccharomyces cerevisiae	Kluyveromyces bulgaricus
282	Growth without pantothenate	Candida kefyr	Kluyveromyces bulgaricus
283	D–Galactose growth.	284	291
284	Nitrate growth.	285	289
285	L–Arabinose growth.	286	Phaffia rhodozyma*
286	L–Sorbose growth.	287	288
287	Starch growth.	Saccharomycopsis vini**	Saccharomycopsis fibuligera**
288	Melezitose growth.	Saccharomycopsis vini**	Pichia mucosa*
289	Sucrose growth.	Hansenula mrakii	290
290	Maltose growth.	Hansenula saturnus	Hansenula beijerinckii
291	Maltose growth.	292	302
292	Lactose growth.	293	295
293	Sucrose growth.	Candida fragicola*	294
294	Melibiose growth.	Kluyveromyces bulgaricus Kluyveromyces lactis	Saccharomyces kluyveri
295	D–Xylose growth.	296	298
296	Nitrate growth.	297	Brettanomyces anomalus
297	Growth without pantothenate	Candida kefyr	Kluyveromyces bulgaricus Kluyveromyces lactis
298	Raffinose growth.	Kluyveromyces lactis Kluyveromyces wickerhamii	299
299	Lactose fermentation.	300	301
300	Growth without biotin.	Kluyveromyces bulgaricus Kluyveromyces lactis	Candida macedoniensis
301	Growth without pantothenate	Candida pseudotropicalis	Kluyveromyces bulgaricus Kluyveromyces lactis
302	Nitrate growth.	303	308

Key No. 4: fermenting yeasts

		Negative	Positive
303	Melibiose growth.	304	Saccharomyces kluyveri
304	Lactose growth.	305	307
305	Cellobiose fermentation.	306	Dekkera intermedia
306	Growth without inositol.	Zygosaccharomyces fermentati**	Kluyveromyces lactis
307	Methyl α−D−glucopyranoside fermentation	Kluyveromyces lactis	Brettanomyces claussenii
308	Lactose growth.	309	253
309	L−Sorbose growth.	Brettanomyces custersii Dekkera intermedia	Hansenula silvicola
310	Glycerol growth.	311	417
311	Raffinose growth.	312	367
312	Maltose growth.	313	333
313	D−Glucitol growth.	314	316
314	L−Rhamnose growth.	315	Torulopsis anatomiae
315	Nitrate growth.	68	Torulopsis halophilus*
316	Methanol growth.	317	Torulopsis sonorensis*
317	Cellobiose growth.	318	332
318	Succinate growth.	26	319
319	D,L−Lactate growth.	320	326
320	Citrate growth.	321	Candida diversa Torulopsis karawaiewi*
321	Trehalose growth.	322	324
322	Growth in 60% D−glucose.	323	Pichia fluxuum Torulaspora delbrueckii**
323	D−δ−Gluconolactone growth	Pichia fluxuum Saccharomyces cerevisiae Torulopsis karawaiewi*	Pichia delftensis Torulopsis karawaiewi*
324	Growth in 60% D−glucose.	325	Torulaspora delbrueckii**
325	2−Ketogluconate growth.	Saccharomyces cerevisiae	Pichia dispora
326	Trehalose growth.	327	330
327	Growth without thiamin.	328	329
328	2−Ketogluconate growth.	Pichia saitoi Saccharomyces cerevisiae	Pichia besseyi*
329	Growth in 60% D−glucose.	Pichia fluxuum Saccharomyces cerevisiae	Pichia fluxuum Torulaspora delbrueckii**
330	Growth in 60% D−glucose.	331	Torulaspora delbrueckii**
331	2−Ketogluconate growth.	Pichia saitoi Saccharomyces cerevisiae	Pichia dispora
332	Sucrose growth.	Brettanomyces naardenensis*	Candida curiosa
333	Ethanol growth.	334	342
334	Cellobiose growth.	335	337
335	D−Xylose growth.	17	336
336	Starch growth.	Torulaspora delbrueckii**	Candida albicans
337	D−Galactose growth.	338	339
338	L−Sorbose growth.	Candida insectamans*	Candida suecica*
339	Citrate growth.	340	341
340	Maltose fermentation.	Metschnikowia bicuspidata	Candida albicans
341	Starch growth.	Candida buinensis*	Candida albicans
342	D−Xylose growth.	343	350
343	Cellobiose growth.	344	345
344	Growth in 60% D−glucose.	Saccharomyces cerevisiae Schwanniomyces occidentalis	Torulaspora delbrueckii** Zygosaccharomyces rouxii**
345	D−Galactose growth.	346	349
346	L−Sorbose growth.	347	348
347	Sucrose fermentation.	Aciculoconidium aculeatum**	Schwanniomyces occidentalis
348	Starch growth.	Pichia spartinae*	Schwanniomyces occidentalis
349	Sucrose fermentation.	Metschnikowia bicuspidata	Schwanniomyces occidentalis
350	L−Rhamnose growth.	351	365
351	Citrate growth.	352	359
352	Sucrose growth.	353	355
353	Cellobiose growth.	Torulaspora delbrueckii**	354

		Negative	Positive
354	Maltose fermentation.	Brettanomyces naardenensis*	Candida tropicalis
355	Cellobiose growth.	356	357
356	Starch growth.	Torulaspora delbrueckii**	Candida albicans Schwanniomyces occidentalis
357	Maltose fermentation.	349	358
358	5-Ketogluconate growth.	Candida albicans Schwanniomyces occidentalis	Candida albicans Candida tropicalis
359	D-Galactose growth.	360	361
360	D,L-Lactate growth.	Schwanniomyces occidentalis	Pichia angophorae
361	D-Galactose fermentation.	362	363
362	Starch growth.	Pichia etchellsii	Candida albicans Schwanniomyces occidentalis
363	Sucrose fermentation.	364	358
364	Starch growth.	Candida buinensis*	Candida albicans Candida tropicalis
365	Sucrose growth.	Brettanomyces naardenensis*	366
366	L-Sorbose growth.	Candida oregonensis	Candida lusitaniae
367	Cellobiose growth.	368	394
368	D-Glucitol growth.	369	381
369	Melibiose growth.	370	375
370	Maltose growth.	371	372
371	Lactose growth.	224	Candida kefyr
372	Starch growth.	373	Saccharomyces cerevisiae Schwanniomyces occidentalis
373	Ethylamine growth.	25	374
374	Growth without niacin.	Kluyveromyces thermotolerans**	Torulaspora delbrueckii**
375	D,L-Lactate growth.	376	237
376	L-Sorbose growth.	377	380
377	Growth in 60% D-glucose.	378	Torulaspora delbrueckii**
378	Starch growth.	379	Saccharomyces cerevisiae Schwanniomyces occidentalis
379	Ethylamine growth.	Saccharomyces cerevisiae	Saccharomyces kluyveri
380	Starch growth.	241	Schwanniomyces occidentalis
381	Melibiose growth.	382	388
382	Maltose growth.	383	385
383	L-Sorbose growth.	384	Kluyveromyces waltii* Torulaspora delbrueckii** Torulopsis apicola
384	Lactose growth.	25	Candida kefyr
385	Nitrate growth.	386	Citeromyces matritensis
386	D,L-Lactate growth.	372	387
387	2-Ketogluconate growth.	25	Torulaspora delbrueckii** Torulaspora pretoriensis**
388	D,L-Lactate growth.	389	237
389	L-Sorbose growth.	390	380
390	Maltose growth.	391	377
391	Growth in 60% D-glucose.	392	Torulaspora delbrueckii**
392	Ethylamine growth.	Saccharomyces cerevisiae	393
393	Growth at 37 degrees.	Zygosaccharomyces mrakii**	Saccharomyces kluyveri
394	Nitrate growth.	395	410
395	Galactitol growth.	396	409
396	D,L-Lactate growth.	397	404
397	Melibiose growth.	398	401
398	D-Ribose growth.	399	Candida pseudointermedia*
399	Lactose growth.	400	Candida intermedia
400	Starch formation.	Schwanniomyces occidentalis	Phaffia rhodozyma*
401	Citrate growth.	402	403

Key No. 4: fermenting yeasts

		Negative	Positive
402	Starch growth.	Saccharomyces kluyveri	Schwanniomyces occidentalis
403	Sucrose fermentation.	Candida melibiosica	Schwanniomyces occidentalis
404	Melibiose growth.	405	408
405	Maltose growth.	406	407
406	D–Xylose growth.	Candida kefyr	Candida pseudotropicalis
407	D–Galactose growth.	Phaffia rhodozyma*	Zygosaccharomyces fermentati**
408	5–Ketogluconate growth.	Saccharomyces kluyveri	Candida salmanticensis
409	D–Ribose growth.	Debaryomyces castellii	Candida pseudointermedia*
410	Maltose growth.	411	414
411	Melibiose growth.	412	413
412	Starch formation.	Candida curiosa	Leucosporidium frigidum
413	Starch formation.	Candida curiosa	Leucosporidium nivalis
414	L–Rhamnose growth.	415	416
415	Methyl α–D–glucopyranoside growth	Candida aquatica	Candida valdiviana*
416	Melibiose growth.	Leucosporidium stokesii	Leucosporidium gelidum
417	Maltose growth.	418	516
418	Methanol growth.	419	509
419	Sucrose growth.	420	452
420	Nitrate growth.	421	445
421	Citrate growth.	422	431
422	D–Xylose growth.	423	425
423	D,L–Lactate growth.	424	202
424	Growth in 60% D–glucose.	Saccharomyces cerevisiae Saccharomycopsis vini** Torulopsis karawaiewi*	Saccharomycopsis vini** Torulaspora delbrueckii** Zygosaccharomyces bailii** Zygosaccharomyces bisporus** Zygosaccharomyces rouxii**
425	D–Galactose growth.	426	428
426	D–Ribose growth.	427	Saccharomycopsis crataegensis*
427	L–Rhamnose growth.	Torulaspora delbrueckii**	Pichia sargentensis*
428	L–Sorbose growth.	207	429
429	Cellobiose growth.	Torulaspora delbrueckii**	430
430	Trehalose growth.	Geotrichum fermentans**	Torulopsis torresii
431	Trehalose growth.	432	437
432	D–Galactose growth.	433	436
433	D–Xylose growth.	434	Pichia pijperi
434	Cellobiose growth.	Candida krissii* Torulopsis karawaiewi*	435
435	2–Ketogluconate growth.	Pichia quercuum	Candida krissii*
436	L–Sorbose growth.	Candida catenulata	Geotrichum fermentans**
437	L–Sorbose growth.	438	440
438	D,L–Lactate growth.	Candida savonica* Candida zeylanoides	439
439	L–Arabinose growth.	Candida catenulata	Pichia abadieae*
440	Cellobiose growth.	441	443
441	D–Xylose growth.	Candida iberica* Candida zeylanoides	442
442	Salicin growth.	Torulopsis fructus*	Candida zeylanoides
443	D,L–Lactate growth.	444	Candida beechii Candida santamariae
444	Trehalose fermentation.	Candida zeylanoides	Candida beechii
445	D–Xylose growth.	446	448
446	L–Sorbose growth.	447	Torulopsis nodaensis*
447	D–Galactose fermentation.	Torulopsis halonitratophila	Torulopsis halophilus*
448	D–Galactose growth.	449	450
449	L–Sorbose growth.	Hansenula mrakii Torulopsis norvegica	Hansenula dimennae
450	L–Rhamnose growth.	Pachysolen tannophilus	451
451	Methyl α–D–glucopyranoside growth	Torulopsis wickerhamii	Hansenula muscicola*
452	D–Glucitol growth.	453	464

		Negative	Positive
453	Lactose growth.	.454	.462
454	Nitrate growth.	.455	.460
455	Melibiose growth.	.456	.236
456	D,L–Lactate growth.	.457	.459
457	Raffinose growth.	.17	.458
458	Growth in 60% D–glucose.	Saccharomyces cerevisiae / Torulopsis bombicola*	Torulaspora delbrueckii** / Torulopsis bombicola*
459	Citrate growth.	.224	Wickerhamia fluorescens
460	L–Sorbose growth.	.461	Hansenula californica
461	D–Xylose growth.	Torulopsis halophilus*	Hansenula saturnus
462	D–Xylose growth.	.463	.257
463	Trehalose growth.	Candida kefyr	Debaryomyces tamarii
464	Ethanol growth.	.465	.479
465	Galactitol growth.	.466	Torulopsis spandovensis*
466	Nitrate growth.	.467	.478
467	Melibiose growth.	.468	.474
468	Raffinose growth.	.469	.470
469	Growth in 60% D–glucose.	Saccharomyces cerevisiae	Torulaspora delbrueckii** / Zygosaccharomyces bailii** / Zygosaccharomyces rouxii**
470	L–Sorbose growth.	.471	.473
471	Lactose growth.	.472	Candida kefyr
472	Cellobiose growth.	.25	Torulopsis gropengiesseri
473	Succinate growth.	Torulaspora delbrueckii** / Torulopsis apicola	Candida bombi* / Torulaspora delbrueckii**
474	L–Sorbose growth.	.475	.60
475	D,L–Lactate growth.	.476	.58
476	Growth in 60% D–glucose.	.477	Torulaspora delbrueckii**
477	D–δ–Gluconolactone growth	Saccharomyces cerevisiae	Zygosaccharomyces mrakii**
478	Trehalose growth.	Torulopsis magnoliae	Candida curiosa
479	Cellobiose growth.	.480	.495
480	Galactitol growth.	.481	Torulopsis spandovensis*
481	Melibiose growth.	.482	.490
482	D,L–Lactate growth.	.483	.488
483	Nitrate growth.	.484	Torulopsis magnoliae
484	Raffinose growth.	.273	.485
485	Growth in 60% D–glucose.	.486	.487
486	Growth at 37 degrees.	Saccharomyces cerevisiae / Saccharomycopsis vini**	Saccharomyces cerevisiae / Torulopsis bombicola*
487	Growth at 37 degrees.	Saccharomycopsis vini** / Torulaspora delbrueckii**	Torulaspora delbrueckii** / Torulopsis bombicola*
488	Lactose growth.	.489	.282
489	Growth with 0.01% cycloheximide .	.25	Kluyveromyces bulgaricus / Kluyveromyces phaseolosporus
490	L–Sorbose growth.	.491	.492
491	D,L–Lactate growth.	.391	.237
492	D,L–Lactate growth.	.241	.493
493	Growth in 60% D–glucose.	.494	Torulaspora delbrueckii**
494	Growth with 0.01% cycloheximide .	.239	Zygosaccharomyces cidri**
495	Nitrate growth.	.496	.506
496	Lactose growth.	.497	.499
497	D–Galactose growth.	Saccharomycopsis vini**	.498
498	Melibiose growth.	Kluyveromyces bulgaricus / Kluyveromyces lactis / Kluyveromyces phaseolosporus	Saccharomyces kluyveri
499	L–Sorbose growth.	.500	.504
500	D–Xylose growth.	.297	.501
501	Lactose fermentation.	.502	.503
502	Growth without biotin.	Kluyveromyces bulgaricus / Kluyveromyces lactis / Kluyveromyces marxianus	Candida macedoniensis
503	Growth without pantothenate	Candida pseudotropicalis / Kluyveromyces marxianus	Kluyveromyces bulgaricus / Kluyveromyces lactis / Kluyveromyces marxianus

Key No. 4: fermenting yeasts

		Negative	Positive
504	L–Arabinose growth.	505	Kluyveromyces bulgaricus Kluyveromyces lactis Kluyveromyces marxianus
505	D–Galactose fermentation.	Kluyveromyces aestuarii	Kluyveromyces bulgaricus Kluyveromyces lactis
506	Methyl α–D–glucopyranoside growth	507	Hansenula californica
507	D–Galactose growth.	508	413
508	Trehalose growth.	Hansenula saturnus	Candida curiosa
509	L–Arabinose growth.	510	513
510	L–Rhamnose growth.	511	512
511	D–Galactose growth.	Hansenula minuta	Torulopsis nodaensis*
512	Cellobiose growth.	Pichia pastoris	Pichia lindnerii* Torulopsis methanolovescens*
513	Ethanol growth.	Torulopsis pignaliae*	514
514	D–Galactose growth.	Torulopsis sonorensis*	515
515	L–Sorbose growth.	Torulopsis nitratophila	Torulopsis nagoyaensis*
516	Raffinose growth.	517	646
517	D–Galactose growth.	518	564
518	L–Rhamnose growth.	519	556
519	D–Xylose growth.	520	539
520	Melezitose growth.	521	530
521	Cellobiose growth.	522	525
522	Methyl α–D–glucopyranoside growth	523	524
523	Growth in 60% D–glucose.	Saccharomyces cerevisiae Saccharomycopsis vini**	Saccharomycopsis vini** Torulaspora delbrueckii** Zygosaccharomyces rouxii**
524	Starch formation.	278	Filobasidium capsuligenum**
525	D–Ribose growth.	526	529
526	Starch growth.	Saccharomycopsis vini**	527
527	Growth at 37 degrees.	528	Saccharomycopsis fibuligera**
528	Growth without biotin.	Aciculoconidium aculeatum**	Filobasidium capsuligenum**
529	Methyl α–D–glucopyranoside growth	Saccharomycopsis malanga*	Filobasidium capsuligenum**
530	Cellobiose growth.	531	532
531	Growth in 60% D–glucose.	Saccharomyces cerevisiae Schwanniomyces occidentalis	Torulaspora delbrueckii**
532	L–Sorbose growth.	533	536
533	Sucrose fermentation.	534	Saccharomycopsis fibuligera** Schwanniomyces occidentalis
534	Starch growth.	Metschnikowia reukaufii	535
535	Growth at 37 degrees.	Aciculoconidium aculeatum**	Saccharomycopsis fibuligera**
536	Citrate growth.	537	348
537	Sucrose fermentation.	538	Schwanniomyces occidentalis
538	Growth without thiamin.	Pichia mucosa*	Metschnikowia reukaufii
539	D,L–Lactate growth.	540	551
540	Cellobiose growth.	541	546
541	Melezitose growth.	542	543
542	Starch growth.	Torulaspora delbrueckii**	Filobasidium capsuligenum**
543	Citrate growth.	544	545
544	Starch growth.	Torulaspora delbrueckii**	Schwanniomyces occidentalis
545	Sucrose fermentation.	Torulopsis musae*	Schwanniomyces occidentalis
546	Melezitose growth.	547	549
547	D–Ribose growth.	548	529

		Negative	Positive
548	Starch growth.	Pichia toletana	Filobasidium capsuligenum**
549	Citrate growth.	537	550
550	Sucrose fermentation.	Pichia toletana	Schwanniomyces occidentalis
551	Nitrate growth.	552	555
552	L–Arabinose growth.	553	Pichia bovis
553	Cellobiose growth.	Torulaspora delbrueckii**	554
554	Growth without pyridoxine	Candida quercuum*	Pichia toletana
555	L–Sorbose growth.	Candida vartiovaarai	Hansenula californica
556	Nitrate growth.	557	561
557	L–Sorbose growth.	558	Candida lusitaniae
558	Growth without pyridoxine	559	560
559	Growth without thiamin.	Pichia rhodanensis Pichia veronae*	Pichia wickerhamii
560	Growth without biotin.	Pichia toletana	Candida freyschussii
561	L–Sorbose growth.	562	563
562	Trehalose growth.	Torulopsis ernobii	Hansenula beckii** Hansenula bimundalis
563	D,L–Lactate growth.	Torulopsis ernobii	Hansenula californica
564	Cellobiose growth.	565	603
565	Ethanol growth.	566	572
566	D–Xylose growth.	567	568
567	Nitrate growth.	17	114
568	L–Arabinose growth.	569	Candida albicans Candida mogii
569	Sucrose growth.	570	571
570	Starch growth.	Torulaspora delbrueckii**	Candida ravautii
571	Starch growth.	Candida sake Torulaspora delbrueckii**	Candida albicans
572	Sucrose growth.	573	585
573	Nitrate growth.	574	584
574	D,L–Lactate growth.	575	577
575	Growth in 60% D–glucose.	576	Torulaspora delbrueckii** Zygosaccharomyces rouxii**
576	2–Ketogluconate growth.	Saccharomyces cerevisiae	Filobasidium capsuligenum**
577	D–Xylose growth.	578	579
578	Citrate growth.	25	Candida catenulata
579	Trehalose growth.	580	583
580	Citrate growth.	581	582
581	Starch growth.	Torulaspora delbrueckii**	Candida brumptii
582	Growth at 37 degrees.	Candida brumptii	Candida catenulata
583	Citrate growth.	570	Candida catenulata Candida ravautii
584	Maltose fermentation.	Torulopsis nodaensis*	Torulopsis etchellsii
585	D–Xylose growth.	586	592
586	Melezitose growth.	587	589
587	Citrate growth.	575	588
588	L–Sorbose growth.	Filobasidium capsuligenum**	Nadsonia fulvescens
589	L–Sorbose growth.	531	590
590	Starch growth.	591	Schwanniomyces occidentalis
591	Growth in 60% D–glucose.	Lodderomyces elongisporus	Torulaspora delbrueckii**
592	L–Arabinose growth.	593	600
593	Starch growth.	594	598
594	Salicin growth.	595	597
595	Growth in 60% D–glucose.	596	Candida sake Torulaspora delbrueckii**
596	Growth at 37 degrees.	Candida sake	Lodderomyces elongisporus
597	Growth at 37 degrees.	Candida sake	Candida maltosa*
598	Melezitose growth.	599	Candida albicans Schwanniomyces occidentalis
599	Growth at 37 degrees.	Filobasidium capsuligenum**	Candida albicans
600	L–Rhamnose growth.	601	Torulopsis haemulonii

Key No. 4: fermenting yeasts

		Negative	Positive
601	Melezitose growth.	.599	.602
602	Starch growth.	Candida parapsilosis	Candida albicans Schwanniomyces occidentalis
603	Nitrate growth.	.604	.642
604	Citrate growth.	.605	.625
605	L–Rhamnose growth.	.606	Candida lusitaniae
606	Sucrose fermentation.	.607	.623
607	Maltose fermentation.	.608	.618
608	L–Sorbose growth.	.609	.614
609	Melezitose growth.	.610	.611
610	D–Galactose fermentation.	Filobasidium capsuligenum**	Kluyveromyces lactis
611	D–Galactose fermentation.	.612	.613
612	Growth at 37 degrees.	Metschnikowia reukaufii	Metschnikowia lunata*
613	Growth without niacin.	Kluyveromyces lactis	Metschnikowia reukaufii
614	D–Galactose fermentation.	.615	.617
615	Growth at 37 degrees.	.616	Metschnikowia lunata* Metschnikowia pulcherrima
616	Growth without thiamin.	Metschnikowia bicuspidata Metschnikowia zobellii	Candida sake Metschnikowia bicuspidata Metschnikowia pulcherrima Metschnikowia reukaufii
617	Growth without niacin.	Kluyveromyces lactis	Candida sake Metschnikowia pulcherrima Metschnikowia reukaufii
618	D–Galactose fermentation.	.619	.621
619	L–Sorbose growth.	.599	.620
620	Starch growth.	Candida sake	Candida albicans
621	Starch growth.	.622	Candida albicans Candida tropicalis
622	Growth without niacin.	Kluyveromyces lactis	Candida sake
623	Starch growth.	.624	.358
624	Growth without niacin.	Candida maltosa* Kluyveromyces lactis	.597
625	Starch growth.	.626	.635
626	L–Rhamnose growth.	.627	Candida lusitaniae
627	Melibiose growth.	.628	Torulopsis navarrensis*
628	L–Arabinose growth.	.629	.634
629	Sucrose fermentation.	.630	.633
630	D–Galactose fermentation.	.631	.632
631	Growth at 37 degrees.	Candida sake Metschnikowia pulcherrima	Metschnikowia lunata* Metschnikowia pulcherrima
632	2–Ketogluconate growth.	Candida buinensis*	Candida oleophila* Candida sake Metschnikowia pulcherrima
633	Trehalose fermentation.	Candida oleophila* Candida sake	.597
634	D–Galactose fermentation.	Pichia etchellsii	Candida buinensis*
635	Lactose growth.	.636	Candida fluviotilis*
636	Melezitose growth.	.637	.639
637	Maltose fermentation.	.638	.599
638	Growth without vitamins.	Filobasidium capsuligenum**	Torulopsis kruisii*
639	Maltose fermentation.	.640	.641
640	Sucrose fermentation.	Torulopsis kruisii*	Schwanniomyces occidentalis
641	Sucrose fermentation.	Candida albicans Candida tropicalis Candida viswanathii	.358
642	D–Glucitol growth.	Torulopsis mannitofaciens*	.643
643	Ethanol growth.	Torulopsis versatilis	.644
644	L–Sorbose growth.	Hansenula muscicola*	.645
645	Starch growth.	Hansenula silvicola	Hansenula holstii
646	Galactitol growth.	.647	.712
647	Nitrate growth.	.648	.703
648	D,L–Lactate growth.	.649	.674
649	Citrate growth.	.650	.659

		Negative	Positive
650	Cellobiose growth.	651	653
651	Melibiose growth.	652	376
652	Melezitose growth.	278	372
653	D–Galactose growth.	654	656
654	L–Arabinose growth.	655	400
655	Melezitose growth.	287	Saccharomycopsis fibuligera** Schwanniomyces occidentalis
656	L–Rhamnose growth.	657	Candida inositophila*
657	Melibiose growth.	658	402
658	Starch growth.	Kluyveromyces lactis	Schwanniomyces occidentalis
659	Melibiose growth.	660	668
660	Ethanol growth.	661	663
661	D–Galactose growth.	Phaffia rhodozyma*	662
662	Methyl α–D–glucopyranoside growth	Candida mogii	Candida pseudointermedia*
663	D–Galactose growth.	664	665
664	L–Arabinose growth.	Saccharomycopsis fibuligera** Schwanniomyces occidentalis	400
665	D–Ribose growth.	666	Candida pseudointermedia*
666	L–Rhamnose growth.	667	Torulopsis haemulonii
667	Melezitose growth.	Pichia ohmeri	Schwanniomyces occidentalis
668	L–Arabinose growth.	669	673
669	L–Sorbose growth.	670	671
670	Starch growth.	Torulopsis navarrensis*	Schwanniomyces occidentalis
671	Sucrose fermentation.	Candida melibiosica Torulopsis pampelonensis*	672
672	Starch growth.	Torulopsis pampelonensis*	Schwanniomyces occidentalis
673	D–Ribose growth.	Schwanniomyces occidentalis	Pichia guilliermondii
674	L–Arabinose growth.	675	693
675	Cellobiose growth.	676	683
676	L–Sorbose growth.	677	679
677	Melibiose growth.	387	678
678	Growth in 60% D–glucose.	379	Torulaspora delbrueckii**
679	Melibiose growth.	680	681
680	Growth without niacin.	Kluyveromyces drosophilarum	Torulaspora delbrueckii**
681	Growth in 60% D–glucose.	682	Torulaspora delbrueckii**
682	Growth with 0.01% cycloheximide	Saccharomyces kluyveri	Zygosaccharomyces cidri**
683	D–Galactose growth.	684	685
684	D–Ribose growth.	Saccharomycopsis fibuligera**	Candida maritima
685	Melibiose growth.	686	Saccharomyces kluyveri
686	Lactose growth.	687	692
687	Melezitose growth.	688	689
688	Citrate growth.	306	Pichia ohmeri
689	Maltose fermentation.	Kluyveromyces drosophilarum Kluyveromyces lactis	690
690	Methyl α–D–glucopyranoside fermentation	306	691
691	Growth without inositol.	Zygosaccharomyces fermentati**	Kluyveromyces dobzhanskii
692	D–Galactose fermentation.	Kluyveromyces aestuarii	Kluyveromyces lactis
693	D–Galactose growth.	694	696
694	L–Rhamnose growth.	695	Pichia rabaulensis*
695	Growth without thiamin.	Pichia onychis	Phaffia rhodozyma*
696	Melibiose growth.	697	702
697	L–Rhamnose growth.	698	701

Key No. 4: fermenting yeasts

	Negative	Positive
698 Methyl α−D−glucopyranoside fermentation	699	700
699 Starch growth.	Kluyveromyces drosophilarum Kluyveromyces lactis	Debaryomyces hansenii
700 Trehalose fermentation.	Debaryomyces hansenii	Kluyveromyces dobzhanskii
701 Starch growth.	Pichia strasburgensis	Debaryomyces hansenii
702 Starch growth.	Pichia guilliermondii	Debaryomyces hansenii
703 D−Glucitol growth.	704	705
704 D−Galactose growth.	Candida utilis	Torulopsis mannitofaciens*
705 D−Galactose growth.	706	710
706 D−Xylose growth.	Citeromyces matritensis	707
707 Citrate growth.	Hansenula beijerinckii	708
708 L−Rhamnose growth.	709	Hansenula petersonii
709 Inulin growth.	Hansenula fabianii	Hansenula jadinii
710 Ethanol growth.	711	415
711 L−Sorbose growth.	Torulopsis versatilis	Candida valdiviana*
712 Melibiose growth.	713	721
713 Ethanol growth.	714	716
714 Cellobiose growth.	Candida mogii	715
715 Citrate growth.	Candida inositophila*	Candida pseudointermedia*
716 Cellobiose growth.	Torulopsis haemulonii	717
717 D,L−Lactate growth.	718	720
718 Citrate growth.	Candida inositophila*	719
719 2−Ketogluconate growth.	Candida steatolytica*	Candida pseudointermedia*
720 Growth without thiamin.	Candida steatolytica*	Debaryomyces hansenii
721 Nitrate growth.	722	Candida valdiviana*
722 Lactose growth.	723	725
723 Sucrose fermentation.	724	702
724 Starch growth.	Candida brassicae*	Debaryomyces hansenii
725 Melibiose fermentation.	Debaryomyces hansenii	Debaryomyces castellii
726 Methanol growth.	727	810
727 Galactitol growth.	728	791
728 Raffinose growth.	729	763
729 Sucrose growth.	730	741
730 D−Mannitol growth.	Candida flavificans*	731
731 Citrate growth.	732	736
732 D−Galactose growth.	733	735
733 L−Sorbose growth.	734	Torulopsis cantarelii
734 Methyl α−D−glucopyranoside growth	Saccharomycopsis malanga*	Saccharomycopsis capsularis**
735 L−Sorbose growth.	Candida cacaoi	Candida conglobata
736 Methyl α−D−glucopyranoside growth	737	739
737 Growth at 37 degrees.	738	Pichia farinosa
738 Growth without vitamins.	Candida boleticola*	Torulopsis schatavii*
739 D−Galactose growth.	Saccharomycopsis capsularis**	740
740 L−Sorbose growth.	Hansenula muscicola*	Pichia acaciae
741 L−Rhamnose growth.	742	752
742 D−Galactose growth.	743	746
743 L−Arabinose growth.	744	Ambrosiozyma cicatricosa* Ambrosiozyma monospora**
744 L−Sorbose growth.	745	Candida incommunis*
745 D−Ribose growth.	Saccharomycopsis fibuligera**	Pichia ambrosiae*
746 Lactose growth.	747	749
747 Starch growth.	748	Pichia nakazawae*
748 2−Ketogluconate growth.	Candida butyri* Candida diddensii	Candida butyri* Pichia philogaea
749 D,L−Lactate growth.	750	Candida chilensis*
750 Maltose fermentation.	751	Trichosporon fennicum*
751 Growth at 37 degrees.	Candida ergatensis*	Candida butyri*
752 D,L−Lactate growth.	753	759
753 Nitrate growth.	754	Hansenula holstii
754 L−Sorbose growth.	755	757

		Negative	Positive

755 Starch growth. ....Candida diddensii 756
Candida insectorum*
Candida tenuis
Candida veronae

756 Growth without vitamins. ....Candida insectorum* ....Candida homilentoma*
Candida naeodendra* Candida tenuis
Candida tenuis

757 Starch growth. ....Candida diddensii 758
Candida insectorum*
Candida tenuis
Candida terebra*

758 Growth without vitamins. ....Candida insectorum* ....Candida homilentoma*
Candida naeodendra* Candida tenuis
Candida tenuis
Pichia nakazawae*

759 Nitrate growth. 760 762
760 D–Galactose growth. ....Ambrosiozyma philentoma* 761
761 Maltose fermentation. ....Candida tenuis ....Pichia stipitis
762 L–Sorbose growth. ....Hormoascus platypodis** ....Candida ishiwadae*
763 Nitrate growth. 764 788
764 Ethanol growth. 765 768
765 Melibiose growth. 766 767
766 Growth at 37 degrees. ....Trichosporon ....Trichosporon fennicum*
melibiosaceum*
767 Maltose fermentation. ....Debaryomyces nepalensis* ....Trichosporon
melibiosaceum*
768 L–Rhamnose growth. 769 783
769 Melibiose growth. 770 777
770 D–Galactose growth. ....Saccharomycopsis 771
fibuligera**
771 Citrate growth. 772 773
772 Starch growth. ....Candida rhagii ....Debaryomyces hansenii
773 D,L–Lactate growth. ....774 ....Debaryomyces hansenii
Debaryomyces marama
774 Lactose growth. 775 776
775 Growth without vitamins. ....Debaryomyces marama ....Hyphopichia burtonii**
776 Growth without vitamins. ....Debaryomyces marama ....Trichosporon fennicum*
777 Lactose growth. 778 781
778 L–Sorbose growth. ....779 ....Debaryomyces hansenii
Debaryomyces marama
Debaryomyces nepalensis*
779 D–Galactose fermentation. ....Debaryomyces hansenii 780
Debaryomyces marama
780 Melibiose fermentation. ....Debaryomyces hansenii ....Candida silvicultrix*
781 D,L–Lactate growth. ....782 ....Debaryomyces hansenii
Debaryomyces marama
782 Starch growth. ....Candida entomophila* ....Debaryomyces marama
783 Succinate growth. ....Pichia scolyti 784
784 D,L–Lactate growth. 785 787
785 Melibiose growth. ....786 ....Candida insectorum*
Candida silvanorum*
786 Maltose fermentation. ....Candida insectorum* ....Wingea robertsii
787 Trehalose fermentation. ....Debaryomyces hansenii ....Wingea robertsii
788 Melibiose growth. ....789 ....Hansenula sydowiorum*
789 Inulin growth. ....790 ....Hansenula lynferdii*
790 Growth without vitamins. ....Hansenula subpelliculosa ....Hansenula anomala
Hansenula ciferrii
791 Raffinose growth. 792 800
792 Ethanol growth. 793 795
793 D,L–Lactate growth. ....Candida terebra* 794
794 Nitrate growth. ....Selenozyma peltata* ....Candida ishiwadae*
795 Melezitose growth. 796 797
796 L–Rhamnose growth. ....Pichia haplophila ....Candida dendronema*
797 Nitrate growth. 798 799
798 L–Sorbose growth. ....Candida entomaea* ....Candida terebra*
799 D,L–Lactate growth. ....Hansenula holstii ....Candida ishiwadae*
800 L–Rhamnose growth. 801 806

Key No. 4: fermenting yeasts

		Negative	Positive
801	Methyl α−D−glucopyranoside growth	Debaryomyces polymorpha**	802
802	D,L−Lactate growth.	803	805
803	Starch growth.	804	Debaryomyces phaffii / Debaryomyces vanriji
804	Inulin growth.	Candida friedrichii	Candida membranaefaciens
805	Starch growth.	Candida membranaefaciens	Debaryomyces hansenii / Debaryomyces vanriji
806	Melibiose growth.	807	809
807	D,L−Lactate growth.	808	Debaryomyces hansenii
808	L−Sorbose growth.	Pichia heimii*	Candida blankii
809	Lactose growth.	Debaryomyces hansenii / Debaryomyces vanriji	Debaryomyces hansenii / Pichia pseudopolymorpha
810	Maltose growth.	811	817
811	Cellobiose growth.	812	814
812	Trehalose growth.	Candida boidinii	813
813	Nitrate growth.	Pichia trehalophila	Hansenula philodendra*
814	Nitrate growth.	815	816
815	D−Galactose growth.	Pichia pinus	Pichia methanolica*
816	Growth without thiamin.	Hansenula glucozyma	Hansenula henricii
817	Sucrose growth.	Hansenula capsulata / Torulopsis molischiana	818
818	D−Ribose growth.	Pichia naganishii*	Hansenula polymorpha

Ascosporogenous yeasts with spherical, oval or reniform ascospores (test No. 76 positive) (Key No. 5)

Key involving physiological tests and microscopical examination

Yeasts in Key No. 5

124 Citeromyces matritensis	213 Kluyveromyces phaffii	312 Saccharomyces cerevisiae
146 Debaryomyces castellii	214 Kluyveromyces phaseolosporus	313 Saccharomyces dairensis
147 Debaryomyces coudertii	215 Kluyveromyces polysporus	314 Saccharomyces exiguus
148 Debaryomyces hansenii	216 Kluyveromyces thermotolerans**	315 Saccharomyces kluyveri
149 Debaryomyces marama	217 Kluyveromyces waltii*	316 Saccharomyces servazzii*
150 Debaryomyces melissophila*	218 Kluyveromyces wickerhamii	317 Saccharomyces telluris
151 Debaryomyces nepalensis*	225 Lipomyces anomalus*	318 Saccharomyces unisporus
152 Debaryomyces phaffii	226 Lipomyces kononenkoae	319 Saccharomycodes ludwigii
153 Debaryomyces polymorpha**	227 Lipomyces lipofer	328 Schizosaccharomyces japonicus
154 Debaryomyces tamarii	228 Lipomyces starkeyi	329 Schizosaccharomyces malidevorans
155 Debaryomyces vanriji	229 Lipomyces tetrasporus*	330 Schizosaccharomyces octosporus
156 Debaryomyces yarrowii*	230 Lodderomyces elongisporus	331 Schizosaccharomyces pombe
164 Guilliermondella selenospora**	237 Nadsonia commutata*	332 Schizosaccharomyces slooffiae*
167 Hanseniaspora osmophila	238 Nadsonia elongata	357 Torulaspora delbrueckii**
168 Hanseniaspora uvarum	239 Nadsonia fulvescens	358 Torulaspora globosa**
170 Hanseniaspora vineae*	242 Pachytichospora transvaalensis	359 Torulaspora pretoriensis**
203 Kluyveromyces aestuarii	244 Pichia abadieae*	429 Wickerhamiella domercqii**
204 Kluyveromyces africanus	254 Pichia etchellsii	430 Wingea robertsii
205 Kluyveromyces blattae*	255 Pichia farinosa	431 Zendera ovetensis
206 Kluyveromyces bulgaricus	261 Pichia humboldtii*	432 Zygosaccharomyces bailii**
207 Kluyveromyces delphensis	263 Pichia kudriavzevii	433 Zygosaccharomyces bisporus**
208 Kluyveromyces dobzhanskii	266 Pichia membranaefaciens	434 Zygosaccharomyces cidri**
209 Kluyveromyces drosophilarum	278 Pichia pseudopolymorpha	435 Zygosaccharomyces fermentati**
210 Kluyveromyces lactis	282 Pichia saitoi	436 Zygosaccharomyces florentinus**
211 Kluyveromyces lodderi	286 Pichia scutulata*	437 Zygosaccharomyces microellipsodes**
212 Kluyveromyces marxianus	290 Pichia terricola	438 Zygosaccharomyces mrakii**
	294 Pichia vini	439 Zygosaccharomyces rouxii**

Tests in Key No. 5

1 D−Glucose fermentation	28 Raffinose growth	55 Growth without thiamin
3 Maltose fermentation	30 Inulin growth	56 Growth without pyridoxine
4 Methyl α−D−glucopyranoside fermentation	32 Glycerol growth	57 Growth without niacin
6 Trehalose fermentation	33 Erythritol growth	60 Growth in 50% D−glucose
7 Melibiose fermentation	35 Galactitol growth	61 Growth in 60% D−glucose
12 D−Galactose growth	37 D−Glucitol growth	63 Growth with 0.01% cycloheximide
13 L−Sorbose growth	39 D−δ−Gluconolactone growth	64 Growth with 0.1% cycloheximide
14 D−Ribose growth	40 2−Ketogluconate growth	65 Starch formation
18 L−Rhamnose growth	42 D,L−Lactate growth	69 Apical budding
22 Methyl α−D−glucopyranoside growth	44 Citrate growth	72 Filamentous
24 Salicin growth	47 Ethylamine growth	74 Septate hyphae
27 Lactose growth	49 Nitrate growth	
	54 Growth without biotin	number of different tests 36

KEY No. 5

		Negative	Positive
1	D–Glucitol growth.	2	101
2	Glycerol growth.	3	53
3	D–Galactose growth.	4	25
4	Raffinose growth.	5	15
5	Salicin growth.	6	14
6	L–Sorbose growth.	7	12
7	Maltose fermentation.	8	10
8	Growth in 60% D–glucose.	Pichia membranaefaciens / Saccharomyces cerevisiae / Saccharomyces telluris	9
9	Starch formation.	Pichia membranaefaciens / Torulaspora delbrueckii** / Zygosaccharomyces bisporus** / Zygosaccharomyces rouxii**	Schizosaccharomyces slooffiae*
10	Starch formation.	11	Schizosaccharomyces octosporus
11	Growth in 60% D–glucose.	Saccharomyces cerevisiae	Torulaspora delbrueckii** / Zygosaccharomyces rouxii**
12	Growth in 50% D–glucose.	13	Pichia membranaefaciens / Torulaspora delbrueckii** / Zygosaccharomyces bisporus** / Zygosaccharomyces rouxii**
13	Apical budding.	Pichia membranaefaciens	Nadsonia elongata
14	2–Ketogluconate growth.	Hanseniaspora osmophila / Hanseniaspora vineae*	Hanseniaspora uvarum
15	L–Sorbose growth.	16	24
16	Maltose fermentation.	17	22
17	D,L–Lactate growth.	18	20
18	Starch formation.	19	Schizosaccharomyces malidevorans
19	Growth in 60% D–glucose.	Saccharomyces cerevisiae	Torulaspora delbrueckii**
20	Growth in 60% D–glucose.	21	Torulaspora delbrueckii**
21	Growth with 0.01% cycloheximide	Saccharomyces cerevisiae	Torulaspora globosa**
22	Starch formation.	19	23
23	Melibiose fermentation.	Schizosaccharomyces pombe	Schizosaccharomyces japonicus
24	Growth in 60% D–glucose.	Zygosaccharomyces florentinus**	Torulaspora delbrueckii**
25	Raffinose growth.	26	36
26	D–Glucose fermentation.	27	28
27	D,L–Lactate growth.	Lipomyces anomalus*	Guilliermondella selenospora**
28	Growth in 60% D–glucose.	29	Torulaspora delbrueckii** / Zygosaccharomyces bisporus** / Zygosaccharomyces rouxii**
29	D,L–Lactate growth.	30	35
30	Growth without niacin.	31	33
31	Growth with 0.1% cycloheximide	Saccharomyces cerevisiae / Saccharomyces dairensis	32
32	Ethylamine growth.	Saccharomyces servazzii*	Saccharomyces unisporus
33	Growth without pyridoxine	34	Saccharomyces cerevisiae / Saccharomyces exiguus
34	Growth without biotin.	Pachytichospora transvaalensis / Saccharomyces cerevisiae	Kluyveromyces africanus / Saccharomyces cerevisiae
35	Septate hyphae.	Saccharomyces cerevisiae / Saccharomyces exiguus	Guilliermondella selenospora**
36	L–Sorbose growth.	37	48
37	Methyl α–D–glucopyranoside growth	38	43
38	Melibiose fermentation.	39	40
39	Growth in 60% D–glucose.	Saccharomyces cerevisiae / Saccharomyces exiguus	Torulaspora delbrueckii**

Key No. 5: ascosporogenous yeasts

		Negative	Positive
40	Growth in 60% D–glucose.	41	Torulaspora delbrueckii**
41	Ethylamine growth.	Saccharomyces cerevisiae	42
42	Growth in 50% D–glucose.	Saccharomyces kluyveri	Zygosaccharomyces microellipsodes**
43	Melibiose fermentation.	44	46
44	Ethylamine growth.	19	45
45	Growth without niacin.	Kluyveromyces thermotolerans**	Torulaspora delbrueckii**
46	Growth in 60% D–glucose.	47	Torulaspora delbrueckii**
47	Ethylamine growth.	Saccharomyces cerevisiae	Saccharomyces kluyveri
48	D,L–Lactate growth.	49	51
49	Melibiose fermentation.	45	50
50	Growth in 50% D–glucose.	Saccharomyces kluyveri	24
51	Growth in 50% D–glucose.	Saccharomyces kluyveri	52
52	Growth in 60% D–glucose.	Zygosaccharomyces microellipsodes**	Torulaspora delbrueckii**
53	Raffinose growth.	54	85
54	D–Galactose growth.	55	68
55	D,L–Lactate growth.	56	63
56	Citrate growth.	57	Pichia membranaefaciens Pichia terricola
57	D–Glucose fermentation.	58	59
58	Apical budding.	Pichia membranaefaciens	Nadsonia commutata*
59	L–Sorbose growth.	60	12
60	Ethylamine growth.	61	Pichia membranaefaciens Torulaspora delbrueckii** Zygosaccharomyces bisporus** Zygosaccharomyces rouxii**
61	Growth with 0.01% cycloheximide	62	Kluyveromyces delphensis Pichia membranaefaciens
62	Growth in 60% D–glucose.	Pichia membranaefaciens Saccharomyces cerevisiae	Pichia membranaefaciens Torulaspora delbrueckii**
63	Citrate growth.	64	67
64	Growth in 60% D–glucose.	65	66
65	Ethylamine growth.	Pichia kudriavzevii Pichia membranaefaciens Saccharomyces cerevisiae	Pichia kudriavzevii Pichia membranaefaciens Pichia scutulata*
66	Filamentous.	Pichia membranaefaciens Torulaspora delbrueckii**	Pichia kudriavzevii Pichia membranaefaciens
67	Growth without thiamin.	Pichia membranaefaciens Pichia terricola	Pichia kudriavzevii Pichia membranaefaciens
68	D–Glucose fermentation.	69	74
69	L–Sorbose growth.	70	73
70	D,L–Lactate growth.	71	72
71	Apical budding.	Pichia humboldtii*	Nadsonia commutata*
72	Septate hyphae.	Pichia humboldtii*	Guilliermondella selenospora**
73	Septate hyphae.	Pichia humboldtii*	Zendera ovetensis
74	D,L–Lactate growth.	75	82
75	Growth in 60% D–glucose.	76	Torulaspora delbrueckii** Zygosaccharomyces bisporus** Zygosaccharomyces rouxii**
76	D–Ribose growth.	77	81
77	Growth without niacin.	78	79
78	Growth with 0.1% cycloheximide	Saccharomyces cerevisiae Saccharomyces dairensis	Saccharomyces servazzii*
79	D–δ–Gluconolactone growth	Kluyveromyces africanus Saccharomyces cerevisiae	80
80	Growth without biotin.	Kluyveromyces phaffii	Kluyveromyces africanus
81	Growth with 0.1% cycloheximide	Kluyveromyces blattae* Saccharomyces dairensis	Saccharomyces servazzii*
82	Lactose growth.	83	Kluyveromyces wickerhamii
83	Growth in 60% D–glucose.	84	Torulaspora delbrueckii**
84	Septate hyphae.	Saccharomyces cerevisiae	Guilliermondella selenospora**
85	D–Galactose growth.	86	93

		Negative	Positive
86	D−Glucose fermentation.	Nadsonia commutata*	87
87	L−Sorbose growth.	88	24
88	D,L−Lactate growth.	89	92
89	Maltose fermentation.	90	91
90	Salicin growth.	19	Saccharomycodes ludwigii
91	Starch formation.	19	Schizosaccharomyces pombe
92	Salicin growth.	20	Saccharomycodes ludwigii
93	Lactose growth.	94	Debaryomyces tamarii
94	D−Glucose fermentation.	Nadsonia commutata*	95
95	Melibiose fermentation.	96	99
96	Methyl α−D−glucopyranoside growth	97	44
97	Growth in 60% D−glucose.	98	Torulaspora delbrueckii**
98	D−δ−Gluconolactone growth	Saccharomyces cerevisiae	Kluyveromyces lodderi Kluyveromyces polysporus
99	L−Sorbose growth.	40	100
100	D,L−Lactate growth.	50	51
101	Erythritol growth.	102	180
102	Raffinose growth.	103	125
103	Glycerol growth.	104	108
104	Citrate growth.	105	Pichia etchellsii
105	D−Glucose fermentation.	Lipomyces anomalus*	106
106	Growth in 60% D−glucose.	107	Torulaspora delbrueckii** Zygosaccharomyces bailii** Zygosaccharomyces bisporus** Zygosaccharomyces rouxii**
107	L−Sorbose growth.	Pichia saitoi Saccharomyces cerevisiae	Nadsonia elongata
108	Citrate growth.	109	120
109	D−Glucose fermentation.	110	111
110	L−Sorbose growth.	Nadsonia commutata*	Lodderomyces elongisporus
111	Lactose growth.	112	Kluyveromyces lactis Kluyveromyces wickerhamii
112	Methyl α−D−glucopyranoside growth	113	118
113	D−Galactose growth.	114	116
114	Growth in 60% D−glucose.	115	Torulaspora delbrueckii** Zygosaccharomyces bailii** Zygosaccharomyces bisporus** Zygosaccharomyces rouxii**
115	L−Sorbose growth.	Saccharomyces cerevisiae	Nadsonia elongata
116	Salicin growth.	117	Kluyveromyces lactis
117	Growth in 60% D−glucose.	Saccharomyces cerevisiae	Torulaspora delbrueckii** Zygosaccharomyces bailii** Zygosaccharomyces bisporus** Zygosaccharomyces rouxii**
118	Growth with 0.01% cycloheximide	19	119
119	Salicin growth.	Lodderomyces elongisporus	Kluyveromyces lactis
120	Methyl α−D−glucopyranoside growth	121	122
121	L−Sorbose growth.	Pichia abadieae*	Wickerhamiella domercqii**
122	Salicin growth.	123	124
123	Apical budding.	Lodderomyces elongisporus	Nadsonia fulvescens
124	Growth without thiamin.	Pichia vini	Pichia etchellsii
125	D,L−Lactate growth.	126	144
126	Galactitol growth.	127	143
127	D−Glucose fermentation.	128	132
128	L−Sorbose growth.	Nadsonia commutata*	129
129	Citrate growth.	130	131
130	Inulin growth.	Debaryomyces yarrowii*	Lipomyces kononenkoae
131	Inulin growth.	Pichia vini	Lipomyces kononenkoae
132	D−Galactose growth.	133	137
133	Nitrate growth.	134	Citeromyces matritensis
134	L−Sorbose growth.	19	135
135	Melibiose fermentation.	136	24

Key No. 5: ascosporogenous yeasts

		Negative	Positive
136	Filamentous.	Torulaspora delbrueckii**	Kluyveromyces waltii*
137	Melibiose fermentation.	138	139
138	Salicin growth.	44	Kluyveromyces lactis
139	L−Sorbose growth.	140	50
140	Growth in 60% D−glucose.	141	Torulaspora delbrueckii**
141	Ethylamine growth.	Saccharomyces cerevisiae	142
142	Growth in 50% D−glucose.	Saccharomyces kluyveri	Zygosaccharomyces mrakii**
143	D−Glucose fermentation.	Lipomyces kononenkoae	Debaryomyces castellii
144	Galactitol growth.	145	179
145	Methyl α−D−glucopyranoside growth	146	153
146	Lactose growth.	147	152
147	Melibiose fermentation.	148	149
148	Growth with 0.01% cycloheximide	19	Kluyveromyces bulgaricus Kluyveromyces lactis Kluyveromyces phaseolosporus
149	L−Sorbose growth.	40	150
150	Growth in 60% D−glucose.	151	Torulaspora delbrueckii**
151	Growth with 0.01% cycloheximide	42	Zygosaccharomyces cidri**
152	Growth with 0.01% cycloheximide	Kluyveromyces aestuarii	Kluyveromyces bulgaricus Kluyveromyces lactis Kluyveromyces marxianus
153	Nitrate growth.	154	Citeromyces matritensis
154	L−Sorbose growth.	155	163
155	Melibiose fermentation.	156	46
156	Salicin growth.	157	158
157	2−Ketogluconate growth.	19	Torulaspora delbrueckii** Torulaspora pretoriensis**
158	Trehalose fermentation.	159	160
159	Growth without niacin.	Kluyveromyces lactis	Debaryomyces hansenii
160	Methyl α−D−glucopyranoside fermentation	161	162
161	Growth without niacin.	Kluyveromyces lactis	Zygosaccharomyces fermentati**
162	Growth without niacin.	Kluyveromyces dobzhanskii	Zygosaccharomyces fermentati**
163	D−Glucose fermentation.	164	165
164	Growth without thiamin.	Pichia vini	Debaryomyces hansenii
165	Trehalose fermentation.	166	173
166	Lactose growth.	167	171
167	Melibiose fermentation.	168	170
168	Growth without niacin.	Kluyveromyces drosophilarum Kluyveromyces lactis	169
169	Salicin growth.	Torulaspora delbrueckii**	Debaryomyces hansenii
170	Growth in 50% D−glucose.	Saccharomyces kluyveri	Torulaspora delbrueckii**
171	Growth without niacin.	172	Debaryomyces hansenii
172	Growth with 0.01% cycloheximide	Kluyveromyces aestuarii	Kluyveromyces lactis
173	Melibiose fermentation.	174	177
174	Growth without niacin.	175	176
175	Methyl α−D−glucopyranoside fermentation	Kluyveromyces lactis	Kluyveromyces dobzhanskii
176	Salicin growth.	Torulaspora delbrueckii**	Zygosaccharomyces fermentati**
177	Growth in 60% D−glucose.	178	Torulaspora delbrueckii**
178	Growth with 0.01% cycloheximide	Saccharomyces kluyveri	Zygosaccharomyces cidri**
179	Melibiose fermentation.	Debaryomyces hansenii	Debaryomyces castellii
180	Galactitol growth.	181	191
181	Glycerol growth.	Lipomyces lipofer Lipomyces tetrasporus*	182
182	Raffinose growth.	183	185
183	D−Glucose fermentation.	184	Pichia farinosa
184	D−Ribose growth.	Debaryomyces melissophila*	Debaryomyces coudertii
185	L−Rhamnose growth.	186	188
186	Inulin growth.	Debaryomyces hansenii Debaryomyces marama Debaryomyces nepalensis*	187

	Negative	Positive
187 Growth in 50% D–glucose.	Debaryomyces marama Lipomyces tetrasporus*	Debaryomyces hansenii Debaryomyces marama
188 D–Glucose fermentation.	189	190
189 Growth in 50% D–glucose.	Lipomyces tetrasporus*	Debaryomyces hansenii
190 Trehalose fermentation.	Debaryomyces hansenii	Wingea robertsii
191 Glycerol growth.	Lipomyces lipofer Lipomyces starkeyi Lipomyces tetrasporus*	192
192 Methyl α–D–glucopyranoside growth	Debaryomyces polymorpha**	193
193 D,L–Lactate growth.	194	196
194 Growth in 50% D–glucose.	195	Debaryomyces phaffii Debaryomyces vanriji
195 Starch formation.	Debaryomyces vanriji Lipomyces tetrasporus*	Lipomyces starkeyi Lipomyces tetrasporus*
196 Lactose growth.	197	198
197 Growth in 50% D–glucose.	Debaryomyces vanriji Lipomyces tetrasporus*	Debaryomyces hansenii Debaryomyces vanriji
198 Growth in 50% D–glucose.	Lipomyces tetrasporus*	Debaryomyces hansenii Pichia pseudopolymorpha

Ascosporogenous yeasts with hat- or Saturn-shaped ascospores (test No. 77 positive) (Key No. 6)

Key involving physiological tests and microscopical examination

Yeasts in Key No. 6

- 4 Ambrosiozyma cicatricosa*
- 5 Ambrosiozyma monospora**
- 6 Ambrosiozyma philentoma*
- 7 Arthroascus javanensis**
- 8 Botryoascus synnaedendrus*
- 157 Dekkera bruxellensis
- 158 Dekkera intermedia
- 165 Hanseniaspora guilliermondii*
- 166 Hanseniaspora occidentalis*
- 168 Hanseniaspora uvarum
- 169 Hanseniaspora valbyensis
- 171 Hansenula anomala
- 172 Hansenula beckii**
- 173 Hansenula beijerinckii
- 174 Hansenula bimundalis
- 175 Hansenula californica
- 176 Hansenula canadensis
- 177 Hansenula capsulata
- 178 Hansenula ciferrii
- 179 Hansenula dimennae
- 180 Hansenula dryadoides*
- 181 Hansenula fabianii
- 182 Hansenula glucozyma
- 183 Hansenula henricii
- 184 Hansenula holstii
- 185 Hansenula jadinii
- 186 Hansenula lynferdii*
- 187 Hansenula minuta
- 188 Hansenula mrakii
- 189 Hansenula muscicola*
- 190 Hansenula nonfermentans
- 191 Hansenula ofunaensis*
- 192 Hansenula petersonii
- 193 Hansenula philodendra*
- 194 Hansenula polymorpha
- 195 Hansenula saturnus
- 196 Hansenula silvicola
- 197 Hansenula subpelliculosa
- 198 Hansenula sydowiorum*
- 199 Hansenula wickerhamii
- 200 Hansenula wingei
- 201 Hormoascus platypodis**
- 202 Hyphopichia burtonii**
- 241 Pachysolen tannophilus
- 244 Pichia abadieae*
- 245 Pichia acaciae
- 246 Pichia ambrosiae*
- 247 Pichia angophorae
- 248 Pichia besseyi*
- 249 Pichia bovis
- 250 Pichia castillae*
- 251 Pichia chambardii
- 252 Pichia delftensis
- 253 Pichia dispora
- 256 Pichia fermentans
- 257 Pichia fluxuum
- 258 Pichia guilliermondii
- 259 Pichia haplophila
- 260 Pichia heimii*
- 262 Pichia kluyveri
- 264 Pichia lindnerii*
- 265 Pichia media
- 266 Pichia membranaefaciens
- 267 Pichia methanolica*
- 268 Pichia mucosa*
- 269 Pichia naganishii*
- 270 Pichia nakazawae*
- 271 Pichia norvegensis*
- 272 Pichia ohmeri
- 273 Pichia onychis
- 274 Pichia pastoris
- 275 Pichia philogaea*
- 276 Pichia pijperi
- 277 Pichia pinus
- 279 Pichia quercuum
- 280 Pichia rabaulensis*
- 281 Pichia rhodanensis
- 282 Pichia saitoi
- 283 Pichia salictaria
- 284 Pichia sargentensis*
- 285 Pichia scolyti
- 287 Pichia spartinae*
- 288 Pichia stipitis
- 289 Pichia strasburgensis
- 291 Pichia toletana
- 292 Pichia trehalophila
- 293 Pichia veronae*
- 295 Pichia wickerhamii
- 320 Saccharomycopsis capsularis**
- 321 Saccharomycopsis crataegensis*
- 322 Saccharomycopsis fibuligera**
- 323 Saccharomycopsis lipolytica**
- 324 Saccharomycopsis malanga*
- 325 Saccharomycopsis vini**
- 333 Schwanniomyces occidentalis
- 348 Stephanoascus ciferrii**
- 428 Wickerhamia fluorescens

Tests in Key No. 6

- 12 D–Galactose growth
- 13 L–Sorbose growth
- 15 D–Xylose growth
- 18 L–Rhamnose growth
- 19 Sucrose growth
- 21 Trehalose growth
- 22 Methyl α–D–glucopyranoside growth
- 23 Cellobiose growth
- 26 Melibiose growth
- 28 Raffinose growth
- 29 Melezitose growth
- 31 Starch growth
- 32 Glycerol growth
- 33 Erythritol growth
- 37 D–Glucitol growth
- 39 D–δ–Gluconolactone growth
- 40 2–Ketogluconate growth
- 44 Citrate growth
- 49 Nitrate growth
- 55 Growth without thiamin
- 56 Growth without pyridoxine
- 62 Growth at 37 degrees
- 73 Pseudohyphae
- 74 Septate hyphae

number of different tests 24

KEY No. 6

	Negative	Positive
1 Erythritol growth.	2	95
2 D–Glucitol growth.	3	28
3 Glycerol growth.	4	11
4 Sucrose growth.	5	8
5 Cellobiose growth.	Pichia membranaefaciens	6
6 2–Ketogluconate growth.	Hanseniaspora valbyensis	7
7 Growth at 37 degrees.	Hanseniaspora uvarum	Hanseniaspora guilliermondii*
8 D–Galactose growth.	9	10
9 Starch growth.	Dekkera bruxellensis	Schwanniomyces occidentalis
10 Starch growth.	Dekkera intermedia	Schwanniomyces occidentalis
11 Sucrose growth.	12	19
12 Cellobiose growth.	13	16
13 D–Xylose growth.	14	Pichia fermentans Pichia membranaefaciens
14 Citrate growth.	15	Pichia kluyveri Pichia membranaefaciens
15 Septate hyphae.	Pichia membranaefaciens	Arthroascus javanensis**
16 Nitrate growth.	17	18
17 D–Galactose growth.	Pichia norvegensis*	Pichia chambardii
18 L–Sorbose growth.	Hansenula mrakii	Hansenula dimennae
19 Methyl α–D–glucopyranoside growth	20	23
20 Melezitose growth.	21	Schwanniomyces occidentalis
21 D–Galactose growth.	22	Wickerhamia fluorescens
22 D–Xylose growth.	Hanseniaspora occidentalis*	Hansenula saturnus
23 D–Galactose growth.	24	10
24 L–Sorbose growth.	25	27
25 Cellobiose growth.	9	26
26 Septate hyphae.	Schwanniomyces occidentalis	Saccharomycopsis fibuligera**
27 Melezitose growth.	Hansenula californica	Schwanniomyces occidentalis
28 Glycerol growth.	29	39
29 Sucrose growth.	30	35
30 Trehalose growth.	31	34
31 Growth without thiamin.	32	33
32 2–Ketogluconate growth.	Pichia saitoi	Pichia besseyi*
33 D–δ–Gluconolactone growth	Pichia fluxuum	Pichia delftensis
34 2–Ketogluconate growth.	Pichia saitoi	Pichia dispora
35 Citrate growth.	8	36
36 L–Sorbose growth.	37	38
37 Starch growth.	Pichia angophorae	Schwanniomyces occidentalis
38 Starch growth.	Pichia spartinae*	Schwanniomyces occidentalis
39 Nitrate growth.	40	74
40 L–Rhamnose growth.	41	63
41 Melezitose growth.	42	51
42 L–Sorbose growth.	43	48
43 Sucrose growth.	44	46
44 Citrate growth.	45	Pichia quercuum
45 Starch growth.	Saccharomycopsis vini**	Saccharomycopsis malanga*
46 D–Xylose growth.	47	Pichia toletana
47 Starch growth.	Saccharomycopsis vini**	Saccharomycopsis fibuligera**
48 D–Xylose growth.	49	50
49 D–Galactose growth.	Saccharomycopsis vini**	Pichia ohmeri
50 Cellobiose growth.	Saccharomycopsis crataegensis*	Pichia pijperi
51 D–Xylose growth.	52	56
52 L–Sorbose growth.	53	54

		Negative	Positive
53	D–Galactose growth.	25	10
54	Citrate growth.	55	38
55	Starch growth.	Pichia mucosa*	Schwanniomyces occidentalis
56	Raffinose growth.	57	60
57	L–Sorbose growth.	58	55
58	Growth without thiamin.	59	Schwanniomyces occidentalis
59	Growth without pyridoxine	Pichia bovis	Pichia toletana
60	D–Galactose growth.	61	62
61	Starch growth.	Pichia onychis	Schwanniomyces occidentalis
62	Starch growth.	Pichia guilliermondii	Schwanniomyces occidentalis
63	Sucrose growth.	64	69
64	Cellobiose growth.	65	67
65	D–Galactose growth.	Pichia pastoris	66
66	L–Sorbose growth.	Pichia abadieae*	Hansenula ofunaensis*
67	Trehalose growth.	68	Pichia lindnerii*
68	Citrate growth.	Pichia sargentensis*	Pichia salictaria
69	Raffinose growth.	70	72
70	Growth without thiamin.	71	Pichia wickerhamii
71	Growth without pyridoxine	Pichia rhodanensis / Pichia veronae*	Pichia toletana
72	D–Galactose growth.	Pichia rabaulensis*	73
73	Melibiose growth.	Pichia strasburgensis	Pichia guilliermondii
74	Melezitose growth.	75	84
75	D–Galactose growth.	76	82
76	L–Sorbose growth.	77	81
77	D–Xylose growth.	78	79
78	Trehalose growth.	Hansenula dryadoides*	Hansenula nonfermentans
79	Sucrose growth.	80	Hansenula saturnus
80	Trehalose growth.	Hansenula mrakii	Hansenula minuta
81	Sucrose growth.	Hansenula dimennae	Hansenula californica
82	L–Sorbose growth.	83	Hansenula silvicola
83	L–Rhamnose growth.	Pachysolen tannophilus	Hansenula muscicola*
84	L–Rhamnose growth.	85	89
85	D–Xylose growth.	86	87
86	D–Galactose growth.	Dekkera bruxellensis	Dekkera intermedia
87	Citrate growth.	Hansenula beijerinckii	88
88	Starch growth.	Hansenula jadinii	Hansenula fabianii
89	D–Galactose growth.	90	94
90	Raffinose growth.	91	93
91	Pseudohyphae.	Hansenula beckii** / Hansenula bimundalis	92
92	Septate hyphae.	Hansenula canadensis	Hansenula wingei
93	Citrate growth.	Hansenula beijerinckii	Hansenula petersonii
94	Starch growth.	Hansenula silvicola	Hansenula holstii
95	Sucrose growth.	96	115
96	Nitrate growth.	97	109
97	D–Galactose growth.	98	104
98	Trehalose growth.	99	101
99	L–Rhamnose growth.	100	Botryoascus synnaedendrus*
100	Citrate growth.	Saccharomycopsis malanga*	Saccharomycopsis lipolytica**
101	Methyl α–D–glucopyranoside growth	102	Saccharomycopsis capsularis**
102	Cellobiose growth.	Pichia trehalophila	103
103	Starch growth.	Pichia pinus	Saccharomycopsis malanga*
104	D–Xylose growth.	105	106
105	Trehalose growth.	Saccharomycopsis lipolytica**	Pichia acaciae
106	Trehalose growth.	Pichia haplophila	107
107	Melibiose growth.	108	Pichia castillae*
108	Starch growth.	Pichia methanolica*	Pichia media
109	Cellobiose growth.	110	111
110	L–Rhamnose growth.	Hansenula philodendra*	Hansenula wickerhamii
111	D–Galactose growth.	112	Hansenula muscicola*

Key No. 7: ascosporogenous yeasts

	Negative	Positive
112 Citrate growth.	113	114
113 Starch growth.	Hansenula henricii	Hansenula capsulata
114 Growth without thiamin.	Hansenula glucozyma	Hansenula henricii
115 Raffinose growth.	116	127
116 D−Galactose growth.	117	122
117 Nitrate growth.	118	121
118 D−Xylose growth.	119	120
119 Starch growth.	Pichia ambrosiae*	Saccharomycopsis fibuligera**
120 L−Rhamnose growth.	Ambrosiozyma cicatricosa* Ambrosiozyma monospora**	Ambrosiozyma philentoma*
121 Growth at 37 degrees.	Hormoascus platypodis**	Hansenula polymorpha
122 Nitrate growth.	123	126
123 L−Rhamnose growth.	124	125
124 Starch growth.	Pichia philogaea*	Pichia nakazawae*
125 D−δ−Gluconolactone growth	Pichia nakazawae*	Pichia stipitis
126 Starch growth.	Hansenula polymorpha	Hansenula holstii
127 Nitrate growth.	128	133
128 L−Rhamnose growth.	129	131
129 D−Xylose growth.	Saccharomycopsis fibuligera**	130
130 Methyl α−D−glucopyranoside growth	Pichia naganishii*	Hyphopichia burtonii**
131 L−Sorbose growth.	132	Stephanoascus ciferrii**
132 Melibiose growth.	Pichia heimii*	Pichia scolyti
133 Melibiose growth.	134	Hansenula sydowiorum*
134 Growth without thiamin.	Hansenula subpelliculosa	135
135 Starch growth.	Hansenula lynferdii*	Hansenula anomala Hansenula ciferrii

---

Ascosporogenous yeasts with ascospores of shapes other than spherical, oval, reniform, hat- or Saturn-shaped (test No. 78 positive) (Key No. 7)

Key involving physiological tests and microscopical examination

Yeasts in Key No. 7
- 231 Metschnikowia bicuspidata
- 232 Metschnikowia krissii
- 233 Metschnikowia lunata*
- 234 Metschnikowia pulcherrima
- 235 Metschnikowia reukaufii
- 236 Metschnikowia zobellii
- 240 Nematospora coryli
- 244 Pichia abadieae*
- 428 Wickerhamia fluorescens

Tests in Key No. 7
- 36 D−Mannitol growth
- 37 D−Glucitol growth
- 38 myo−Inositol growth
- 55 Growth without thiamin
- 71 Cells of other shapes

number of different tests 5

KEY No. 7

	Negative	Positive
1 Glucitol growth.	2	4
2 Mannitol growth.	Nematospora coryli	3
3 Growth without thiamin.	Metschnikowia krissii	Wickerhamia fluorescens
4 myo−Inositol growth.	5	Pichia abadieae*
5 Growth without thiamin.	6	7
6 Cells of other shapes.	Metschnikowia bicuspidata Metschnikowia zobellii	Metschnikowia lunata*
7 Cells of other shapes.	Metschnikowia bicuspidata Metschnikowia pulcherrima Metschnikowia reukaufii	Metschnikowia lunata*

Basidiomycetous yeasts (Key No. 8)

Key involving physiological tests and microscopical examination

Yeasts in Key No. 8
- 159 Filobasidium capsuligenum**
- 219 Leucosporidium antarcticum
- 220 Leucosporidium frigidum
- 221 Leucosporidium gelidum
- 222 Leucosporidium nivalis
- 223 Leucosporidium scottii
- 224 Leucosporidium stokesii
- 296 Rhodosporidium bisporidiis*
- 297 Rhodosporidium capitatum*
- 298 Rhodosporidium dacryoidum*
- 299 Rhodosporidium infirmo–miniatum**
- 300 Rhodosporidium malvinellum*
- 304 Rhodotorula glutinis
- 335 Sporidibolus johnsonii
- 336 Sporidibolus ruinenii

Tests in Key No. 8
- 26 Melibiose growth
- 29 Melezitose growth
- 38 myo–Inositol growth
- 49 Nitrate growth
- 51 Growth without vitamins
- 66 Pink colonies
- 74 Septate hyphae

number of different tests 7

KEY No. 8

	Negative	Positive
1 myo–Inositol growth.	2	13
2 Melezitose growth.	3	7
3 Nitrate growth.	Rhodosporidium dacryoidum*	4
4 Growth without vitamins.	5	6
5 Pink colonies.	Leucosporidium frigidum	Rhodosporidium malvinellum*
6 Pink colonies.	Leucosporidium antarcticum	Sporidibolus ruinenii
7 Melibiose growth.	8	Leucosporidium gelidum
8 Nitrate growth.	Rhodosporidium dacryoidum*	9
9 Growth without vitamins.	10	11
10 Pink colonies.	Leucosporidium stokesii	Rhodotorula glutinis
11 Pink colonies.	Leucosporidium scottii	12
12 Septate hyphae.	Rhodotorula glutinis	Sporidibolus johnsonii
13 Melibiose growth.	14	17
14 Melezitose growth.	15	16
15 Nitrate growth.	Filobasidium capsuligenum**	Leucosporidium frigidum
16 Pink colonies.	Leucosporidium stokesii	Rhodosporidium capitatum* Rhodosporidium infirmo–miniatum**
17 Melezitose growth.	Leucosporidium nivalis	18
18 Pink colonies.	Leucosporidium gelidum	Rhodosporidium bisporidiis*

Yeasts with pink colonies (test No. 66 positive) (Key No. 9)

Key involving physiological tests and microscopical examination

Yeasts in Key No. 9
- 3 Aessosporon salmonicolor**
- 134 Cryptococcus hungaricus
- 137 Cryptococcus laurentii
- 139 Cryptococcus macerans
- 243 Phaffia rhodozyma*
- 296 Rhodosporidium bisporidiis*
- 297 Rhodosporidium capitatum*
- 298 Rhodosporidium dacryoidum*
- 299 Rhodosporidium infirmo–miniatum**
- 300 Rhodosporidium malvinellum*
- 301 Rhodotorula acheniorum*
- 302 Rhodotorula araucariae*
- 303 Rhodotorula aurantiaca
- 304 Rhodotorula glutinis
- 305 Rhodotorula graminis
- 306 Rhodotorula lactosa
- 307 Rhodotorula marina
- 308 Rhodotorula minuta
- 309 Rhodotorula pallida
- 310 Rhodotorula pilimanae
- 311 Rhodotorula rubra
- 335 Sporidibolus johnsonii
- 336 Sporidibolus ruinenii
- 337 Sporobolomyces albo–rubescens
- 339 Sporobolomyces gracilis
- 340 Sporobolomyces hispanicus
- 341 Sporobolomyces holsaticus
- 342 Sporobolomyces odorus
- 343 Sporobolomyces pararoseus
- 344 Sporobolomyces puniceus*
- 345 Sporobolomyces roseus
- 346 Sporobolomyces salmonicolor
- 374 Torulopsis fujisanensis

Tests in Key No. 9
- 15 D–Xylose growth
- 16 L–Arabinose growth
- 19 Sucrose growth
- 20 Maltose growth
- 26 Melibiose growth
- 29 Melezitose growth
- 31 Starch growth
- 33 Erythritol growth
- 35 Galactitol growth
- 38 myo–Inositol growth
- 40 2–Ketogluconate growth
- 49 Nitrate growth
- 51 Growth without vitamins
- 59 Growth without p–aminobenzoate
- 72 Filamentous
- 74 Septate hyphae

number of different tests 16

KEY No. 9

	Negative	Positive
1 myo–Inositol growth.	2	39
2 Nitrate growth.	3	18
3 Galactitol growth.	4	14
4 L–Arabinose growth.	5	11
5 Maltose growth.	6	8
6 D–Xylose growth.	7	Rhodotorula pallida / Sporobolomyces gracilis
7 Filamentous.	Rhodotorula pallida	Rhodosporidium dacryoidum*
8 D–Xylose growth.	9	10
9 Starch growth.	Rhodosporidium dacryoidum*	Sporobolomyces pararoseus
10 Starch growth.	Rhodotorula rubra	Sporobolomyces pararoseus
11 Maltose growth.	12	13
12 Melezitose growth.	Rhodotorula pilimanae	Rhodotorula minuta
13 2–Ketogluconate growth.	Rhodotorula rubra / Sporobolomyces albo–rubescens	Phaffia rhodozyma*
14 Melezitose growth.	15	17
15 L–Arabinose growth.	Rhodotorula pallida	16
16 Sucrose growth.	Torulopsis fujisanensis	Rhodotorula pilimanae
17 Growth without p–aminobenzoate	Rhodotorula marina	Rhodotorula rubra
18 Melezitose growth.	19	31
19 Galactitol growth.	20	28
20 Sucrose growth.	21	23
21 Filamentous.	Rhodotorula araucariae*	22
22 Septate hyphae.	Sporobolomyces odorus	Sporobolomyces hispanicus
23 Maltose growth.	24	26
24 2–Ketogluconate growth.	25	Rhodotorula graminis
25 Septate hyphae.	Sporobolomyces odorus	Aessosporon salmonicolor** / Sporobolomyces salmonicolor
26 Growth without vitamins.	Rhodosporidium malvinellum*	27
27 2–Ketogluconate growth.	Sporobolomyces holsaticus	Rhodotorula graminis
28 Sucrose growth.	Rhodotorula araucariae*	29
29 Growth without vitamins.	Rhodosporidium malvinellum*	30
30 2–Ketogluconate growth.	Sporidibolus ruinenii	Rhodotorula graminis
31 Erythritol growth.	32	Rhodotorula acheniorum*
32 Melibiose growth.	33	Rhodotorula lactosa
33 D–Xylose growth.	34	38
34 2–Ketogluconate growth.	35	37
35 Septate hyphae.	36	Sporidibolus johnsonii / Sporobolomyces holsaticus
36 Starch growth.	Rhodotorula glutinis	Sporobolomyces roseus
37 Starch growth.	Rhodotorula glutinis	Sporobolomyces puniceus*
38 2–Ketogluconate growth.	35	Rhodotorula aurantiaca / Rhodotorula glutinis
39 Erythritol growth.	40	42
40 Nitrate growth.	Cryptococcus hungaricus	41
41 Melibiose growth.	Rhodosporidium capitatum* / Rhodosporidium infirmo–miniatum**	Rhodosporidium bisporidiis*
42 Melibiose growth.	Cryptococcus macerans	Cryptococcus laurentii

Yeasts that use hydrocarbons (Keys No. 10 and No. 11)

Yeasts in Keys No. 10 and No. 11

- 19 Candida aaseri
- 20 Candida albicans
- 23 Candida beechii
- 25 Candida blankii
- 35 Candida cacaoi
- 36 Candida catenulata
- 38 Candida chiropterorum*
- 40 Candida conglobata
- 42 Candida curvata
- 43 Candida dendronema*
- 44 Candida diddensii
- 47 Candida edax*
- 49 Candida entomophila*
- 53 Candida foliarum
- 56 Candida friedrichii
- 57 Candida glaebosa
- 60 Candida humicola
- 61 Candida hydrocarbofumarica*
- 64 Candida incommunis*
- 65 Candida inositophila*
- 68 Candida intermedia
- 70 Candida javanica
- 74 Candida lusitaniae
- 76 Candida maltosa*
- 79 Candida melibiosica
- 81 Candida membranaefaciens

82 Candida mesenterica	202 Hyphopichia burtonii**	342 Sporobolomyces odorus
84 Candida mogii	223 Leucosporidium scottii	343 Sporobolomyces pararoseus
89 Candida oregonensis	230 Lodderomyces elongisporus	346 Sporobolomyces salmonicolor
90 Candida parapsilosis	233 Metschnikowia lunata*	348 Stephanoascus ciferrii**
95 Candida ravautii	234 Metschnikowia pulcherrima	362 Torulopsis apis
96 Candida rhagii	235 Metschnikowia reukaufii	366 Torulopsis bombicola*
98 Candida rugosa	250 Pichia castillae*	376 Torulopsis gropengiesseri
99 Candida sake	254 Pichia etchellsii	388 Torulopsis magnoliae
101 Candida santamariae	255 Pichia farinosa	390 Torulopsis maris
104 Candida silvanorum*	258 Pichia guilliermondii	398 Torulopsis nitratophila
110 Candida tenuis	302 Rhodotorula araucariae*	400 Torulopsis norvegica
113 Candida tropicalis	304 Rhodotorula glutinis	404 Torulopsis pinus
116 Candida valdiviana*	305 Rhodotorula graminis	414 Torulopsis torresii
119 Candida veronae	310 Rhodotorula pilimanae	415 Torulopsis vanderwaltii
123 Candida zeylanoides	334 Selenozyma peltata*	418 Torulopsis xestobii*
146 Debaryomyces castellii	335 Sporidibolus johnsonii	419 Trichosporon aquatile*
148 Debaryomyces hansenii	336 Sporidibolus ruinenii	423 Trichosporon fennicum*
149 Debaryomyces marama	338 Sporobolomyces antarcticus*	424 Trichosporon melibiosaceum*
152 Debaryomyces phaffii	340 Sporobolomyces hispanicus	
155 Debaryomyces vanriji	341 Sporobolomyces holsaticus	

Key involving physiological tests and microscopical examination (Key No. 10)

Tests in Key No. 10

6 Trehalose fermentation	28 Raffinose growth	46 Ethanol growth
7 Melibiose fermentation	29 Melezitose growth	49 Nitrate growth
12 D–Galactose growth	30 Inulin growth	62 Growth at 37 degrees
16 L–Arabinose growth	31 Starch growth	65 Starch formation
18 L–Rhamnose growth	33 Erythritol growth	66 Pink colonies
19 Sucrose growth	35 Galactitol growth	71 Cells of other shapes
20 Maltose growth	36 D–Mannitol growth	72 Filamentous
24 Salicin growth	42 D,L–Lactate growth	74 Septate hyphae
27 Lactose growth	43 Succinate growth	
	44 Citrate growth	number of different tests 27

KEY No. 10

	Negative	Positive
1 Erythritol growth.	2	90
2 Nitrate growth.	3	55
3 Raffinose growth.	4	34
4 D–Mannitol growth.	5	6
5 D,L–Lactate growth.	Torulopsis xestobii*	Trichosporon aquatile*
6 Sucrose growth.	7	17
7 D,L–Lactate growth.	8	13
8 D–Galactose growth.	9	11
9 L–Rhamnose growth.	10	Torulopsis maris
10 Trehalose fermentation.	Candida zeylanoides	Candida beechii
11 Maltose growth.	12	Candida tropicalis
12 Citrate growth.	Torulopsis torresii	Candida zeylanoides
13 Maltose growth.	14	16
14 D–Galactose growth.	Candida beechii / Candida santamariae	15
15 Trehalose fermentation.	Candida catenulata / Candida rugosa	Candida santamariae
16 Melezitose growth.	Candida catenulata / Candida ravautii	Candida tropicalis
17 L–Arabinose growth.	18	29
18 L–Rhamnose growth.	19	28
19 Trehalose fermentation.	20	24
20 Growth at 37 degrees.	Candida sake / Metschnikowia pulcherrima / Metschnikowia reukaufii	21
21 Starch growth.	22	Candida albicans
22 Salicin growth.	Lodderomyces elongisporus	23
23 Cells of other shapes.	Metschnikowia pulcherrima	Metschnikowia lunata*
24 Starch growth.	25	Candida albicans / Candida tropicalis
25 Growth at 37 degrees.	Candida sake	26
26 Salicin growth.	Lodderomyces elongisporus	27
27 Cells of other shapes.	Candida maltosa*	Metschnikowia lunata*

Key No. 10: hydrocarbon-using yeasts

		Negative	Positive
28	Starch growth.	Candida lusitaniae	Candida oregonensis
29	Ethanol growth.	30	31
30	Septate hyphae.	Candida mogii	Candida albicans
31	L–Rhamnose growth.	32	Candida lusitaniae
32	Starch growth.	33	Candida albicans
33	Salicin growth.	Candida parapsilosis	Pichia etchellsii
34	Ethanol growth.	35	42
35	D–Mannitol growth.	36	37
36	Starch growth.	Torulopsis xestobii*	Sporobolomyces pararoseus
37	L–Arabinose growth.	38	40
38	Maltose growth.	39	Sporobolomyces pararoseus
39	Salicin growth.	Torulopsis apis	Torulopsis gropengiesseri
40	L–Rhamnose growth.	41	Candida inositophila*
41	Maltose growth.	Rhodotorula pilimanae	Candida mogii
42	Lactose growth.	43	51
43	D–Mannitol growth.	36	44
44	L–Arabinose growth.	45	47
45	Maltose growth.	Torulopsis bombicola*	46
46	Trehalose fermentation.	Sporobolomyces pararoseus	Candida melibiosica
47	Maltose growth.	Rhodotorula pilimanae	48
48	D,L–Lactate growth.	49	50
49	Citrate growth.	Candida inositophila*	Pichia guilliermondii
50	Starch growth.	Pichia guilliermondii	Debaryomyces hansenii
51	Galactitol growth.	52	54
52	D,L–Lactate growth.	Candida intermedia	53
53	Starch growth.	Candida glaebosa	Debaryomyces hansenii
54	Melibiose fermentation.	Debaryomyces hansenii	Debaryomyces castellii
55	Sucrose growth.	56	69
56	Maltose growth.	57	Candida javanica
57	D–Galactose growth.	58	64
58	Salicin growth.	59	63
59	Growth at 37 degrees.	60	62
60	Filamentous.	Rhodotorula araucariae*	61
61	Septate hyphae.	Sporobolomyces odorus	Sporobolomyces hispanicus
62	Pink colonies.	Candida foliarum	Sporobolomyces odorus
63	Pink colonies.	Torulopsis norvegica	Rhodotorula araucariae*
64	Ethanol growth.	65	66
65	L–Arabinose growth.	Sporobolomyces hispanicus	Torulopsis vanderwaltii
66	L–Rhamnose growth.	67	Torulopsis nitratophila
67	Growth at 37 degrees.	68	Candida foliarum
68	Filamentous.	Rhodotorula araucariae*	Sporobolomyces hispanicus
69	Galactitol growth.	70	84
70	Maltose growth.	71	77
71	D–Galactose growth.	72	74
72	Pink colonies.	Torulopsis magnoliae	73
73	Septate hyphae.	Sporobolomyces odorus	Sporobolomyces salmonicolor
74	Pink colonies.	75	76
75	L–Arabinose growth.	Torulopsis magnoliae	Torulopsis vanderwaltii
76	Septate hyphae.	Rhodotorula graminis	Sporobolomyces salmonicolor
77	Raffinose growth.	78	79
78	Septate hyphae.	Rhodotorula glutinis	Sporidibolus johnsonii Sporobolomyces holsaticus
79	Melezitose growth.	80	82
80	Pink colonies.	Candida valdiviana*	81
81	Septate hyphae.	Rhodotorula graminis	Sporobolomyces holsaticus
82	Pink colonies.	Candida valdiviana*	83
83	Septate hyphae.	Rhodotorula glutinis	Sporobolomyces holsaticus
84	Melezitose growth.	85	88
85	Raffinose growth.	Torulopsis vanderwaltii	86
86	Pink colonies.	Candida valdiviana*	87
87	Septate hyphae.	Rhodotorula graminis	Sporidibolus ruinenii
88	Succinate growth.	Leucosporidium scottii	89
89	Pink colonies.	Candida valdiviana*	Rhodotorula glutinis
90	Galactitol growth.	91	133
91	Raffinose growth.	92	106
92	Sucrose growth.	93	96

		Negative	Positive
93	Citrate growth.	94	95
94	Growth at 37 degrees.	Candida conglobata	Candida cacaoi
95	D−Galactose growth.	Torulopsis pinus	Pichia farinosa
96	D−Galactose growth.	97	99
97	Nitrate growth.	98	Candida incommunis*
98	Starch formation.	Candida mesenterica	Candida humicola
99	L−Rhamnose growth.	100	104
100	Starch formation.	101	Candida humicola
101	Lactose growth.	102	103
102	Cells of other shapes.	Candida aaseri	Candida diddensii
103	Starch growth.	Candida aaseri	Trichosporon fennicum*
104	Starch formation.	105	Candida humicola
105	Cells of other shapes.	Candida tenuis Candida veronae	Candida diddensii
106	Ethanol growth.	107	110
107	Nitrate growth.	108	Sporobolomyces antarcticus*
108	Starch formation.	109	Candida humicola
109	Growth at 37 degrees.	Trichosporon melibiosaceum*	Trichosporon fennicum*
110	Lactose growth.	111	124
111	L−Rhamnose growth.	112	120
112	Sucrose growth.	Pichia castillae*	113
113	Citrate growth.	114	116
114	Starch formation.	115	Candida humicola
115	Starch growth.	Candida rhagii	Debaryomyces hansenii
116	D,L−Lactate growth.	117	119
117	Starch formation.	118	Candida humicola
118	Filamentous.	Debaryomyces marama	Hyphopichia burtonii**
119	Starch formation.	Debaryomyces hansenii Debaryomyces marama	Candida humicola
120	Nitrate growth.	121	Candida edax*
121	D,L−Lactate growth.	122	123
122	Starch formation.	Candida silvanorum*	Candida humicola
123	Starch formation.	Debaryomyces hansenii	Candida humicola
124	Nitrate growth.	125	132
125	L−Rhamnose growth.	126	130
126	D,L−Lactate growth.	127	119
127	Starch formation.	128	Candida humicola
128	Starch growth.	Candida entomophila*	129
129	Growth at 37 degrees.	Debaryomyces marama	Trichosporon fennicum*
130	Starch formation.	Debaryomyces hansenii	131
131	Septate hyphae.	Candida curvata	Candida humicola
132	D−Galactose growth.	Sporobolomyces antarcticus*	Candida edax*
133	Ethanol growth.	134	139
134	Lactose growth.	135	138
135	Succinate growth.	136	137
136	Starch formation.	Candida chiropterorum*	Candida humicola
137	Starch formation.	Selenozyma peltata*	Candida humicola
138	Starch formation.	Candida hydrocarbofumarica*	Candida humicola
139	L−Rhamnose growth.	140	148
140	Sucrose growth.	Pichia castillae*	141
141	D,L−Lactate growth.	142	146
142	Inulin growth.	143	145
143	Starch formation.	144	Candida humicola
144	Starch growth.	Candida friedrichii	Debaryomyces vanrijii
145	Starch growth.	Candida membranaefaciens	Debaryomyces phaffii Debaryomyces vanriji
146	Starch formation.	147	Candida humicola
147	Starch growth.	Candida membranaefaciens	Debaryomyces hansenii Debaryomyces vanriji
148	D,L−Lactate growth.	149	155
149	Lactose growth.	150	154
150	Succinate growth.	136	151
151	Raffinose growth.	152	153
152	Starch formation.	Candida dendronema*	Candida humicola

Key No. 11: hydrocarbon-using yeasts

	Negative	Positive
153 Starch formation.	...Debaryomyces vanriji	...Candida humicola
154 Starch formation.	...Candida blankii Candida hydrocarbofumarica*	...Candida humicola
155 Nitrate growth.	...156	...Candida edax*
156 Succinate growth.	...136	...157
157 Melezitose growth.	...158	...159
158 Starch formation.	...Stephanoascus ciferrii**	...Candida humicola
159 Starch formation.	...Debaryomyces hansenii Debaryomyces vanriji	...Candida humicola

Key involving physiological tests only (Key No. 11)

Tests in Key No. 11

5 Sucrose fermentation	26 Melibiose growth	43 Succinate growth
6 Trehalose fermentation	27 Lactose growth	44 Citrate growth
7 Melibiose fermentation	28 Raffinose growth	46 Ethanol growth
12 D−Galactose growth	29 Melezitose growth	49 Nitrate growth
16 L−Arabinose growth	30 Inulin growth	51 Growth without vitamins
18 L−Rhamnose growth	31 Starch growth	62 Growth at 37 degrees
19 Sucrose growth	33 Erythritol growth	65 Starch formation
20 Maltose growth	35 Galactitol growth	
21 Trehalose growth	36 D−Mannitol growth	number of different tests 28
24 Salicin growth	40 2−Ketogluconate growth	
	42 D,L−Lactate growth	

KEY No. 11

	Negative	Positive
1 Erythritol growth.	...2	...85
2 Nitrate growth.	...3	...59
3 Raffinose growth.	...4	...35
4 D−Mannitol growth.	...5	...6
5 D,L−Lactate growth.	...Torulopsis xestobii*	...Trichosporon aquatile*
6 Sucrose growth.	...7	...18
7 D,L−Lactate growth.	...8	...13
8 D−Galactose growth.	...9	...11
9 L−Rhamnose growth.	...10	...Torulopsis maris
10 Trehalose fermentation.	...Candida zeylanoides	...Candida beechii
11 Maltose growth.	...12	...Candida tropicalis
12 Citrate growth.	...Torulopsis torresii	...Candida zeylanoides
13 Maltose growth.	...14	...17
14 D−Galactose growth.	...Candida beechii Candida santamariae	...15
15 Trehalose growth.	...Candida catenulata Candida rugosa	...16
16 Trehalose fermentation.	...Candida catenulata	...Candida santamariae
17 Melezitose growth.	...Candida catenulata Candida ravautii	...Candida tropicalis
18 L−Arabinose growth.	...19	...31
19 L−Rhamnose growth.	...20	...30
20 Trehalose fermentation.	...21	...24
21 Growth at 37 degrees.	...Candida sake Metschnikowia pulcherrima Metschnikowia reukaufii	...22
22 Starch growth.	...23	...Candida albicans
23 Salicin growth.	...Lodderomyces elongisporus	...Metschnikowia lunata* Metschnikowia pulcherrima
24 Starch growth.	...25	...Candida albicans Candida tropicalis
25 Sucrose fermentation.	...26	...28
26 Growth at 37 degrees.	...Candida sake	...27
27 Salicin growth.	...Lodderomyces elongisporus	...Metschnikowia lunata*
28 Growth at 37 degrees.	...Candida sake	...29
29 Salicin growth.	...Lodderomyces elongisporus	...Candida maltosa*
30 Starch growth.	...Candida lusitaniae	...Candida oregonensis

		Negative	Positive
31	Ethanol growth.	Candida albicans / Candida mogii	32
32	L–Rhamnose growth.	33	Candida lusitaniae
33	Starch growth.	34	Candida albicans
34	Salicin growth.	Candida parapsilosis	Pichia etchellsii
35	Ethanol growth.	36	43
36	D–Mannitol growth.	37	38
37	Starch growth.	Torulopsis xestobii*	Sporobolomyces pararoseus
38	L–Arabinose growth.	39	41
39	Maltose growth.	40	Sporobolomyces pararoseus
40	Sucrose fermentation.	Torulopsis apis	Torulopsis gropengiesseri
41	L–Rhamnose growth.	42	Candida inositophila*
42	Maltose growth.	Rhodotorula pilimanae	Candida mogii
43	Melibiose growth.	44	50
44	D–Mannitol growth.	37	45
45	Maltose growth.	46	47
46	L–Arabinose growth.	Torulopsis bombicola*	Rhodotorula pilimanae
47	D,L–Lactate growth.	48	Debaryomyces hansenii
48	Lactose growth.	49	Candida intermedia
49	L–Arabinose growth.	Sporobolomyces pararoseus	Candida inositophila*
50	D–Mannitol growth.	Torulopsis xestobii*	51
51	Galactitol growth.	52	57
52	L–Arabinose growth.	53	54
53	Lactose growth.	Candida melibiosica	Candida glaebosa
54	Lactose growth.	55	56
55	Starch growth.	Pichia guilliermondii	Debaryomyces hansenii
56	Starch growth.	Candida glaebosa	Debaryomyces hansenii
57	Lactose growth.	55	58
58	Melibiose fermentation.	Debaryomyces hansenii	Debaryomyces castellii
59	Galactitol growth.	60	75
60	Sucrose growth.	61	66
61	Ethanol growth.	62	63
62	L–Arabinose growth.	Sporobolomyces hispanicus / Sporobolomyces odorus	Torulopsis vanderwaltii
63	Trehalose growth.	Torulopsis norvegica	64
64	L–Rhamnose growth.	65	Torulopsis nitratophila
65	Growth at 37 degrees.	Rhodotorula araucariae* / Sporobolomyces hispanicus / Sporobolomyces odorus	Candida foliarum / Sporobolomyces odorus
66	Maltose growth.	67	72
67	Trehalose growth.	Torulopsis magnoliae	68
68	D–Galactose growth.	Sporobolomyces odorus / Sporobolomyces salmonicolor	69
69	Raffinose growth.	70	71
70	2–Ketogluconate growth.	Sporobolomyces salmonicolor	Torulopsis vanderwaltii
71	2–Ketogluconate growth.	Sporobolomyces salmonicolor	Rhodotorula graminis
72	Melibiose growth.	73	Candida valdiviana*
73	Melezitose growth.	74	Rhodotorula glutinis / Sporidibolus johnsonii / Sporobolomyces holsaticus
74	2–Ketogluconate growth.	Sporobolomyces holsaticus	Rhodotorula graminis
75	Ethanol growth.	76	80
76	Melibiose growth.	77	Candida valdiviana*
77	Melezitose growth.	78	Rhodotorula glutinis
78	Raffinose growth.	Torulopsis vanderwaltii	79
79	2–Ketogluconate growth.	Sporidibolus ruinenii	Rhodotorula graminis
80	Sucrose growth.	81	82
81	Maltose growth.	Rhodotorula araucariae*	Candida javanica
82	Melibiose growth.	83	Candida valdiviana*
83	Melezitose growth.	Rhodotorula graminis	84
84	Succinate growth.	Leucosporidium scottii	Rhodotorula glutinis
85	Galactitol growth.	86	126
86	Raffinose growth.	87	99
87	Sucrose growth.	88	91
88	Citrate growth.	89	90

Key No. 11: hydrocarbon-using yeasts

		Negative	Positive
89	Growth without vitamins.	Candida conglobata	Candida cacaoi
90	D–Galactose growth.	Torulopsis pinus	Pichia farinosa
91	D–Galactose growth.	92	94
92	Melibiose growth.	93	Candida humicola
93	Nitrate growth.	Candida mesenterica	Candida incommunis*
94	L–Rhamnose growth.	95	98
95	Melibiose growth.	96	Candida humicola
96	Lactose growth.	Candida aaseri Candida diddensii	97
97	Sucrose fermentation.	Candida aaseri	Trichosporon fennicum*
98	Melibiose growth.	Candida diddensii Candida tenuis Candida veronae	Candida humicola
99	Ethanol growth.	100	104
100	Nitrate growth.	101	Sporobolomyces antarcticus*
101	Melibiose growth.	102	103
102	Growth at 37 degrees.	Trichosporon melibiosaceum*	Trichosporon fennicum*
103	Sucrose fermentation.	Candida humicola	Trichosporon melibiosaceum*
104	Melibiose growth.	105	114
105	Lactose growth.	106	110
106	Citrate growth.	107	108
107	Starch growth.	Candida rhagii	Debaryomyces hansenii
108	D,L–Lactate growth.	109	Debaryomyces hansenii Debaryomyces marama
109	Growth without vitamins.	Debaryomyces marama	Hyphopichia burtonii**
110	L–Rhamnose growth.	111	113
111	D,L–Lactate growth.	112	Debaryomyces hansenii Debaryomyces marama
112	Growth without vitamins.	Debaryomyces marama	Trichosporon fennicum*
113	Starch formation.	Debaryomyces hansenii	Candida curvata
114	Nitrate growth.	115	125
115	Sucrose growth.	Pichia castillae*	116
116	L–Rhamnose growth.	117	122
117	D,L–Lactate growth.	118	121
118	Sucrose fermentation.	119	120
119	Starch formation.	Debaryomyces marama	Candida humicola
120	Starch growth.	Candida entomophila*	Debaryomyces marama
121	Starch formation.	Debaryomyces hansenii Debaryomyces marama	Candida humicola
122	D,L–Lactate growth.	123	124
123	Starch formation.	Candida silvanorum*	Candida humicola
124	Starch formation.	Debaryomyces hansenii	Candida humicola
125	D–Galactose growth.	Sporobolomyces antarcticus*	Candida edax*
126	Melibiose growth.	127	134
127	Ethanol growth.	128	130
128	Lactose growth.	129	Candida hydrocarbofumarica*
129	Succinate growth.	Candida chiropterorum*	Selenozyma peltata*
130	Lactose growth.	131	133
131	Succinate growth.	Candida chiropterorum*	132
132	Raffinose growth.	Candida dendronema*	Debaryomyces hansenii
133	D,L–Lactate growth.	Candida blankii Candida hydrocarbofumarica*	Debaryomyces hansenii
134	Melezitose growth.	135	137
135	Sucrose growth.	Pichia castillae*	136
136	Starch formation.	Stephanoascus ciferrii**	Candida humicola
137	Nitrate growth.	138	Candida edax*
138	D,L–Lactate growth.	139	143
139	Inulin growth.	140	142
140	Starch formation.	141	Candida humicola
141	Starch growth.	Candida friedrichii	Debaryomyces vanriji
142	Starch growth.	Candida membranaefaciens	Debaryomyces phaffii Debaryomyces vanriji

	Negative	Positive
143 Starch formation.	144	.Candida humicola
144 Starch growth.	Candida membranaefaciens	Debaryomyces hansenii Debaryomyces vanriji

Yeasts that use methanol (test No. 45 positive) (Key No. 12)

Key involving physiological tests only

Yeasts in Key No. 12
- 27 Candida boidinii
- 177 Hansenula capsulata
- 182 Hansenula glucozyma
- 183 Hansenula henricii
- 187 Hansenula minuta
- 190 Hansenula nonfermentans
- 191 Hansenula ofunaensis*
- 193 Hansenula philodendra*
- 194 Hansenula polymorpha
- 199 Hansenula wickerhamii
- 237 Nadsonia commutata*
- 264 Pichia lindnerii*
- 267 Pichia methanolica*
- 269 Pichia naganishii*
- 274 Pichia pastoris
- 277 Pichia pinus
- 292 Pichia trehalophila
- 338 Sporobolomyces antarcticus*
- 364 Torulopsis austromarina*
- 391 Torulopsis methanolovescens*
- 392 Torulopsis molischiana
- 395 Torulopsis nagoyaensis*
- 397 Torulopsis nemodendra*
- 398 Torulopsis nitratophila
- 399 Torulopsis nodaensis*
- 403 Torulopsis pignaliae*
- 404 Torulopsis pinus
- 409 Torulopsis sonorensis*
- 419 Trichosporon aquatile*
- 426 Trichosporon terrestre*
- 431 Zendera ovetensis

The authors do not know whether or not the following yeasts utilize methanol:

Nadsonia commutata
Sporobolomyces antarcticus
Torulopsis austromarina
Torulopsis nodaensis
Trichosporon aquatile
Trichosporon terrestre
Zendera ovetensis

See also Sahm (1977).

Torulopsis maris has been omitted incorrectly; it keys out at 6 with *Torulopsis austromarina*.

Tests in Key No. 12
- 1 D–Glucose fermentation
- 12 D–Galactose growth
- 15 D–Xylose growth
- 20 Maltose growth
- 23 Cellobiose growth
- 33 Erythritol growth
- 42 D,L–Lactate growth
- 49 Nitrate growth
- 55 Growth without thiamin

number of different tests 9

KEY No. 12

	Negative	Positive
1 Erythritol growth.	2	15
2 Cellobiose growth.	3	10
3 D–Xylose growth.	4	8
4 D–Glucose fermentation.	5	7
5 Maltose growth.	6	Nadsonia commutata*
6 D,L–Lactate growth.	Torulopsis austromarina*	Zendera ovetensis
7 D–Galactose growth.	Pichia pastoris	Torulopsis nodaensis*
8 D–Galactose growth.	Torulopsis pignaliae*	9
9 Nitrate growth.	Hansenula ofunaensis*	Torulopsis nitratophila
10 Nitrate growth.	11	13
11 D–Galactose growth.	12	Trichosporon aquatile*
12 Growth without thiamin.	Torulopsis sonorensis*	Pichia lindnerii* Torulopsis methanolovescens*
13 D–Galactose growth.	14	Torulopsis nagoyaensis*
14 D–Xylose growth.	Hansenula nonfermentans	Hansenula minuta
15 Maltose growth.	16	26
16 Cellobiose growth.	17	22
17 Nitrate growth.	18	19
18 D–Glucose fermentation.	Torulopsis pinus	Pichia trehalophila
19 D–Glucose fermentation.	20	21
20 Growth without thiamin.	Hansenula philodendra*	Hansenula wickerhamii
21 Growth without thiamin.	Hansenula philodendra*	Candida boidinii
22 D–Galactose growth.	23	25
23 Nitrate growth.	Pichia pinus	24
24 Growth without thiamin.	Hansenula glucozyma	Hansenula henricii
25 D–Glucose fermentation.	Torulopsis nemodendra*	Pichia methanolica*
26 D–Glucose fermentation.	27	28
27 D–Galactose growth.	Sporobolomyces antarcticus*	Trichosporon terrestre*
28 Nitrate growth.	Pichia naganishii*	29

	Negative	Positive
29 Growth without thiamin.	Hansenula capsulata Torulopsis molischiana	Hansenula polymorpha

Yeasts most commonly isolated clinically (Key No. 13)

Key involving physiological tests only

Yeasts in Key No. 13
- 20 Candida albicans
- 31 Candida brumptii
- 36 Candida catenulata
- 90 Candida parapsilosis
- 95 Candida ravautii
- 113 Candida tropicalis
- 125 Cryptococcus albidus
- 137 Cryptococcus laurentii
- 142 Cryptococcus neoformans
- 145 Cryptococcus uniguttulatus
- 148 Debaryomyces hansenii
- 161 Geotrichum capitatum**
- 212 Kluyveromyces marxianus
- 255 Pichia farinosa
- 258 Pichia guilliermondii
- 263 Pichia kudriavzevii
- 271 Pichia norvegensis*
- 304 Rhodotorula glutinis
- 311 Rhodotorula rubra
- 312 Saccharomyces cerevisiae
- 317 Saccharomyces telluris
- 375 Torulopsis glabrata
- 421 Trichosporon cutaneum

Note: Medically important yeasts have been reviewed by Ahearn (1978).

Tests in Key No. 13
- 1 D−Glucose fermentation
- 15 D−Xylose growth
- 19 Sucrose growth
- 21 Trehalose growth
- 23 Cellobiose growth
- 27 Lactose growth
- 28 Raffinose growth
- 31 Starch growth
- 33 Erythritol growth
- 37 D−Glucitol growth
- 38 myo−Inositol growth
- 44 Citrate growth
- 49 Nitrate growth
- 51 Growth without vitamins
- 60 Growth in 50% D−glucose
- 62 Growth at 37 degrees

number of different tests 16

KEY No. 13

	Negative	Positive
1 Growth at 37 degrees.	2	16
2 myo−Inositol growth.	3	12
3 Sucrose growth.	4	7
4 D−Xylose growth.	Saccharomyces cerevisiae	5
5 Lactose growth.	6	Trichosporon cutaneum
6 Trehalose growth.	Candida brumptii	Candida ravautii
7 Nitrate growth.	8	Rhodotorula glutinis
8 D−Xylose growth.	Saccharomyces cerevisiae	9
9 Lactose growth.	10	11
10 Starch growth.	Rhodotorula rubra	Debaryomyces hansenii
11 Growth in 50x D−glucose.	Trichosporon cutaneum	Debaryomyces hansenii
12 Erythritol growth.	13	15
13 Lactose growth.	Cryptococcus uniguttulatus	14
14 Nitrate growth.	Trichosporon cutaneum	Cryptococcus albidus
15 Nitrate growth.	Cryptococcus laurentii Trichosporon cutaneum	Cryptococcus albidus
16 D−Glucitol growth.	17	27
17 D−Glucose fermentation.	18	23
18 Sucrose growth.	19	21
19 Lactose growth.	20	Trichosporon cutaneum
20 Cellobiose growth.	Geotrichum capitatum**	Pichia norvegensis*
21 Lactose growth.	22	Trichosporon cutaneum
22 Nitrate growth.	Rhodotorula rubra	Rhodotorula glutinis
23 Trehalose growth.	24	26
24 Cellobiose growth.	25	Pichia norvegensis*
25 Growth without vitamins.	Saccharomyces cerevisiae Saccharomyces telluris	Pichia kudriavzevii Saccharomyces cerevisiae
26 D−Xylose growth.	Saccharomyces cerevisiae Torulopsis glabrata	Candida albicans
27 myo−Inositol growth.	28	47
28 Erythritol growth.	29	45
29 Raffinose growth.	30	40
30 D−Glucose fermentation.	31	33
31 Lactose growth.	32	Trichosporon cutaneum
32 Sucrose growth.	Candida catenulata Candida ravautii	Rhodotorula glutinis

		Negative	Positive

		Negative	Positive
33	Sucrose growth.	34	37
34	Cellobiose growth.	35	Candida tropicalis
35	D–Xylose growth.	36	Candida catenulata / Candida ravautii
36	Citrate growth.	Saccharomyces cerevisiae	Candida catenulata
37	D–Xylose growth.	Saccharomyces cerevisiae	38
38	Cellobiose growth.	39	Candida albicans / Candida tropicalis
39	Starch growth.	Candida parapsilosis	Candida albicans
40	D–Glucose fermentation.	41	42
41	Nitrate growth.	9	Rhodotorula glutinis
42	D–Xylose growth.	Saccharomyces cerevisiae	43
43	Trehalose growth.	Kluyveromyces marxianus	44
44	Starch growth.	Pichia guilliermondii	Debaryomyces hansenii
45	D–Glucose fermentation.	11	46
46	Sucrose growth.	Pichia farinosa	Debaryomyces hansenii
47	Lactose growth.	Cryptococcus neoformans	Cryptococcus laurentii / Trichosporon cutaneum

Extended list of yeasts isolated clinically (Key No. 14)

Key involving physiological tests only

Yeasts in Key No. 14

19	Candida aaseri	125	Cryptococcus albidus	303	Rhodotorula aurantiaca
20	Candida albicans	126	Cryptococcus ater	304	Rhodotorula glutinis
31	Candida brumptii	131	Cryptococcus gastricus	308	Rhodotorula minuta
36	Candida catenulata	135	Cryptococcus kuetzingii	311	Rhodotorula rubra
42	Candida curvata	137	Cryptococcus laurentii	312	Saccharomyces cerevisiae
60	Candida humicola	142	Cryptococcus neoformans	317	Saccharomyces telluris
68	Candida intermedia	145	Cryptococcus uniguttulatus	323	Saccharomycopsis lipolytica**
74	Candida lusitaniae	148	Debaryomyces hansenii	326	Sarcinosporon inkin**
75	Candida macedoniensis	149	Debaryomyces marama	327	Schizoblastosporion starkeyi–henricii
79	Candida melibiosica	161	Geotrichum capitatum**	345	Sporobolomyces roseus
81	Candida membranaefaciens	169	Hanseniaspora valbyensis	346	Sporobolomyces salmonicolor
90	Candida parapsilosis	185	Hansenula jadinii	350	Sterigmatomyces elviae*
93	Candida pseudotropicalis	192	Hansenula petersonii	351	Sterigmatomyces halophilus
95	Candida ravautii	202	Hyphopichia burtonii**	371	Torulopsis etchellsii
98	Candida rugosa	210	Kluyveromyces lactis	375	Torulopsis glabrata
103	Candida silvae	212	Kluyveromyces marxianus	380	Torulopsis holmii
108	Candida steatolytica*	254	Pichia etchellsii	382	Torulopsis inconspicua
110	Candida tenuis	255	Pichia farinosa	388	Torulopsis magnoliae
113	Candida tropicalis	258	Pichia guilliermondii	400	Torulopsis norvegica
115	Candida utilis	263	Pichia kudriavzevii	421	Trichosporon cutaneum
122	Candida viswanathii	266	Pichia membranaefaciens	432	Zygosaccharomyces bailii**
123	Candida zeylanoides	271	Pichia norvegensis*		
		294	Pichia vini		

Tests in Key No. 14

1	D–Glucose fermentation	28	Raffinose growth	51	Growth without vitamins
11	Raffinose fermentation	29	Melezitose growth	53	Growth without pantothenate
12	D–Galactose growth	31	Starch growth	54	Growth without biotin
15	D–Xylose growth	33	Erythritol growth	55	Growth without thiamin
16	L–Arabinose growth	37	D–Glucitol growth	59	Growth without p–aminobenzoate
18	L–Rhamnose growth	38	myo–Inositol growth	60	Growth in 50% D–glucose
19	Sucrose growth	39	D–δ–Gluconolactone growth	62	Growth at 37 degrees
21	Trehalose growth	42	D,L–Lactate growth	65	Starch formation
23	Cellobiose growth	44	Citrate growth		
26	Melibiose growth	46	Ethanol growth	number of different tests 31	
27	Lactose growth	47	Ethylamine growth		
		49	Nitrate growth		

KEY No. 14

	Negative	Positive
1 Growth at 37 degrees.	..2	.74
2 Erythritol growth.	..3	.57
3 myo–Inositol growth.	..4	.50
4 D–Glucitol growth.	..5	.21
5 Ethanol growth.	..6	.12
6 D–Glucose fermentation.	..7	.10
7 Sucrose growth.	Schizoblastosporion starkeyi–henricii	..8
8 Nitrate growth.	..9	Rhodotorula glutinis
9 Raffinose growth.	Rhodotorula minuta	Rhodotorula rubra
10 Cellobiose growth.	..11	Hanseniaspora valbyensis
11 D–δ–Gluconolactone growth	Saccharomyces cerevisiae	Torulopsis holmii
12 D–Xylose growth.	..13	..16
13 Nitrate growth.	..14	Rhodotorula glutinis
14 D–Glucose fermentation.	..15	Pichia membranaefaciens Saccharomyces cerevisiae
15 D–Galactose growth.	Pichia membranaefaciens	Schizoblastosporion starkeyi–henricii
16 Nitrate growth.	..17	Rhodotorula glutinis
17 Sucrose growth.	..18	..19
18 Lactose growth.	Pichia membranaefaciens	Trichosporon cutaneum
19 Lactose growth.	..9	..20
20 Growth without p–aminobenzoate	Rhodotorula minuta	Trichosporon cutaneum
21 Nitrate growth.	..22	..44
22 Sucrose growth.	..23	..32
23 D–Xylose growth.	..24	..29
24 D–Glucose fermentation.	..25	..27
25 Citrate growth.	..26	Candida zeylanoides
26 D–Galactose growth.	Candida silvae	Schizoblastosporion starkeyi–henricii
27 Citrate growth.	..28	Candida zeylanoides
28 Ethylamine growth.	Saccharomyces cerevisiae	Zygosaccharomyces bailii**
29 Lactose growth.	..30	Trichosporon cutaneum
30 Trehalose growth.	Candida brumptii	..31
31 D,L–Lactate growth.	Candida zeylanoides	Candida ravautii
32 D–Glucose fermentation.	..33	..40
33 Lactose growth.	..34	..38
34 Raffinose growth.	..35	..36
35 Citrate growth.	Rhodotorula minuta	Pichia vini
36 Starch growth.	Rhodotorula rubra	..37
37 Growth without thiamin.	Pichia vini	Debaryomyces hansenii
38 Raffinose growth.	..20	..39
39 Growth in 50% D–glucose.	Trichosporon cutaneum	Debaryomyces hansenii
40 Cellobiose growth.	..28	..41
41 D,L–Lactate growth.	..42	..43
42 Starch growth.	Kluyveromyces lactis	Candida intermedia
43 Starch growth.	Kluyveromyces lactis	Debaryomyces hansenii
44 Melezitose growth.	..45	..48
45 Sucrose growth.	..46	..47
46 D–Galactose growth.	Torulopsis norvegica	Torulopsis etchellsii
47 D–Glucose fermentation.	Sporobolomyces salmonicolor	Torulopsis magnoliae
48 Raffinose growth.	Rhodotorula aurantiaca Rhodotorula glutinis	..49
49 Starch growth.	Rhodotorula glutinis	Sporobolomyces roseus
50 Nitrate growth.	..51	..56
51 Ethanol growth.	..52	..55
52 L–Rhamnose growth.	..53	..54
53 D–δ–Gluconolactone growth	Cryptococcus uniguttulatus	Cryptococcus gastricus
54 D–δ–Gluconolactone growth	Cryptococcus ater	Cryptococcus gastricus
55 Lactose growth.	..53	Cryptococcus gastricus Trichosporon cutaneum
56 Lactose growth.	Cryptococcus kuetzingii	Cryptococcus albidus
57 myo–Inositol growth.	..58	..72
58 Raffinose growth.	..59	..63
59 Sucrose growth.	..60	..62

		Negative	Positive
60	D–Xylose growth.	Saccharomycopsis lipolytica**	.61
61	Nitrate growth.	Trichosporon cutaneum	Sterigmatomyces halophilus
62	D–Glucose fermentation.	Candida humicola Trichosporon cutaneum	Candida tenuis
63	D–Glucose fermentation.	.64	.70
64	Melibiose growth.	.65	.69
65	Lactose growth.	.66	.68
66	D,L–Lactate growth.	.67	Debaryomyces hansenii Debaryomyces marama
67	Growth without vitamins.	Debaryomyces marama	Hyphopichia burtonii**
68	Growth in 50% D–glucose.	Debaryomyces marama Trichosporon cutaneum	Debaryomyces hansenii Debaryomyces marama
69	Starch formation.	.68	Candida humicola Trichosporon cutaneum
70	Melibiose growth.	.66	.71
71	Starch growth.	Candida membranaefaciens	Debaryomyces hansenii Debaryomyces marama
72	Nitrate growth.	.73	Cryptococcus albidus
73	Melibiose growth.	Candida curvata Trichosporon cutaneum	Candida humicola Cryptococcus laurentii Trichosporon cutaneum
74	Erythritol growth.	.75	.133
75	D–Glucitol growth.	.76	.90
76	Sucrose growth.	.77	.84
77	D–Glucose fermentation.	.78	.81
78	Lactose growth.	.79	Trichosporon cutaneum
79	Cellobiose growth.	.80	Pichia norvegensis*
80	Growth without biotin.	Pichia membranaefaciens Torulopsis inconspicua	Geotrichum capitatum** Pichia membranaefaciens
81	Cellobiose growth.	.82	Pichia norvegensis*
82	Trehalose growth.	.83	Saccharomyces cerevisiae Torulopsis glabrata
83	Growth without vitamins.	Pichia membranaefaciens Saccharomyces cerevisiae Saccharomyces telluris	Pichia kudriavzevii Pichia membranaefaciens Saccharomyces cerevisiae
84	D–Glucose fermentation.	.85	.86
85	Nitrate growth.	.19	Rhodotorula glutinis
86	Lactose growth.	.87	.89
87	Nitrate growth.	.88	Candida utilis
88	D–Xylose growth.	.11	Candida albicans
89	Growth without pantothenate	Candida pseudotropicalis	Candida macedoniensis
90	myo–Inositol growth.	.91	.131
91	Melezitose growth.	.92	.106
92	Lactose growth.	.93	.103
93	Sucrose growth.	.94	.99
94	D–Glucose fermentation.	.95	.97
95	D–Galactose growth.	Candida silvae	.96
96	Trehalose growth.	Candida catenulata Candida rugosa	Candida catenulata Candida ravautii
97	D–Xylose growth.	.98	Candida catenulata Candida ravautii
98	Citrate growth.	.28	Candida catenulata
99	Nitrate growth.	.100	.47
100	D–Xylose growth.	.101	.102
101	Cellobiose growth.	.28	Kluyveromyces lactis
102	Starch growth.	Kluyveromyces lactis	Candida albicans
103	D–Glucose fermentation.	Trichosporon cutaneum	.104
104	Growth without biotin.	.105	Candida macedoniensis
105	Growth without pantothenate	Candida pseudotropicalis Kluyveromyces marxianus	Kluyveromyces lactis Kluyveromyces marxianus
106	D–Glucose fermentation.	.107	.115
107	Nitrate growth.	.108	.113
108	Raffinose growth.	.109	.112
109	Lactose growth.	.110	.20
110	Citrate growth.	Rhodotorula minuta	.111
111	Starch growth.	Pichia etchellsii	Pichia vini
112	Lactose growth.	.36	.39

Key No. 14: clinical yeasts

		Negative	Positive
113	D−Galactose growth.	114	Rhodotorula aurantiaca / Rhodotorula glutinis
114	Raffinose fermentation.	Rhodotorula glutinis	Hansenula jadinii
115	Raffinose growth.	116	124
116	Cellobiose growth.	117	119
117	D−Xylose growth.	Saccharomyces cerevisiae	118
118	Starch growth.	Candida parapsilosis	Candida albicans
119	L−Rhamnose growth.	120	Candida lusitaniae
120	Citrate growth.	121	122
121	Starch growth.	Kluyveromyces lactis	Candida albicans / Candida tropicalis
122	L−Arabinose growth.	Candida albicans / Candida tropicalis / Candida viswanathii	123
123	Starch growth.	Pichia etchellsii	Candida albicans / Candida viswanathii
124	Nitrate growth.	125	130
125	L−Arabinose growth.	126	128
126	Cellobiose growth.	Saccharomyces cerevisiae	127
127	Melibiose growth.	42	Candida melibiosica
128	Melibiose growth.	41	129
129	Starch growth.	Pichia guilliermondii	Debaryomyces hansenii
130	L−Rhamnose growth.	Hansenula jadinii	Hansenula petersonii
131	D−Glucose fermentation.	132	Candida steatolytica*
132	Lactose growth.	Cryptococcus neoformans	Trichosporon cutaneum
133	myo−Inositol growth.	134	147
134	Raffinose growth.	135	139
135	D−Glucose fermentation.	136	138
136	D−Xylose growth.	Saccharomycopsis lipolytica**	137
137	Melibiose growth.	Candida aaseri / Trichosporon cutaneum	Candida humicola / Trichosporon cutaneum
138	L−Rhamnose growth.	Pichia farinosa	Candida tenuis
139	Melibiose growth.	140	143
140	Lactose growth.	141	142
141	D,L−Lactate growth.	Hyphopichia burtonii**	Debaryomyces hansenii
142	Melezitose growth.	Sterigmatomyces elviae* / Trichosporon cutaneum	39
143	D−Glucose fermentation.	144	146
144	Growth in 50% D−glucose.	Candida humicola / Trichosporon cutaneum	145
145	Starch formation.	Debaryomyces hansenii	Candida humicola
146	Starch growth.	Candida membranaefaciens	Debaryomyces hansenii
147	Melibiose growth.	148	Candida humicola / Cryptococcus laurentii / Trichosporon cutaneum
148	Lactose growth.	Cryptococcus neoformans	149
149	L−Rhamnose growth.	Sarcinosporon inkin** / Trichosporon cutaneum	Candida curvata / Trichosporon cutaneum

Yeasts associated with food (Key No. 15)

Key involving physiological tests and microscopical examination

Yeasts in Key No. 15

27	Candida boidinii	60	Candida humicola	99	Candida sake
28	Candida boleticola*	63	Candida iberica*	101	Candida santamariae
34	Candida butyri*	68	Candida intermedia	103	Candida silvae
35	Candida cacaoi	71	Candida kefyr	106	Candida solani
36	Candida catenulata	73	Candida lambica	107	Candida sorboxylosa*
39	Candida citrea*	74	Candida lusitaniae	110	Candida tenuis
41	Candida curiosa	75	Candida macedoniensis	113	Candida tropicalis
42	Candida curvata	78	Candida maritima	115	Candida utilis
44	Candida diddensii	82	Candida mesenterica	117	Candida valida
54	Candida fragicola*	87	Candida norvegensis	121	Candida vini
56	Candida friedrichii	90	Candida parapsilosis	123	Candida zeylanoides
57	Candida glaebosa	93	Candida pseudotropicalis	124	Citeromyces matritensis
		98	Candida rugosa	125	Cryptococcus albidus

130	Cryptococcus flavus
135	Cryptococcus kuetzingii
137	Cryptococcus laurentii
142	Cryptococcus neoformans
144	Cryptococcus terreus
145	Cryptococcus uniguttulatus
148	Debaryomyces hansenii
153	Debaryomyces polymorpha**
154	Debaryomyces tamarii
161	Geotrichum capitatum**
162	Geotrichum fermentans**
167	Hanseniaspora osmophila
168	Hanseniaspora uvarum
169	Hanseniaspora valbyensis
170	Hanseniaspora vineae*
171	Hansenula anomala
175	Hansenula californica
181	Hansenula fabianii
184	Hansenula holstii
187	Hansenula minuta
197	Hansenula subpelliculosa
202	Hyphopichia burtonii**
206	Kluyveromyces bulgaricus
207	Kluyveromyces delphensis
210	Kluyveromyces lactis
212	Kluyveromyces marxianus
216	Kluyveromyces thermotolerans**
223	Leucosporidium scottii
234	Metschnikowia pulcherrima
236	Metschnikowia zobellii
254	Pichia etchellsii
255	Pichia farinosa
256	Pichia fermentans
258	Pichia guilliermondii
262	Pichia kluyveri
263	Pichia kudriavzevii
266	Pichia membranaefaciens
272	Pichia ohmeri
273	Pichia onychis
276	Pichia pijperi
290	Pichia terricola
299	Rhodosporidium infirmo−miniatum**
303	Rhodotorula aurantiaca
304	Rhodotorula glutinis
307	Rhodotorula marina
308	Rhodotorula minuta
309	Rhodotorula pallida
311	Rhodotorula rubra
312	Saccharomyces cerevisiae
313	Saccharomyces dairensis
314	Saccharomyces exiguus
318	Saccharomyces unisporus
322	Saccharomycopsis fibuligera**
323	Saccharomycopsis lipolytica**
324	Saccharomycopsis malanga*
325	Saccharomycopsis vini**
329	Schizosaccharomyces malidevorans
330	Schizosaccharomyces octosporus
331	Schizosaccharomyces pombe
341	Sporobolomyces holsaticus
343	Sporobolomyces pararoseus
344	Sporobolomyces puniceus*
346	Sporobolomyces salmonicolor
357	Torulaspora delbrueckii**
361	Torulopsis apicola
371	Torulopsis etchellsii
372	Torulopsis fragaria*
373	Torulopsis fructus*
375	Torulopsis glabrata
376	Torulopsis gropengiesseri
377	Torulopsis haemulonii
380	Torulopsis holmii
382	Torulopsis inconspicua
387	Torulopsis lactis−condensi
388	Torulopsis magnoliae
392	Torulopsis molischiana
394	Torulopsis musae*
412	Torulopsis stellata
416	Torulopsis versatilis
420	Trichosporon brassicae*
421	Trichosporon cutaneum
425	Trichosporon pullulans
432	Zygosaccharomyces bailii**
433	Zygosaccharomyces bisporus**
439	Zygosaccharomyces rouxii**

Tests in Key No. 15

1	D−Glucose fermentation
12	D−Galactose growth
13	L−Sorbose growth
15	D−Xylose growth
16	L−Arabinose growth
18	L−Rhamnose growth
20	Maltose growth
21	Trehalose growth
22	Methyl α−D−glucopyranoside growth
23	Cellobiose growth
26	Melibiose growth
27	Lactose growth
28	Raffinose growth
31	Starch growth
33	Erythritol growth
37	D−Glucitol growth
39	D−δ−Gluconolactone growth
40	2−Ketogluconate growth
42	D,L−Lactate growth
43	Succinate growth
44	Citrate growth
47	Ethylamine growth
49	Nitrate growth
53	Growth without pantothenate
54	Growth without biotin
55	Growth without thiamin
56	Growth without pyridoxine
57	Growth without niacin
59	Growth without p−aminobenzoate
61	Growth in 60% D−glucose
62	Growth at 37 degrees
63	Growth with 0.01% cycloheximide
65	Starch formation
66	Pink colonies
68	Splitting cells
71	Cells of other shapes
72	Filamentous
74	Septate hyphae

number of different tests 38

KEY No. 15

		Negative	Positive
1	D−Glucitol growth.	2	114
2	Erythritol growth.	3	105
3	Maltose growth.	4	80
4	D,L−Lactate growth.	5	48
5	D−Galactose growth.	6	37
6	D−Glucose fermentation.	7	13
7	D−Xylose growth.	8	10
8	Citrate growth.	9	Pichia membranaefaciens / Pichia terricola
9	Splitting cells.	Candida valida / Pichia membranaefaciens	Geotrichum capitatum**
10	L−Arabinose growth.	11	12
11	Lactose growth.	Candida valida / Pichia membranaefaciens	Trichosporon cutaneum
12	Growth without p−aminobenzoate.	Rhodotorula minuta	Trichosporon cutaneum
13	Cellobiose growth.	14	36
14	Raffinose growth.	15	32
15	Nitrate growth.	16	Torulopsis lactis−condensi
16	Succinate growth.	17	28
17	Trehalose growth.	18	24
18	Ethylamine growth.	19	23
19	Growth without pyridoxine	20	21
20	Growth with 0.01% cycloheximide.	Pichia membranaefaciens / Saccharomyces cerevisiae	Kluyveromyces delphensis / Pichia membranaefaciens

Key No. 15: yeasts of food

	Negative	Positive
21 Growth in 60% D−glucose.	Candida valida Pichia membranaefaciens Saccharomyces cerevisiae	.22
22 Filamentous.	Pichia membranaefaciens Torulaspora delbrueckii**	Candida valida Pichia membranaefaciens
23 Filamentous.	Pichia membranaefaciens Torulaspora delbrueckii** Zygosaccharomyces bisporus** Zygosaccharomyces rouxii**	Candida valida Pichia membranaefaciens Zygosaccharomyces bisporus** Zygosaccharomyces rouxii**
24 Growth without pyridoxine	.25	.27
25 Growth without niacin.	Saccharomyces cerevisiae Torulopsis glabrata	.26
26 Growth with 0.01% cycloheximide	Saccharomyces cerevisiae	Kluyveromyces delphensis
27 Growth in 60% D−glucose.	Saccharomyces cerevisiae	Torulaspora delbrueckii** Zygosaccharomyces rouxii**
28 Citrate growth.	.29	.31
29 D−δ−Gluconolactone growth	.21	.30
30 Growth without thiamin.	Candida citrea* Candida valida Pichia membranaefaciens	.22
31 D−δ−Gluconolactone growth	Pichia membranaefaciens Pichia terricola	Candida citrea* Pichia membranaefaciens
32 Nitrate growth.	.33	Torulopsis lactis−condensi
33 Starch formation.	.34	Schizosaccharomyces malidevorans
34 Growth without thiamin.	Saccharomyces cerevisiae Torulopsis stellata	.35
35 Growth in 60% D−glucose.	Saccharomyces cerevisiae	Torulaspora delbrueckii**
36 2−Ketogluconate growth.	Hanseniaspora valbyensis	Hanseniaspora uvarum
37 D−Glucose fermentation.	.38	.40
38 D−Xylose growth.	.39	.12
39 Trehalose growth.	Geotrichum capitatum**	Debaryomyces tamarii
40 Cellobiose growth.	.41	Debaryomyces tamarii
41 Trehalose growth.	.42	.44
42 Growth in 60% D−glucose.	.43	Torulaspora delbrueckii** Zygosaccharomyces bisporus** Zygosaccharomyces rouxii**
43 Ethylamine growth.	Saccharomyces cerevisiae Saccharomyces dairensis	Saccharomyces unisporus
44 Growth in 60% D−glucose.	.45	Torulaspora delbrueckii** Zygosaccharomyces rouxii**
45 Raffinose growth.	.46	.47
46 Growth without pantothenate	Saccharomyces cerevisiae Saccharomyces dairensis	Saccharomyces cerevisiae Saccharomyces exiguus
47 D−δ−Gluconolactone growth	Saccharomyces cerevisiae Saccharomyces exiguus	Saccharomyces exiguus Torulopsis holmii
48 D−Xylose growth.	.49	.66
49 D−Galactose growth.	.50	.60
50 Citrate growth.	.51	.56
51 D−Glucose fermentation.	.52	.53
52 Growth without pyridoxine	Pichia membranaefaciens Torulopsis inconspicua	.9
53 Growth without pyridoxine	Candida sorboxylosa* Pichia membranaefaciens Saccharomyces cerevisiae	.54
54 Growth in 60% D−glucose.	Candida valida Pichia kudriavzevii Pichia membranaefaciens Saccharomyces cerevisiae	.55
55 Filamentous.	Pichia membranaefaciens Torulaspora delbrueckii**	Candida valida Pichia kudriavzevii Pichia membranaefaciens
56 Cellobiose growth.	.57	Candida norvegensis
57 Growth without thiamin.	.58	.59
58 Growth without pyridoxine	Candida sorboxylosa* Pichia membranaefaciens	Pichia membranaefaciens Pichia terricola

		Negative	Positive
59	Growth without biotin.	Pichia kluyveri Pichia membranaefaciens	Pichia kudriavzevii Pichia membranaefaciens
60	Trehalose growth.	61	65
61	D–Glucose fermentation.	Geotrichum capitatum**	62
62	Lactose growth.	63	Candida kefyr
63	D–δ–Gluconolactone growth	35	64
64	Growth without thiamin.	Candida fragicola*	Torulaspora delbrueckii**
65	Growth in 60% D–glucose.	47	Torulaspora delbrueckii**
66	Lactose growth.	67	78
67	Cellobiose growth.	68	75
68	L–Arabinose growth.	69	74
69	D–Galactose growth.	70	64
70	Citrate growth.	71	73
71	Growth without pyridoxine	Candida sorboxylosa* Pichia membranaefaciens	72
72	Filamentous.	Pichia membranaefaciens Torulaspora delbrueckii**	Candida lambica Candida valida Pichia membranaefaciens
73	Growth without pyridoxine	Candida sorboxylosa* Pichia membranaefaciens	Candida lambica Pichia fermentans Pichia membranaefaciens
74	D–Glucose fermentation.	Rhodotorula minuta	Candida fragicola*
75	Methyl α–D–glucopyranoside growth	76	Hansenula californica
76	Trehalose growth.	77	Rhodotorula minuta
77	D–Galactose growth.	Candida norvegensis	Candida fragicola*
78	D–Glucose fermentation.	12	79
79	Growth without pantothenate	Candida pseudotropicalis	Candida macedoniensis
80	D–Glucose fermentation.	81	94
81	Nitrate growth.	82	91
82	L–Arabinose growth.	83	87
83	Lactose growth.	84	Trichosporon cutaneum
84	D–Xylose growth.	85	86
85	Pink colonies.	Saccharomycopsis fibuligera**	Sporobolomyces pararoseus
86	Starch growth.	Rhodotorula rubra	Sporobolomyces pararoseus
87	Lactose growth.	88	Trichosporon cutaneum
88	L–Sorbose growth.	89	90
89	2–Ketogluconate growth.	Rhodotorula rubra	Cryptococcus uniguttulatus
90	Raffinose growth.	Trichosporon brassicae*	Rhodotorula rubra
91	Starch formation.	92	93
92	Succinate growth.	Leucosporidium scottii	Rhodotorula glutinis
93	Pink colonies.	Cryptococcus albidus	Rhodosporidium infirmo–miniatum**
94	Cellobiose growth.	95	101
95	Raffinose growth.	96	97
96	Starch formation.	27	Schizosaccharomyces octosporus
97	D–Galactose growth.	98	99
98	Starch formation.	35	Schizosaccharomyces pombe
99	Ethylamine growth.	35	100
100	Growth without niacin.	Kluyveromyces thermotolerans**	Torulaspora delbrueckii**
101	L–Sorbose growth.	102	104
102	Nitrate growth.	103	Candida utilis
103	Methyl α–D–glucopyranoside growth	Hanseniaspora osmophila Hanseniaspora vineae*	Saccharomycopsis fibuligera**
104	Nitrate growth.	Candida solani	Hansenula californica
105	D–Xylose growth.	106	109
106	Maltose growth.	Saccharomycopsis lipolytica**	107
107	D–Galactose growth.	108	Trichosporon pullulans
108	Starch growth.	Candida mesenterica	Saccharomycopsis fibuligera**
109	Nitrate growth.	110	113
110	Melibiose growth.	111	112
111	Splitting cells.	Candida curvata	Trichosporon cutaneum
112	Splitting cells.	Candida humicola	Trichosporon cutaneum

Key No. 15: yeasts of food

		Negative	Positive
113	Splitting cells.	...Cryptococcus albidus	...Trichosporon pullulans
114	Erythritol growth.	...115	...272
115	Nitrate growth.	...116	...250
116	Maltose growth.	...117	...179
117	Cellobiose growth.	...118	...151
118	D–Glucose fermentation.	...119	...132
119	Trehalose growth.	...120	...124
120	D–Galactose growth.	...121	...123
121	D–Xylose growth.	...122	...Trichosporon cutaneum
122	Septate hyphae.	...Candida silvae / Candida vini	...Saccharomycopsis vini**
123	Lactose growth.	...Candida catenulata / Candida rugosa	...Trichosporon cutaneum
124	D,L–Lactate growth.	...125	...129
125	D–Xylose growth.	...126	...127
126	Citrate growth.	...Saccharomycopsis vini**	...Candida iberica* / Candida zeylanoides
127	Lactose growth.	...128	...12
128	Citrate growth.	...Rhodotorula minuta	...Candida zeylanoides
129	L–Arabinose growth.	...130	...12
130	Lactose growth.	...131	...Trichosporon cutaneum
131	Growth without p–aminobenzoate	...Rhodotorula pallida	...Candida catenulata
132	D,L–Lactate growth.	...133	...145
133	Citrate growth.	...134	...143
134	Raffinose growth.	...135	...139
135	Growth in 60% D–glucose.	...136	...137
136	Septate hyphae.	...Saccharomyces cerevisiae	...Saccharomycopsis vini**
137	Filamentous.	...Torulaspora delbrueckii** / Zygosaccharomyces bailii** / Zygosaccharomyces bisporus** / Zygosaccharomyces rouxii**	...138
138	Septate hyphae.	...Zygosaccharomyces bailii** / Zygosaccharomyces bisporus** / Zygosaccharomyces rouxii**	...Saccharomycopsis vini**
139	L–Sorbose growth.	...140	...142
140	Growth in 60% D–glucose.	...136	...141
141	Filamentous.	...Torulaspora delbrueckii**	...Saccharomycopsis vini**
142	Filamentous.	...Torulaspora delbrueckii** / Torulopsis apicola	...Saccharomycopsis vini**
143	D–Xylose growth.	...Candida iberica* / Candida zeylanoides	...144
144	Growth without pyridoxine	...Torulopsis fructus*	...Candida zeylanoides
145	Raffinose growth.	...146	...147
146	Citrate growth.	...63	...Candida catenulata
147	Lactose growth.	...148	...150
148	Growth in 60% D–glucose.	...149	...Torulaspora delbrueckii**
149	Ethylamine growth.	...Saccharomyces cerevisiae	...Kluyveromyces bulgaricus
150	Growth without pantothenate	...Candida kefyr	...Kluyveromyces bulgaricus
151	Raffinose growth.	...152	...171
152	D–Glucose fermentation.	...153	...163
153	L–Arabinose growth.	...154	...159
154	D,L–Lactate growth.	...155	...157
155	Lactose growth.	...156	...Trichosporon cutaneum
156	Citrate growth.	...Saccharomycopsis vini**	...Candida zeylanoides
157	Lactose growth.	...158	...Trichosporon cutaneum
158	Trehalose growth.	...Geotrichum fermentans**	...Rhodotorula pallida
159	L–Rhamnose growth.	...160	...162
160	Trehalose growth.	...161	...127
161	Lactose growth.	...Geotrichum fermentans**	...Trichosporon cutaneum
162	Growth without p–aminobenzoate	...Rhodotorula marina	...Trichosporon cutaneum
163	Citrate growth.	...164	...168
164	D–Galactose growth.	...Saccharomycopsis vini**	...165
165	L–Sorbose growth.	...166	...167
166	Growth without thiamin.	...Candida fragicola*	...Kluyveromyces lactis
167	Growth without biotin.	...Kluyveromyces lactis	...Geotrichum fermentans**
168	Trehalose growth.	...169	...170

		Negative	Positive
169	D−Galactose growth.	Pichia pijperi	Geotrichum fermentans**
170	D,L−Lactate growth.	Candida zeylanoides	Candida santamariae
171	D−Xylose growth.	172	176
172	D,L−Lactate growth.	173	175
173	D−Galactose growth.	Saccharomycopsis vini**	174
174	Succinate growth.	Torulopsis gropengiesseri	Kluyveromyces lactis
175	Growth without pantothenate	Candida kefyr	Kluyveromyces bulgaricus Kluyveromyces lactis
176	D−Glucose fermentation.	162	177
177	Growth without biotin.	178	Candida macedoniensis
178	Growth without pantothenate	Candida pseudotropicalis Kluyveromyces marxianus	Kluyveromyces bulgaricus Kluyveromyces lactis Kluyveromyces marxianus
179	Cellobiose growth.	180	203
180	D−Glucose fermentation.	181	189
181	Lactose growth.	182	Trichosporon cutaneum
182	L−Arabinose growth.	183	188
183	D−Galactose growth.	184	185
184	D−Xylose growth.	Saccharomycopsis vini**	Rhodotorula rubra
185	L−Rhamnose growth.	186	187
186	Raffinose growth.	Candida catenulata	Rhodotorula rubra
187	2−Ketogluconate growth.	Rhodotorula rubra	Cryptococcus neoformans
188	L−Rhamnose growth.	88	187
189	D−Xylose growth.	190	196
190	Citrate growth.	191	Candida catenulata
191	D−Galactose growth.	192	195
192	Growth in 60% D−glucose.	136	193
193	Filamentous.	Torulaspora delbrueckii** Zygosaccharomyces rouxii**	194
194	Septate hyphae.	Zygosaccharomyces rouxii**	Saccharomycopsis vini**
195	Methyl α−D−glucopyranoside growth	27	99
196	L−Arabinose growth.	197	202
197	D−Galactose growth.	198	199
198	Citrate growth.	Torulaspora delbrueckii**	Torulopsis musae*
199	L−Sorbose growth.	200	201
200	Citrate growth.	Torulaspora delbrueckii**	Candida catenulata
201	Filamentous.	Torulaspora delbrueckii**	Candida sake
202	L−Rhamnose growth.	Candida parapsilosis	Torulopsis haemulonii
203	D−Glucose fermentation.	204	229
204	L−Arabinose growth.	205	215
205	Lactose growth.	206	214
206	D−Galactose growth.	207	212
207	D−Xylose growth.	208	211
208	Methyl α−D−glucopyranoside growth	209	210
209	Starch growth.	Saccharomycopsis vini**	Saccharomycopsis malanga*
210	Starch growth.	Saccharomycopsis vini**	85
211	Raffinose growth.	Saccharomycopsis malanga*	86
212	L−Rhamnose growth.	213	187
213	Raffinose growth.	Metschnikowia zobellii	86
214	Growth without pantothenate	Candida glaebosa	Trichosporon cutaneum
215	Lactose growth.	216	225
216	L−Rhamnose growth.	217	222
217	Raffinose growth.	218	220
218	L−Sorbose growth.	219	Pichia etchellsii
219	Growth at 37 degrees.	Cryptococcus uniguttulatus	Saccharomycopsis malanga*
220	D,L−Lactate growth.	89	221
221	Starch growth.	Rhodotorula rubra	Debaryomyces hansenii
222	Growth without thiamin.	223	Debaryomyces hansenii
223	2−Ketogluconate growth.	224	Cryptococcus neoformans
224	Growth without p−aminobenzoate	Rhodotorula marina	Rhodotorula rubra
225	L−Rhamnose growth.	226	228
226	Growth without pantothenate	Candida glaebosa	227
227	Splitting cells.	Debaryomyces hansenii	Trichosporon cutaneum
228	Growth without p−aminobenzoate	Rhodotorula marina	227
229	D−Galactose growth.	230	236
230	L−Rhamnose growth.	231	235

Key No. 15: yeasts of food

		Negative	Positive
231	D−Xylose growth.	232	234
232	Methyl α−D−glucopyranoside growth	209	233
233	Starch growth.	Saccharomycopsis vini**	Saccharomycopsis fibuligera**
234	Methyl α−D−glucopyranoside growth	Saccharomycopsis malanga*	Pichia onychis
235	L−Sorbose growth.	Candida maritima	Candida lusitaniae
236	Raffinose growth.	237	243
237	L−Rhamnose growth.	238	Candida lusitaniae
238	L−Arabinose growth.	239	242
239	Starch growth.	240	Candida tropicalis
240	Growth without thiamin.	Metschnikowia zobellii	241
241	Growth without niacin.	Kluyveromyces lactis	Candida sake Metschnikowia pulcherrima
242	Citrate growth.	Kluyveromyces lactis	Pichia etchellsii
243	D−Xylose growth.	244	245
244	Citrate growth.	Kluyveromyces lactis	Pichia ohmeri
245	Melibiose growth.	246	249
246	D,L−Lactate growth.	247	248
247	Starch growth.	Kluyveromyces lactis	Candida intermedia
248	Starch growth.	Kluyveromyces lactis	Debaryomyces hansenii
249	Starch growth.	Pichia guilliermondii	Debaryomyces hansenii
250	D−Glucose fermentation.	251	264
251	Maltose growth.	252	255
252	Cellobiose growth.	Sporobolomyces salmonicolor	253
253	L−Sorbose growth.	254	Cryptococcus terreus
254	L−Arabinose growth.	Hansenula minuta	Cryptococcus kuetzingii
255	Raffinose growth.	256	260
256	Citrate growth.	257	258
257	Starch formation.	Rhodotorula aurantiaca Rhodotorula glutinis	Cryptococcus terreus
258	2−Ketogluconate growth.	259	Rhodotorula aurantiaca Rhodotorula glutinis
259	Septate hyphae.	Rhodotorula glutinis	Sporobolomyces holsaticus
260	L−Rhamnose growth.	261	91
261	Lactose growth.	262	263
262	Growth without biotin.	Sporobolomyces puniceus*	259
263	Starch formation.	Torulopsis fragaria*	Cryptococcus albidus
264	Methyl α−D−glucopyranoside growth	265	269
265	Maltose growth.	266	268
266	Trehalose growth.	Torulopsis magnoliae	267
267	Filamentous.	Hansenula minuta	Candida curiosa
268	Trehalose growth.	Torulopsis etchellsii	Torulopsis versatilis
269	Raffinose growth.	270	271
270	D−Galactose growth.	Hansenula californica	Hansenula holstii
271	D−Xylose growth.	Citeromyces matritensis	Hansenula fabianii
272	Raffinose growth.	273	295
273	Maltose growth.	274	280
274	D−Glucose fermentation.	275	277
275	Lactose growth.	276	Trichosporon cutaneum
276	Trehalose growth.	Saccharomycopsis lipolytica**	Rhodotorula pallida
277	Citrate growth.	278	279
278	D−Galactose growth.	Candida boidinii	Candida cacaoi
279	Growth at 37 degrees.	Candida boleticola*	Pichia farinosa
280	D−Galactose growth.	281	287
281	D−Xylose growth.	282	284
282	Methyl α−D−glucopyranoside growth	283	108
283	Starch growth.	Candida mesenterica	Saccharomycopsis malanga*
284	Nitrate growth.	285	Torulopsis molischiana
285	Methyl α−D−glucopyranoside growth	286	112
286	Lactose growth.	Saccharomycopsis malanga*	Trichosporon cutaneum
287	D−Glucose fermentation.	288	291

		Negative	Positive
288	Melibiose growth.	289	112
289	Lactose growth.	290	Trichosporon cutaneum
290	2–Ketogluconate growth.	Candida diddensii	Cryptococcus neoformans
291	Nitrate growth.	292	Hansenula holstii
292	L–Rhamnose growth.	293	294
293	Cells of other shapes.	Candida butyri*	Candida diddensii
294	Cells of other shapes.	Candida tenuis	Candida diddensii
295	Nitrate growth.	296	313
296	Melibiose growth.	297	306
297	L–Rhamnose growth.	298	303
298	D–Xylose growth.	Saccharomycopsis fibuligera**	299
299	Methyl α–D–glucopyranoside growth	300	301
300	D–Glucose fermentation.	Trichosporon cutaneum	Debaryomyces polymorpha**
301	Lactose growth.	302	227
302	D,L–Lactate growth.	Hyphopichia burtonii**	Debaryomyces hansenii
303	Lactose growth.	304	305
304	Growth without thiamin.	Cryptococcus neoformans	Debaryomyces hansenii
305	Starch formation.	227	111
306	D–Glucose fermentation.	307	311
307	Starch formation.	308	310
308	Growth without thiamin.	309	227
309	Splitting cells.	Cryptococcus flavus	Trichosporon cutaneum
310	Splitting cells.	Candida humicola / Cryptococcus laurentii	Trichosporon cutaneum
311	Methyl α–D–glucopyranoside growth	Debaryomyces polymorpha**	312
312	D,L–Lactate growth.	Candida friedrichii	Debaryomyces hansenii
313	D–Glucose fermentation.	113	314
314	Growth without thiamin.	Hansenula subpelliculosa	Hansenula anomala

Yeasts associated with wine and wine-making (Key No. 16)

Key involving physiological tests and microscopical examination

Yeasts in Key No. 16

- 11 Brettanomyces claussenii
- 13 Brettanomyces custersii
- 14 Brettanomyces lambicus
- 20 Candida albicans
- 26 Candida bogoriensis
- 27 Candida boidinii
- 31 Candida brumptii
- 36 Candida catenulata
- 46 Candida diversa
- 60 Candida humicola
- 64 Candida incommunis*
- 68 Candida intermedia
- 80 Candida melinii
- 90 Candida parapsilosis
- 98 Candida rugosa
- 99 Candida sake
- 106 Candida solani
- 108 Candida steatolytica*
- 110 Candida tenuis
- 113 Candida tropicalis
- 115 Candida utilis
- 117 Candida valida
- 119 Candida veronae
- 121 Candida vini
- 123 Candida zeylanoides
- 124 Citeromyces matritensis
- 125 Cryptococcus albidus
- 137 Cryptococcus laurentii
- 138 Cryptococcus luteolus
- 142 Cryptococcus neoformans
- 148 Debaryomyces hansenii
- 153 Debaryomyces polymorpha**
- 157 Dekkera bruxellensis
- 158 Dekkera intermedia
- 159 Filobasidium capsuligenum**
- 162 Geotrichum fermentans**
- 166 Hanseniaspora occidentalis*
- 167 Hanseniaspora osmophila
- 168 Hanseniaspora uvarum
- 169 Hanseniaspora valbyensis
- 170 Hanseniaspora vineae*
- 171 Hansenula anomala
- 175 Hansenula californica
- 195 Hansenula saturnus
- 196 Hansenula silvicola
- 197 Hansenula subpelliculosa
- 202 Hyphopichia burtonii**
- 210 Kluyveromyces lactis
- 212 Kluyveromyces marxianus
- 216 Kluyveromyces thermotolerans**
- 223 Leucosporidium scottii
- 228 Lipomyces starkeyi
- 230 Lodderomyces elongisporus
- 234 Metschnikowia pulcherrima
- 235 Metschnikowia reukaufii
- 238 Nadsonia elongata
- 242 Pachytichospora transvaalensis
- 254 Pichia etchellsii
- 255 Pichia farinosa
- 256 Pichia fermentans
- 258 Pichia guilliermondii
- 261 Pichia humboldtii*
- 263 Pichia kudriavzevii
- 266 Pichia membranaefaciens
- 294 Pichia vini
- 303 Rhodotorula aurantiaca
- 304 Rhodotorula glutinis
- 308 Rhodotorula minuta
- 309 Rhodotorula pallida
- 311 Rhodotorula rubra
- 312 Saccharomyces cerevisiae
- 314 Saccharomyces exiguus
- 315 Saccharomyces kluyveri
- 318 Saccharomyces unisporus
- 319 Saccharomycodes ludwigii
- 322 Saccharomycopsis fibuligera**
- 323 Saccharomycopsis lipolytica**
- 328 Schizosaccharomyces japonicus
- 329 Schizosaccharomyces malidevorans
- 330 Schizosaccharomyces octosporus
- 331 Schizosaccharomyces pombe
- 343 Sporobolomyces pararoseus
- 345 Sporobolomyces roseus
- 346 Sporobolomyces salmonicolor
- 357 Torulaspora delbrueckii**
- 358 Torulaspora globosa**
- 361 Torulopsis apicola
- 367 Torulopsis cantarelii
- 375 Torulopsis glabrata
- 382 Torulopsis inconspicua
- 400 Torulopsis norvegica
- 412 Torulopsis stellata
- 415 Torulopsis vanderwaltii
- 416 Torulopsis versatilis
- 421 Trichosporon cutaneum

Key No. 16: yeasts of wine

425 Trichosporon pullulans
429 Wickerhamiella domercqii**
432 Zygosaccharomyces bailii**
433 Zygosaccharomyces bisporus**
436 Zygosaccharomyces florentinus**
437 Zygosaccharomyces microellipsodes**
439 Zygosaccharomyces rouxii**

Tests in Key No. 16
- 1 D−Glucose fermentation
- 8 Lactose fermentation
- 9 Cellobiose fermentation
- 12 D−Galactose growth
- 13 L−Sorbose growth
- 15 D−Xylose growth
- 16 L−Arabinose growth
- 19 Sucrose growth
- 20 Maltose growth
- 21 Trehalose growth
- 22 Methyl α−D−glucopyranoside growth
- 23 Cellobiose growth
- 26 Melibiose growth
- 27 Lactose growth
- 28 Raffinose growth
- 30 Inulin growth
- 31 Starch growth
- 33 Erythritol growth
- 37 D−Glucitol growth
- 38 myo−Inositol growth
- 40 2−Ketogluconate growth
- 42 D,L−Lactate growth
- 44 Citrate growth
- 47 Ethylamine growth
- 49 Nitrate growth
- 55 Growth without thiamin
- 56 Growth without pyridoxine
- 57 Growth without niacin
- 61 Growth in 60% D−glucose
- 62 Growth at 37 degrees
- 63 Growth with 0.01% cycloheximide
- 65 Starch formation
- 66 Pink colonies
- 68 Splitting cells
- 69 Apical budding
- 72 Filamentous

number of different tests 36

KEY No. 16

	Negative	Positive
1 D−Glucitol growth.	2	101
2 Maltose growth.	3	55
3 Sucrose growth.	4	28
4 Erythritol growth.	5	27
5 D−Glucose fermentation.	6	9
6 Lactose growth.	7	Trichosporon cutaneum
7 D−Galactose growth.	8	Pichia humboldtii*
8 Growth without pyridoxine	Pichia membranaefaciens / Torulopsis inconspicua	Candida valida / Pichia membranaefaciens
9 Cellobiose growth.	10	26
10 D−Galactose growth.	11	24
11 D,L−Lactate growth.	12	20
12 Trehalose growth.	13	18
13 L−Sorbose growth.	14	16
14 Growth in 60% D−glucose.	Candida valida / Pichia membranaefaciens / Saccharomyces cerevisiae	15
15 Filamentous.	Pichia membranaefaciens / Torulaspora delbrueckii** / Zygosaccharomyces bisporus** / Zygosaccharomyces rouxii**	Candida valida / Pichia membranaefaciens / Zygosaccharomyces bisporus** / Zygosaccharomyces rouxii**
16 Growth in 60% D−glucose.	17	Pichia membranaefaciens / Torulaspora delbrueckii** / Zygosaccharomyces bisporus** / Zygosaccharomyces rouxii**
17 Apical budding.	Pichia membranaefaciens	Nadsonia elongata
18 Growth without pyridoxine	Saccharomyces cerevisiae / Torulopsis glabrata	19
19 Growth in 60% D−glucose.	Saccharomyces cerevisiae	Torulaspora delbrueckii** / Zygosaccharomyces rouxii**
20 Citrate growth.	21	23
21 Growth in 60% D−glucose.	Candida valida / Pichia kudriavzevii / Pichia membranaefaciens / Saccharomyces cerevisiae	22
22 Filamentous.	Pichia membranaefaciens / Torulaspora delbrueckii**	Candida valida / Pichia kudriavzevii / Pichia membranaefaciens
23 D−Xylose growth.	Pichia kudriavzevii / Pichia membranaefaciens	Pichia fermentans / Pichia membranaefaciens
24 Growth in 60% D−glucose.	25	Torulaspora delbrueckii** / Zygosaccharomyces bisporus** / Zygosaccharomyces rouxii**

		Negative	Positive
25	Ethylamine growth.	Pachytichospora transvaalensis	Saccharomyces unisporus
		Saccharomyces cerevisiae	
26	2-Ketogluconate growth.	Hanseniaspora valbyensis	Hanseniaspora uvarum
27	D-Xylose growth.	Saccharomycopsis lipolytica**	Trichosporon cutaneum
28	Cellobiose growth.	29	49
29	D-Glucose fermentation.	30	31
30	Pink colonies.	Trichosporon cutaneum	Rhodotorula minuta
31	Melibiose growth.	32	41
32	D-Galactose growth.	33	40
33	Raffinose growth.	19	34
34	D,L-Lactate growth.	35	38
35	Starch formation.	36	Schizosaccharomyces malidevorans
36	Growth without thiamin.	Saccharomyces cerevisiae	37
		Torulopsis stellata	
37	Growth in 60% D-glucose.	Saccharomyces cerevisiae	Torulaspora delbrueckii**
38	Growth in 60% D-glucose.	39	Torulaspora delbrueckii**
39	Growth with 0.01% cycloheximide	Saccharomyces cerevisiae	Torulaspora globosa**
40	Growth in 60% D-glucose.	Saccharomyces cerevisiae	Torulaspora delbrueckii**
		Saccharomyces exiguus	Zygosaccharomyces rouxii**
41	L-Sorbose growth.	42	45
42	Growth in 60% D-glucose.	43	Torulaspora delbrueckii**
43	Ethylamine growth.	Saccharomyces cerevisiae	44
44	Growth at 37 degrees.	Zygosaccharomyces microellipsodes**	Saccharomyces kluyveri
45	D,L-Lactate growth.	46	48
46	Growth in 60% D-glucose.	47	Torulaspora delbrueckii**
47	Growth at 37 degrees.	Zygosaccharomyces florentinus**	Saccharomyces kluyveri
48	Growth in 60% D-glucose.	44	Torulaspora delbrueckii**
49	Raffinose growth.	50	52
50	D-Glucose fermentation.	30	51
51	L-Sorbose growth.	Hanseniaspora occidentalis*	Hansenula californica
52	D-Glucose fermentation.	Trichosporon cutaneum	53
53	D-Galactose growth.	54	Saccharomyces kluyveri
54	D-Xylose growth.	Saccharomycodes ludwigii	Hansenula saturnus
55	D-Glucose fermentation.	56	72
56	Erythritol growth.	57	67
57	myo-Inositol growth.	58	64
58	Nitrate growth.	59	63
59	Lactose growth.	60	Trichosporon cutaneum
60	D-Xylose growth.	61	62
61	Pink colonies.	Saccharomycopsis fibuligera**	Sporobolomyces pararoseus
62	Starch growth.	Rhodotorula rubra	Sporobolomyces pararoseus
63	Pink colonies.	Leucosporidium scottii	Rhodotorula glutinis
64	D-Xylose growth.	Saccharomycopsis fibuligera**	65
65	Nitrate growth.	66	Cryptococcus albidus
66	Splitting cells.	Cryptococcus luteolus	Trichosporon cutaneum
67	Nitrate growth.	68	71
68	D-Xylose growth.	Saccharomycopsis fibuligera**	69
69	Splitting cells.	70	Trichosporon cutaneum
70	Filamentous.	Cryptococcus luteolus	Candida humicola
71	Splitting cells.	Cryptococcus albidus	Trichosporon pullulans
72	Cellobiose growth.	73	92
73	Raffinose growth.	74	82
74	Methyl α-D-glucopyranoside growth	75	78
75	D-Xylose growth.	76	77
76	Starch formation.	19	Schizosaccharomyces octosporus
77	Starch growth.	Torulaspora delbrueckii**	Candida albicans
78	D-Xylose growth.	79	77

Key No. 16: yeasts of wine

	Negative	Positive
79 Nitrate growth.	80	Brettanomyces lambicus Dekkera bruxellensis
80 Growth in 60% D-glucose.	81	Torulaspora delbrueckii**
81 Ethylamine growth.	Saccharomyces cerevisiae	Dekkera bruxellensis
82 Melibiose growth.	83	89
83 Trehalose growth.	84	86
84 Starch formation.	37	85
85 Filamentous.	Schizosaccharomyces pombe	Schizosaccharomyces japonicus
86 D-Galactose growth.	80	87
87 Ethylamine growth.	37	88
88 Growth without niacin.	Kluyveromyces thermotolerans**	Torulaspora delbrueckii**
89 L-Sorbose growth.	90	46
90 Growth in 60% D-glucose.	91	Torulaspora delbrueckii**
91 Ethylamine growth.	Saccharomyces cerevisiae	Saccharomyces kluyveri
92 D-Galactose growth.	93	97
93 L-Sorbose growth.	94	96
94 Nitrate growth.	95	Candida utilis
95 Methyl α-D-glucopyranoside growth	Hanseniaspora osmophila Hanseniaspora vineae*	Saccharomycopsis fibuligera**
96 Nitrate growth.	Candida solani	Hansenula californica
97 Melibiose growth.	98	Saccharomyces kluyveri
98 D-Xylose growth.	99	Candida albicans
99 Lactose growth.	Brettanomyces custersii Dekkera intermedia	100
100 Lactose fermentation.	Brettanomyces custersii	Brettanomyces claussenii
101 Erythritol growth.	102	219
102 myo-Inositol growth.	103	210
103 Nitrate growth.	104	192
104 Cellobiose growth.	105	148
105 Maltose growth.	106	123
106 D-Glucose fermentation.	107	115
107 Trehalose growth.	108	110
108 Lactose growth.	109	Trichosporon cutaneum
109 D-Galactose growth.	Candida vini	Candida catenulata Candida rugosa
110 Lactose growth.	111	30
111 L-Arabinose growth.	112	114
112 D,L-Lactate growth.	Candida zeylanoides	113
113 Growth at 37 degrees.	Rhodotorula pallida	Candida catenulata
114 Sucrose growth.	Candida zeylanoides	Rhodotorula minuta
115 Citrate growth.	116	121
116 Raffinose growth.	117	119
117 Growth in 60% D-glucose.	118	Torulaspora delbrueckii** Zygosaccharomyces bailii** Zygosaccharomyces bisporus** Zygosaccharomyces rouxii**
118 L-Sorbose growth.	Saccharomyces cerevisiae	Nadsonia elongata
119 Melibiose growth.	120	41
120 L-Sorbose growth.	37	Torulaspora delbrueckii** Torulopsis apicola
121 D,L-Lactate growth.	122	Candida catenulata
122 Trehalose growth.	Candida diversa	Candida zeylanoides
123 D-Xylose growth.	124	130
124 Raffinose growth.	125	129
125 Methyl α-D-glucopyranoside growth	126	127
126 Citrate growth.	19	Candida catenulata
127 L-Sorbose growth.	80	128
128 Growth in 60% D-glucose.	Lodderomyces elongisporus	Torulaspora delbrueckii**
129 Melibiose growth.	86	89
130 D-Glucose fermentation.	131	138
131 Sucrose growth.	132	134
132 Lactose growth.	133	Trichosporon cutaneum
133 Growth at 37 degrees.	Candida brumptii	Candida catenulata
134 Lactose growth.	135	Trichosporon cutaneum

		Negative	Positive
135	Raffinose growth.	...136	...137
136	Starch growth.	...Lodderomyces elongisporus	...Pichia vini
137	Starch growth.	...Rhodotorula rubra	...Pichia vini
138	Sucrose growth.	...139	...141
139	Citrate growth.	...140	...133
140	Starch growth.	...Torulaspora delbrueckii**	...Candida brumptii
141	Melibiose growth.	...142	...46
142	L–Arabinose growth.	...143	...147
143	Starch growth.	...144	...Candida albicans
144	Growth in 60% D–glucose.	...145	...146
145	Growth at 37 degrees.	...Candida sake	...Lodderomyces elongisporus
146	Filamentous.	...Torulaspora delbrueckii**	...Candida sake
147	Starch growth.	...Candida parapsilosis	...Candida albicans
148	Raffinose growth.	...149	...173
149	Maltose growth.	...150	...157
150	Sucrose growth.	...151	...154
151	Lactose growth.	...152	...Trichosporon cutaneum
152	Trehalose growth.	...Geotrichum fermentans**	...153
153	D,L–Lactate growth.	...Candida zeylanoides	...Rhodotorula pallida
154	D–Glucose fermentation.	...155	...Kluyveromyces lactis
155	L–Arabinose growth.	...156	...30
156	Lactose growth.	...Rhodotorula pallida	...Trichosporon cutaneum
157	D–Glucose fermentation.	...158	...162
158	D–Xylose growth.	...Saccharomycopsis fibuligera**	...159
159	Lactose growth.	...160	...Trichosporon cutaneum
160	Sucrose growth.	...Candida bogoriensis	...161
161	Starch growth.	...Pichia etchellsii	...Pichia vini
162	D–Xylose growth.	...163	...167
163	D–Galactose growth.	...164	...165
164	Starch growth.	...Metschnikowia reukaufii	...Saccharomycopsis fibuligera**
165	Cellobiose fermentation.	...166	...Dekkera intermedia
166	Growth without niacin.	...Kluyveromyces lactis	...Metschnikowia reukaufii
167	L–Arabinose growth.	...168	...170
168	Starch growth.	...169	...Candida albicans Candida tropicalis
169	Growth without niacin.	...Kluyveromyces lactis	...Candida sake Metschnikowia pulcherrima Metschnikowia reukaufii
170	Citrate growth.	...171	...172
171	Starch growth.	...Kluyveromyces lactis	...Candida albicans
172	Starch growth.	...Pichia etchellsii	...Candida albicans
173	D–Glucose fermentation.	...174	...179
174	D–Xylose growth.	...61	...175
175	Lactose growth.	...176	...178
176	Growth without thiamin.	...137	...177
177	L–Arabinose growth.	...Sporobolomyces pararoseus	...Debaryomyces hansenii
178	Splitting cells.	...Debaryomyces hansenii	...Trichosporon cutaneum
179	D–Xylose growth.	...180	...185
180	D–Galactose growth.	...Saccharomycopsis fibuligera**	...181
181	Melibiose growth.	...182	...Saccharomyces kluyveri
182	Lactose growth.	...183	...184
183	Cellobiose fermentation.	...Kluyveromyces lactis	...Dekkera intermedia
184	Cellobiose fermentation.	...Kluyveromyces lactis	...Brettanomyces claussenii
185	Melibiose growth.	...186	...190
186	Maltose growth.	...Kluyveromyces lactis Kluyveromyces marxianus	...187
187	D,L–Lactate growth.	...188	...189
188	Starch growth.	...Kluyveromyces lactis	...Candida intermedia
189	Starch growth.	...Kluyveromyces lactis	...Debaryomyces hansenii
190	L–Arabinose growth.	...Saccharomyces kluyveri	...191
191	Starch growth.	...Pichia guilliermondii	...Debaryomyces hansenii
192	D–Glucose fermentation.	...193	...201
193	Maltose growth.	...194	...197
194	Trehalose growth.	...195	...196
195	L–Sorbose growth.	...Torulopsis norvegica	...Wickerhamiella domercqii**

Key No. 16: yeasts of wine

	Negative	Positive
196 2–Ketogluconate growth.	Sporobolomyces salmonicolor	Torulopsis vanderwaltii
197 Raffinose growth.	198	200
198 D–Galactose growth.	199	Rhodotorula aurantiaca Rhodotorula glutinis
199 Pink colonies.	Candida melinii	Rhodotorula glutinis
200 Starch growth.	63	Sporobolomyces roseus
201 D–Xylose growth.	202	207
202 Cellobiose growth.	203	206
203 Methyl α–D–glucopyranoside growth	Torulopsis versatilis	204
204 Raffinose growth.	Brettanomyces lambicus Dekkera bruxellensis	205
205 Growth in 60% D–glucose.	Dekkera bruxellensis	Citeromyces matritensis
206 Methyl α–D–glucopyranoside growth	Torulopsis versatilis	99
207 L–Sorbose growth.	208	209
208 Sucrose growth.	Torulopsis norvegica	Hansenula saturnus
209 D–Galactose growth.	Hansenula californica	Hansenula silvicola
210 Lactose growth.	211	218
211 D–Galactose growth.	212	213
212 Growth at 37 degrees.	Filobasidium capsuligenum**	Saccharomycopsis fibuligera**
213 D–Glucose fermentation.	214	217
214 Raffinose growth.	215	216
215 Growth at 37 degrees.	Filobasidium capsuligenum**	Cryptococcus neoformans
216 Growth at 37 degrees.	Cryptococcus luteolus	Cryptococcus neoformans
217 L–Sorbose growth.	Filobasidium capsuligenum**	Candida steatolytica*
218 D–Glucose fermentation.	65	Candida steatolytica*
219 myo–Inositol growth.	220	243
220 Raffinose growth.	221	231
221 D–Glucose fermentation.	222	227
222 D–Xylose growth.	223	225
223 Maltose growth.	224	Saccharomycopsis fibuligera**
224 Trehalose growth.	Saccharomycopsis lipolytica**	Rhodotorula pallida
225 Maltose growth.	156	226
226 Splitting cells.	Candida humicola	Trichosporon cutaneum
227 D–Galactose growth.	228	230
228 D–Xylose growth.	229	Candida boidinii
229 L–Sorbose growth.	Saccharomycopsis fibuligera**	Torulopsis cantarelii
230 Sucrose growth.	Pichia farinosa	Candida tenuis Candida veronae
231 Nitrate growth.	232	241
232 D–Xylose growth.	Saccharomycopsis fibuligera**	233
233 Methyl α–D–glucopyranoside growth	234	235
234 D–Glucose fermentation.	Trichosporon cutaneum	Debaryomyces polymorpha**
235 Melibiose growth.	236	238
236 Lactose growth.	237	178
237 D,L–Lactate growth.	Hyphopichia burtonii**	Debaryomyces hansenii
238 D,L–Lactate growth.	239	240
239 Inulin growth.	226	Lipomyces starkeyi
240 Starch formation.	178	226
241 D–Glucose fermentation.	Trichosporon pullulans	242
242 Growth without thiamin.	Hansenula subpelliculosa	Hansenula anomala
243 Nitrate growth.	244	250
244 Melibiose growth.	245	247
245 D–Xylose growth.	Saccharomycopsis fibuligera**	246
246 Lactose growth.	216	66
247 Inulin growth.	248	Lipomyces starkeyi
248 Splitting cells.	249	Trichosporon cutaneum

	Negative	Positive
249 Filamentous.	Cryptococcus laurentii Cryptococcus luteolus	Candida humicola Cryptococcus laurentii
250 Raffinose growth.	Candida incommunis*	71

Yeasts found in brewing (Keys No. 17 and No. 18)

Yeasts in Keys No. 17 and No. 18

10	Brettanomyces anomalus	117	Candida valida	263	Pichia kudriavzevii
11	Brettanomyces claussenii	118	Candida vartiovaarai	266	Pichia membranaefaciens
12	Brettanomyces custersianus	148	Debaryomyces hansenii	272	Pichia ohmeri
13	Brettanomyces custersii	149	Debaryomyces marama	273	Pichia onychis
14	Brettanomyces lambicus	157	Dekkera bruxellensis	312	Saccharomyces cerevisiae
23	Candida beechii	158	Dekkera intermedia	314	Saccharomyces exiguus
68	Candida intermedia	159	Filobasidium capsuligenum**	318	Saccharomyces unisporus
73	Candida lambica	168	Hanseniaspora uvarum	331	Schizosaccharomyces pombe
75	Candida macedoniensis	169	Hanseniaspora valbyensis	357	Torulaspora delbrueckii**
88	Candida oleophila*	170	Hanseniaspora vineae*	370	Torulopsis ernobii
90	Candida parapsilosis	171	Hansenula anomala	381	Torulopsis humilis*
93	Candida pseudotropicalis	181	Hansenula fabianii	400	Torulopsis norvegica
99	Candida sake	197	Hansenula subpelliculosa	412	Torulopsis stellata
106	Candida solani	206	Kluyveromyces bulgaricus	416	Torulopsis versatilis
110	Candida tenuis	212	Kluyveromyces marxianus	432	Zygosaccharomyces bailii**
113	Candida tropicalis	255	Pichia farinosa	439	Zygosaccharomyces rouxii**
115	Candida utilis	256	Pichia fermentans		
		258	Pichia guilliermondii		

Key involving physiological tests and microscopical examination (Key No. 17)

Tests in Key No. 17

5	Sucrose fermentation	28	Raffinose growth	54	Growth without biotin
8	Lactose fermentation	31	Starch growth	55	Growth without thiamin
12	D–Galactose growth	33	Erythritol growth	61	Growth in 60% D–glucose
15	D–Xylose growth	37	D–Glucitol growth	63	Growth with 0.01% cycloheximide
16	L–Arabinose growth	40	2–Ketogluconate growth	67	Budding cells
20	Maltose growth	42	D,L–Lactate growth	72	Filamentous
21	Trehalose growth	44	Citrate growth	74	Septate hyphae
23	Cellobiose growth	49	Nitrate growth		
		53	Growth without pantothenate	number of different tests 24	

KEY No. 17

	Negative	Positive
1 D–Glucitol growth.	2	41
2 Cellobiose growth.	3	32
3 Trehalose growth.	4	22
4 D,L–Lactate growth.	5	17
5 Raffinose growth.	6	11
6 D–Galactose growth.	7	9
7 Growth in 60% D–glucose.	Candida valida Pichia membranaefaciens Saccharomyces cerevisiae	8
8 Filamentous.	Pichia membranaefaciens Torulaspora delbrueckii** Zygosaccharomyces rouxii**	Candida valida Pichia membranaefaciens Zygosaccharomyces rouxii**
9 Growth in 60% D–glucose.	10	Torulaspora delbrueckii** Zygosaccharomyces rouxii**
10 Growth with 0.01% cycloheximide	Saccharomyces cerevisiae	Saccharomyces unisporus
11 Maltose growth.	12	14
12 Growth without thiamin.	Saccharomyces cerevisiae Torulopsis stellata	13
13 Growth in 60% D–glucose.	Saccharomyces cerevisiae	Torulaspora delbrueckii**
14 Growth in 60% D–glucose.	15	16
15 Budding cells.	Schizosaccharomyces pombe	Saccharomyces cerevisiae
16 Budding cells.	Schizosaccharomyces pombe	Torulaspora delbrueckii**
17 D–Xylose growth.	18	20

Key No. 17: yeasts of brewing

		Negative	Positive
18	Growth in 60% D−glucose.	Candida valida Pichia kudriavzevii Pichia membranaefaciens Saccharomyces cerevisiae	.19
19	Filamentous.	Pichia membranaefaciens Torulaspora delbrueckii**	Candida valida Pichia kudriavzevii Pichia membranaefaciens
20	Citrate growth.	21	Candida lambica Pichia fermentans Pichia membranaefaciens
21	Filamentous.	Pichia membranaefaciens Torulaspora delbrueckii**	Candida lambica Candida valida Pichia membranaefaciens
22	Maltose growth.	23	.27
23	Growth in 60% D−glucose.	24	Torulaspora delbrueckii** Zygosaccharomyces rouxii**
24	Sucrose fermentation.	25	Saccharomyces cerevisiae Saccharomyces exiguus
25	Growth with 0.01% cycloheximide	Saccharomyces cerevisiae	.26
26	Filamentous.	Torulopsis humilis*	Brettanomyces custersianus
27	Nitrate growth.	28	Brettanomyces lambicus Dekkera bruxellensis
28	Growth in 60% D−glucose.	29	Torulaspora delbrueckii** Zygosaccharomyces rouxii**
29	Sucrose fermentation.	30	.31
30	Growth with 0.01% cycloheximide	Saccharomyces cerevisiae	Brettanomyces custersianus
31	Growth with 0.01% cycloheximide	Saccharomyces cerevisiae	Dekkera bruxellensis
32	D−Galactose growth.	33	.37
33	Maltose growth.	34	.35
34	2−Ketogluconate growth.	Hanseniaspora valbyensis	Hanseniaspora uvarum
35	Raffinose growth.	36	Candida utilis
36	D−Xylose growth.	Hanseniaspora vineae*	Candida solani
37	D−Xylose growth.	38	.40
38	Lactose fermentation.	Brettanomyces custersii Dekkera intermedia	.39
39	Septate hyphae.	Brettanomyces claussenii	Brettanomyces anomalus
40	Lactose fermentation.	Candida macedoniensis	Candida pseudotropicalis
41	Erythritol growth.	42	.89
42	D−Xylose growth.	43	.65
43	Cellobiose growth.	44	.55
44	Nitrate growth.	45	.53
45	D,L−Lactate growth.	46	.49
46	Growth in 60% D−glucose.	47	Torulaspora delbrueckii** Zygosaccharomyces bailii** Zygosaccharomyces rouxii**
47	Sucrose fermentation.	48	.31
48	2−Ketogluconate growth.	Saccharomyces cerevisiae	Filobasidium capsuligenum**
49	Sucrose fermentation.	50	.51
50	Growth in 60% D−glucose.	30	Torulaspora delbrueckii**
51	Growth with 0.01% cycloheximide	13	.52
52	D−Galactose growth.	Dekkera bruxellensis	Kluyveromyces bulgaricus
53	D−Galactose growth.	Brettanomyces lambicus Dekkera bruxellensis	.54
54	Filamentous.	Torulopsis versatilis	Brettanomyces lambicus
55	Nitrate growth.	56	.62
56	Maltose growth.	57	.58
57	D−Galactose growth.	Candida beechii	Kluyveromyces bulgaricus
58	Raffinose growth.	59	.60
59	Sucrose fermentation.	Filobasidium capsuligenum**	Dekkera intermedia
60	Citrate growth.	61	Pichia ohmeri
61	Lactose fermentation.	Dekkera intermedia	Brettanomyces claussenii
62	Lactose fermentation.	63	.64
63	Filamentous.	Torulopsis versatilis	Brettanomyces custersii Dekkera intermedia
64	Filamentous.	Torulopsis versatilis	.39
65	Raffinose growth.	66	.76

	Negative	Positive
66 Nitrate growth.	67	74
67 Cellobiose growth.	68	72
68 L−Arabinose growth.	69	71
69 Starch growth.	70	Filobasidium capsuligenum**
70 Filamentous.	Torulaspora delbrueckii**	Candida sake
71 Starch growth.	Candida parapsilosis	Filobasidium capsuligenum**
72 Starch growth.	Candida oleophila* / Candida sake	73
73 Growth without biotin.	Candida tropicalis	Filobasidium capsuligenum**
74 Maltose growth.	Torulopsis norvegica	75
75 Trehalose growth.	Torulopsis ernobii	Candida vartiovaarai
76 Maltose growth.	77	82
77 L−Arabinose growth.	78	79
78 Growth in 60% D−glucose.	Kluyveromyces bulgaricus	Torulaspora delbrueckii**
79 Lactose fermentation.	80	81
80 Growth without biotin.	Kluyveromyces bulgaricus / Kluyveromyces marxianus	Candida macedoniensis
81 Growth without pantothenate	Candida pseudotropicalis / Kluyveromyces marxianus	Kluyveromyces bulgaricus / Kluyveromyces marxianus
82 D−Galactose growth.	83	85
83 L−Arabinose growth.	84	Pichia onychis
84 Cellobiose growth.	Torulaspora delbrueckii**	Hansenula fabianii
85 Cellobiose growth.	Torulaspora delbrueckii**	86
86 D,L−Lactate growth.	87	88
87 Starch growth.	Pichia guilliermondii	Candida intermedia
88 Starch growth.	Pichia guilliermondii	Debaryomyces hansenii
89 Raffinose growth.	90	91
90 Maltose growth.	Pichia farinosa	Candida tenuis
91 Nitrate growth.	Debaryomyces hansenii / Debaryomyces marama	92
92 Growth without thiamin.	Hansenula subpelliculosa	Hansenula anomala

Key involving physiological tests only (Key No. 18)

Tests in Key No. 18

5 Sucrose fermentation	28 Raffinose growth	49 Nitrate growth
8 Lactose fermentation	31 Starch growth	53 Growth without pantothenate
12 D−Galactose growth	33 Erythritol growth	54 Growth without biotin
15 D−Xylose growth	37 D−Glucitol growth	55 Growth without thiamin
16 L−Arabinose growth	40 2−Ketogluconate growth	61 Growth in 60% D−glucose
20 Maltose growth	42 D,L−Lactate growth	63 Growth with 0.01% cycloheximide
21 Trehalose growth	43 Succinate growth	65 Starch formation
23 Cellobiose growth	44 Citrate growth	
	46 Ethanol growth	number of different tests 24

KEY No. 18

	Negative	Positive
1 D−Glucitol growth.	2	43
2 Ethanol growth.	3	17
3 Cellobiose growth.	4	14
4 D−Galactose growth.	5	11
5 Raffinose growth.	6	7
6 Growth in 60% D−glucose.	Saccharomyces cerevisiae	Torulaspora delbrueckii** / Zygosaccharomyces rouxii**
7 Maltose growth.	8	10
8 Growth without thiamin.	Saccharomyces cerevisiae / Torulopsis stellata	9
9 Growth in 60% D−glucose.	Saccharomyces cerevisiae	Torulaspora delbrueckii**
10 Starch formation.	9	Schizosaccharomyces pombe
11 Growth in 60% D−glucose.	12	Torulaspora delbrueckii** / Zygosaccharomyces rouxii**

Key No. 18: yeasts of brewing

	Negative	Positive
12 Trehalose growth.	13	Saccharomyces cerevisiae Saccharomyces exiguus
13 Growth with 0.01%. cycloheximide	Saccharomyces cerevisiae	Saccharomyces unisporus
14 Maltose growth.	15	16
15 2-Ketogluconate growth.	Hanseniaspora valbyensis	Hanseniaspora uvarum
16 Raffinose growth.	Hanseniaspora vineae*	Candida utilis
17 Cellobiose growth.	18	37
18 Trehalose growth.	19	26
19 D-Xylose growth.	20	25
20 D-Galactose growth.	21	24
21 D,L-Lactate growth.	22	23
22 Growth in 60% D-glucose.	Candida valida Pichia membranaefaciens Saccharomyces cerevisiae	Candida valida Pichia membranaefaciens Torulaspora delbrueckii** Zygosaccharomyces rouxii**
23 Growth in 60% D-glucose.	Candida valida Pichia kudriavzevii Pichia membranaefaciens Saccharomyces cerevisiae	Candida valida Pichia kudriavzevii Pichia membranaefaciens Torulaspora delbrueckii**
24 Growth in 60% D-glucose.	13	Torulaspora delbrueckii** zaygosaccharomyces rouxii**
25 Citrate growth.	Candida lambica Candida valida Pichia membranaefaciens Torulaspora delbrueckii**	Candida lambica Pichia fermentans Pichia membranaefaciens
26 Maltose growth.	27	33
27 Growth in 60% D-glucose.	28	Torulaspora delbrueckii** Zygosaccharomyces rouxii**
28 Succinate growth.	29	31
29 Sucrose fermentation.	30	Saccharomyces cerevisiae Saccharomyces exiguus
30 Growth with 0.01%. cycloheximide	Saccharomyces cerevisiae	Torulopsis humilis*
31 Sucrose fermentation.	32	Saccharomyces cerevisiae Saccharomyces exiguus
32 Growth with 0.01%. cycloheximide	Saccharomyces cerevisiae	Brettanomyces custersianus
33 Nitrate growth.	34	Brettanomyces lambicus Dekkera bruxellensis
34 Growth in 60% D-glucose.	35	Torulaspora delbrueckii** Zygosaccharomyces rouxii**
35 Sucrose fermentation.	32	36
36 Growth with 0.01%. cycloheximide	Saccharomyces cerevisiae	Dekkera bruxellensis
37 D-Xylose growth.	38	40
38 D-Galactose growth.	Candida utilis	39
39 Lactose fermentation.	Brettanomyces custersii Dekkera intermedia	Brettanomyces anomalus Brettanomyces claussenii
40 D-Galactose growth.	41	42
41 Raffinose growth.	Candida solani	Candida utilis
42 Lactose fermentation.	Candida macedoniensis	Candida pseudotropicalis
43 Erythritol growth.	44	88
44 D-Xylose growth.	45	65
45 Cellobiose growth.	46	56
46 Nitrate growth.	47	55
47 D,L-Lactate growth.	48	51
48 Growth in 60% D-glucose.	49	Torulaspora delbrueckii** Zygosaccharomyces bailii** Zygosaccharomyces rouxii**
49 Sucrose fermentation.	50	36
50 2-Ketogluconate growth.	Saccharomyces cerevisiae	Filobasidium capsuligenum**
51 Sucrose fermentation.	52	53
52 Growth in 60% D-glucose.	32	Torulaspora delbrueckii**
53 Growth with 0.01%. cycloheximide	9	54
54 D-Galactose growth.	Dekkera bruxellensis	Kluyveromyces bulgaricus
55 Ethanol growth.	Torulopsis versatilis	Brettanomyces lambicus Dekkera bruxellensis
56 Ethanol growth.	Torulopsis versatilis	57
57 Maltose growth.	58	60
58 D-Galactose growth.	Candida beechii	59

		Negative	Positive
59	Nitrate growth.	Kluyveromyces bulgaricus	Brettanomyces anomalus
60	Nitrate growth.	61	39
61	Raffinose growth.	62	63
62	Sucrose fermentation.	Filobasidium capsuligenum**	Dekkera intermedia
63	Citrate growth.	64	Pichia ohmeri
64	Lactose fermentation.	Dekkera intermedia	Brettanomyces claussenii
65	Raffinose growth.	66	75
66	Nitrate growth.	67	73
67	Cellobiose growth.	68	71
68	L−Arabinose growth.	69	70
69	Starch growth.	Candida sake Torulaspora delbrueckii**	Filobasidium capsuligenum**
70	Starch growth.	Candida parapsilosis	Filobasidium capsuligenum**
71	Starch growth.	Candida oleophila* Candida sake	72
72	Growth without biotin.	Candida tropicalis	Filobasidium capsuligenum**
73	Maltose growth.	Torulopsis norvegica	74
74	Trehalose growth.	Torulopsis ernobii	Candida vartiovaarai
75	Maltose growth.	76	81
76	L−Arabinose growth.	77	78
77	Growth in 60% D−glucose.	Kluyveromyces bulgaricus	Torulaspora delbrueckii**
78	Lactose fermentation.	79	80
79	Growth without biotin.	Kluyveromyces bulgaricus Kluyveromyces marxianus	Candida macedoniensis
80	Growth without pantothenate	Candida pseudotropicalis Kluyveromyces marxianus	Kluyveromyces bulgaricus Kluyveromyces marxianus
81	D−Galactose growth.	82	84
82	L−Arabinose growth.	83	Pichia onychis
83	Cellobiose growth.	Torulaspora delbrueckii**	Hansenula fabianii
84	Cellobiose growth.	Torulaspora delbrueckii**	85
85	D,L−Lactate growth.	86	87
86	Starch growth.	Pichia guilliermondii	Candida intermedia
87	Starch growth.	Pichia guilliermondii	Debaryomyces hansenii
88	Raffinose growth.	89	90
89	Maltose growth.	Pichia farinosa	Candida tenuis
90	Nitrate growth.	Debaryomyces hansenii Debaryomyces marama	91
91	Growth without thiamin.	Hansenula subpelliculosa	Hansenula anomala

8
The identification of particular species

The table below provides lists of tests to check the identity of each species. The species are arranged alphabetically and the tests are numbered as in Chapter 6. Each list constitutes a minimal group of tests to distinguish each species from the other species in this book. However, new species might give some of the same sets of results and thus might not be distinguished.

A clinician may wish to know whether or not an isolate is *Candida albicans*, a food yeast manufacturer may want to ascertain that the cultured yeast is still uncontaminated *Saccharomycopsis lipolytica*, or an academic research worker may need to be sure that the organism under study remains *Candida utilis*. This table may help such people and it can also be used to confirm identifications obtained with one of the keys. This could be particularly important for the specialist keys, because correct identification of a given yeast must depend on its species having been included in the key.

The use of the table is exemplified by the need to check whether a strain of yeast is *Candida utilis*. First, the entry for *Candida utilis* (species arranged alphabetically) is found on p. 204, as shown below.

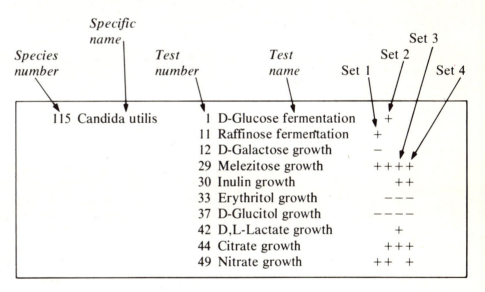

Thus for *Candida utilis* four sets of test results are listed; these are as follows.

Set 1 (5 tests)
11 Raffinose fermentation +
12 D-Galactose growth −
29 Melezitose growth +
37 D-Glucitol growth −
49 Nitrate growth +

Set 2 (6 tests)
1 D-Glucose fermentation +
29 Melezitose growth +
33 Erythritol growth −
37 D-Glucitol growth −
44 Citrate growth +
49 Nitrate growth +

Set 3 (6 tests)
29 Melezitose growth +
30 Inulin growth +
33 Erythritol growth −
37 D-Glucitol growth −
42 DL-Lactate growth +
44 Citrate growth +

Set 4 (6 tests)
29 Melezitose growth +
30 Inulin growth +
33 Erythritol growth −
37 D-Glucitol growth −
44 Citrate growth +
49 Nitrate growth +

Although any one of these four sets distinguishes *Candida utilis* from the other species listed in this book, the different sets are offered to allow for individual preferences for tests. For example, Set 1 is of five tests only, but it includes a fermentation test; on the other hand, Set 3 and 4 are each of six tests, but neither includes a fermentation test. Set 4, but not Set 3, includes the test for nitrate growth.

Only the physiological tests have been used for these sets, but with some yeasts such tests are insufficient for complete discrimination. The example given below of *Candida albicans* is taken from p. 182. Four additional species are listed that give the same sets of results as *Candida albicans*.

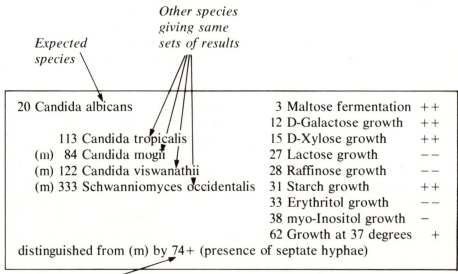

Although most strains of *Candida albicans* are negative for sucrose fermentation (test number 5), some give a very weak positive response to the test so that the response is given in this book as V (variable) and could not be used by the computer program to distinguish this species from *Candida tropicalis* which has a positive response to sucrose fermentation. Such information may be obtained from *The Yeasts* (Lodder, 1970). The letter '(m)', preceding the other three alternative specific names, indicates that they may be distinguished from *Candida albicans* by an additional test, a micro- or macroscopical examination. As written below the lists, the additional test is number 74; this is positive for *Candida albicans* which (unlike species 84, 122 and 333) can form septate hyphae. Furthermore the illustrations in *The Yeasts* (Lodder, 1970) show that the cells of these three species have different shapes from those of *Candida albicans*; in addition, *Schwanniomyces occidentalis* can produce ascospores. For some yeasts, two species are distinguished only by the ability of one to form ascospores or by the shape of the ascospores; here the letter '(a)' precedes the alternative specific name. An example is *Hansenula capsulata* (species number 177, p. 217) which is distinguished from '(a) 392 *Torulopsis molischiana*' by test 75+ (formation of ascospores).

When as with *Hansenula canadensis* (p. 217), several species give the same sets of results, more than one additional test may be listed. These tests are in numerical order rather than in the order of the species that they distinguish. Thus, the first test '66− (colonies not pink)' distinguishes *Hansenula canadensis* from the last of the species marked '(m)', namely *Rhodotorula glutinis*. Such details can best be clarified by reference to *The Yeasts* (Lodder, 1970).

Table 8.1. *Sets of test responses for identifying each yeast species*

Species	Tests	Responses
1 Aciculoconidium aculeatum**	5 Sucrose fermentation	− − −
	15 D−Xylose growth	− − − − −
	19 Sucrose growth	+ + +
	23 Cellobiose growth	+ + + + +
	28 Raffinose growth	−
	31 Starch growth	+ + + + +
	33 Erythritol growth	− −
	38 myo−Inositol growth	− − − −
	49 Nitrate growth	− − − −
	54 Growth sans biotin	−
	55 Growth sans thiamin	− −
	62 Growth at 37 degrees	− − − − −
2 Aessosporon dendrophilum*	26 Melibiose growth	− −
	28 Raffinose growth	− − −
(m) 421 Trichosporon cutaneum	31 Starch growth	+
	33 Erythritol growth	− − −
	35 Galactitol growth	+ + + +
	38 myo−Inositol growth	− − − −
	46 Ethanol growth	+
	49 Nitrate growth	− −
	65 Starch formation	+ + +
distinguished from (m) by 68− (cells not splitting)		
3 Aessosporon salmonicolor**	1 D−Glucose fermentation	− −
	19 Sucrose growth	+ + +
346 Sporobolomyces salmonicolor	20 Maltose growth	− − −
(m) 342 Sporobolomyces odorus	21 Trehalose growth	+ + +
	23 Cellobiose growth	− −
	35 Galactitol growth	−
	37 D−Glucitol growth	+
	40 2−Ketogluconate growth	− − −
	49 Nitrate growth	+ + +
distinguished from (m) by 74+ (presence of septate hyphae)		
4 Ambrosiozyma cicatricosa*	1 D−Glucose fermentation	+ +
	5 Sucrose fermentation	+
5 Ambrosiozyma monospora**	12 D−Galactose growth	− − −
	16 L−Arabinose growth	+ + +
	22 Me α−D−glucoside grth	+
	29 Melezitose growth	+
	33 Erythritol growth	+ + +
	49 Nitrate growth	− − −
5 Ambrosiozyma monospora**	12 D−Galactose growth	− −
	16 L−Arabinose growth	+ +
4 Ambrosiozyma cicatricosa*	22 Me α−D−glucoside grth	+
	26 Melibiose growth	− −
	27 Lactose growth	−
	29 Melezitose growth	+
	33 Erythritol growth	+ +
	49 Nitrate growth	− −

The identification of particular species

Species	Tests	Responses
6 Ambrosiozyma philentoma*	12 D−Galactose growth	− −
	18 L−Rhamnose growth	+ +
	22 Me α−D−glucoside grth	+
	26 Melibiose growth	− −
	27 Lactose growth	− −
	29 Melezitose growth	+
	33 Erythritol growth	+ +
	49 Nitrate growth	− −

Species	Tests	Responses
7 Arthroascus javanensis**	1 D−Glucose fermentation	− −
	12 D−Galactose growth	− −
(m) 117 Candida valida	15 D−Xylose growth	−
(m) 237 Nadsonia commutata*	19 Sucrose growth	− −
(m) 266 Pichia membranaefaciens	27 Lactose growth	−
	37 D−Glucitol growth	− −
	42 D,L−Lactate growth	− −
	44 Citrate growth	− −
	49 Nitrate growth	− −
	62 Growth at 37 degrees	− −

distinguished from (m) by 74+ (presence of septate hyphae)

Species	Tests	Responses
8 Botryoascus synnaedendrus*	19 Sucrose growth	−
	21 Trehalose growth	− −
	22 Me α−D−glucoside grth	−
	25 Arbutin growth	− −
	27 Lactose growth	− − −
	33 Erythritol growth	+ + +
	38 myo−Inositol growth	+ + +

Species	Tests	Responses
9 Brettanomyces abstinens*	1 D−Glucose fermentation	+ +
	12 D−Galactose growth	+ + +
	19 Sucrose growth	− − −
	21 Trehalose growth	−
	36 D−Mannitol growth	− − −
	37 D−Glucitol growth	− − −
	49 Nitrate growth	+ + + +
	51 Growth sans vitamins	−
	62 Growth at 37 degrees	+

Species	Tests	Responses
10 Brettanomyces anomalus	8 Lactose fermentation	+ +
	46 Ethanol growth	+
(m) 11 Brettanomyces claussenii	49 Nitrate growth	+ +
	62 Growth at 37 degrees	+

distinguished from (m) by 74+ (presence of septate hyphae)

Species	Tests	Responses
11 Brettanomyces claussenii	4 Me α−D−glucoside fermn	+
	8 Lactose fermentation	+ +
(m) 10 Brettanomyces anomalus	9 Cellobiose fermentn	+
	36 D−Mannitol growth	− −
	46 Ethanol growth	+

distinguished from (m) by 74− (absence of septate hyphae)

Species	Tests	Responses
12 Brettanomyces custersianus	2 Galactose fermentation	− −
	15 D−Xylose growth	− − −
(m) 298 Rhodosporidium dacryoidum*	21 Trehalose growth	+ + +
	23 Cellobiose growth	− − −
	28 Raffinose growth	− −
	29 Melezitose growth	− −
	32 Glycerol growth	+
	33 Erythritol growth	− − −
	36 D−Mannitol growth	− − −
	42 D,L−Lactate growth	+ + +
	43 Succinate growth	+ +
	44 Citrate growth	− − −
	63 0.01% cyclohex. growth	+ + +

distinguished from (m) by 66− (colonies not pink)

Species	Tests	Responses
13 Brettanomyces custersii	3 Maltose fermentation	+ +
	8 Lactose fermentation	− − − −
(a) 158 Dekkera intermedia	9 Cellobiose fermentn	+ +
	15 D−Xylose growth	− −
	22 Me α−D−glucoside grth	+ + +
	23 Cellobiose growth	+ +
	33 Erythritol growth	−
	36 D−Mannitol growth	− − −
	46 Ethanol growth	+
	49 Nitrate growth	+ + + +

distinguished from (a) by 75− (inability to form ascospores)

Species	Tests	Responses
14 Brettanomyces lambicus	1 D−Glucose fermentation	+
	4 Me α−D−glucoside fermn	+
(a) 157 Dekkera bruxellensis	10 Melezitose fermentn	+ +
	22 Me α−D−glucoside grth	+
	23 Cellobiose growth	− − − −
	33 Erythritol growth	−
	36 D−Mannitol growth	− − −
	49 Nitrate growth	+ + + +

distinguished from (a) by 75− (inability to form ascospores)

Species	Tests	Responses
15 Brettanomyces naardenensis*	3 Maltose fermentation	− − −
	15 D−Xylose growth	+
	16 L−Arabinose growth	− −
	19 Sucrose growth	− − − −
	23 Cellobiose growth	+ + + + +
	25 Arbutin growth	− − −
	27 Lactose growth	− − −
	32 Glycerol growth	− − − −
	37 D−Glucitol growth	+
	38 myo−Inositol growth	−
	43 Succinate growth	+ + +
	44 Citrate growth	−
	46 Ethanol growth	+
	49 Nitrate growth	− − −
	62 Growth at 37 degrees	− −

182 The identification of particular species

Species   Tests   Responses

16 Bullera alba   18 L−Rhamnose growth   +
                                                                 27 Lactose growth   +
   126 Cryptococcus ater   28 Raffinose growth   + +
   130 Cryptococcus flavus   30 Inulin growth   − −
   137 Cryptococcus laurentii   38 myo−Inositol growth   + +
   138 Cryptococcus luteolus   46 Ethanol growth   − −
   140 Cryptococcus magnus**   49 Nitrate growth   − −
(m) 60 Candida humicola   51 Growth sans vitamins   − −
(m) 134 Cryptococcus hungaricus   62 Growth at 37 degrees   −
(m) 355 Sterigmatomyces polyborus*
distinguished from (m) by 66− (colonies not pink)
                67+ (formation of budding cells)
                72− (absence of filaments)

17 Bullera piricola*   22 Me α−D−glucoside grth   − −
                                                                 29 Melezitose growth   +
                                                                 31 Starch growth   + +
                                                                 33 Erythritol growth   −
                                                                 38 myo−Inositol growth   + + +
                                                                 49 Nitrate growth   + + +
                                                                 65 Starch formation   − − −

18 Bullera tsugae   12 D−Galactose growth   − − − −
                                                                 15 D−Xylose growth   − −
                                                                 27 Lactose growth   + + + +
                                                                 28 Raffinose growth   − − −
                                                                 32 Glycerol growth   +
                                                                 34 Ribitol growth   −   −
                                                                 38 myo−Inositol growth   −
                                                                 42 D,L−Lactate growth   −
                                                                 49 Nitrate growth   +   +
                                                                 51 Growth sans vitamins   + +

19 Candida aaseri   1 D−Glucose fermentation   − − −
                                                                 12 D−Galactose growth   + +
(m) 44 Candida diddensii   13 L−Sorbose growth   +   +
(m) 421 Trichosporon cutaneum   16 L−Arabinose growth   + + + +
                                                                 17 D−Arabinose growth   −
                                                                 18 L−Rhamnose growth   − − − −
                                                                 19 Sucrose growth   +
                                                                 26 Melibiose growth   − − − −
                                                                 28 Raffinose growth   − − − −
                                                                 29 Melezitose growth   + + +
                                                                 31 Starch growth   −
                                                                 33 Erythritol growth   + + + +
                                                                 46 Ethanol growth   +
                                                                 60 Growth in 50% glucose   −
distinguished from (m) by 71− (cells spherical, oval or cylindrical only)
                73+ (presence of pseudohyphae)

20 Candida albicans   3 Maltose fermentation   + +
                                                                 12 D−Galactose growth   + +
   113 Candida tropicalis   15 D−Xylose growth   + +
(m) 84 Candida mogii   27 Lactose growth   − −
(m) 122 Candida viswanathii   28 Raffinose growth   − −
(m) 333 Schwanniomyces occidentalis   31 Starch growth   + +
                                                                 33 Erythritol growth   − −
                                                                 38 myo−Inositol growth   −
                                                                 62 Growth at 37 degrees   +

distinguished from (m) by 74+ (presence of septate hyphae)

The identification of particular species

Species	Tests	Responses
21 Candida amylolenta*	26 Melibiose growth	+ +
	27 Lactose growth	− −
(m) 60 Candida humicola	29 Melezitose growth	+
	30 Inulin growth	− −
	31 Starch growth	+ +
	38 myo−Inositol growth	+ +
	42 D,L−Lactate growth	−
	46 Ethanol growth	+ +
	49 Nitrate growth	− −

distinguished from (m) by 74− (absence of septate hyphae)

Species	Tests	Responses
22 Candida aquatica	18 L−Rhamnose growth	−
	26 Melibiose growth	+
	27 Lactose growth	+ +
	29 Melezitose growth	+
	31 Starch growth	+ +
	33 Erythritol growth	− − −
	34 Ribitol growth	+
	38 myo−Inositol growth	− − −
	46 Ethanol growth	+
	49 Nitrate growth	+ + +

Species	Tests	Responses
23 Candida beechii	6 Trehalose fermentation	+ + +
	12 D−Galactose growth	− −
101 Candida santamariae	13 L−Sorbose growth	+ + +
	15 D−Xylose growth	−
	19 Sucrose growth	− − −
	23 Cellobiose growth	+ + +
	33 Erythritol growth	− −
	44 Citrate growth	+ + +
	45 Methanol growth	−

Species	Tests	Responses
24 Candida berthetii	12 D−Galactose growth	− − − −
	15 D−Xylose growth	− − − −
	19 Sucrose growth	− − − − −
	23 Cellobiose growth	+ + +
	24 Salicin growth	+ + +
	36 D−Mannitol growth	− − − −
	37 D−Glucitol growth	− −
	42 D,L−Lactate growth	− −
	49 Nitrate growth	+ + + +
	51 Growth sans vitamins	+ +

Species	Tests	Responses
25 Candida blankii	13 L−Sorbose growth	+
	14 D−Ribose growth	−
61 Candida hydrocarbofumarica*	26 Melibiose growth	− − −
(m) 421 Trichosporon cutaneum	27 Lactose growth	+ + +
	30 Inulin growth	− − −
	33 Erythritol growth	+ +
	35 Galactitol growth	+ +
	42 D,L−Lactate growth	− − −
	51 Growth sans vitamins	+
	62 Growth at 37 degrees	+ + +

distinguished from (m) by 73+ (presence of pseudohyphae)

The identification of particular species

Species	Tests	Responses
26 Candida bogoriensis	16 L−Arabinose growth	+ + +
	19 Sucrose growth	− − −
	27 Lactose growth	− − −
	29 Melezitose growth	+ + +
	34 Ribitol growth	+
	43 Succinate growth	+
	44 Citrate growth	+ +
	49 Nitrate growth	− −
27 Candida boidinii	18 L−Rhamnose growth	−
	21 Trehalose growth	− − −
	23 Cellobiose growth	− −
	33 Erythritol growth	+ +
	34 Ribitol growth	+
	45 Methanol growth	+ + +
28 Candida boleticola*	1 D−Glucose fermentation	+ + + +
	12 D−Galactose growth	+ + + + +
	14 D−Ribose growth	+
	16 L−Arabinose growth	− −
	19 Sucrose growth	− − − −
	27 Lactose growth	−
	33 Erythritol growth	+ + + + +
	44 Citrate growth	+ + +
	51 Growth sans vitamins	− −
	54 Growth sans biotin	− − −
	62 Growth at 37 degrees	− − − − −
29 Candida bombi*	5 Sucrose fermentation	+
(m) 357 Torulaspora delbrueckii**	11 Raffinose fermentation	+ +
	12 D−Galactose growth	− − −
	13 L−Sorbose growth	+ + + +
	15 D−Xylose growth	−
	22 Me α−D−glucoside grth	−
	26 Melibiose growth	− − −
	28 Raffinose growth	+
	34 Ribitol growth	−
	43 Succinate growth	+ + + +
	46 Ethanol growth	− −
	49 Nitrate growth	− −
	51 Growth sans vitamins	+ +

distinguished from (m) by 72+ (presence of filaments)

30 Candida brassicae*	1 D−Glucose fermentation	+ + + +
	5 Sucrose fermentation	− −
	22 Me α−D−glucoside grth	+ +
	26 Melibiose growth	+ +
	30 Inulin growth	+
	31 Starch growth	− − − − −
	33 Erythritol growth	− − − − −
	35 Galactitol growth	+ + + + +
	38 myo−Inositol growth	− −
	51 Growth sans vitamins	+ + +

Species	Tests	Responses
31 Candida brumptii	15 D−Xylose growth	+ +
	16 L−Arabinose growth	−
	19 Sucrose growth	− −
	21 Trehalose growth	− −
	27 Lactose growth	− −
	31 Starch growth	+ +
	36 D−Mannitol growth	+
	38 myo−Inositol growth	− −
	62 Growth at 37 degrees	− −
32 Candida buffonii	14 D−Ribose growth	−
	19 Sucrose growth	− − −
	29 Melezitose growth	+ + +
	33 Erythritol growth	−
	38 myo−Inositol growth	− −
	45 Methanol growth	−
	49 Nitrate growth	+ + +
33 Candida buinensis*	1 D−Glucose fermentation	+ +
	2 Galactose fermentation	+ +
	12 D−Galactose growth	+ +
	17 D−Arabinose growth	+ + + +
	28 Raffinose growth	− −
	34 Ribitol growth	− − − −
	38 myo−Inositol growth	− − −
	42 D,L−Lactate growth	− −
34 Candida butyri*	1 D−Glucose fermentation	+ +
	12 D−Galactose growth	+
(m) 44 Candida diddensii	17 D−Arabinose growth	+
(a) 275 Pichia philogaea*	18 L−Rhamnose growth	− −
	19 Sucrose growth	+
	28 Raffinose growth	− −
	31 Starch growth	− −
	33 Erythritol growth	+ +
	45 Methanol growth	− −
	62 Growth at 37 degrees	+ +

distinguished from (m) by 71− (cells spherical, oval or cylindrical only)
distinguished from (a) by 75− (inability to form ascospores)

Species	Tests	Responses
35 Candida cacaoi	1 D−Glucose fermentation	+ +
	2 Galactose fermentation	−
	6 Trehalose fermentation	+
	12 D−Galactose growth	+ +
	19 Sucrose growth	− − −
	23 Cellobiose growth	+ +
	27 Lactose growth	−
	29 Melezitose growth	−
	33 Erythritol growth	+ + +
	44 Citrate growth	− − − −
	45 Methanol growth	− − − −
	51 Growth sans vitamins	+ + + +

The identification of particular species

Species	Tests	Responses
36 Candida catenulata	12 D−Galactose growth	+ +
	13 L−Sorbose growth	− − −
95 Candida ravautii	19 Sucrose growth	− − −
98 Candida rugosa	22 Me α−D−glucoside grth	−
	23 Cellobiose growth	−
	27 Lactose growth	− − − −
	32 Glycerol growth	+ +
	33 Erythritol growth	− − − −
	36 D−Mannitol growth	+
	37 D−Glucitol growth	+
	44 Citrate growth	+ + + +
	49 Nitrate growth	− − − −
	62 Growth at 37 degrees	+ + + +
37 Candida chilensis*	1 D−Glucose fermentation	+ +
	16 L−Arabinose growth	+ + +
	19 Sucrose growth	+
	22 Me α−D−glucoside grth	+
	27 Lactose growth	+ + + + +
	28 Raffinose growth	− − −
	33 Erythritol growth	+ + + +
	42 D,L−Lactate growth	+ +
	45 Methanol growth	− −
	49 Nitrate growth	+ + + + +
38 Candida chiropterorum*	18 L−Rhamnose growth	+
	26 Melibiose growth	− −
142 Cryptococcus neoformans	27 Lactose growth	− − −
	30 Inulin growth	− −
	33 Erythritol growth	+
	35 Galactitol growth	+ + +
	38 myo−Inositol growth	+ +
	43 Succinate growth	− − −
	49 Nitrate growth	−
	65 Starch formation	−
39 Candida citrea*	1 D−Glucose fermentation	+ + + +
	12 D−Galactose growth	−
117 Candida valida	19 Sucrose growth	− − −
(m) 385 Torulopsis karawaiewi*	21 Trehalose growth	− −
(a) 266 Pichia membranaefaciens	36 D−Mannitol growth	− − −
	37 D−Glucitol growth	− − − −
	39 Gluconolactone growth	+ + + +
	42 D,L−Lactate growth	− − − −
	43 Succinate growth	+ + + +
	49 Nitrate growth	−
	55 Growth sans thiamin	− − − −
	62 Growth at 37 degrees	− −

distinguished from (m) by 72+ (presence of filaments)
distinguished from (a) by 75− (inability to form ascospores)

Species	Tests	Responses
40 Candida conglobata	1 D–Glucose fermentation	+
	6 Trehalose fermentation	+ +
	12 D–Galactose growth	+ +
	13 L–Sorbose growth	+
	19 Sucrose growth	– – –
	20 Maltose growth	–
	23 Cellobiose growth	+ +
	27 Lactose growth	–
	33 Erythritol growth	+ + + +
	44 Citrate growth	– – – –
	45 Methanol growth	– –
	48 D–Glucosamine growth	+ +
	55 Growth sans thiamin	+
	62 Growth at 37 degrees	–

Species	Tests	Responses
41 Candida curiosa	1 D–Glucose fermentation	+
	5 Sucrose fermentation	+ +
	19 Sucrose growth	+ +
	20 Maltose growth	– – –
	22 Me α–D–glucoside grth	–
	29 Melezitose growth	–
	34 Ribitol growth	+ + + +
	36 D–Mannitol growth	+ + +
	49 Nitrate growth	+ + + +
	54 Growth sans biotin	–
	62 Growth at 37 degrees	–
	65 Starch formation	– – – –

Species	Tests	Responses
42 Candida curvata	12 D–Galactose growth	+ +
(m) 421 Trichosporon cutaneum	17 D–Arabinose growth	– –
	26 Melibiose growth	– – – – – –
	27 Lactose growth	+ + + +
	28 Raffinose growth	+ + +
	30 Inulin growth	– – – – – –
	33 Erythritol growth	+ + + +
	34 Ribitol growth	+ + +
	35 Galactitol growth	– – – – – –
	38 myo–Inositol growth	+ + + + + +
	46 Ethanol growth	+ + + + + +
	49 Nitrate growth	– – – – – –

distinguished from (m) by 73+ (presence of pseudohyphae)

Species	Tests	Responses
43 Candida dendronema*	1 D–Glucose fermentation	+ + +
	18 L–Rhamnose growth	+ + +
	19 Sucrose growth	+
	27 Lactose growth	–
	29 Melezitose growth	– – – –
	33 Erythritol growth	+ +
	35 Galactitol growth	+ + + +
	38 myo–Inositol growth	–
	45 Methanol growth	– – –
	54 Growth sans biotin	–
	62 Growth at 37 degrees	–

Species	Tests	Responses
44 Candida diddensii	12 D–Galactose growth	+ +
	16 L–Arabinose growth	+
(m) 19 Candida aaseri	19 Sucrose growth	+ +
(m) 34 Candida butyri*	26 Melibiose growth	–
(m) 67 Candida insectorum*	27 Lactose growth	– –
(m) 110 Candida tenuis	28 Raffinose growth	– –
(m) 112 Candida terebra*	31 Starch growth	– –
(m) 119 Candida veronae	33 Erythritol growth	+ +
	35 Galactitol growth	–
	40 2–Ketogluconate growth	– –
	46 Ethanol growth	+ +
	54 Growth sans biotin	–
	55 Growth sans thiamin	–

distinguished from (m) by 71+ (presence of cells not spherical, oval or cylindrical)

Species	Tests	Responses
45 Candida diffluens	1 D–Glucose fermentation	– –
	15 D–Xylose growth	– –
	19 Sucrose growth	+
	23 Cellobiose growth	– –
	28 Raffinose growth	– – – –
	33 Erythritol growth	+ + + + +
	45 Methanol growth	– –
	49 Nitrate growth	+ + + + +
	51 Growth sans vitamins	+ + +

Species	Tests	Responses
46 Candida diversa	1 D–Glucose fermentation	+ + + +
	13 L–Sorbose growth	–
385 Torulopsis karawaiewi*	19 Sucrose growth	– – –
	21 Trehalose growth	– –
	23 Cellobiose growth	– –
	32 Glycerol growth	– – – –
	34 Ribitol growth	+ + +
	37 D–Glucitol growth	+ +
	42 D,L–Lactate growth	–
	44 Citrate growth	+ + + +
	56 Grth sans pyridoxine	–

Species	Tests	Responses
47 Candida edax*	12 D–Galactose growth	+
	14 D–Ribose growth	+
(m) 349 Sterigmatomyces aphidis*	26 Melibiose growth	+ + +
	33 Erythritol growth	+ +
	34 Ribitol growth	+ +
	49 Nitrate growth	+ + +
	62 Growth at 37 degrees	+ + +

distinguished from (m) by 74+ (presence of septate hyphae)

Species	Tests	Responses
48 Candida entomaea*	1 D–Glucose fermentation	+ +
	2 Galactose fermentation	+ +
	13 L–Sorbose growth	– –
	26 Melibiose growth	–
	27 Lactose growth	+ + +
	28 Raffinose growth	– – – –
	33 Erythritol growth	+ +
	35 Galactitol growth	+ + + + +
	38 myo–Inositol growth	– – –
	60 Growth in 50% glucose	+

The identification of particular species

Species	Tests	Responses
49 Candida entomophila*	3 Maltose fermentation	−
	5 Sucrose fermentation	+
	11 Raffinose fermentation	+ + +
	18 L−Rhamnose growth	−
	26 Melibiose growth	+
	27 Lactose growth	+ + + +
	31 Starch growth	− −
	33 Erythritol growth	+ + + +
	35 Galactitol growth	− −
	42 D,L−Lactate growth	− −
	43 Succinate growth	+ + +
	46 Ethanol growth	+ + +
50 Candida ergatensis*	1 D−Glucose fermentation	+ + + +
	18 L−Rhamnose growth	−
	19 Sucrose growth	+ +
	27 Lactose growth	+ + + +
	28 Raffinose growth	−
	29 Melezitose growth	− − −
	33 Erythritol growth	+ + +
	44 Citrate growth	+
	45 Methanol growth	−
	49 Nitrate growth	− −
	62 Growth at 37 degrees	− − − −
51 Candida flavificans*	19 Sucrose growth	− − −
	20 Maltose growth	−
	27 Lactose growth	− − −
	33 Erythritol growth	+ + + +
	34 Ribitol growth	− − −
	36 D−Mannitol growth	− − −
	38 myo−Inositol growth	−
	42 D,L−Lactate growth	− −
	44 Citrate growth	−
	56 Grth sans pyridoxine	−
52 Candida fluviotilis*	13 L−Sorbose growth	−
	14 D−Ribose growth	−
(m) 421 Trichosporon cutaneum	16 L−Arabinose growth	− −
	22 Me α−D−glucoside grth	+ +
	27 Lactose growth	+ + +
	28 Raffinose growth	− −
	33 Erythritol growth	− −
	42 D,L−Lactate growth	+
	44 Citrate growth	+ +
	51 Growth sans vitamins	+

distinguished from (m) by 73+ (presence of pseudohyphae)

Species	Tests	Responses
53 Candida foliarum	1 D−Glucose fermentation	− −
	13 L−Sorbose growth	− −
(m) 342 Sporobolomyces odorus	19 Sucrose growth	− − − −
	21 Trehalose growth	+ + + +
	24 Salicin growth	− −
	45 Methanol growth	− − − −
	46 Ethanol growth	+ +
	49 Nitrate growth	+ + + +
	62 Growth at 37 degrees	+ + + +

distinguished from (m) by 66− (colonies not pink)

Species	Tests	Responses
54 Candida fragicola*	1 D−Glucose fermentation	+
	2 Galactose fermentation	+ + +
	12 D−Galactose growth	+
	18 L−Rhamnose growth	−
	19 Sucrose growth	− − − −
	21 Trehalose growth	−
	27 Lactose growth	−
	38 myo−Inositol growth	+ + + + +
	42 D,L−Lactate growth	+ +
	44 Citrate growth	− −
55 Candida freyschussii	1 D−Glucose fermentation	+ +
	9 Cellobiose fermentn	+
	16 L−Arabinose growth	−
	18 L−Rhamnose growth	+ + +
	28 Raffinose growth	− −
	32 Glycerol growth	+ +
	34 Ribitol growth	− − −
	44 Citrate growth	+
	49 Nitrate growth	− − −
	54 Growth sans biotin	+ + +
56 Candida friedrichii	1 D−Glucose fermentation	+ + + + +
	17 D−Arabinose growth	+ + +
	26 Melibiose growth	+ + + +
	28 Raffinose growth	+
	30 Inulin growth	− − − − −
	31 Starch growth	− −
	33 Erythritol growth	+ + +
	35 Galactitol growth	+ + + + +
	38 myo−Inositol growth	− −
	42 D,L−Lactate growth	− − −
57 Candida glaebosa	1 D−Glucose fermentation	− − − −
	26 Melibiose growth	+ + + +
	27 Lactose growth	+ + + +
	33 Erythritol growth	− − − − −
	34 Ribitol growth	+ +
	35 Galactitol growth	− − − −
	37 D−Glucitol growth	+ +
	38 myo−Inositol growth	− − − − −
	44 Citrate growth	+ +
	52 Growth sans inositol	− − −
	53 Grth sans pantothenate	− − −
58 Candida graminis*	1 D−Glucose fermentation	−
	13 L−Sorbose growth	+
	16 L−Arabinose growth	− − −
	18 L−Rhamnose growth	− −
	28 Raffinose growth	−
	33 Erythritol growth	+ + + +
	34 Ribitol growth	+
	38 myo−Inositol growth	−
	42 D,L−Lactate growth	+
	46 Ethanol growth	− −
	49 Nitrate growth	+ + + +
	54 Growth sans biotin	+

Species	Tests	Responses

59 Candida homilentoma*
 	
 (m) 110 Candida tenuis
	1 D−Glucose fermentation
	 2 Galactose fermentation
	 18 L−Rhamnose growth
	 19 Sucrose growth
	 22 Me α−D−glucoside grth
	 26 Melibiose growth
	 27 Lactose growth
	 28 Raffinose growth
	 33 Erythritol growth
	 42 D,L−Lactate growth
	 45 Methanol growth
	 51 Growth sans vitamins
		+ +
		    +
		    + +
		 +
		 +
		    −
		 − − −
		 − − − −
		 + + + +
		 −
		 −
		 + + + +

distinguished from (m) by 71+ (presence of cells not spherical, oval or cylindrical)

60 Candida humicola 137 Cryptococcus laurentii (m) 16 Bullera alba (m) 21 Candida amylolenta* (m) 138 Cryptococcus luteolus (m) 354 Sterigmatomyces penicillatus* (m) 421 Trichosporon cutaneum	1 D−Glucose fermentation 14 D−Ribose growth 26 Melibiose growth 30 Inulin growth 33 Erythritol growth 49 Nitrate growth 65 Starch formation	− + + − + − +

distinguished from (m) by 68− (cells not splitting)
                        72+ (presence of filaments)
                        74+ (presence of septate hyphae)

61 Candida hydrocarbofumarica* 25 Candida blankii (m) 421 Trichosporon cutaneum	26 Melibiose growth 27 Lactose growth 30 Inulin growth 33 Erythritol growth 35 Galactitol growth 38 myo−Inositol growth 42 D,L−Lactate growth 51 Growth sans vitamins 62 Growth at 37 degrees 65 Starch formation	− − − − + + + − − − − + + + + + + + + + − − − + + + + + + + + + −

distinguished from (m) by 73+ (presence of pseudohyphae)

62 Candida hylophila*	1 D−Glucose fermentation 12 D−Galactose growth 13 L−Sorbose growth 15 D−Xylose growth 19 Sucrose growth 21 Trehalose growth 33 Erythritol growth 36 D−Mannitol growth 37 D−Glucitol growth 38 myo−Inositol growth 42 D,L−Lactate growth 44 Citrate growth 60 Growth in 50% glucose	− − − − − − − − − − + + + + + + − − − − − − − − − − − − − − − − − − + + + + + + − − − − + + + + + + − − − −

Species	Tests	Responses

63 Candida iberica*
 	
123 Candida zeylanoides
(m) 298 Rhodosporidium dacryoidum*
	5 Sucrose fermentation
	13 L−Sorbose growth
	15 D−Xylose growth
	19 Sucrose growth
	21 Trehalose growth
	22 Me α−D−glucoside grth
	23 Cellobiose growth
	32 Glycerol growth
	33 Erythritol growth
	38 myo−Inositol growth
	42 D,L−Lactate growth
	44 Citrate growth
	49 Nitrate growth
		− −
		+ +
		− − − − −
		− − −
		+ + + + +
		− − −
		− − − − −
		      + + +
		− − − − −
		           −
		− − − − −
		+ + + + +
		− − − − −

distinguished from (m) by 66− (colonies not pink)

Species	Tests	Responses
64 Candida incommunis*	12 D−Galactose growth 16 L−Arabinose growth 22 Me α−D−glucoside grth 23 Cellobiose growth 28 Raffinose growth 33 Erythritol growth 34 Ribitol growth 38 myo−Inositol growth 49 Nitrate growth	− − − − + + − − − + + + − − + + + + + + + + +
65 Candida inositophila*	1 D−Glucose fermentation 13 L−Sorbose growth 18 L−Rhamnose growth 27 Lactose growth 33 Erythritol growth 38 myo−Inositol growth 40 2−Ketogluconate growth 44 Citrate growth 49 Nitrate growth 62 Growth at 37 degrees	+ + + + + + + + + − − + + + + + − − − − − − − − − + +
66 Candida insectamans*	14 D−Ribose growth 18 L−Rhamnose growth 19 Sucrose growth 22 Me α−D−glucoside grth 44 Citrate growth 46 Ethanol growth 62 Growth at 37 degrees	+ − − − − + + + + + − − − + +
67 Candida insectorum* 110 Candida tenuis 112 Candida terebra* 119 Candida veronae (m) 44 Candida diddensii (m) 86 Candida naeodendra* (m) 104 Candida silvanorum* (m) 270 Pichia nakazawae*	1 D−Glucose fermentation 5 Sucrose fermentation 11 Raffinose fermentation 12 D−Galactose growth 18 L−Rhamnose growth 33 Erythritol growth 35 Galactitol growth 42 D,L−Lactate growth 43 Succinate growth 45 Methanol growth 49 Nitrate growth 51 Growth sans vitamins	+ + + + − − − − + + + + + + + + + + − − − − − − − − + + + + − − − − − − − − − −

distinguished from (m) by 71− (cells spherical, oval or cylindrical only)
                       74− (absence of septate hyphae)

Species	Tests	Responses
68 Candida intermedia	1 D−Glucose fermentation	+ + +
	2 Galactose fermentation	+
	13 L−Sorbose growth	+
	26 Melibiose growth	−
	27 Lactose growth	+ + + +
	31 Starch growth	+
	32 Glycerol growth	− − −
	34 Ribitol growth	− − −
	42 D,L−Lactate growth	− − −
	46 Ethanol growth	+
	49 Nitrate growth	− −
69 Candida ishiwadae*	1 D−Glucose fermentation	+ + +
	13 L−Sorbose growth	+ + +
	17 D−Arabinose growth	+ +
	18 L−Rhamnose growth	+ + + +
	28 Raffinose growth	− −
	31 Starch growth	+
	33 Erythritol growth	+ + + +
	42 D,L−Lactate growth	+ + + + +
	49 Nitrate growth	+ + + + +
70 Candida javanica	19 Sucrose growth	− − −
	26 Melibiose growth	+ + +
	27 Lactose growth	−
	28 Raffinose growth	+
	33 Erythritol growth	−
	35 Galactitol growth	+ + +
	38 myo−Inositol growth	−
	46 Ethanol growth	+
	49 Nitrate growth	+ + +
71 Candida kefyr	1 D−Glucose fermentation	+
	9 Cellobiose fermentn	−
	12 D−Galactose growth	+
	15 D−Xylose growth	− − − −
	21 Trehalose growth	− − −
	27 Lactose growth	+ + + +
	42 D,L−Lactate growth	+ +
	53 Grth sans pantothenate	− − − −
72 Candida krissii*	12 D−Galactose growth	− − − −
(m) 385 Torulopsis karawaiewi*	15 D−Xylose growth	− − −
	19 Sucrose growth	− − − −
	21 Trehalose growth	− − − −
	24 Salicin growth	+
	27 Lactose growth	− − − −
	33 Erythritol growth	− − −
	35 Galactitol growth	− −
	36 D−Mannitol growth	+ +
	37 D−Glucitol growth	+ + +
	38 myo−Inositol growth	− −
	40 2−Ketogluconate growth	+
	44 Citrate growth	+ + + + +
	49 Nitrate growth	− − − −
	55 Growth sans thiamin	+ + + +

distinguished from (m) by 72+ (presence of filaments)

Species	Tests	Responses
73 Candida lambica	1 D–Glucose fermentation	+ + + +
	12 D–Galactose growth	– – – – –
117 Candida valida	13 L–Sorbose growth	–
(m) 357 Torulaspora delbrueckii**	15 D–Xylose growth	+ + + + +
(a) 256 Pichia fermentans	19 Sucrose growth	– – – –
(a) 266 Pichia membranaefaciens	23 Cellobiose growth	– – –
	27 Lactose growth	–
	36 D–Mannitol growth	– – – –
	37 D–Glucitol growth	–
	38 myo–Inositol growth	–
	49 Nitrate growth	– –
	56 Grth sans pyridoxine	+ + + + +

distinguished from (m) by 72+ (presence of filaments)
distinguished from (a) by 75– (inability to form ascospores)

Species	Tests	Responses
74 Candida lusitaniae	1 D–Glucose fermentation	+ +
	13 L–Sorbose growth	+ + + +
	18 L–Rhamnose growth	+ + + +
	19 Sucrose growth	+
	27 Lactose growth	– –
	28 Raffinose growth	– – – –
	33 Erythritol growth	– – – –
	35 Galactitol growth	–
	38 myo–Inositol growth	–
	49 Nitrate growth	– – – –
	62 Growth at 37 degrees	+ + +

Species	Tests	Responses
75 Candida macedoniensis	1 D–Glucose fermentation	+ +
	5 Sucrose fermentation	+
	11 Raffinose fermentation	+
	15 D–Xylose growth	+ + +
	16 L–Arabinose growth	+
	20 Maltose growth	– – –
	27 Lactose growth	+ + +
	29 Melezitose growth	–
	35 Galactitol growth	–
	38 myo–Inositol growth	–
	44 Citrate growth	–
	54 Growth sans biotin	+ + + +

Species	Tests	Responses
76 Candida maltosa*	1 D–Glucose fermentation	+
	5 Sucrose fermentation	+ + + +
(a) 210 Kluyveromyces lactis	15 D–Xylose growth	+ + + + +
	16 L–Arabinose growth	– – – –
	18 L–Rhamnose growth	– – – –
	24 Salicin growth	+ + + + +
	26 Melibiose growth	– – – – –
	28 Raffinose growth	– – – – –
	29 Melezitose growth	+
	31 Starch growth	– – – – –
	33 Erythritol growth	– –
	34 Ribitol growth	+ + +
	36 D–Mannitol growth	+ +
	41 5–Ketogluconate growth	+
	45 Methanol growth	– –
	62 Growth at 37 degrees	+ + + +

distinguished from (a) by 75– (inability to form ascospores)

The identification of particular species

Species	Tests	Responses
77 Candida marina	12 D–Galactose growth	−
	13 L–Sorbose growth	+ +
	16 L–Arabinose growth	+ + +
	19 Sucrose growth	− − −
	20 Maltose growth	− −
	21 Trehalose growth	+
	22 Me α–D–glucoside grth	−
	23 Cellobiose growth	− −
	27 Lactose growth	− − − −
	33 Erythritol growth	+ + +
	38 myo–Inositol growth	+ + + + + +
	43 Succinate growth	−
	46 Ethanol growth	+
	56 Grth sans pyridoxine	−
78 Candida maritima	1 D–Glucose fermentation	+ + + +
	12 D–Galactose growth	− − −
	14 D–Ribose growth	+ + + +
	16 L–Arabinose growth	−
	18 L–Rhamnose growth	+ + +
	23 Cellobiose growth	+
	28 Raffinose growth	+ + + +
	32 Glycerol growth	+
	33 Erythritol growth	−
	34 Ribitol growth	− −
	38 myo–Inositol growth	−
	49 Nitrate growth	− − −
	56 Grth sans pyridoxine	−
79 Candida melibiosica	1 D–Glucose fermentation	+ +
(m) 401 Torulopsis pampelonensis*	5 Sucrose fermentation	− − − − −
	9 Cellobiose fermentn	+ + +
	13 L–Sorbose growth	+ + + + + +
	16 L–Arabinose growth	− − − − −
	23 Cellobiose growth	+ +
	26 Melibiose growth	+ + + + + +
	33 Erythritol growth	− −
	35 Galactitol growth	− − − −
	36 D–Mannitol growth	+ + + +
	37 D–Glucitol growth	+ +
	38 myo–Inositol growth	− − − −
	49 Nitrate growth	− −

distinguished from (m) by 72+ (presence of filaments)

Species	Tests	Responses
80 Candida melinii	1 D–Glucose fermentation	−
(m) 172 Hansenula beckii**	13 L–Sorbose growth	−
(m) 174 Hansenula bimundalis	14 D–Ribose growth	
(m) 200 Hansenula wingei	18 L–Rhamnose growth	+ + + +
(m) 304 Rhodotorula glutinis	19 Sucrose growth	+ +
(a) 176 Hansenula canadensis	28 Raffinose growth	− − − −
	29 Melezitose growth	+ +
	32 Glycerol growth	+
	34 Ribitol growth	− − −
	46 Ethanol growth	+
	49 Nitrate growth	+ + + +

distinguished from (m) by 66− (colonies not pink)
74− (absence of septate hyphae)
distinguished from (a) by 75− (inability to form ascospores)

Species	Tests	Responses
81 Candida membranaefaciens	30 Inulin growth	+ +
	31 Starch growth	− −
	33 Erythritol growth	+ +
	35 Galactitol growth	+
	49 Nitrate growth	−
82 Candida mesenterica	12 D−Galactose growth	− −
	15 D−Xylose growth	− − − −
	19 Sucrose growth	+ + + +
	21 Trehalose growth	+ +
	31 Starch growth	− −
	33 Erythritol growth	+ + + +
	49 Nitrate growth	− − − −
	55 Growth sans thiamin	+ + + +
	62 Growth at 37 degrees	− −
83 Candida milleri*	2 Galactose fermentation	+ + +
	7 Melibiose fermentation	−
380 Torulopsis holmii	11 Raffinose fermentation	+ + +
(a) 211 Kluyveromyces lodderi	12 D−Galactose growth	+
(a) 215 Kluyveromyces polysporus	23 Cellobiose growth	−
(a) 312 Saccharomyces cerevisiae	27 Lactose growth	−
(a) 314 Saccharomyces exiguus	28 Raffinose growth	+
	29 Melezitose growth	−
	36 D−Mannitol growth	− − −
	37 D−Glucitol growth	− − −
	60 Growth in 50% glucose	− − − −
	62 Growth at 37 degrees	− − −

distinguished from (a) by 75− (inability to form ascospores)

Species	Tests	Responses
84 Candida mogii	1 D−Glucose fermentation	+ +
	3 Maltose fermentation	+ +
(m) 20 Candida albicans	13 L−Sorbose growth	+ + +
	15 D−Xylose growth	+
	16 L−Arabinose growth	+ +
	23 Cellobiose growth	−
	25 Arbutin growth	− −
	29 Melezitose growth	−
	31 Starch growth	+ +
	33 Erythritol growth	− −
	46 Ethanol growth	− −
	51 Growth sans vitamins	+

distinguished from (m) by 74− (absence of septate hyphae)

Species	Tests	Responses
85 Candida muscorum	1 D−Glucose fermentation	−
	16 L−Arabinose growth	− − − −
	22 Me α−D−glucoside grth	+ + +
	26 Melibiose growth	−
	27 Lactose growth	+ + + +
	28 Raffinose growth	+
	33 Erythritol growth	−
	38 myo−Inositol growth	− −
	42 D,L−Lactate growth	− −
	46 Ethanol growth	− −
	49 Nitrate growth	+ +
	51 Growth sans vitamins	−

Species	Tests	Responses
86 Candida naeodendra*	18 L–Rhamnose growth	+ +
	26 Melibiose growth	– – –
(m) 67 Candida insectorum*	27 Lactose growth	– –
(m) 110 Candida tenuis	28 Raffinose growth	– – –
(m) 270 Pichia nakazawae*	29 Melezitose growth	+
	31 Starch growth	+ + +
	32 Glycerol growth	+
	33 Erythritol growth	+ +
	35 Galactitol growth	– –
	38 myo–Inositol growth	– –
	49 Nitrate growth	– – –
	51 Growth sans vitamins	– – –
	60 Growth in 50% glucose	+

distinguished from (m) by 71+ (presence of cells not spherical, oval or cylindrical)
74– (absence of septate hyphae)

Species	Tests	Responses
87 Candida norvegensis	12 D–Galactose growth	– – – – –
	19 Sucrose growth	– – – – –
(a) 271 Pichia norvegensis*	23 Cellobiose growth	+ + +
	24 Salicin growth	+ +
	27 Lactose growth	– – –
	33 Erythritol growth	– – –
	36 D–Mannitol growth	– – – – –
	42 D,L–Lactate growth	+ + + +
	44 Citrate growth	+ + +
	49 Nitrate growth	–
	56 Grth sans pyridoxine	– –

distinguished from (a) by 75– (inability to form ascospores)

Species	Tests	Responses
88 Candida oleophila*	2 Galactose fermentation	+ + + +
	6 Trehalose fermentation	– – –
99 Candida sake	13 L–Sorbose growth	+ + +
(a) 234 Metschnikowia pulcherrima	16 L–Arabinose growth	–
	17 D–Arabinose growth	– –
	18 L–Rhamnose growth	– – – –
	23 Cellobiose growth	+ + +
	26 Melibiose growth	–
	28 Raffinose growth	– – – –
	29 Melezitose growth	+ + + +
	31 Starch growth	– – – –
	33 Erythritol growth	– –
	40 2–Ketogluconate growth	+
	44 Citrate growth	+ + + +
	62 Growth at 37 degrees	–

distinguished from (a) by 75– (inability to form ascospores)

Species	Tests	Responses
89 Candida oregonensis	1 D–Glucose fermentation	+ +
	12 D–Galactose growth	– –
	13 L–Sorbose growth	– –
	14 D–Ribose growth	–
	18 L–Rhamnose growth	+ + +
	19 Sucrose growth	+ + +
	22 Me α–D–glucoside grth	+
	27 Lactose growth	–
	31 Starch growth	+
	32 Glycerol growth	– –
	49 Nitrate growth	– –

Species	Tests	Responses
90 Candida parapsilosis	1 D–Glucose fermentation	+ +
	16 L–Arabinose growth	+ + +
	18 L–Rhamnose growth	–
	22 Me α–D–glucoside grth	+ + +
	23 Cellobiose growth	– –
	24 Salicin growth	
	27 Lactose growth	–
	28 Raffinose growth	– – –
	31 Starch growth	– – –
	36 D–Mannitol growth	+
	37 D–Glucitol growth	+
	49 Nitrate growth	–
	62 Growth at 37 degrees	+

Species	Tests	Responses
91 Candida podzolica*	18 L–Rhamnose growth	+
	33 Erythritol growth	– –
(m) 140 Cryptococcus magnus**	35 Galactitol growth	+
(m) 421 Trichosporon cutaneum	36 D–Mannitol growth	– –
	49 Nitrate growth	– –
	51 Growth sans vitamins	+ +

distinguished from (m) by 72+ (presence of filaments)
                              73+ (presence of pseudohyphae)

Species	Tests	Responses
92 Candida pseudointermedia*	1 D–Glucose fermentation	+
	2 Galactose fermentation	+ +
	14 D–Ribose growth	+ + + +
	22 Me α–D–glucoside grth	+ +
	26 Melibiose growth	– –
	27 Lactose growth	–
	28 Raffinose growth	+ + + +
	31 Starch growth	+ +
	33 Erythritol growth	– – – –
	36 D–Mannitol growth	+
	38 myo–Inositol growth	– – –
	42 D,L–Lactate growth	– – – –
	44 Citrate growth	+
	49 Nitrate growth	–
	54 Growth sans biotin	–
	55 Growth sans thiamin	+

Species	Tests	Responses
93 Candida pseudotropicalis	8 Lactose fermentation	+ +
	15 D–Xylose growth	+ +
(a) 212 Kluyveromyces marxianus	20 Maltose growth	–
	21 Trehalose growth	–
	27 Lactose growth	+ +
	34 Ribitol growth	–
	53 Grth sans pantothenate	– – –

distinguished from (a) by 75– (inability to form ascospores)

Species	Tests	Responses
94 Candida quercuum*	12 D−Galactose growth	− − −
	13 L−Sorbose growth	−
	15 D−Xylose growth	+ + +
	16 L−Arabinose growth	− − −
	18 L−Rhamnose growth	− − −
	19 Sucrose growth	+ +
	28 Raffinose growth	− −
	29 Melezitose growth	+
	42 D,L−Lactate growth	+
	44 Citrate growth	+
	49 Nitrate growth	−
	56 Grth sans pyridoxine	− − −
	62 Growth at 37 degrees	+ +
95 Candida ravautii	15 D−Xylose growth	+ + +
36 Candida catenulata	16 L−Arabinose growth	−
	19 Sucrose growth	− − −
	21 Trehalose growth	+ + +
	22 Me α−D−glucoside grth	−
	23 Cellobiose growth	− −
	27 Lactose growth	− − −
	31 Starch growth	+ + +
	32 Glycerol growth	+
	34 Ribitol growth	− −
	38 myo−Inositol growth	− −
96 Candida rhagii	1 D−Glucose fermentation	+
	5 Sucrose fermentation	+ +
	13 L−Sorbose growth	−
	18 L−Rhamnose growth	−
	22 Me α−D−glucoside grth	+
	26 Melibiose growth	− − −
	27 Lactose growth	−
	28 Raffinose growth	+ + + +
	31 Starch growth	− − −
	33 Erythritol growth	+ + + +
	44 Citrate growth	− −
	45 Methanol growth	−
	46 Ethanol growth	+ + + +
	49 Nitrate growth	− −
97 Candida rugopelliculosa*	13 L−Sorbose growth	−
	15 D−Xylose growth	− − − − −
(m) 357 Torulaspora delbrueckii**	19 Sucrose growth	− − − −
(a) 266 Pichia membranaefaciens	21 Trehalose growth	−
	32 Glycerol growth	− − − − −
	36 D−Mannitol growth	− − −
	37 D−Glucitol growth	− −
	39 Gluconolactone growth	+ +
	42 D,L−Lactate growth	+ + + +
	43 Succinate growth	+
	48 D−Glucosamine growth	+ + +

distinguished from (m) by 72+ (presence of filaments)
distinguished from (a) by 75− (inability to form ascospores)

Species	Tests	Responses
98 Candida rugosa	1 D–Glucose fermentation	− − − −
	12 D–Galactose growth	+ + + +
36 Candida catenulata	19 Sucrose growth	− −
	21 Trehalose growth	− − − −
	23 Cellobiose growth	− −
	24 Salicin growth	− −
	27 Lactose growth	− − − −
	33 Erythritol growth	− − − −
	36 D–Mannitol growth	+ +
	37 D–Glucitol growth	+ +
	42 D,L–Lactate growth	+ + + +
	62 Growth at 37 degrees	+ + + +

Species	Tests	Responses
99 Candida sake	1 D–Glucose fermentation	+ + +
	12 D–Galactose growth	+ + + +
88 Candida oleophila*	13 L–Sorbose growth	+
(m) 357 Torulaspora delbrueckii**	15 D–Xylose growth	+ + + +
(a) 231 Metschnikowia bicuspidata	16 L–Arabinose growth	− − −
(a) 234 Metschnikowia pulcherrima	17 D–Arabinose growth	− −
(a) 235 Metschnikowia reukaufii	19 Sucrose growth	+ + + +
	26 Melibiose growth	− − −
	27 Lactose growth	−
	28 Raffinose growth	−
	31 Starch growth	− − − −
	33 Erythritol growth	− − − −
	35 Galactitol growth	− −
	36 D–Mannitol growth	+ + +
	37 D–Glucitol growth	+
	40 2–Ketogluconate growth	+ +
	49 Nitrate growth	−
	55 Growth sans thiamin	+ + + +
	57 Growth sans niacin	+ + + +
	62 Growth at 37 degrees	− − − −

distinguished from (m) by 72+ (presence of filaments)
distinguished from (a) by 75− (inability to form ascospores)

Species	Tests	Responses
100 Candida salmanticensis	1 D–Glucose fermentation	+
	5 Sucrose fermentation	+
	14 D–Ribose growth	−
	16 L–Arabinose growth	
	17 D–Arabinose growth	+ + + +
	22 Me α–D–glucoside grth	+
	26 Melibiose growth	+ +
	30 Inulin growth	+
	32 Glycerol growth	− − − −
	33 Erythritol growth	− −
	38 myo–Inositol growth	−
	46 Ethanol growth	+
	49 Nitrate growth	−
	56 Grth sans pyridoxine	−

Species	Tests	Responses
101 Candida santamariae	6 Trehalose fermentation	+ + + + +
	13 L–Sorbose growth	+ +
23 Candida beechii	16 L–Arabinose growth	− − −
	19 Sucrose growth	− − − − −
	22 Me α–D–glucoside grth	− −
	29 Melezitose growth	−
	36 D–Mannitol growth	+
	37 D–Glucitol growth	+
	42 D,L–Lactate growth	+ + + + +
	44 Citrate growth	+ + + +
	45 Methanol growth	−
	62 Growth at 37 degrees	− −

The identification of particular species

Species	Tests	Responses
102 Candida savonica*	12 D−Galactose growth	+ + + +
	13 L−Sorbose growth	− − − −
123 Candida zeylanoides	18 L−Rhamnose growth	−
	19 Sucrose growth	− − −
	20 Maltose growth	−
	24 Salicin growth	+
	27 Lactose growth	− − − −
	29 Melezitose growth	−
	32 Glycerol growth	+
	33 Erythritol growth	− − − −
	34 Ribitol growth	+ + +
	38 myo−Inositol growth	− − −
	41 5−Ketogluconate growth	−
	42 D,L−Lactate growth	− − − −
	44 Citrate growth	+ + +
	49 Nitrate growth	− − − −

Species	Tests	Responses
103 Candida silvae	1 D−Glucose fermentation	− − − −
	12 D−Galactose growth	− − − −
121 Candida vini	13 L−Sorbose growth	− − − −
(m) 325 Saccharomycopsis vini**	15 D−Xylose growth	−
	18 L−Rhamnose growth	− −
	20 Maltose growth	− − −
	21 Trehalose growth	− − − −
	24 Salicin growth	−
	27 Lactose growth	− − −
	28 Raffinose growth	−
	32 Glycerol growth	+ + + +
	33 Erythritol growth	−
	36 D−Mannitol growth	+ + +
	37 D−Glucitol growth	+
	44 Citrate growth	− − −
	46 Ethanol growth	+ + +
	49 Nitrate growth	− − − −

distinguished from (m) by 74− (absence of septate hyphae)

Species	Tests	Responses
104 Candida silvanorum*	1 D−Glucose fermentation	+ + +
	2 Galactose fermentation	+
(m) 67 Candida insectorum*	18 L−Rhamnose growth	+ + + + +
	26 Melibiose growth	+ + + +
	27 Lactose growth	− −
	28 Raffinose growth	+
	30 Inulin growth	−
	31 Starch growth	+
	33 Erythritol growth	+ + +
	35 Galactitol growth	− − − −
	42 D,L−Lactate growth	− − − −
	43 Succinate growth	+ +
	49 Nitrate growth	− − − −
	65 Starch formation	−

distinguished from (m) by 71+ (presence of cells not spherical, oval or cylindrical)

Species	Tests	Responses
105 Candida silvicultrix*	2 Galactose fermentation	+
	7 Melibiose fermentation	+ + + +
	13 L–Sorbose growth	– –
	17 D–Arabinose growth	+ + +
	18 L–Rhamnose growth	
	27 Lactose growth	–
	33 Erythritol growth	+ + + +
	35 Galactitol growth	– –
	42 D,L–Lactate growth	+ + +
	46 Ethanol growth	+ + + +
	49 Nitrate growth	–
106 Candida solani	1 D–Glucose fermentation	+ + + +
	12 D–Galactose growth	– – – –
	13 L–Sorbose growth	+ + + +
	23 Cellobiose growth	+ +
	24 Salicin growth	+ +
	36 D–Mannitol growth	– – – –
	37 D–Glucitol growth	– –
	42 D,L–Lactate growth	+ +
107 Candida sorboxylosa*	12 D–Galactose growth	– – –
	19 Sucrose growth	–
(a) 266 Pichia membranaefaciens	21 Trehalose growth	– – –
(a) 312 Saccharomyces cerevisiae	23 Cellobiose growth	– –
	36 D–Mannitol growth	– –
	37 D–Glucitol growth	–
	42 D,L–Lactate growth	+ + +
	54 Growth sans biotin	+ + +
	56 Grth sans pyridoxine	– – –

distinguished from (a) by 75– (inability to form ascospores)

Species	Tests	Responses
108 Candida steatolytica*	1 D–Glucose fermentation	+ +
	5 Sucrose fermentation	+ +
	12 D–Galactose growth	+
	13 L–Sorbose growth	+ +
	18 L–Rhamnose growth	+
	33 Erythritol growth	– – –
	35 Galactitol growth	+
	38 myo–Inositol growth	+ + + +
	44 Citrate growth	+ + + +
	49 Nitrate growth	– – –
	51 Growth sans vitamins	–
109 Candida suecica*	12 D–Galactose growth	– – – –
	13 L–Sorbose growth	+
	14 D–Ribose growth	– – –
	18 L–Rhamnose growth	– –
	22 Me α–D–glucoside grth	+ +
	23 Cellobiose growth	+ + + +
	29 Melezitose growth	– –
	32 Glycerol growth	–
	33 Erythritol growth	–
	34 Ribitol growth	+
	43 Succinate growth	– – –
	46 Ethanol growth	– – – –
	49 Nitrate growth	– – –

The identification of particular species

Species	Tests	Responses
110 Candida tenuis	1 D–Glucose fermentation	+
	2 Galactose fermentation	+ +
67 Candida insectorum*	3 Maltose fermentation	– – –
112 Candida terebra*	9 Cellobiose fermentn	–
119 Candida veronae	12 D–Galactose growth	+
(m) 44 Candida diddensii	18 L–Rhamnose growth	+ + +
(m) 59 Candida homilentoma*	19 Sucrose growth	+
(m) 86 Candida naeodendra*	28 Raffinose growth	– – –
(m) 270 Pichia nakazawae*	33 Erythritol growth	+ +
	35 Galactitol growth	– – –
	49 Nitrate growth	– – –

distinguished from (m) by 71– (cells spherical, oval or cylindrical only)
74– (absence of septate hyphae)

Species	Tests	Responses
111 Candida tepae*	1 D–Glucose fermentation	– – – – –
	14 D–Ribose growth	– –
	15 D–Xylose growth	+ + + + +
	16 L–Arabinose growth	–
	19 Sucrose growth	+ + + +
	20 Maltose growth	+ + + + +
	23 Cellobiose growth	+
	27 Lactose growth	– – – – –
	28 Raffinose growth	– – – –
	29 Melezitose growth	– – – –
	32 Glycerol growth	+ +
	33 Erythritol growth	– – –
	38 myo–Inositol growth	–
	42 D,L–Lactate growth	+ + + + +
	44 Citrate growth	– – – – –
	49 Nitrate growth	– –

Species	Tests	Responses
112 Candida terebra*	13 L–Sorbose growth	+ + + + +
	16 L–Arabinose growth	+
67 Candida insectorum*	17 D–Arabinose growth	+
110 Candida tenuis	18 L–Rhamnose growth	+ + + + +
(m) 44 Candida diddensii	26 Melibiose growth	– – – –
	27 Lactose growth	– – – – –
	28 Raffinose growth	– – – –
	29 Melezitose growth	+ + +
	31 Starch growth	– – – – –
	33 Erythritol growth	+ + + + +
	38 myo–Inositol growth	– – –
	42 D,L–Lactate growth	– – – – –
	45 Methanol growth	– – – –
	54 Growth sans biotin	–
	62 Growth at 37 degrees	+ +

distinguished from (m) by 71– (cells spherical, oval or cylindrical only)

Species	Tests	Responses
113 Candida tropicalis	2 Galactose fermentation	+ +
	3 Maltose fermentation	+ +
20 Candida albicans	12 D–Galactose growth	+
(m) 122 Candida viswanathii	18 L–Rhamnose growth	–
	22 Me α–D–glucoside grth	+
	27 Lactose growth	–
	28 Raffinose growth	– – –
	31 Starch growth	+ + +
	33 Erythritol growth	– –
	38 myo–Inositol growth	–
	41 5–Ketogluconate growth	+ + +
	51 Growth sans vitamins	–
	62 Growth at 37 degrees	+

distinguished from (m) by 74+ (presence of septate hyphae)

Species	Tests	Responses
114 Candida tsukubaensis*	24 Salicin growth	− −
	27 Lactose growth	+
	33 Erythritol growth	+
	36 D−Mannitol growth	− −
	37 D−Glucitol growth	− − −
	38 myo−Inositol growth	+ + +
	49 Nitrate growth	+ + + + +
	62 Growth at 37 degrees	+ + +
115 Candida utilis	1 D−Glucose fermentation	+
	11 Raffinose fermentation	+
	12 D−Galactose growth	−
	29 Melezitose growth	+ + + +
	30 Inulin growth	+ +
	33 Erythritol growth	− − −
	37 D−Glucitol growth	− − − −
	42 D,L−Lactate growth	+
	44 Citrate growth	+ + +
	49 Nitrate growth	+ + +
116 Candida valdiviana*	14 D−Ribose growth	− −
	22 Me α−D−glucoside grth	+ + + +
	26 Melibiose growth	+ +
	33 Erythritol growth	− − −
	38 myo−Inositol growth	+ + +
	40 2−Ketogluconate growth	− −
	49 Nitrate growth	+ + + +
	65 Starch formation	− −
117 Candida valida	12 D−Galactose growth	− −
39 Candida citrea*	13 L−Sorbose growth	− −
73 Candida lambica	19 Sucrose growth	−
(m) 7 Arthroascus javanensis**	20 Maltose growth	− −
(m) 161 Geotrichum capitatum**	21 Trehalose growth	− −
(m) 357 Torulaspora delbrueckii**	27 Lactose growth	− −
(m) 385 Torulopsis karawaiewi*	28 Raffinose growth	−
(a) 263 Pichia kudriavzevii	32 Glycerol growth	+ +
(a) 266 Pichia membranaefaciens	37 D−Glucitol growth	− −
(a) 286 Pichia scutulata*	44 Citrate growth	− −
(a) 312 Saccharomyces cerevisiae	49 Nitrate growth	− −
(a) 433 Zygosaccharomyces bisporus**	56 Grth sans pyridoxine	+ +
(a) 439 Zygosaccharomyces rouxii**		

distinguished from (m) by 67+ (formation of budding cells)
                       72+ (presence of filaments)
                       73+ (presence of pseudohyphae)
distinguished from (a) by 75− (inability to form ascospores)

Species	Tests	Responses
118 Candida vartiovaarai	1 D−Glucose fermentation	+
	5 Sucrose fermentation	+ + + +
	15 D−Xylose growth	+ +
	22 Me α−D−glucoside grth	+
	28 Raffinose growth	− − − − −
	29 Melezitose growth	+
	33 Erythritol growth	−
	34 Ribitol growth	− −
	36 D−Mannitol growth	+ +
	42 D,L−Lactate growth	+ + +
	44 Citrate growth	−
	49 Nitrate growth	+ + + + +
	51 Growth sans vitamins	+

Species	Tests	Responses
119 Candida veronae	1 D–Glucose fermentation	+ + +
	12 D–Galactose growth	+
67 Candida insectorum*	13 L–Sorbose growth	– – – –
110 Candida tenuis	16 L–Arabinose growth	+
(m) 44 Candida diddensii	18 L–Rhamnose growth	+ + + +
	19 Sucrose growth	+
	26 Melibiose growth	–
	27 Lactose growth	– –
	28 Raffinose growth	– –
	31 Starch growth	– – – –
	33 Erythritol growth	+ + + +
	35 Galactitol growth	– –
	42 D,L–Lactate growth	– –
	45 Methanol growth	– –
	46 Ethanol growth	+
	49 Nitrate growth	–
	54 Growth sans biotin	–

distinguished from (m) by 71– (cells spherical, oval or cylindrical only)

Species	Tests	Responses
120 Candida vinaria*	1 D–Glucose fermentation	– – –
	12 D–Galactose growth	+ + +
	13 L–Sorbose growth	+ +
	16 L–Arabinose growth	– –
	21 Trehalose growth	– – –
	27 Lactose growth	– – –
	32 Glycerol growth	+
	33 Erythritol growth	– – –
	38 myo–Inositol growth	–
	42 D,L–Lactate growth	–
	43 Succinate growth	– – –
	49 Nitrate growth	–
	51 Growth sans vitamins	– – –
	62 Growth at 37 degrees	– – –

Species	Tests	Responses
121 Candida vini	1 D–Glucose fermentation	– – – –
	12 D–Galactose growth	– – – –
103 Candida silvae	13 L–Sorbose growth	– – – –
(m) 325 Saccharomycopsis vini**	15 D–Xylose growth	– –
(a) 252 Pichia delftensis	18 L–Rhamnose growth	– –
(a) 257 Pichia fluxuum	19 Sucrose growth	– – – –
	20 Maltose growth	– – – –
	21 Trehalose growth	– – – –
	23 Cellobiose growth	– –
	27 Lactose growth	– –
	33 Erythritol growth	– –
	36 D–Mannitol growth	+ +
	37 D–Glucitol growth	+ +
	44 Citrate growth	– – – –
	49 Nitrate growth	– – – –

distinguished from (m) by 74– (absence of septate hyphae)
distinguished from (a) by 75– (inability to form ascospores)

Species	Tests	Responses
122 Candida viswanathii	3 Maltose fermentation	+ + + +
	5 Sucrose fermentation	− − − −
(m) 20 Candida albicans	6 Trehalose fermentation	+
(m) 113 Candida tropicalis	12 D−Galactose growth	+ + +
	15 D−Xylose growth	+
	27 Lactose growth	− −
	28 Raffinose growth	− − − −
	29 Melezitose growth	+
	31 Starch growth	+ + + +
	33 Erythritol growth	− − − −
	34 Ribitol growth	+ +
	38 myo−Inositol growth	− − −
	44 Citrate growth	+

distinguished from (m) by 74− (absence of septate hyphae)

Species	Tests	Responses
123 Candida zeylanoides	6 Trehalose fermentation	− − − −
	18 L−Rhamnose growth	− − −
63 Candida iberica*	19 Sucrose growth	− −
102 Candida savonica*	20 Maltose growth	− −
	21 Trehalose growth	+ + + +
	24 Salicin growth	+ + + +
	27 Lactose growth	− − − −
	28 Raffinose growth	−
	29 Melezitose growth	− −
	33 Erythritol growth	− − − −
	42 D,L−Lactate growth	− −
	44 Citrate growth	+ + + +
	45 Methanol growth	−
	46 Ethanol growth	+ + + +
	49 Nitrate growth	− − − −
	55 Growth sans thiamin	+ +

Species	Tests	Responses
124 Citeromyces matritensis	1 D−Glucose fermentation	+
	11 Raffinose fermentation	+ +
	22 Me α−D−glucoside grth	+ + +
	23 Cellobiose growth	− − − −
	28 Raffinose growth	+
	29 Melezitose growth	− −
	33 Erythritol growth	−
	36 D−Mannitol growth	+ + +
	37 D−Glucitol growth	+
	43 Succinate growth	− − − −
	45 Methanol growth	−
	49 Nitrate growth	+ + + + +

Species	Tests	Responses
125 Cryptococcus albidus	19 Sucrose growth	+ +
	23 Cellobiose growth	+ +
132 Cryptococcus heveanensis*	27 Lactose growth	+ + + +
160 Filobasidium floriforme*	28 Raffinose growth	+ +
(m) 299 Rhodosporidium infirmo−miniatum**	38 myo−Inositol growth	+ + + +
(m) 425 Trichosporon pullulans	49 Nitrate growth	+ + + +
	54 Growth sans biotin	+ + + +
	62 Growth at 37 degrees	− −
	65 Starch formation	+ + + +

distinguished from (m) by 66− (colonies not pink)
68− (cells not splitting)

The identification of particular species

Species	Tests	Responses
126 Cryptococcus ater	18 L–Rhamnose growth	+
	22 Me α–D–glucoside grth	+ + +
16 Bullera alba	26 Melibiose growth	– – –
	27 Lactose growth	+ +
	33 Erythritol growth	–
	34 Ribitol growth	– –
	35 Galactitol growth	– – –
	38 myo–Inositol growth	+
	46 Ethanol growth	– – –
	49 Nitrate growth	– – –
127 Cryptococcus bhutanensis*	15 D–Xylose growth	+
	22 Me α–D–glucoside grth	+
	26 Melibiose growth	–
	31 Starch growth	+ +
	32 Glycerol growth	– –
	33 Erythritol growth	–
	34 Ribitol growth	– –
	38 myo–Inositol growth	– – –
	46 Ethanol growth	– –
	49 Nitrate growth	+ + + +
	65 Starch formation	+ +
128 Cryptococcus cereanus*	18 L–Rhamnose growth	– –
	19 Sucrose growth	– – –
	22 Me α–D–glucoside grth	–
	27 Lactose growth	– – –
	33 Erythritol growth	+ + +
	34 Ribitol growth	+
	38 myo–Inositol growth	+ + + +
	43 Succinate growth	–
	46 Ethanol growth	+
	62 Growth at 37 degrees	+ +
129 Cryptococcus dimennae	18 L–Rhamnose growth	+
	19 Sucrose growth	+
(m) 421 Trichosporon cutaneum	20 Maltose growth	– –
	27 Lactose growth	+
	28 Raffinose growth	+
	29 Melezitose growth	–
	33 Erythritol growth	–
	38 myo–Inositol growth	+ +
	46 Ethanol growth	+ +
	49 Nitrate growth	– –
	65 Starch formation	+

distinguished from (m) by 68– (cells not splitting)

130 Cryptococcus flavus	12 D–Galactose growth	+
	18 L–Rhamnose growth	+ +
16 Bullera alba	22 Me α–D–glucoside grth	+
(m) 355 Sterigmatomyces polyborus*	26 Melibiose growth	+ + +
	32 Glycerol growth	– –
	33 Erythritol growth	+ +
	38 myo–Inositol growth	+
	46 Ethanol growth	– – –
	65 Starch formation	– – –

distinguished from (m) by 67+ (formation of budding cells)

Species	Tests	Responses
131 Cryptococcus gastricus	13 L−Sorbose growth	−
	14 D−Ribose growth	− −
(m) 421 Trichosporon cutaneum	22 Me α−D−glucoside grth	− −
	28 Raffinose growth	− − −
	29 Melezitose growth	+ +
	32 Glycerol growth	−
	34 Ribitol growth	−
	38 myo−Inositol growth	+ + +
distinguished from (m) by 68− (cells not splitting)	39 Gluconolactone growth	+ + +

132 Cryptococcus heveanensis*	1 D−Glucose fermentation	−
	26 Melibiose growth	+ + +
125 Cryptococcus albidus	27 Lactose growth	+ + +
	33 Erythritol growth	− − −
	38 myo−Inositol growth	+ +
	40 2−Ketogluconate growth	+
	42 D,L−Lactate growth	+ +
	49 Nitrate growth	+ + +
	65 Starch formation	+ +

133 Cryptococcus himalayensis*	19 Sucrose growth	− − −
	28 Raffinose growth	−
	32 Glycerol growth	−
	34 Ribitol growth	−
	35 Galactitol growth	+ +
	38 myo−Inositol growth	+ + +
	44 Citrate growth	+ + + +
	46 Ethanol growth	−
	49 Nitrate growth	+ + +

134 Cryptococcus hungaricus	19 Sucrose growth	+ +
	28 Raffinose growth	+
(m) 16 Bullera alba	33 Erythritol growth	− − −
(m) 138 Cryptococcus luteolus	36 D−Mannitol growth	+ +
	38 myo−Inositol growth	+ + +
	42 D,L−Lactate growth	+ + +
	46 Ethanol growth	− − −
	49 Nitrate growth	− − −
	51 Growth sans vitamins	−
distinguished from (m) by 66+ (pink colonies)	62 Growth at 37 degrees	− − −

135 Cryptococcus kuetzingii	1 D−Glucose fermentation	− −
	12 D−Galactose growth	− −
	13 L−Sorbose growth	−
	20 Maltose growth	− −
	27 Lactose growth	−
	28 Raffinose growth	+
	29 Melezitose growth	−
	38 myo−Inositol growth	+ + +
	46 Ethanol growth	+
	49 Nitrate growth	+ +
	54 Growth sans biotin	+

Species	Tests	Responses
136 Cryptococcus lactativorus	1 D−Glucose fermentation	−
	12 D−Galactose growth	− −
	13 L−Sorbose growth	+
	16 L−Arabinose growth	−
	19 Sucrose growth	− −
	27 Lactose growth	− − −
	36 D−Mannitol growth	−
	37 D−Glucitol growth	− −
	38 myo−Inositol growth	+ + +
137 Cryptococcus laurentii	26 Melibiose growth	+
	27 Lactose growth	+
16 Bullera alba	28 Raffinose growth	+
60 Candida humicola	30 Inulin growth	−
138 Cryptococcus luteolus	33 Erythritol growth	+
(m) 421 Trichosporon cutaneum	38 myo−Inositol growth	+
	49 Nitrate growth	−
	65 Starch formation	+
distinguished from (m) by 68− (cells not splitting)		
138 Cryptococcus luteolus	22 Me α−D−glucoside grth	+
	28 Raffinose growth	+ +
16 Bullera alba	29 Melezitose growth	+
137 Cryptococcus laurentii	31 Starch growth	− −
140 Cryptococcus magnus**	35 Galactitol growth	+ +
(m) 60 Candida humicola	49 Nitrate growth	− −
(m) 134 Cryptococcus hungaricus	62 Growth at 37 degrees	− −
(m) 421 Trichosporon cutaneum	65 Starch formation	+ +
distinguished from (m) by 66− (colonies not pink)		
72− (absence of filaments)		
139 Cryptococcus macerans	16 L−Arabinose growth	+
	22 Me α−D−glucoside grth	−
	23 Cellobiose growth	+
	26 Melibiose growth	−
	33 Erythritol growth	+ + +
	38 myo−Inositol growth	+ +
	49 Nitrate growth	+ + +
	54 Growth sans biotin	− − −
	65 Starch formation	+
140 Cryptococcus magnus**	1 D−Glucose fermentation	−
	19 Sucrose growth	+
16 Bullera alba	26 Melibiose growth	+ + +
138 Cryptococcus luteolus	27 Lactose growth	+
(m) 91 Candida podzolica*	29 Melezitose growth	+
(m) 421 Trichosporon cutaneum	33 Erythritol growth	− − −
	37 D−Glucitol growth	+
	38 myo−Inositol growth	+ +
	42 D,L−Lactate growth	− − −
	45 Methanol growth	−
	49 Nitrate growth	− − −
distinguished from (m) by 72− (absence of filaments)		

Species	Tests	Responses
141 Cryptococcus melibiosum	1 D–Glucose fermentation	–
	12 D–Galactose growth	+
	15 D–Xylose growth	– – –
	19 Sucrose growth	–
	26 Melibiose growth	+ + +
	36 D–Mannitol growth	– –
	38 myo–Inositol growth	+ + +
	43 Succinate growth	–
142 Cryptococcus neoformans 38 Candida chiropterorum*	1 D–Glucose fermentation	– –
	12 D–Galactose growth	+ +
	26 Melibiose growth	– – –
	27 Lactose growth	– – –
	29 Melezitose growth	+ +
	30 Inulin growth	– – –
	35 Galactitol growth	+
	38 myo–Inositol growth	+ + +
	40 2–Ketogluconate growth	+
	49 Nitrate growth	–
	62 Growth at 37 degrees	+ + +
143 Cryptococcus skinneri	18 L–Rhamnose growth	+ + +
	19 Sucrose growth	– –
	27 Lactose growth	– – –
	28 Raffinose growth	–
	29 Melezitose growth	–
	33 Erythritol growth	– –
	38 myo–Inositol growth	+ +
	44 Citrate growth	+ +
	49 Nitrate growth	–
	65 Starch formation	+
144 Cryptococcus terreus	1 D–Glucose fermentation	–
	13 L–Sorbose growth	+
	19 Sucrose growth	– –
	28 Raffinose growth	–
	32 Glycerol growth	–
	38 myo–Inositol growth	+ + +
	44 Citrate growth	– – –
	46 Ethanol growth	–
	49 Nitrate growth	+ +
145 Cryptococcus uniguttulatus	13 L–Sorbose growth	–
	15 D–Xylose growth	+ + +
	16 L–Arabinose growth	+
	27 Lactose growth	– – – –
	29 Melezitose growth	+ + +
	33 Erythritol growth	– – – –
	34 Ribitol growth	
	38 myo–Inositol growth	+ + + +
	39 Gluconolactone growth	– – – –
	42 D,L–Lactate growth	–
	49 Nitrate growth	– – –

Species	Tests	Responses
146 Debaryomyces castellii	5 Sucrose fermentation	+
	7 Melibiose fermentation	+ + +
	14 D−Ribose growth	−
	18 L−Rhamnose growth	+
	22 Me α−D−glucoside grth	+
	27 Lactose growth	+ +
	33 Erythritol growth	− −
	35 Galactitol growth	+ +
	38 myo−Inositol growth	−
147 Debaryomyces coudertii	1 D−Glucose fermentation	−
	12 D−Galactose growth	+ + +
	14 D−Ribose growth	+
	19 Sucrose growth	− − − −
	20 Maltose growth	+ +
	23 Cellobiose growth	+
	24 Salicin growth	− −
	27 Lactose growth	−
	33 Erythritol growth	+ + + +
	44 Citrate growth	− − − −
	46 Ethanol growth	+
	60 Growth in 50% glucose	+ + +
148 Debaryomyces hansenii	6 Trehalose fermentation	− − −
	7 Melibiose fermentation	− − −
149 Debaryomyces marama	15 D−Xylose growth	+
151 Debaryomyces nepalensis*	22 Me α−D−glucoside grth	+ + +
155 Debaryomyces vanriji	28 Raffinose growth	+ + +
278 Pichia pseudopolymorpha	29 Melezitose growth	+
	31 Starch growth	+ + +
	34 Ribitol growth	+ +
	38 myo−Inositol growth	− −
	42 D,L−Lactate growth	+ + +
	46 Ethanol growth	+ + +
	49 Nitrate growth	− − −
	55 Growth sans thiamin	+ + +
	60 Growth in 50% glucose	+ + +
	65 Starch formation	− − −
149 Debaryomyces marama	2 Galactose fermentation	− −
	12 D−Galactose growth	+ +
148 Debaryomyces hansenii	17 D−Arabinose growth	− −
151 Debaryomyces nepalensis*	18 L−Rhamnose growth	− − − −
229 Lipomyces tetrasporus*	19 Sucrose growth	+
(m) 421 Trichosporon cutaneum	28 Raffinose growth	+ + + +
	31 Starch growth	+ + +
	33 Erythritol growth	+ + + +
	35 Galactitol growth	− −
	44 Citrate growth	+
	46 Ethanol growth	+ +
	49 Nitrate growth	− − − −
	51 Growth sans vitamins	− − − −
	62 Growth at 37 degrees	− −
	65 Starch formation	− − − −

distinguished from (m) by 68− (cells not splitting)

Species	Tests	Responses
150 Debaryomyces melissophila*	1 D−Glucose fermentation	−
	12 D−Galactose growth	+ + +
	15 D−Xylose growth	− − −
	19 Sucrose growth	+ +
	21 Trehalose growth	− −
	22 Me α−D−glucoside grth	+
	33 Erythritol growth	+ +
	43 Succinate growth	− −
	49 Nitrate growth	−
151 Debaryomyces nepalensis*	3 Maltose fermentation	− − −
148 Debaryomyces hansenii	13 L−Sorbose growth	+ + +
149 Debaryomyces marama	17 D−Arabinose growth	− −
	18 L−Rhamnose growth	− − − −
	19 Sucrose growth	+ +
	24 Salicin growth	+ +
	26 Melibiose growth	+ + + +
	27 Lactose growth	− − − −
	30 Inulin growth	− − −
	33 Erythritol growth	+ + + +
	35 Galactitol growth	− − −
	49 Nitrate growth	− −
	65 Starch formation	− −
152 Debaryomyces phaffii	14 D−Ribose growth	+
155 Debaryomyces vanriji	22 Me α−D−glucoside grth	+ + + +
	30 Inulin growth	+ + + +
	31 Starch growth	+ + + +
	33 Erythritol growth	+
	35 Galactitol growth	+ +
	42 D,L−Lactate growth	− − −
	51 Growth sans vitamins	+
	60 Growth in 50% glucose	+ + + +
	62 Growth at 37 degrees	+
153 Debaryomyces polymorpha**	1 D−Glucose fermentation	+
	11 Raffinose fermentation	+ +
	22 Me α−D−glucoside grth	− − −
	28 Raffinose growth	+
	30 Inulin growth	+
	33 Erythritol growth	+ + + +
	34 Ribitol growth	+
	35 Galactitol growth	+ + +
	46 Ethanol growth	+
154 Debaryomyces tamarii	15 D−Xylose growth	− − −
	19 Sucrose growth	+
	26 Melibiose growth	+ + +
	27 Lactose growth	+ + +
	33 Erythritol growth	−
	36 D−Mannitol growth	+
	37 D−Glucitol growth	−
	38 myo−Inositol growth	−
	49 Nitrate growth	− −

The identification of particular species

Species	Tests	Responses
155 Debaryomyces vanriji	13 L−Sorbose growth	+
	17 D−Arabinose growth	− −
148 Debaryomyces hansenii	22 Me α−D−glucoside grth	+ +
152 Debaryomyces phaffii	26 Melibiose growth	+
229 Lipomyces tetrasporus*	27 Lactose growth	− −
	33 Erythritol growth	+ +
	35 Galactitol growth	+ +
	38 myo−Inositol growth	−
	45 Methanol growth	−
	46 Ethanol growth	+ +
	65 Starch formation	− −

Species	Tests	Responses
156 Debaryomyces yarrowii*	1 D−Glucose fermentation	− − − −
	12 D−Galactose growth	+
	13 L−Sorbose growth	+
	21 Trehalose growth	− − − −
	25 Arbutin growth	+
	27 Lactose growth	−
	28 Raffinose growth	+ +
	29 Melezitose growth	+ + + +
	33 Erythritol growth	−
	38 myo−Inositol growth	−
	46 Ethanol growth	+
	49 Nitrate growth	−
	51 Growth sans vitamins	+ +
	60 Growth in 50% glucose	+ + + +

Species	Tests	Responses
157 Dekkera bruxellensis	5 Sucrose fermentation	+
	10 Melezitose fermentn	+ +
(a) 14 Brettanomyces lambicus	13 L−Sorbose growth	−
	15 D−Xylose growth	−
	23 Cellobiose growth	− −
	24 Salicin growth	−
	29 Melezitose growth	+
	36 D−Mannitol growth	− − −
	44 Citrate growth	−
	63 0.01% cyclohex. growth	+ + +

distinguished from (a) by 75+ (formation of ascospores)

Species	Tests	Responses
158 Dekkera intermedia	2 Galactose fermentation	+ +
	7 Melibiose fermentation	−
(a) 13 Brettanomyces custersii	9 Cellobiose fermentn	+ + + +
	12 D−Galactose growth	+ +
	15 D−Xylose growth	− − −
	16 L−Arabinose growth	−
	22 Me α−D−glucoside grth	+ + +
	26 Melibiose growth	− − −
	27 Lactose growth	− − − −
	29 Melezitose growth	+
	36 D−Mannitol growth	− − − −

distinguished from (a) by 75+ (formation of ascospores)

Species	Tests	Responses
159 Filobasidium capsuligenum**	13 L–Sorbose growth	−
	20 Maltose growth	+ +
	27 Lactose growth	− − −
	29 Melezitose growth	− − −
	32 Glycerol growth	+
	33 Erythritol growth	− − −
	38 myo–Inositol growth	+ + +
	41 5–Ketogluconate growth	−
	62 Growth at 37 degrees	−
	65 Starch formation	+ +
160 Filobasidium floriforme*	14 D–Ribose growth	+
	22 Me α–D–glucoside grth	+ + +
125 Cryptococcus albidus	26 Melibiose growth	− −
	32 Glycerol growth	+
	33 Erythritol growth	− −
	34 Ribitol growth	+
	38 myo–Inositol growth	+ + +
	42 D,L–Lactate growth	− −
	49 Nitrate growth	+ + +
161 Geotrichum capitatum**	1 D–Glucose fermentation	− − − −
	15 D–Xylose growth	− − − −
(m) 117 Candida valida	19 Sucrose growth	− − − −
(m) 266 Pichia membranaefaciens	21 Trehalose growth	− − − − −
	23 Cellobiose growth	−
	36 D–Mannitol growth	− − − −
	37 D–Glucitol growth	−
	44 Citrate growth	− − − − −
	51 Growth sans vitamins	− − − −
	54 Growth sans biotin	+ + + +
	55 Growth sans thiamin	−
	56 Grth sans pyridoxine	+
	62 Growth at 37 degrees	+ + + + +

distinguished from (m) by 68+ (formation of splitting cells)

Species	Tests	Responses
162 Geotrichum fermentans**	5 Sucrose fermentation	−
	12 D–Galactose growth	+
	13 L–Sorbose growth	+ +
	19 Sucrose growth	− − − −
	21 Trehalose growth	− − − −
	23 Cellobiose growth	+ + + +
	24 Salicin growth	+
	27 Lactose growth	− − − − −
	33 Erythritol growth	− − −
	34 Ribitol growth	+
	42 D,L–Lactate growth	+ +
	46 Ethanol growth	+
	49 Nitrate growth	− −
	51 Growth sans vitamins	+ + + + +
163 Geotrichum penicillatum**	12 D–Galactose growth	+ + +
	13 L–Sorbose growth	+ +
	15 D–Xylose growth	+ + +
	19 Sucrose growth	− − −
	27 Lactose growth	− − −
	36 D–Mannitol growth	− − −
	49 Nitrate growth	−
	51 Growth sans vitamins	+ + +
	60 Growth in 50% glucose	− −
	61 Growth in 60% glucose	−

Species	Tests	Responses
164 Guilliermondella selenospora**	16 L−Arabinose growth	+ + +
	19 Sucrose growth	− −
	21 Trehalose growth	−
	27 Lactose growth	− − −
	33 Erythritol growth	−
	34 Ribitol growth	+ +
	36 D−Mannitol growth	− −
	37 D−Glucitol growth	− −
	38 myo−Inositol growth	− − −
	46 Ethanol growth	+ +
165 Hanseniaspora guilliermondii*	1 D−Glucose fermentation	+ +
	12 D−Galactose growth	− −
	19 Sucrose growth	− −
	23 Cellobiose growth	+ +
	24 Salicin growth	+ +
	32 Glycerol growth	− − − −
	36 D−Mannitol growth	− − − −
	46 Ethanol growth	− −
	62 Growth at 37 degrees	+ + + +
166 Hanseniaspora occidentalis*	5 Sucrose fermentation	+ + +
	12 D−Galactose growth	− −
	23 Cellobiose growth	+ +
	24 Salicin growth	+
	28 Raffinose growth	− −
	36 D−Mannitol growth	− − −
	46 Ethanol growth	− − −
	52 Growth sans inositol	−
167 Hanseniaspora osmophila	1 D−Glucose fermentation	+ + +
170 Hanseniaspora vineae*	12 D−Galactose growth	− −
	20 Maltose growth	+ + + +
	23 Cellobiose growth	+ + +
	24 Salicin growth	+
	28 Raffinose growth	− −
	29 Melezitose growth	−
	32 Glycerol growth	− − −
	36 D−Mannitol growth	− − − −
	46 Ethanol growth	− − −
168 Hanseniaspora uvarum	1 D−Glucose fermentation	+ +
	12 D−Galactose growth	−
	19 Sucrose growth	− −
	23 Cellobiose growth	+ + +
	32 Glycerol growth	− −
	36 D−Mannitol growth	− − −
	40 2−Ketogluconate growth	+ + +
	43 Succinate growth	−
	46 Ethanol growth	−
	62 Growth at 37 degrees	− − −

Species	Tests	Responses
169 Hanseniaspora valbyensis	12 D−Galactose growth	− −
	15 D−Xylose growth	−
	19 Sucrose growth	−
	20 Maltose growth	− − −
	23 Cellobiose growth	+ + +
	32 Glycerol growth	− −
	36 D−Mannitol growth	− − −
	40 2−Ketogluconate growth	− − −
	43 Succinate growth	− −
	44 Citrate growth	−
	46 Ethanol growth	− −
170 Hanseniaspora vineae*	1 D−Glucose fermentation	+
167 Hanseniaspora osmophila	12 D−Galactose growth	− −
	15 D−Xylose growth	−
	20 Maltose growth	+ + +
	22 Me α−D−glucoside grth	−
	23 Cellobiose growth	+ + +
	29 Melezitose growth	− −
	32 Glycerol growth	−
	36 D−Mannitol growth	− −
	37 D−Glucitol growth	−
	43 Succinate growth	−
	46 Ethanol growth	− −
171 Hansenula anomala	5 Sucrose fermentation	+
178 Hansenula ciferrii	11 Raffinose fermentation	+
	17 D−Arabinose growth	−
	26 Melibiose growth	− − −
	30 Inulin growth	−
	31 Starch growth	+ +
	33 Erythritol growth	+ +
	38 myo−Inositol growth	−
	49 Nitrate growth	+ + +
	51 Growth sans vitamins	+ + +
172 Hansenula beckii**	13 L−Sorbose growth	−
174 Hansenula bimundalis	16 L−Arabinose growth	− −
(m) 80 Candida melinii	18 L−Rhamnose growth	+ + +
(m) 176 Hansenula canadensis	21 Trehalose growth	+ +
(m) 200 Hansenula wingei	28 Raffinose growth	− − −
(m) 304 Rhodotorula glutinis	29 Melezitose growth	+ +
	33 Erythritol growth	− −
	44 Citrate growth	+
	45 Methanol growth	−
	49 Nitrate growth	+ + +
	62 Growth at 37 degrees	+

distinguished from (m) by 73− (absence of pseudohyphae)
74+ (presence of septate hyphae)

Species	Tests	Responses
173 Hansenula beijerinckii	11 Raffinose fermentation	+ +
	12 D−Galactose growth	− −
	15 D−Xylose growth	+ +
	17 D−Arabinose growth	−
	20 Maltose growth	+
	29 Melezitose growth	+ +
	30 Inulin growth	+
	44 Citrate growth	− − −
	49 Nitrate growth	+ + +

174 Hansenula bimundalis

    172 Hansenula beckii**
(m) 80 Candida melinii
(m) 176 Hansenula canadensis
(m) 200 Hansenula wingei
(m) 304 Rhodotorula glutinis

13 L−Sorbose growth	−
18 L−Rhamnose growth	+ + +
19 Sucrose growth	+
28 Raffinose growth	− − −
29 Melezitose growth	+ +
32 Glycerol growth	+
34 Ribitol growth	− − −
46 Ethanol growth	+
49 Nitrate growth	+ + +

distinguished from (m) by 73− (absence of pseudohyphae)
                            74+ (presence of septate hyphae)

175 Hansenula californica

1 D−Glucose fermentation	+ +
12 D−Galactose growth	− −
13 L−Sorbose growth	+ +
15 D−Xylose growth	+
19 Sucrose growth	+ + +
22 Me α−D−glucoside grth	+
23 Cellobiose growth	+
28 Raffinose growth	−
29 Melezitose growth	−
34 Ribitol growth	− − −
36 D−Mannitol growth	+
38 myo−Inositol growth	−
42 D,L−Lactate growth	+
49 Nitrate growth	+ +

176 Hansenula canadensis

(m) 172 Hansenula beckii**
(m) 174 Hansenula bimundalis
(m) 200 Hansenula wingei
(m) 304 Rhodotorula glutinis
(a) 80 Candida melinii

13 L−Sorbose growth	−
18 L−Rhamnose growth	+ + +
19 Sucrose growth	+
28 Raffinose growth	− − −
29 Melezitose growth	+ +
32 Glycerol growth	+
34 Ribitol growth	− − −
46 Ethanol growth	+
49 Nitrate growth	+ + +

distinguished from (m) by 66− (colonies not pink)
                            74− (absence of septate hyphae)
distinguished from (a) by 75+ (formation of ascospores)

177 Hansenula capsulata

(a) 392 Torulopsis molischiana

15 D−Xylose growth	+
19 Sucrose growth	− −
20 Maltose growth	+ +
31 Starch growth	+
34 Ribitol growth	+ +
45 Methanol growth	+ + +

distinguished from (a) by 75+ (formation of ascospores)

178 Hansenula ciferrii

    171 Hansenula anomala

1 D−Glucose fermentation	+ +
5 Sucrose fermentation	+
12 D−Galactose growth	+ +
26 Melibiose growth	− − − −
30 Inulin growth	−
31 Starch growth	+ + +
33 Erythritol growth	+ +
38 myo−Inositol growth	−
44 Citrate growth	+
49 Nitrate growth	+ + + +
51 Growth sans vitamins	+ + + +

Species	Tests	Responses
179 Hansenula dimennae	13 L−Sorbose growth	+ + +
	14 D−Ribose growth	−
	15 D−Xylose growth	+
	19 Sucrose growth	− − −
	21 Trehalose growth	−
	23 Cellobiose growth	+ +
	32 Glycerol growth	+
	34 Ribitol growth	− −
	46 Ethanol growth	+
	49 Nitrate growth	+ + +
180 Hansenula dryadoides*	1 D−Glucose fermentation	−
	15 D−Xylose growth	− − −
	19 Sucrose growth	−
	21 Trehalose growth	− − −
	23 Cellobiose growth	+
	28 Raffinose growth	−
	36 D−Mannitol growth	+
	38 myo−Inositol growth	−
	42 D,L−Lactate growth	+ +
	44 Citrate growth	+ +
	49 Nitrate growth	+ + +
181 Hansenula fabianii	1 D−Glucose fermentation	+
	5 Sucrose fermentation	+
	31 Starch growth	+ + +
	33 Erythritol growth	− − −
	34 Ribitol growth	− −
	42 D,L−Lactate growth	+
	49 Nitrate growth	+ + +
	51 Growth sans vitamins	−
	62 Growth at 37 degrees	+
182 Hansenula glucozyma	19 Sucrose growth	−
	20 Maltose growth	− −
	23 Cellobiose growth	+ +
	24 Salicin growth	+
	33 Erythritol growth	+ + +
	44 Citrate growth	+
	45 Methanol growth	+ + +
	55 Growth sans thiamin	− − −
183 Hansenula henricii	12 D−Galactose growth	− −
	19 Sucrose growth	− − −
	23 Cellobiose growth	+ +
	24 Salicin growth	+
	33 Erythritol growth	+ +
	45 Methanol growth	+ +
	49 Nitrate growth	+ + +
	55 Growth sans thiamin	+ + +

The identification of particular species

Species	Tests	Responses
184 Hansenula holstii	1 D−Glucose fermentation	+ +
	12 D−Galactose growth	+ + +
	17 D−Arabinose growth	+
	18 L−Rhamnose growth	+
	22 Me α−D−glucoside grth	+
	26 Melibiose growth	−
	28 Raffinose growth	− − −
	31 Starch growth	+ + +
	32 Glycerol growth	+
	42 D,L−Lactate growth	− − − −
	45 Methanol growth	−
	49 Nitrate growth	+ + + +
	51 Growth sans vitamins	−
185 Hansenula jadinii	11 Raffinose fermentation	+
	12 D−Galactose growth	−
	18 L−Rhamnose growth	−
	21 Trehalose growth	+
	29 Melezitose growth	+ +
	30 Inulin growth	+ +
	33 Erythritol growth	− −
	37 D−Glucitol growth	+ + +
	44 Citrate growth	+ + +
	49 Nitrate growth	+ +
	51 Growth sans vitamins	+ +
	62 Growth at 37 degrees	+
186 Hansenula lynferdii*	13 L−Sorbose growth	−
	15 D−Xylose growth	−
	16 L−Arabinose growth	−
	30 Inulin growth	+ +
	33 Erythritol growth	+ +
187 Hansenula minuta	15 D−Xylose growth	+ + +
	16 L−Arabinose growth	− −
	18 L−Rhamnose growth	−
	23 Cellobiose growth	+
	33 Erythritol growth	− − −
	42 D,L−Lactate growth	−
	45 Methanol growth	+ + +
	49 Nitrate growth	+ +
188 Hansenula mrakii	1 D−Glucose fermentation	+ + +
(a) 400 Torulopsis norvegica	12 D−Galactose growth	−
	13 L−Sorbose growth	− − − −
	15 D−Xylose growth	+ + +
	19 Sucrose growth	− − −
	21 Trehalose growth	− −
	23 Cellobiose growth	+
	28 Raffinose growth	−
	34 Ribitol growth	− − −
	42 D,L−Lactate growth	+ +
	49 Nitrate growth	+ + + +

distinguished from (a) by 75+ (formation of ascospores)

Species	Tests	Responses
189 Hansenula muscicola*	12 D−Galactose growth	+
	16 L−Arabinose growth	+
	19 Sucrose growth	− −
	22 Me α−D−glucoside grth	+ +
	27 Lactose growth	−
	38 myo−Inositol growth	−
	42 D,L−Lactate growth	+
	49 Nitrate growth	+
190 Hansenula nonfermentans	15 D−Xylose growth	− − −
	18 L−Rhamnose growth	−
	23 Cellobiose growth	+
	33 Erythritol growth	− − −
	34 Ribitol growth	+ +
	45 Methanol growth	+ + +
	49 Nitrate growth	+ +
191 Hansenula ofunaensis*	33 Erythritol growth	− −
	35 Galactitol growth	+
	38 myo−Inositol growth	+ +
	42 D,L−Lactate growth	−
	45 Methanol growth	+ + +
192 Hansenula petersonii	11 Raffinose fermentation	+ +
	12 D−Galactose growth	− − −
	14 D−Ribose growth	−
	16 L−Arabinose growth	−
	18 L−Rhamnose growth	+ + + +
	22 Me α−D−glucoside grth	+ + + +
	30 Inulin growth	+ +
	44 Citrate growth	+ + + +
	49 Nitrate growth	+ +
193 Hansenula philodendra*	17 D−Arabinose growth	−
	19 Sucrose growth	− − −
	23 Cellobiose growth	− −
	24 Salicin growth	−
	33 Erythritol growth	+ +
	43 Succinate growth	+
	44 Citrate growth	+ +
	45 Methanol growth	+ +
	49 Nitrate growth	+ +
194 Hansenula polymorpha	16 L−Arabinose growth	− −
	19 Sucrose growth	+ + + +
	28 Raffinose growth	− −
	33 Erythritol growth	+ +
	34 Ribitol growth	+ +
	45 Methanol growth	+ + + +

Species	Tests	Responses
195 Hansenula saturnus	5 Sucrose fermentation	+ +
	12 D-Galactose growth	−
	15 D-Xylose growth	+
	21 Trehalose growth	− − −
	23 Cellobiose growth	+
	28 Raffinose growth	+
	29 Melezitose growth	− − −
	42 D,L-Lactate growth	+ +
	49 Nitrate growth	+ +
196 Hansenula silvicola	1 D-Glucose fermentation	+ + +
	12 D-Galactose growth	+
	13 L-Sorbose growth	+ +
	16 L-Arabinose growth	+ +
	17 D-Arabinose growth	− − −
	19 Sucrose growth	+
	22 Me α-D-glucoside grth	+ + +
	26 Melibiose growth	−
	28 Raffinose growth	− −
	34 Ribitol growth	+
	45 Methanol growth	−
	49 Nitrate growth	+ + +
197 Hansenula subpelliculosa	1 D-Glucose fermentation	+
	5 Sucrose fermentation	+ +
	18 L-Rhamnose growth	−
	24 Salicin growth	+
	25 Arbutin growth	+
	28 Raffinose growth	+ + +
	33 Erythritol growth	+ + + +
	38 myo-Inositol growth	−
	42 D,L-Lactate growth	+
	49 Nitrate growth	+ + + +
	51 Growth sans vitamins	− − − −
198 Hansenula sydowiorum*	1 D-Glucose fermentation	+
	2 Galactose fermentation	+ +
	22 Me α-D-glucoside grth	+
	24 Salicin growth	+
	26 Melibiose growth	+ + + +
	32 Glycerol growth	+
	33 Erythritol growth	+ + +
	38 myo-Inositol growth	−
	49 Nitrate growth	+ + + +
	65 Starch formation	−
199 Hansenula wickerhamii	1 D-Glucose fermentation	− −
	17 D-Arabinose growth	+
	18 L-Rhamnose growth	+
	19 Sucrose growth	−
	23 Cellobiose growth	− −
	24 Salicin growth	−
	33 Erythritol growth	+ +
	38 myo-Inositol growth	−
	44 Citrate growth	−
	45 Methanol growth	+ + +

Species	Tests	Responses
200 Hansenula wingei	1 D−Glucose fermentation	− −
	12 D−Galactose growth	−
(m) 80 Candida melinii	14 D−Ribose growth	−
(m) 172 Hansenula beckii**	18 L−Rhamnose growth	+ + +
(m) 174 Hansenula bimundalis	19 Sucrose growth	+ +
(m) 176 Hansenula canadensis	28 Raffinose growth	− −
(m) 304 Rhodotorula glutinis	29 Melezitose growth	+
	33 Erythritol growth	− −
	42 D,L−Lactate growth	+
distinguished from (m) by 73+ (presence of pseudohyphae)	49 Nitrate growth	+ + +
74+ (presence of septate hyphae)		

201 Hormoascus platypodis**	12 D−Galactose growth	− −
	13 L−Sorbose growth	− −
	17 D−Arabinose growth	−
	18 L−Rhamnose growth	+ +
	19 Sucrose growth	+
	28 Raffinose growth	−
	33 Erythritol growth	+ + +
	38 myo−Inositol growth	− −
	42 D,L−Lactate growth	+
	45 Methanol growth	− −
	49 Nitrate growth	+ + +

202 Hyphopichia burtonii**	12 D−Galactose growth	+ +
	15 D−Xylose growth	+
	18 L−Rhamnose growth	− − − −
	22 Me α−D−glucoside grth	+
	26 Melibiose growth	− − − −
	27 Lactose growth	− − − −
	28 Raffinose growth	+ +
	30 Inulin growth	− −
	31 Starch growth	+
	33 Erythritol growth	+ + +
	35 Galactitol growth	−
	42 D,L−Lactate growth	− − − −
	43 Succinate growth	+
	44 Citrate growth	+ + +
	46 Ethanol growth	+ + + +
	51 Growth sans vitamins	+ + + +
	60 Growth in 50% glucose	+ +

203 Kluyveromyces aestuarii	1 D−Glucose fermentation	+
	2 Galactose fermentation	− −
	6 Trehalose fermentation	−
	11 Raffinose fermentation	+
	13 L−Sorbose growth	+
	26 Melibiose growth	−
	27 Lactose growth	+ + +
	33 Erythritol growth	−
	34 Ribitol growth	− −
	36 D−Mannitol growth	+
	44 Citrate growth	−
	46 Ethanol growth	+
	57 Growth sans niacin	−
	63 0.01% cyclohex. growth	−

Species	Tests	Responses
204 Kluyveromyces africanus	2 Galactose fermentation	+ +
	12 D−Galactose growth	+ +
312 Saccharomyces cerevisiae	15 D−Xylose growth	− −
	36 D−Mannitol growth	− − − −
	43 Succinate growth	− −
	46 Ethanol growth	− − − −
	52 Growth sans inositol	− − − −
	53 Grth sans pantothenate	+ +
	54 Growth sans biotin	+ +
205 Kluyveromyces blattae*	2 Galactose fermentation	+ +
	14 D−Ribose growth	+ + +
313 Saccharomyces dairensis	15 D−Xylose growth	−
	19 Sucrose growth	−
	23 Cellobiose growth	−
	33 Erythritol growth	− −
	36 D−Mannitol growth	−
	37 D−Glucitol growth	− −
	38 myo−Inositol growth	−
	43 Succinate growth	−
	47 Ethylamine growth	−
	62 Growth at 37 degrees	+
	63 0.01% cyclohex. growth	−
	64 0.1% cyclohex. growth	− −
206 Kluyveromyces bulgaricus	2 Galactose fermentation	+
	20 Maltose growth	−
210 Kluyveromyces lactis	22 Me α−D−glucoside grth	−
212 Kluyveromyces marxianus	26 Melibiose growth	−
214 Kluyveromyces phaseolosporus	28 Raffinose growth	+
	30 Inulin growth	+
	37 D−Glucitol growth	+ +
	42 D,L−Lactate growth	+
	49 Nitrate growth	− −
	53 Grth sans pantothenate	+ +
	54 Growth sans biotin	− −
	63 0.01% cyclohex. growth	+ +
207 Kluyveromyces delphensis	1 D−Glucose fermentation	+ +
	12 D−Galactose growth	− −
266 Pichia membranaefaciens	19 Sucrose growth	− − −
	22 Me α−D−glucoside grth	−
	32 Glycerol growth	+ +
	36 D−Mannitol growth	− −
	37 D−Glucitol growth	− −
	42 D,L−Lactate growth	− −
	43 Succinate growth	− −
	46 Ethanol growth	+ +
	49 Nitrate growth	− −
	56 Grth sans pyridoxine	− − − −
	57 Growth sans niacin	+ + + +
	62 Growth at 37 degrees	+ + + +
	63 0.01% cyclohex. growth	+ + + +

The identification of particular species

Species	Tests	Responses
208 Kluyveromyces dobzhanskii	4 Me α−D−glucoside fermn	+ + +
	11 Raffinose fermentation	+ +
	12 D−Galactose growth	+ + +
	18 L−Rhamnose growth	−
	23 Cellobiose growth	+ +
	26 Melibiose growth	− − −
	28 Raffinose growth	+
	33 Erythritol growth	−
	36 D−Mannitol growth	+ + +
	57 Growth sans niacin	− − −
	63 0.01% cyclohex. growth	+
209 Kluyveromyces drosophilarum	2 Galactose fermentation	+
210 Kluyveromyces lactis	11 Raffinose fermentation	+ +
	13 L−Sorbose growth	+ + +
	22 Me α−D−glucoside grth	+ +
	26 Melibiose growth	− −
	28 Raffinose growth	+
	29 Melezitose growth	+
	33 Erythritol growth	−
	38 myo−Inositol growth	−
	42 D,L−Lactate growth	+ +
	57 Growth sans niacin	− −
	60 Growth in 50% glucose	−
	62 Growth at 37 degrees	+ + +
	63 0.01% cyclohex. growth	+
210 Kluyveromyces lactis	2 Galactose fermentation	+ + + +
	4 Me α−D−glucoside fermn	− − − −
206 Kluyveromyces bulgaricus	9 Cellobiose fermentn	− − − −
209 Kluyveromyces drosophilarum	18 L−Rhamnose growth	−
212 Kluyveromyces marxianus	23 Cellobiose growth	+ + +
214 Kluyveromyces phaseolosporus	24 Salicin growth	+
218 Kluyveromyces wickerhamii	26 Melibiose growth	− − − −
(a) 76 Candida maltosa*	32 Glycerol growth	+ + + +
	33 Erythritol growth	−
	37 D−Glucitol growth	+
	44 Citrate growth	− − − −
	46 Ethanol growth	+ + +
	49 Nitrate growth	− − −
	53 Grth sans pantothenate	+ + + +
	54 Growth sans biotin	− − − −
	57 Growth sans niacin	− − − −

distinguished from (a) by 75+ (formation of ascospores)

211 Kluyveromyces lodderi	2 Galactose fermentation	+ + + +
	5 Sucrose fermentation	+ +
215 Kluyveromyces polysporus	7 Melibiose fermentation	−
(a) 83 Candida milleri*	22 Me α−D−glucoside grth	−
	23 Cellobiose growth	− −
	24 Salicin growth	−
	26 Melibiose growth	− − −
	27 Lactose growth	− −
	28 Raffinose growth	+ +
	32 Glycerol growth	+ + + +
	36 D−Mannitol growth	− − −
	37 D−Glucitol growth	− −
	46 Ethanol growth	+ + + +
	62 Growth at 37 degrees	− −
	63 0.01% cyclohex. growth	+ + + +

distinguished from (a) by 75+ (formation of ascospores)

Species	Tests	Responses
212 Kluyveromyces marxianus	1 D−Glucose fermentation	+
	15 D−Xylose growth	+ + +
206 Kluyveromyces bulgaricus	16 L−Arabinose growth	+ +
210 Kluyveromyces lactis	21 Trehalose growth	−
(a) 93 Candida pseudotropicalis	27 Lactose growth	+ +
	29 Melezitose growth	− −
	30 Inulin growth	+ +
	32 Glycerol growth	+
	54 Growth sans biotin	− − −

distinguished from (a) by 75+ (formation of ascospores)

Species	Tests	Responses
213 Kluyveromyces phaffii	2 Galactose fermentation	+ + + +
	5 Sucrose fermentation	− −
	28 Raffinose growth	− −
	32 Glycerol growth	+ + + +
	36 D−Mannitol growth	− − − −
	37 D−Glucitol growth	−
	39 Gluconolactone growth	+ + + +
	46 Ethanol growth	− − − −
	49 Nitrate growth	−
	53 Grth sans pantothenate	− −
	54 Growth sans biotin	− − − −
	57 Growth sans niacin	+ + + +
	60 Growth in 50% glucose	− −

Species	Tests	Responses
214 Kluyveromyces phaseolosporus	7 Melibiose fermentation	−
	11 Raffinose fermentation	+
206 Kluyveromyces bulgaricus	12 D−Galactose growth	+ + +
210 Kluyveromyces lactis	13 L−Sorbose growth	+
	15 D−Xylose growth	+ + +
	16 L−Arabinose growth	− −
	20 Maltose growth	−
	22 Me α−D−glucoside grth	− −
	26 Melibiose growth	− − −
	27 Lactose growth	− − −
	28 Raffinose growth	+ + +
	29 Melezitose growth	− −
	42 D,L−Lactate growth	+
	57 Growth sans niacin	− − − −
	62 Growth at 37 degrees	+

Species	Tests	Responses
215 Kluyveromyces polysporus	2 Galactose fermentation	+ + +
	11 Raffinose fermentation	+ + +
211 Kluyveromyces lodderi	23 Cellobiose growth	− − −
(a) 83 Candida milleri*	26 Melibiose growth	− −
(a) 380 Torulopsis holmii	32 Glycerol growth	+ + +
	36 D−Mannitol growth	− − −
	39 Gluconolactone growth	+ + +
	42 D,L−Lactate growth	− −
	60 Growth in 50% glucose	− −
	61 Growth in 60% glucose	−
	62 Growth at 37 degrees	−

distinguished from (a) by 75+ (formation of ascospores)

Species	Tests	Responses
216 Kluyveromyces thermotolerans**	7 Melibiose fermentation	−
	11 Raffinose fermentation	+ + + +
	12 D−Galactose growth	+
	22 Me α−D−glucoside grth	+
	23 Cellobiose growth	− − − −
	26 Melibiose growth	−
	29 Melezitose growth	+ + +
	42 D,L−Lactate growth	− − −
	47 Ethylamine growth	+ + + +
	57 Growth sans niacin	− − − −
	60 Growth in 50% glucose	+

Species	Tests	Responses
217 Kluyveromyces waltii*	1 D−Glucose fermentation	+
	11 Raffinose fermentation	+
(m) 357 Torulaspora delbrueckii**	12 D−Galactose growth	− − −
(m) 361 Torulopsis apicola	13 L−Sorbose growth	+ + + +
	21 Trehalose growth	− − − −
	26 Melibiose growth	−
	27 Lactose growth	− −
	28 Raffinose growth	+ + +
	29 Melezitose growth	− − − −
	32 Glycerol growth	− − − −
	35 Galactitol growth	−
	43 Succinate growth	−

distinguished from (m) by 72+ (presence of filaments)

Species	Tests	Responses
218 Kluyveromyces wickerhamii	1 D−Glucose fermentation	+ + +
210 Kluyveromyces lactis	15 D−Xylose growth	+ + +
	16 L−Arabinose growth	−
	27 Lactose growth	+ + + +
	28 Raffinose growth	− − − −
	33 Erythritol growth	−
	34 Ribitol growth	−
	36 D−Mannitol growth	− −
	49 Nitrate growth	−
	57 Growth sans niacin	−

Species	Tests	Responses
219 Leucosporidium antarcticum	1 D−Glucose fermentation	− − −
	5 Sucrose fermentation	−
	23 Cellobiose growth	− − −
	24 Salicin growth	−
	37 D−Glucitol growth	− − − −
	43 Succinate growth	− − − −
	49 Nitrate growth	+ + + +
	51 Growth sans vitamins	+ + +
	55 Growth sans thiamin	+

Species	Tests	Responses
220 Leucosporidium frigidum	12 D−Galactose growth	+
	20 Maltose growth	−
	26 Melibiose growth	−
	27 Lactose growth	+ +
	29 Melezitose growth	− −
	32 Glycerol growth	−
	34 Ribitol growth	+ +
	44 Citrate growth	+ +
	46 Ethanol growth	+
	49 Nitrate growth	+ + +
	65 Starch formation	+ + +

Species	Tests	Responses
221 Leucosporidium gelidum	18 L–Rhamnose growth	+ +
	26 Melibiose growth	+ + +
(m) 296 Rhodosporidium bisporidiis*	27 Lactose growth	– –
	29 Melezitose growth	+ +
	31 Starch growth	+
	32 Glycerol growth	– –
	33 Erythritol growth	–
	35 Galactitol growth	–
	49 Nitrate growth	+ +
	54 Growth sans biotin	– –

distinguished from (m) by 69+ (apical budding)

222 Leucosporidium nivalis	20 Maltose growth	–
	26 Melibiose growth	+ + +
	27 Lactose growth	–
	29 Melezitose growth	– –
	33 Erythritol growth	–
	36 D–Mannitol growth	+
	49 Nitrate growth	+ +
	65 Starch formation	+ + +

223 Leucosporidium scottii	15 D–Xylose growth	+
	22 Me α–D–glucoside grth	+ +
	29 Melezitose growth	+
	33 Erythritol growth	–
	35 Galactitol growth	+ + +
	38 myo–Inositol growth	– – –
	43 Succinate growth	– – –
	45 Methanol growth	– –
	49 Nitrate growth	+ + +

224 Leucosporidium stokesii	15 D–Xylose growth	+
	18 L–Rhamnose growth	+ +
(m) 299 Rhodosporidium infirmo–miniatum**	22 Me α–D–glucoside grth	–
	26 Melibiose growth	– – –
	27 Lactose growth	– –
	28 Raffinose growth	+ +
	29 Melezitose growth	+ +
	33 Erythritol growth	–
	35 Galactitol growth	–
	43 Succinate growth	– – –
	49 Nitrate growth	+
	54 Growth sans biotin	–
	65 Starch formation	+

distinguished from (m) by 69+ (apical budding)

225 Lipomyces anomalus*	15 D–Xylose growth	+ +
	16 L–Arabinose growth	+ +
	21 Trehalose growth	– –
	23 Cellobiose growth	+
	32 Glycerol growth	– –
	33 Erythritol growth	–
	36 D–Mannitol growth	– –
	38 myo–Inositol growth	– – –
	43 Succinate growth	– –
	46 Ethanol growth	– – –

Species	Tests	Responses
226 Lipomyces kononenkoae	1 D–Glucose fermentation	– – – –
	16 L–Arabinose growth	– – –
	26 Melibiose growth	+ + + +
	27 Lactose growth	– – –
	30 Inulin growth	+
	33 Erythritol growth	– – – –
	34 Ribitol growth	– – –
	36 D–Mannitol growth	+
	38 myo–Inositol growth	–
	43 Succinate growth	– –
227 Lipomyces lipofer	14 D–Ribose growth	– – –
228 Lipomyces starkeyi	26 Melibiose growth	+
229 Lipomyces tetrasporus*	32 Glycerol growth	– –
(m) 421 Trichosporon cutaneum	33 Erythritol growth	+ + +
	34 Ribitol growth	– –
	38 myo–Inositol growth	– – –
	43 Succinate growth	–
	49 Nitrate growth	– – –

distinguished from (m) by 68– (cells not splitting)

Species	Tests	Responses
228 Lipomyces starkeyi	26 Melibiose growth	+
	30 Inulin growth	+ +
227 Lipomyces lipofer	33 Erythritol growth	+
229 Lipomyces tetrasporus*	42 D,L–Lactate growth	–
	65 Starch formation	+ +
229 Lipomyces tetrasporus*	1 D–Glucose fermentation	– –
	12 D–Galactose growth	+
149 Debaryomyces marama	13 L–Sorbose growth	+
155 Debaryomyces vanriji	30 Inulin growth	+ + +
227 Lipomyces lipofer	31 Starch growth	+
228 Lipomyces starkeyi	33 Erythritol growth	+ + +
	46 Ethanol growth	+ +
	49 Nitrate growth	– –
	60 Growth in 50% glucose	– – –
230 Lodderomyces elongisporus	12 D–Galactose growth	+ +
	16 L–Arabinose growth	– – – –
	22 Me α–D–glucoside grth	+ +
	24 Salicin growth	– –
	25 Arbutin growth	– –
	27 Lactose growth	– – – –
	28 Raffinose growth	– – – –
	29 Melezitose growth	+ +
	31 Starch growth	– – – –
	36 D–Mannitol growth	+ +
	38 myo–Inositol growth	– –
	49 Nitrate growth	– – – –
	62 Growth at 37 degrees	+ + + +
	63 0.01% cyclohex. growth	+ + + +

The identification of particular species

Species	Tests	Responses
231 Metschnikowia bicuspidata	1 D–Glucose fermentation	+ + +
	2 Galactose fermentation	– – –
234 Metschnikowia pulcherrima	5 Sucrose fermentation	– – –
235 Metschnikowia reukaufii	12 D–Galactose growth	+ + +
236 Metschnikowia zobellii	13 L–Sorbose growth	+ +
(a) 99 Candida sake	16 L–Arabinose growth	– –
	19 Sucrose growth	+ + +
	23 Cellobiose growth	+ + +
	27 Lactose growth	–
	33 Erythritol growth	–
	36 D–Mannitol growth	+
	38 myo–Inositol growth	–
	42 D,L–Lactate growth	–
	44 Citrate growth	– – –
	49 Nitrate growth	–
	62 Growth at 37 degrees	– – –

distinguished from (a) by 75+ (formation of ascospores)

Species	Tests	Responses
232 Metschnikowia krissii	1 D–Glucose fermentation	– – –
	12 D–Galactose growth	–
	15 D–Xylose growth	– – – –
	16 L–Arabinose growth	– – – –
	22 Me α–D–glucoside grth	+ + +
	23 Cellobiose growth	+
	29 Melezitose growth	+
	31 Starch growth	– – – –
	33 Erythritol growth	– – – –
	37 D–Glucitol growth	– – – –
	39 Gluconolactone growth	–
	49 Nitrate growth	– – –

Species	Tests	Responses
233 Metschnikowia lunata*	2 Galactose fermentation	– – –
	5 Sucrose fermentation	– – –
(m) 234 Metschnikowia pulcherrima	12 D–Galactose growth	+ + +
	16 L–Arabinose growth	– – –
	18 L–Rhamnose growth	– – –
	19 Sucrose growth	+ +
	23 Cellobiose growth	+ +
	24 Salicin growth	+
	26 Melibiose growth	– – –
	27 Lactose growth	– – –
	28 Raffinose growth	– – –
	29 Melezitose growth	+
	31 Starch growth	– – –
	49 Nitrate growth	– – –
	62 Growth at 37 degrees	+ + +

distinguished from (m) by 70– (cells not spherical, oval or cylindrical)

Species	Tests	Responses
234 Metschnikowia pulcherrima	1 D–Glucose fermentation	+ + +
	5 Sucrose fermentation	– – –
231 Metschnikowia bicuspidata	9 Cellobiose fermentn	–
235 Metschnikowia reukaufii	12 D–Galactose growth	+ + +
(m) 233 Metschnikowia lunata*	16 L–Arabinose growth	– – –
(a) 88 Candida oleophila*	18 L–Rhamnose growth	–
(a) 99 Candida sake	19 Sucrose growth	+ + +
	23 Cellobiose growth	+ +
	24 Salicin growth	+
	26 Melibiose growth	– – –
	31 Starch growth	– – –
	33 Erythritol growth	– – –
	34 Ribitol growth	+ + +
	36 D–Mannitol growth	+ +
	37 D–Glucitol growth	+
	55 Growth sans thiamin	+ + +
	57 Growth sans niacin	+ + +

distinguished from (m) by 70+ (presence of cells spherical, oval or cylindrical)
distinguished from (a) by 75+ (formation of ascospores)

Species	Tests	Responses
235 Metschnikowia reukaufii	1 D–Glucose fermentation	+ +
	5 Sucrose fermentation	– –
231 Metschnikowia bicuspidata	16 L–Arabinose growth	– –
234 Metschnikowia pulcherrima	18 L–Rhamnose growth	– –
(a) 99 Candida sake	19 Sucrose growth	+
	23 Cellobiose growth	+ + +
	27 Lactose growth	–
	28 Raffinose growth	–
	29 Melezitose growth	+ + +
	31 Starch growth	–
	32 Glycerol growth	+ +
	33 Erythritol growth	–
	34 Ribitol growth	+
	36 D–Mannitol growth	+
	37 D–Glucitol growth	+
	44 Citrate growth	– – –
	49 Nitrate growth	–
	55 Growth sans thiamin	+ + +
	57 Growth sans niacin	+ + +
	62 Growth at 37 degrees	– – –

distinguished from (a) by 75+ (formation of ascospores)

Species	Tests	Responses
236 Metschnikowia zobellii	12 D–Galactose growth	+ + +
	13 L–Sorbose growth	+
231 Metschnikowia bicuspidata	14 D–Ribose growth	–
	16 L–Arabinose growth	–
	17 D–Arabinose growth	–
	19 Sucrose growth	+
	22 Me α–D–glucoside grth	+ +
	27 Lactose growth	– – – –
	28 Raffinose growth	– – – –
	29 Melezitose growth	+ + +
	32 Glycerol growth	+ + +
	33 Erythritol growth	– –
	34 Ribitol growth	+ + +
	39 Gluconolactone growth	+
	44 Citrate growth	– – –
	49 Nitrate growth	–
	55 Growth sans thiamin	– – – –
	62 Growth at 37 degrees	– – – –

The identification of particular species

Species	Tests	Responses
237 Nadsonia commutata*	1 D−Glucose fermentation	− − −
	15 D−Xylose growth	− −
(m) 7 Arthroascus javanensis**	16 L−Arabinose growth	−
(m) 298 Rhodosporidium dacryoidum*	20 Maltose growth	+ + +
(m) 325 Saccharomycopsis vini**	22 Me α−D−glucoside grth	− −
	23 Cellobiose growth	−
	27 Lactose growth	−
	29 Melezitose growth	− −
	32 Glycerol growth	+
	33 Erythritol growth	− − −
	34 Ribitol growth	−
	42 D,L−Lactate growth	− − −
	49 Nitrate growth	− − −
	62 Growth at 37 degrees	− − −

distinguished from (m) by 69+ (apical budding)
74− (absence of septate hyphae)

Species	Tests	Responses
238 Nadsonia elongata	1 D−Glucose fermentation	+ +
	12 D−Galactose growth	− −
(m) 266 Pichia membranaefaciens	13 L−Sorbose growth	+ + +
	15 D−Xylose growth	− −
	19 Sucrose growth	− −
	21 Trehalose growth	−
	36 D−Mannitol growth	− −
	42 D,L−Lactate growth	− − −
	51 Growth sans vitamins	+ + +
	60 Growth in 50% glucose	− − −

distinguished from (m) by 69+ (apical budding)

Species	Tests	Responses
239 Nadsonia fulvescens	5 Sucrose fermentation	+
	12 D−Galactose growth	+
	13 L−Sorbose growth	+
	15 D−Xylose growth	−
	21 Trehalose growth	− − −
	22 Me α−D−glucoside grth	+ +
	25 Arbutin growth	−
	28 Raffinose growth	−
	29 Melezitose growth	−
	44 Citrate growth	+ + +
	56 Grth sans pyridoxine	−

Species	Tests	Responses
240 Nematospora coryli	2 Galactose fermentation	−
	21 Trehalose growth	+ + + +
(m) 312 Saccharomyces cerevisiae	22 Me α−D−glucoside grth	− − −
	24 Salicin growth	− −
	26 Melibiose growth	− −
	27 Lactose growth	−
	29 Melezitose growth	−
	34 Ribitol growth	− − − −
	36 D−Mannitol growth	− −
	42 D,L−Lactate growth	− − − −
	43 Succinate growth	+ + + +
	52 Growth sans inositol	− − − −
	62 Growth at 37 degrees	+ + + +

distinguished from (m) by 74+ (presence of septate hyphae)

Species	Tests	Responses
241 Pachysolen tannophilus	1 D–Glucose fermentation	+ +
	12 D–Galactose growth	+
	15 D–Xylose growth	+
	18 L–Rhamnose growth	– – –
	19 Sucrose growth	– – –
	21 Trehalose growth	–
	32 Glycerol growth	+
	34 Ribitol growth	+ +
	45 Methanol growth	– –
	49 Nitrate growth	+ + +
242 Pachytichospora transvaalensis	2 Galactose fermentation	+ +
	12 D–Galactose growth	+ +
312 Saccharomyces cerevisiae	14 D–Ribose growth	–
	15 D–Xylose growth	– –
	16 L–Arabinose growth	–
	19 Sucrose growth	– – – –
	32 Glycerol growth	– – –
	36 D–Mannitol growth	– – –
	49 Nitrate growth	– – – –
	52 Growth sans inositol	– – – –
	53 Grth sans pantothenate	– –
	54 Growth sans biotin	– –
243 Phaffia rhodozyma*	12 D–Galactose growth	– –
	13 L–Sorbose growth	– –
	14 D–Ribose growth	–
	16 L–Arabinose growth	+ +
	27 Lactose growth	– – –
	28 Raffinose growth	+
	29 Melezitose growth	+
	33 Erythritol growth	– –
	34 Ribitol growth	– –
	35 Galactitol growth	–
	38 myo–Inositol growth	– – –
	49 Nitrate growth	–
	65 Starch formation	+ + +
244 Pichia abadieae*	18 L–Rhamnose growth	+
	19 Sucrose growth	– – –
(m) 421 Trichosporon cutaneum	26 Melibiose growth	+ +
	27 Lactose growth	+
	38 myo–Inositol growth	+
	42 D,L–Lactate growth	+ +
	45 Methanol growth	–

distinguished from (m) by 68– (cells not splitting)

Species	Tests	Responses
245 Pichia acaciae	1 D–Glucose fermentation	+
	12 D–Galactose growth	+
	13 L–Sorbose growth	+ +
	15 D–Xylose growth	– –
	19 Sucrose growth	– – –
	22 Me α–D–glucoside grth	+ + +
	33 Erythritol growth	+ +
	44 Citrate growth	+

The identification of particular species

Species	Tests	Responses
246 Pichia ambrosiae*	13 L–Sorbose growth	–
	14 D–Ribose growth	+ + +
	15 D–Xylose growth	– –
	19 Sucrose growth	+ + +
	22 Me α–D–glucoside grth	+ +
	28 Raffinose growth	–
	33 Erythritol growth	+ + +
	49 Nitrate growth	–
	54 Growth sans biotin	–
	55 Growth sans thiamin	– – –
247 Pichia angophorae	3 Maltose fermentation	+
	5 Sucrose fermentation	+ +
	12 D–Galactose growth	– –
	13 L–Sorbose growth	– –
	15 D–Xylose growth	+
	22 Me α–D–glucoside grth	+ +
	28 Raffinose growth	–
	32 Glycerol growth	– – –
	34 Ribitol growth	+
	42 D,L–Lactate growth	+ +
	44 Citrate growth	+ +
	55 Growth sans thiamin	–
	62 Growth at 37 degrees	–
248 Pichia besseyi*	12 D–Galactose growth	– –
	19 Sucrose growth	– –
	21 Trehalose growth	– – – –
	32 Glycerol growth	– – – –
	36 D–Mannitol growth	+ +
	37 D–Glucitol growth	+ +
	40 2–Ketogluconate growth	+ + + +
	42 D,L–Lactate growth	+ + + +
	56 Grth sans pyridoxine	– – – –
249 Pichia bovis	1 D–Glucose fermentation	+ +
	12 D–Galactose growth	– – –
	16 L–Arabinose growth	+ + +
	18 L–Rhamnose growth	– – –
	19 Sucrose growth	+
	22 Me α–D–glucoside grth	+
	27 Lactose growth	–
	28 Raffinose growth	– – –
	33 Erythritol growth	– –
	34 Ribitol growth	–
	42 D,L–Lactate growth	+ + +
	44 Citrate growth	+
	49 Nitrate growth	–
250 Pichia castillae*	19 Sucrose growth	– – –
	26 Melibiose growth	+ +
	27 Lactose growth	– – –
	28 Raffinose growth	+
	33 Erythritol growth	+ +
	44 Citrate growth	+

Species	Tests	Responses
251 Pichia chambardii	1 D–Glucose fermentation	−
	12 D–Galactose growth	+ + +
	15 D–Xylose growth	− −
	19 Sucrose growth	− − −
	21 Trehalose growth	− − −
	23 Cellobiose growth	+
	27 Lactose growth	−
	33 Erythritol growth	− − −
	36 D–Mannitol growth	− −
	38 myo–Inositol growth	−
	44 Citrate growth	+ +

Species	Tests	Responses
252 Pichia delftensis	12 D–Galactose growth	−
	15 D–Xylose growth	− − −
(a) 121 Candida vini	19 Sucrose growth	− − −
(a) 385 Torulopsis karawaiewi*	21 Trehalose growth	− −
	32 Glycerol growth	− − −
	34 Ribitol growth	+ + +
	39 Gluconolactone growth	+ + +
	42 D,L–Lactate growth	− −
	44 Citrate growth	− − −
	49 Nitrate growth	−
	60 Growth in 50% glucose	− −
	61 Growth in 60% glucose	−

distinguished from (a) by 75+ (formation of ascospores)

Species	Tests	Responses
253 Pichia dispora	1 D–Glucose fermentation	+ +
	15 D–Xylose growth	−
	19 Sucrose growth	− − −
	21 Trehalose growth	+ + +
	23 Cellobiose growth	− −
	32 Glycerol growth	− − −
	34 Ribitol growth	+ + +
	49 Nitrate growth	−
	56 Grth sans pyridoxine	−
	60 Growth in 50% glucose	−
	61 Growth in 60% glucose	−

Species	Tests	Responses
254 Pichia etchellsii	16 L–Arabinose growth	+ + +
	18 L–Rhamnose growth	− − −
	19 Sucrose growth	+ +
	22 Me α–D–glucoside grth	+
	23 Cellobiose growth	+ +
	24 Salicin growth	+
	27 Lactose growth	− − −
	28 Raffinose growth	− − −
	31 Starch growth	− − −
	33 Erythritol growth	− − −
	34 Ribitol growth	+ + +
	36 D–Mannitol growth	+ +
	37 D–Glucitol growth	+
	44 Citrate growth	+ + +
	49 Nitrate growth	− − −

Species	Tests	Responses
255 Pichia farinosa	1 D−Glucose fermentation	+
	19 Sucrose growth	− −
	20 Maltose growth	−
	22 Me α−D−glucoside grth	− −
	27 Lactose growth	− −
	28 Raffinose growth	−
	33 Erythritol growth	+ + +
	42 D,L−Lactate growth	− −
	44 Citrate growth	+ + +
	45 Methanol growth	− − −
	62 Growth at 37 degrees	+ + +
256 Pichia fermentans	1 D−Glucose fermentation	+
	12 D−Galactose growth	−
266 Pichia membranaefaciens	13 L−Sorbose growth	−
(a) 73 Candida lambica	15 D−Xylose growth	+ + +
	19 Sucrose growth	− − −
	27 Lactose growth	− −
	36 D−Mannitol growth	− − −
	37 D−Glucitol growth	−
	38 myo−Inositol growth	−
	44 Citrate growth	+ + +
	56 Grth sans pyridoxine	+ + +

distinguished from (a) by 75+ (formation of ascospores)

Species	Tests	Responses
257 Pichia fluxuum	12 D−Galactose growth	− − −
	13 L−Sorbose growth	−
(a) 121 Candida vini	15 D−Xylose growth	− −
(a) 312 Saccharomyces cerevisiae	19 Sucrose growth	− − −
(a) 357 Torulaspora delbrueckii**	21 Trehalose growth	− − −
(a) 385 Torulopsis karawaiewi*	27 Lactose growth	−
	32 Glycerol growth	− − −
	36 D−Mannitol growth	+
	37 D−Glucitol growth	+ + +
	39 Gluconolactone growth	− − −
	43 Succinate growth	+ + +
	44 Citrate growth	−
	54 Growth sans biotin	−
	55 Growth sans thiamin	+ +

distinguished from (a) by 75+ (formation of ascospores)
                77+ (ascospores hat− or saturn−shaped)

Species	Tests	Responses
258 Pichia guilliermondii	5 Sucrose fermentation	+ +
	16 L−Arabinose growth	+ +
	17 D−Arabinose growth	+ +
	22 Me α−D−glucoside grth	+ +
	26 Melibiose growth	+ + + +
	27 Lactose growth	− −
	31 Starch growth	− − − −
	33 Erythritol growth	− − − +
	36 D−Mannitol growth	+
	38 myo−Inositol growth	−
	49 Nitrate growth	− −
	54 Growth sans biotin	− −

Species	Tests	Responses
259 Pichia haplophila	19 Sucrose growth	− − −
	21 Trehalose growth	− − −
	23 Cellobiose growth	−
	27 Lactose growth	− − −
	28 Raffinose growth	−
	33 Erythritol growth	+ + +
	35 Galactitol growth	+ + + +
	43 Succinate growth	−
260 Pichia heimii*	1 D−Glucose fermentation	+
	2 Galactose fermentation	+ +
	13 L−Sorbose growth	− − −
	17 D−Arabinose growth	−
	27 Lactose growth	−
	28 Raffinose growth	+
	33 Erythritol growth	+
	35 Galactitol growth	+ + +
	42 D,L−Lactate growth	− − −
	51 Growth sans vitamins	+
261 Pichia humboldtii*	1 D−Glucose fermentation	− −
	12 D−Galactose growth	+ +
	15 D−Xylose growth	− −
	19 Sucrose growth	−
	23 Cellobiose growth	−
	36 D−Mannitol growth	−
	37 D−Glucitol growth	−
	49 Nitrate growth	− −
	51 Growth sans vitamins	+ +
262 Pichia kluyveri	1 D−Glucose fermentation	+ +
266 Pichia membranaefaciens	12 D−Galactose growth	− −
	15 D−Xylose growth	−
	19 Sucrose growth	−
	21 Trehalose growth	−
	23 Cellobiose growth	− − −
	36 D−Mannitol growth	−
	37 D−Glucitol growth	−
	42 D,L−Lactate growth	+ + +
	44 Citrate growth	+ + +
	54 Growth sans biotin	− − −
263 Pichia kudriavzevii	1 D−Glucose fermentation	+ +
266 Pichia membranaefaciens	12 D−Galactose growth	− −
286 Pichia scutulata*	15 D−Xylose growth	− − − −
312 Saccharomyces cerevisiae	19 Sucrose growth	− − −
(m) 357 Torulaspora delbrueckii**	28 Raffinose growth	−
(a) 117 Candida valida	32 Glycerol growth	+ + + +
	36 D−Mannitol growth	−
	37 D−Glucitol growth	− − −
	42 D,L−Lactate growth	+ + + +
	51 Growth sans vitamins	+ + + +

distinguished from (m) by 72+ (presence of filaments)
distinguished from (a) by 75+ (formation of ascospores)

Species	Tests	Responses
264 Pichia lindnerii*	12 D−Galactose growth	− −
	13 L−Sorbose growth	−
(a) 391 Torulopsis methanolovescens*	16 L−Arabinose growth	−
	18 L−Rhamnose growth	+ + +
	33 Erythritol growth	− − − −
	34 Ribitol growth	+ +
	44 Citrate growth	+ +
	45 Methanol growth	+ + + +
	49 Nitrate growth	−

distinguished from (a) by 75+ (formation of ascospores)

Species	Tests	Responses
265 Pichia media	12 D−Galactose growth	+
	13 L−Sorbose growth	+
	14 D−Ribose growth	− − −
	19 Sucrose growth	− − −
	20 Maltose growth	+ +
	22 Me α−D−glucoside grth	−
	27 Lactose growth	− − −
	31 Starch growth	+
	33 Erythritol growth	+ + +
	38 myo−Inositol growth	−
	46 Ethanol growth	+

Species	Tests	Responses
266 Pichia membranaefaciens	12 D−Galactose growth	− − −
	19 Sucrose growth	− −
207 Kluyveromyces delphensis	20 Maltose growth	− − −
256 Pichia fermentans	21 Trehalose growth	− − −
262 Pichia kluyveri	23 Cellobiose growth	− −
263 Pichia kudriavzevii	24 Salicin growth	− −
286 Pichia scutulata*	25 Arbutin growth	−
290 Pichia terricola	27 Lactose growth	− − −
312 Saccharomyces cerevisiae	28 Raffinose growth	−
317 Saccharomyces telluris	33 Erythritol growth	− − −
357 Torulaspora delbrueckii**	36 D−Mannitol growth	−
433 Zygosaccharomyces bisporus**	37 D−Glucitol growth	− − −
439 Zygosaccharomyces rouxii**	38 myo−Inositol growth	− − −
(m) 7 Arthroascus javanensis**	46 Ethanol growth	+ + +
(m) 161 Geotrichum capitatum**	49 Nitrate growth	− − −
(m) 238 Nadsonia elongata		
(a) 39 Candida citrea*		
(a) 73 Candida lambica		
(a) 97 Candida rugopelliculosa*		
(a) 107 Candida sorboxylosa*		
(a) 117 Candida valida		
(a) 382 Torulopsis inconspicua		
(a) 385 Torulopsis karawaiewi*		

distinguished from (m) by 67+ (formation of budding cells)
                69− (budding not apical)
distinguished from (a) by 75+ (formation of ascospores)

Species	Tests	Responses
267 Pichia methanolica*	1 D−Glucose fermentation	+ +
	12 D−Galactose growth	+ + +
	14 D−Ribose growth	+
	18 L−Rhamnose growth	−
	19 Sucrose growth	− −
	33 Erythritol growth	+
	34 Ribitol growth	+
	45 Methanol growth	+ + +
	49 Nitrate growth	−
	62 Growth at 37 degrees	+

Species	Tests	Responses
268 Pichia mucosa*	1 D–Glucose fermentation	+ + +
	5 Sucrose fermentation	– – –
	12 D–Galactose growth	– – – –
	13 L–Sorbose growth	+ + + +
	19 Sucrose growth	+
	23 Cellobiose growth	+ + +
	28 Raffinose growth	–
	29 Melezitose growth	+ + + +
	32 Glycerol growth	+ + +
	33 Erythritol growth	–
	34 Ribitol growth	– – – –
	37 D–Glucitol growth	+ + +
	38 myo–Inositol growth	– –
	42 D,L–Lactate growth	–
	44 Citrate growth	– –
	56 Grth sans pyridoxine	–
269 Pichia naganishii*	1 D–Glucose fermentation	+
	14 D–Ribose growth	–
	28 Raffinose growth	+ +
	33 Erythritol growth	+
	34 Ribitol growth	+
	38 myo–Inositol growth	–
	45 Methanol growth	+ + +
	49 Nitrate growth	–
270 Pichia nakazawae*	1 D–Glucose fermentation	+ +
	12 D–Galactose growth	+
(m) 67 Candida insectorum*	15 D–Xylose growth	+ +
(m) 86 Candida naeodendra*	19 Sucrose growth	+
(m) 110 Candida tenuis	26 Melibiose growth	–
	27 Lactose growth	–
	28 Raffinose growth	– – –
	29 Melezitose growth	+
	31 Starch growth	+ + +
	33 Erythritol growth	+ + +
	35 Galactitol growth	– –
	42 D,L–Lactate growth	–
	49 Nitrate growth	– – –
	51 Growth sans vitamins	– –
	55 Growth sans thiamin	–

distinguished from (m) by 71– (cells spherical, oval or cylindrical only)
74+ (presence of septate hyphae)

Species	Tests	Responses
271 Pichia norvegensis*	12 D–Galactose growth	– – –
	19 Sucrose growth	– – –
(a) 87 Candida norvegensis	23 Cellobiose growth	+ + +
	27 Lactose growth	– –
	33 Erythritol growth	–
	36 D–Mannitol growth	– – –
	42 D,L–Lactate growth	+ + +
	48 D–Glucosamine growth	+
	49 Nitrate growth	–
	56 Grth sans pyridoxine	–

distinguished from (a) by 75+ (formation of ascospores)

Species	Tests	Responses
272 Pichia ohmeri	5 Sucrose fermentation	+
	11 Raffinose fermentation	+
	12 D–Galactose growth	+ +
	13 L–Sorbose growth	+
	15 D–Xylose growth	− −
	16 L–Arabinose growth	−
	22 Me α–D–glucoside grth	+ + +
	23 Cellobiose growth	+ + +
	28 Raffinose growth	+
	29 Melezitose growth	− − −
	33 Erythritol growth	− −
	38 myo–Inositol growth	−
	44 Citrate growth	+ + +
	49 Nitrate growth	−
273 Pichia onychis	1 D–Glucose fermentation	+ +
	5 Sucrose fermentation	+
	12 D–Galactose growth	− − − −
	16 L–Arabinose growth	+ + + +
	18 L–Rhamnose growth	− − −
	28 Raffinose growth	+ + + +
	33 Erythritol growth	− −
	42 D,L–Lactate growth	+
	45 Methanol growth	−
	49 Nitrate growth	−
	56 Grth sans pyridoxine	− − −
	62 Growth at 37 degrees	+
274 Pichia pastoris	1 D–Glucose fermentation	+
	12 D–Galactose growth	−
	14 D–Ribose growth	−
	18 L–Rhamnose growth	+ + +
	34 Ribitol growth	− − −
	38 myo–Inositol growth	−
	44 Citrate growth	− −
	45 Methanol growth	+ + + +
275 Pichia philogaea*	1 D–Glucose fermentation	+ +
(a) 34 Candida butyri*	12 D–Galactose growth	+
	16 L–Arabinose growth	+
	18 L–Rhamnose growth	− − −
	19 Sucrose growth	+ +
	26 Melibiose growth	−
	27 Lactose growth	−
	28 Raffinose growth	− − −
	29 Melezitose growth	+
	31 Starch growth	− − −
	33 Erythritol growth	+ + +
	38 myo–Inositol growth	−
	40 2–Ketogluconate growth	+ + +
	42 D,L–Lactate growth	− −
	60 Growth in 50% glucose	+

distinguished from (a) by 75+ (formation of ascospores)

Species	Tests	Responses
276 Pichia pijperi	1 D−Glucose fermentation	+ + +
	12 D−Galactose growth	− − − −
	13 L−Sorbose growth	+ + + +
	21 Trehalose growth	− − −
	23 Cellobiose growth	+ + +
	27 Lactose growth	−
	33 Erythritol growth	
	34 Ribitol growth	−
	36 D−Mannitol growth	+
	42 D,L−Lactate growth	+ + + +
	44 Citrate growth	+ +
	49 Nitrate growth	− − − −
277 Pichia pinus	12 D−Galactose growth	− − −
	19 Sucrose growth	− −
	23 Cellobiose growth	+ +
	24 Salicin growth	+
	28 Raffinose growth	−
	33 Erythritol growth	+ + +
	45 Methanol growth	+ + +
	49 Nitrate growth	− − −
278 Pichia pseudopolymorpha	18 L−Rhamnose growth	+ +
148 Debaryomyces hansenii	22 Me α−D−glucoside grth	+
	26 Melibiose growth	+
	27 Lactose growth	+ + +
	33 Erythritol growth	+ + +
	35 Galactitol growth	+ + +
	42 D,L−Lactate growth	+ +
	46 Ethanol growth	+ + +
	60 Growth in 50% glucose	+ + +
	65 Starch formation	− − −
279 Pichia quercuum	15 D−Xylose growth	− − −
	19 Sucrose growth	− −
	21 Trehalose growth	− − −
	22 Me α−D−glucoside grth	−
	23 Cellobiose growth	+ + +
	27 Lactose growth	− −
	33 Erythritol growth	− − −
	36 D−Mannitol growth	+
	37 D−Glucitol growth	+ +
	38 myo−Inositol growth	−
	40 2−Ketogluconate growth	−
	44 Citrate growth	+ + +
	49 Nitrate growth	−
	55 Growth sans thiamin	− −
280 Pichia rabaulensis*	1 D−Glucose fermentation	+ +
	11 Raffinose fermentation	+
	12 D−Galactose growth	− − − −
	14 D−Ribose growth	−
	16 L−Arabinose growth	+ + + +
	18 L−Rhamnose growth	+ + + +
	22 Me α−D−glucoside grth	+ +
	28 Raffinose growth	+ + +
	38 myo−Inositol growth	−
	49 Nitrate growth	−
	56 Grth sans pyridoxine	−

Species	Tests	Responses
281 Pichia rhodanensis	1 D−Glucose fermentation	+
	14 D−Ribose growth	− −
293 Pichia veronae*	16 L−Arabinose growth	− −
	18 L−Rhamnose growth	+ + +
	19 Sucrose growth	+ +
	28 Raffinose growth	−
	32 Glycerol growth	+
	33 Erythritol growth	−
	34 Ribitol growth	− − −
	49 Nitrate growth	− − −
	56 Grth sans pyridoxine	− − −
282 Pichia saitoi	1 D−Glucose fermentation	+ +
	12 D−Galactose growth	− −
312 Saccharomyces cerevisiae	15 D−Xylose growth	− − −
	19 Sucrose growth	− −
	23 Cellobiose growth	−
	32 Glycerol growth	− − −
	36 D−Mannitol growth	+ + +
	37 D−Glucitol growth	+
	38 myo−Inositol growth	−
	40 2−Ketogluconate growth	− − − −
	42 D,L−Lactate growth	+ + + +
	49 Nitrate growth	− −
	55 Growth sans thiamin	−
	56 Grth sans pyridoxine	− − −
283 Pichia salictaria	1 D−Glucose fermentation	−
	18 L−Rhamnose growth	+ + +
	19 Sucrose growth	− −
	21 Trehalose growth	− − −
	27 Lactose growth	− − −
	29 Melezitose growth	−
	34 Ribitol growth	− −
	38 myo−Inositol growth	−
	44 Citrate growth	+ +
	49 Nitrate growth	− − −
284 Pichia sargentensis*	1 D−Glucose fermentation	+ +
	18 L−Rhamnose growth	+ + +
	19 Sucrose growth	−
	21 Trehalose growth	− −
	27 Lactose growth	−
	32 Glycerol growth	+
	33 Erythritol growth	−
	34 Ribitol growth	− −
	44 Citrate growth	− −
	45 Methanol growth	−
	49 Nitrate growth	− − −
285 Pichia scolyti	13 L−Sorbose growth	− − −
	16 L−Arabinose growth	+
(m) 421 Trichosporon cutaneum	22 Me α−D−glucoside grth	+
	26 Melibiose growth	+ +
	27 Lactose growth	+
	28 Raffinose growth	+
	33 Erythritol growth	+ +
	43 Succinate growth	− − −
	54 Growth sans biotin	−
	65 Starch formation	− −

distinguished from (m) by 68− (cells not splitting)

The identification of particular species

Species	Tests	Responses
286 Pichia scutulata*	12 D–Galactose growth	– – –
	15 D–Xylose growth	– – –
263 Pichia kudriavzevii	19 Sucrose growth	– –
266 Pichia membranaefaciens	21 Trehalose growth	–
(a) 117 Candida valida	32 Glycerol growth	+ + +
	36 D–Mannitol growth	–
	37 D–Glucitol growth	– –
	42 D,L–Lactate growth	+ + +
	44 Citrate growth	–
	47 Ethylamine growth	+ + +
	51 Growth sans vitamins	+ + +
	60 Growth in 50% glucose	– –
	61 Growth in 60% glucose	–

distinguished from (a) by 75+ (formation of ascospores)

Species	Tests	Responses
287 Pichia spartinae*	1 D–Glucose fermentation	+ +
	12 D–Galactose growth	– – – – –
	15 D–Xylose growth	– – – –
	16 L–Arabinose growth	–
	22 Me α–D–glucoside grth	+ +
	29 Melezitose growth	+ + +
	31 Starch growth	– – – –
	33 Erythritol growth	– – – – –
	34 Ribitol growth	+ + + +
	43 Succinate growth	+
	44 Citrate growth	+ + + + +
	46 Ethanol growth	+
	49 Nitrate growth	– – –

Species	Tests	Responses
288 Pichia stipitis	3 Maltose fermentation	+ + + +
	12 D–Galactose growth	+
	13 L–Sorbose growth	+
	18 L–Rhamnose growth	+ +
	27 Lactose growth	+ + +
	28 Raffinose growth	– – – –
	33 Erythritol growth	+ +
	42 D,L–Lactate growth	+ + +
	46 Ethanol growth	+
	49 Nitrate growth	–

Species	Tests	Responses
289 Pichia strasburgensis	12 D–Galactose growth	+ + +
	13 L–Sorbose growth	–
	17 D–Arabinose growth	– –
	18 L–Rhamnose growth	+ + +
	19 Sucrose growth	+
	26 Melibiose growth	–
	27 Lactose growth	–
	28 Raffinose growth	+ +
	33 Erythritol growth	– – –
	38 myo–Inositol growth	–
	46 Ethanol growth	+
	56 Grth sans pyridoxine	– – –

Species	Tests	Responses
290 Pichia terricola	12 D−Galactose growth	−
	13 L−Sorbose growth	−
266 Pichia membranaefaciens	15 D−Xylose growth	− − −
	19 Sucrose growth	− −
	21 Trehalose growth	−
	23 Cellobiose growth	−
	33 Erythritol growth	− − −
	36 D−Mannitol growth	− − −
	44 Citrate growth	+ + +
	51 Growth sans vitamins	− −
	55 Growth sans thiamin	−
	56 Grth sans pyridoxine	+ + +
	62 Growth at 37 degrees	+ + +
291 Pichia toletana	5 Sucrose fermentation	−
	12 D−Galactose growth	− −
	15 D−Xylose growth	+ +
	16 L−Arabinose growth	−
	19 Sucrose growth	+ +
	27 Lactose growth	− −
	34 Ribitol growth	− −
	44 Citrate growth	+ +
	49 Nitrate growth	− −
	54 Growth sans biotin	− −
	55 Growth sans thiamin	−
	56 Grth sans pyridoxine	+ +
292 Pichia trehalophila	1 D−Glucose fermentation	+ +
	17 D−Arabinose growth	+
	19 Sucrose growth	−
	23 Cellobiose growth	− − −
	24 Salicin growth	−
	33 Erythritol growth	+ +
	34 Ribitol growth	+
	43 Succinate growth	+ +
	45 Methanol growth	+ + + +
	49 Nitrate growth	− − −
293 Pichia veronae*	18 L−Rhamnose growth	+ +
	19 Sucrose growth	+
281 Pichia rhodanensis	22 Me α−D−glucoside grth	+
	28 Raffinose growth	− −
	32 Glycerol growth	+
	34 Ribitol growth	− −
	49 Nitrate growth	− −
	56 Grth sans pyridoxine	− −
294 Pichia vini	1 D−Glucose fermentation	− −
	13 L−Sorbose growth	+ +
	17 D−Arabinose growth	−
	19 Sucrose growth	+
	22 Me α−D−glucoside grth	+
	27 Lactose growth	− −
	31 Starch growth	+ +
	32 Glycerol growth	+
	33 Erythritol growth	− −
	34 Ribitol growth	+
	35 Galactitol growth	− −
	49 Nitrate growth	−
	55 Growth sans thiamin	− −

Species	Tests	Responses
295 Pichia wickerhamii	1 D−Glucose fermentation	+ + +
	12 D−Galactose growth	−
	13 L−Sorbose growth	− − − −
	14 D−Ribose growth	−
	16 L−Arabinose growth	− − −
	18 L−Rhamnose growth	+ + + + +
	19 Sucrose growth	+
	27 Lactose growth	−
	28 Raffinose growth	− −
	32 Glycerol growth	+ + +
	33 Erythritol growth	− − − −
	34 Ribitol growth	+ + + + +
	42 D,L−Lactate growth	+
	45 Methanol growth	− − −
	49 Nitrate growth	− − − −
	56 Grth sans pyridoxine	−
	62 Growth at 37 degrees	+

Species	Tests	Responses
296 Rhodosporidium bisporidiis*	1 D−Glucose fermentation	−
	12 D−Galactose growth	+
(m) 221 Leucosporidium gelidum	18 L−Rhamnose growth	+ +
	22 Me α−D−glucoside grth	− −
	26 Melibiose growth	+ + + +
	27 Lactose growth	− − − −
	29 Melezitose growth	+ + +
	33 Erythritol growth	− −
	38 myo−Inositol growth	+ + +
	44 Citrate growth	− −
	46 Ethanol growth	+ +

distinguished from (m) by 66+ (pink colonies)

Species	Tests	Responses
297 Rhodosporidium capitatum*	1 D−Glucose fermentation	−
	17 D−Arabinose growth	+
299 Rhodosporidium infirmo−miniatum**	22 Me α−D−glucoside grth	− − −
	26 Melibiose growth	− − −
	32 Glycerol growth	+
	33 Erythritol growth	− − −
	35 Galactitol growth	+ +
	38 myo−Inositol growth	+ +
	46 Ethanol growth	− −
	49 Nitrate growth	+
	54 Growth sans biotin	− − −

Species	Tests	Responses
298 Rhodosporidium dacryoidum*	1 D−Glucose fermentation	− − −
	12 D−Galactose growth	+ + +
(m) 12 Brettanomyces custersianus	15 D−Xylose growth	− − −
(m) 63 Candida iberica*	21 Trehalose growth	+ + +
(m) 237 Nadsonia commutata*	22 Me α−D−glucoside grth	− −
(m) 309 Rhodotorula pallida	24 Salicin growth	− −
(m) 327 Schizoblastosporion starkeyi−henricii	28 Raffinose growth	−
	33 Erythritol growth	−
	34 Ribitol growth	− −
	49 Nitrate growth	− − −
	54 Growth sans biotin	+ + +
	62 Growth at 37 degrees	− − −

distinguished from (m) by 66+ (pink colonies)
72+ (presence of filaments)

The identification of particular species

Species	Tests	Responses
299 Rhodosporidium infirmo−miniatum**	1 D−Glucose fermentation	− −
	18 L−Rhamnose growth	+ +
297 Rhodosporidium capitatum*	19 Sucrose growth	+ +
(m) 125 Cryptococcus albidus	22 Me α−D−glucoside grth	− − − −
(m) 224 Leucosporidium stokesii	26 Melibiose growth	− − − −
	28 Raffinose growth	+ +
	29 Melezitose growth	+ +
	33 Erythritol growth	− − − −
	38 myo−Inositol growth	+ + + +
	49 Nitrate growth	+ + + +

distinguished from (m) by 66+ (pink colonies)

Species	Tests	Responses
300 Rhodosporidium malvinellum*	1 D−Glucose fermentation	− − − −
	14 D−Ribose growth	−
	19 Sucrose growth	+ + +
	24 Salicin growth	+
	27 Lactose growth	−
	28 Raffinose growth	+ + +
	29 Melezitose growth	− − − −
	32 Glycerol growth	+ + +
	38 myo−Inositol growth	− − − −
	49 Nitrate growth	+ + + +
	51 Growth sans vitamins	− − − −

Species	Tests	Responses
301 Rhodotorula acheniorum*	1 D−Glucose fermentation	−
	12 D−Galactose growth	+ + +
	16 L−Arabinose growth	+
	24 Salicin growth	− −
	28 Raffinose growth	+ +
	33 Erythritol growth	+ + +
	38 myo−Inositol growth	− − −
	49 Nitrate growth	+ + +
	51 Growth sans vitamins	−

Species	Tests	Responses
302 Rhodotorula araucariae*	1 D−Glucose fermentation	− −
	14 D−Ribose growth	− −
(m) 340 Sporobolomyces hispanicus	19 Sucrose growth	− − − −
(m) 342 Sporobolomyces odorus	21 Trehalose growth	+
	29 Melezitose growth	−
	32 Glycerol growth	+ + + +
	33 Erythritol growth	− −
	34 Ribitol growth	+ + +
	43 Succinate growth	+ + +
	45 Methanol growth	− −
	46 Ethanol growth	+ + + +
	48 D−Glucosamine growth	− −
	49 Nitrate growth	+ + + +
	62 Growth at 37 degrees	− −

distinguished from (m) by 72− (absence of filaments)

Species	Tests	Responses
303 Rhodotorula aurantiaca	1 D−Glucose fermentation	−
	12 D−Galactose growth	+ +
304 Rhodotorula glutinis	15 D−Xylose growth	+
	18 L−Rhamnose growth	−
	28 Raffinose growth	− − −
	29 Melezitose growth	+ + +
	33 Erythritol growth	− − −
	36 D−Mannitol growth	+
	43 Succinate growth	+
	49 Nitrate growth	+ + +
	51 Growth sans vitamins	− − −

Species	Tests	Responses

304 Rhodotorula glutinis
 	
    303 Rhodotorula aurantiaca
(m) 80 Candida melinii
(m) 172 Hansenula beckii**
(m) 174 Hansenula bimundalis
(m) 176 Hansenula canadensis
(m) 200 Hansenula wingei
(m) 335 Sporidibolus johnsonii
(m) 341 Sporobolomyces holsaticus
 	
distinguished from (m) by 66+ (pink colonies)
                           74− (absence of septate hyphae)
	1 D−Glucose fermentation
	11 Raffinose fermentation
	26 Melibiose growth
	27 Lactose growth
	29 Melezitose growth
	31 Starch growth
	33 Erythritol growth
	38 myo−Inositol growth
	41 5−Ketogluconate growth
	43 Succinate growth
	49 Nitrate growth
		− −
		− −
		−
		− −
		+ +
		
		− −
		− −
		−
		+
		+ +

305 Rhodotorula graminis	1 D−Glucose fermentation
	15 D−Xylose growth
	19 Sucrose growth
	21 Trehalose growth
	27 Lactose growth
	28 Raffinose growth
	29 Melezitose growth
	33 Erythritol growth
	38 myo−Inositol growth
	40 2−Ketogluconate growth
	42 D,L−Lactate growth
	43 Succinate growth
	49 Nitrate growth
	51 Growth sans vitamins
		−
		+
		    +
		    +
		
		+ +
		− − −
		
		−
		+ + +
		−
		+ + +
		+ +
		+ + +

306 Rhodotorula lactosa	1 D−Glucose fermentation 5 Sucrose fermentation 12 D−Galactose growth 26 Melibiose growth 29 Melezitose growth 38 myo−Inositol growth 49 Nitrate growth	− − − − − + + + + − − − + + +

307 Rhodotorula marina	16 L−Arabinose growth
	18 L−Rhamnose growth
	29 Melezitose growth
	33 Erythritol growth
	35 Galactitol growth
	38 myo−Inositol growth
	46 Ethanol growth
	49 Nitrate growth
	59 Growth sans PABA
		+ +
		  + +
		    +
		− − −
		+ +
		− − −
		− − −
		− −
		− − −

308 Rhodotorula minuta	16 L−Arabinose growth
	17 D−Arabinose growth
	19 Sucrose growth
	20 Maltose growth
	28 Raffinose growth
	29 Melezitose growth
	32 Glycerol growth
	35 Galactitol growth
	44 Citrate growth
	49 Nitrate growth
	59 Growth sans PABA
		+ +
		    +
		+
		− −
		−
		  + +
		+
		− −
		−
		−
		− − −

Species	Tests	Responses

309 Rhodotorula pallida

    339 Sporobolomyces gracilis
(m) 298 Rhodosporidium dacryoidum*
(m) 321 Saccharomycopsis crataegensis*

	Tests	Responses
	1 D−Glucose fermentation	− −
	16 L−Arabinose growth	− −
	20 Maltose growth	− −
	21 Trehalose growth	+ +
	27 Lactose growth	−
	32 Glycerol growth	+ +
	36 D−Mannitol growth	+ +
	38 myo−Inositol growth	−
	42 D,L−Lactate growth	+ +
	49 Nitrate growth	− −
	59 Growth sans PABA	−
	62 Growth at 37 degrees	− −

distinguished from (m) by 72− (absence of filaments)

310 Rhodotorula pilimanae

Tests	Responses
1 D−Glucose fermentation	− −
12 D−Galactose growth	+
15 D−Xylose growth	+
16 L−Arabinose growth	+ +
18 L−Rhamnose growth	− −
20 Maltose growth	− − − −
27 Lactose growth	− − − −
28 Raffinose growth	+ + + +
29 Melezitose growth	− −
34 Ribitol growth	+
49 Nitrate growth	− − − −
55 Growth sans thiamin	− −

311 Rhodotorula rubra

    337 Sporobolomyces albo−rubescens
(m) 418 Torulopsis xestobii*

Tests	Responses
1 D−Glucose fermentation	− −
15 D−Xylose growth	+
21 Trehalose growth	+
26 Melibiose growth	−
27 Lactose growth	− − −
28 Raffinose growth	+ + +
29 Melezitose growth	+ + +
31 Starch growth	− −
33 Erythritol growth	− − −
38 myo−Inositol growth	− −
40 2−Ketogluconate growth	−
49 Nitrate growth	− − −
54 Growth sans biotin	+
55 Growth sans thiamin	− −
56 Grth sans pyridoxine	+
59 Growth sans PABA	+ + +

distinguished from (m) by 66+ (pink colonies)

Species	Tests	Responses
312 Saccharomyces cerevisiae	1 D−Glucose fermentation	+ +
	13 L−Sorbose growth	− −
204 Kluyveromyces africanus	14 D−Ribose growth	− −
242 Pachytichospora transvaalensis	15 D−Xylose growth	− −
263 Pichia kudriavzevii	18 L−Rhamnose growth	−
266 Pichia membranaefaciens	23 Cellobiose growth	− −
282 Pichia saitoi	27 Lactose growth	− −
313 Saccharomyces dairensis	39 Gluconolactone growth	− −
314 Saccharomyces exiguus	40 2−Ketogluconate growth	− −
317 Saccharomyces telluris	44 Citrate growth	− −
(m) 240 Nematospora coryli	45 Methanol growth	−
(m) 325 Saccharomycopsis vini**	47 Ethylamine growth	− −
(a) 83 Candida milleri*	49 Nitrate growth	− −
(a) 107 Candida sorboxylosa*	61 Growth in 60% glucose	− −
(a) 117 Candida valida	63 0.01% cyclohex. growth	− −
(a) 257 Pichia fluxuum	65 Starch formation	− −
(a) 333 Schwanniomyces occidentalis		
(a) 366 Torulopsis bombicola*		
(a) 375 Torulopsis glabrata		
(a) 385 Torulopsis karawaiewi*		
(a) 412 Torulopsis stellata		

distinguished from (m) by 74− (absence of septate hyphae)
distinguished from (a) by 75+ (formation of ascospores)
                   76+ (ascospores spherical, oval or reniform)

Species	Tests	Responses
313 Saccharomyces dairensis	2 Galactose fermentation	+ +
	12 D−Galactose growth	+ +
205 Kluyveromyces blattae*	15 D−Xylose growth	− −
312 Saccharomyces cerevisiae	19 Sucrose growth	− − − −
	36 D−Mannitol growth	− −
	37 D−Glucitol growth	− −
	43 Succinate growth	− −
	49 Nitrate growth	− − − −
	57 Growth sans niacin	− − − −
	64 0.1% cyclohex. growth	− − − −

Species	Tests	Responses
314 Saccharomyces exiguus	2 Galactose fermentation	+ +
	5 Sucrose fermentation	+ + + +
312 Saccharomyces cerevisiae	7 Melibiose fermentation	−
(a) 83 Candida milleri*	9 Cellobiose fermentn	−
(a) 380 Torulopsis holmii	12 D−Galactose growth	+ +
	21 Trehalose growth	+ +
	23 Cellobiose growth	−
	26 Melibiose growth	−
	32 Glycerol growth	− − − −
	36 D−Mannitol growth	− − − −
	60 Growth in 50% glucose	− − − −
	62 Growth at 37 degrees	− −

distinguished from (a) by 75+ (formation of ascospores)

Species	Tests	Responses
315 Saccharomyces kluyveri	1 D−Glucose fermentation	+
	11 Raffinose fermentation	+
	17 D−Arabinose growth	− − −
	26 Melibiose growth	+ + +
	27 Lactose growth	− −
	31 Starch growth	− − −
	33 Erythritol growth	− −
	34 Ribitol growth	−
	44 Citrate growth	− − −
	46 Ethanol growth	+
	47 Ethylamine growth	+ + +
	49 Nitrate growth	−
	60 Growth in 50% glucose	− − −
	62 Growth at 37 degrees	+
	64 0.1% cyclohex. growth	− − −
316 Saccharomyces servazzii*	2 Galactose fermentation	+ + +
	13 L−Sorbose growth	−
	15 D−Xylose growth	−
	19 Sucrose growth	−
	28 Raffinose growth	−
	36 D−Mannitol growth	− −
	37 D−Glucitol growth	−
	42 D,L−Lactate growth	− −
	43 Succinate growth	−
	47 Ethylamine growth	− − −
	49 Nitrate growth	− − −
	56 Grth sans pyridoxine	−
	64 0.1% cyclohex. growth	+ + +
317 Saccharomyces telluris	1 D−Glucose fermentation	+
266 Pichia membranaefaciens	2 Galactose fermentation	−
312 Saccharomyces cerevisiae	12 D−Galactose growth	− − −
	15 D−Xylose growth	− −
	19 Sucrose growth	− − − −
	21 Trehalose growth	− − − −
	27 Lactose growth	−
	32 Glycerol growth	− − − −
	36 D−Mannitol growth	− − −
	43 Succinate growth	− − − −
	49 Nitrate growth	−
	60 Growth in 50% glucose	− − − −
	62 Growth at 37 degrees	+ + + +
318 Saccharomyces unisporus	2 Galactose fermentation	+ +
	12 D−Galactose growth	+ +
	15 D−Xylose growth	−
	16 L−Arabinose growth	−
	19 Sucrose growth	− − −
	23 Cellobiose growth	−
	28 Raffinose growth	
	32 Glycerol growth	− −
	36 D−Mannitol growth	− −
	37 D−Glucitol growth	−
	47 Ethylamine growth	+ + + +
	49 Nitrate growth	− − − −
	57 Growth sans niacin	− − −
	64 0.1% cyclohex. growth	+

Species	Tests	Responses
319 Saccharomycodes ludwigii	11 Raffinose fermentation	+
	12 D–Galactose growth	–
	23 Cellobiose growth	+ +
	28 Raffinose growth	+
	29 Melezitose growth	–
	36 D–Mannitol growth	–
	43 Succinate growth	–
	46 Ethanol growth	– –
	49 Nitrate growth	–
320 Saccharomycopsis capsularis**	12 D–Galactose growth	– –
	13 L–Sorbose growth	–
	15 D–Xylose growth	– –
	19 Sucrose growth	– – –
	22 Me α–D–glucoside grth	+ + +
	27 Lactose growth	+ + –
	33 Erythritol growth	+ + +
	45 Methanol growth	–
321 Saccharomycopsis crataegensis*	12 D–Galactose growth	– –
	13 L–Sorbose growth	+
(m) 309 Rhodotorula pallida	14 D–Ribose growth	+ + +
	15 D–Xylose growth	+ + +
	19 Sucrose growth	– – –
	23 Cellobiose growth	– –
	24 Salicin growth	–
	27 Lactose growth	– – –
	33 Erythritol growth	–
	38 myo–Inositol growth	– –
	44 Citrate growth	– – –
	49 Nitrate growth	– – –
distinguished from (m) by 72+ (presence of filaments)		
322 Saccharomycopsis fibuligera**	12 D–Galactose growth	– – –
	13 L–Sorbose growth	– – –
(m) 333 Schwanniomyces occidentalis	14 D–Ribose growth	– –
	15 D–Xylose growth	– – – –
	19 Sucrose growth	+ + + +
	23 Cellobiose growth	+ + +
	24 Salicin growth	+
	31 Starch growth	+ +
	49 Nitrate growth	– – – –
	51 Growth sans vitamins	– – – –
	62 Growth at 37 degrees	+ + + +
distinguished from (m) by 74+ (presence of septate hyphae)		
323 Saccharomycopsis lipolytica**	1 D–Glucose fermentation	– –
	15 D–Xylose growth	– – – –
	19 Sucrose growth	– –
	21 Trehalose growth	– – – –
	22 Me α–D–glucoside grth	– –
	33 Erythritol growth	+ + + +
	38 myo–Inositol growth	– –
	42 D,L–Lactate growth	+ +
	44 Citrate growth	+ +

Species	Tests	Responses
324 Saccharomycopsis malanga*	12 D−Galactose growth	− −
	14 D−Ribose growth	+ +
	19 Sucrose growth	−
	20 Maltose growth	+
	22 Me α−D−glucoside grth	− −
	23 Cellobiose growth	+
	27 Lactose growth	− −
	31 Starch growth	+
	44 Citrate growth	−
	49 Nitrate growth	− −
	62 Growth at 37 degrees	+
325 Saccharomycopsis vini**	7 Melibiose fermentation	−
	12 D−Galactose growth	− − −
(m) 103 Candida silvae	15 D−Xylose growth	− − −
(m) 121 Candida vini	18 L−Rhamnose growth	−
(m) 237 Nadsonia commutata*	26 Melibiose growth	−
(m) 312 Saccharomyces cerevisiae	27 Lactose growth	− − −
(m) 357 Torulaspora delbrueckii**	29 Melezitose growth	− −
(m) 385 Torulopsis karawaiewi*	31 Starch growth	− − −
(m) 432 Zygosaccharomyces bailii**	32 Glycerol growth	+ + +
(m) 433 Zygosaccharomyces bisporus**	33 Erythritol growth	− − −
(m) 439 Zygosaccharomyces rouxii**	37 D−Glucitol growth	+ + +
	42 D,L−Lactate growth	− − −
	44 Citrate growth	− − −
	45 Methanol growth	− −
	46 Ethanol growth	+ + +
	49 Nitrate growth	− − −
	51 Growth sans vitamins	− − −
	62 Growth at 37 degrees	− − −

distinguished from (m) by 72+ (presence of filaments)
74+ (presence of septate hyphae)

Species	Tests	Responses
326 Sarcinosporon inkin**	12 D−Galactose growth	+
	13 L−Sorbose growth	− −
421 Trichosporon cutaneum	16 L−Arabinose growth	−
	26 Melibiose growth	− − − −
	27 Lactose growth	+ + +
	28 Raffinose growth	− − −
	33 Erythritol growth	+ + +
	34 Ribitol growth	− − −
	38 myo−Inositol growth	+ + +
327 Schizoblastosporion starkeyi−henricii	1 D−Glucose fermentation	− − −
	12 D−Galactose growth	+ + +
(m) 298 Rhodosporidium dacryoidum*	15 D−Xylose growth	− −
(m) 364 Torulopsis austromarina*	16 L−Arabinose growth	−
	19 Sucrose growth	−
	20 Maltose growth	− − −
	27 Lactose growth	− −
	32 Glycerol growth	+
	42 D,L−Lactate growth	− − −
	43 Succinate growth	+ + +
	44 Citrate growth	− − −
	49 Nitrate growth	− − −
	51 Growth sans vitamins	− − −
	62 Growth at 37 degrees	− − −

distinguished from (m) by 69+ (apical budding)

Species	Tests	Responses
328 Schizosaccharomyces japonicus	7 Melibiose fermentation	+ +
	15 D–Xylose growth	−
	36 D–Mannitol growth	−
	37 D–Glucitol growth	−
	46 Ethanol growth	−
	60 Growth in 50% glucose	−
	62 Growth at 37 degrees	+
	65 Starch formation	+ + +
329 Schizosaccharomyces malidevorans	11 Raffinose fermentation	+
	15 D–Xylose growth	− −
	20 Maltose growth	− − − −
	28 Raffinose growth	+ + +
	32 Glycerol growth	− − −
	37 D–Glucitol growth	−
	46 Ethanol growth	− −
	65 Starch formation	+ + + +
330 Schizosaccharomyces octosporus	3 Maltose fermentation	+ + +
	11 Raffinose fermentation	−
	12 D–Galactose growth	−
	28 Raffinose growth	− −
	32 Glycerol growth	−
	36 D–Mannitol growth	− −
	43 Succinate growth	−
	46 Ethanol growth	− −
	65 Starch formation	+ + +
331 Schizosaccharomyces pombe	1 D–Glucose fermentation	+ +
	7 Melibiose fermentation	− − − −
	11 Raffinose fermentation	+ +
	20 Maltose growth	+ + + +
	21 Trehalose growth	− −
	23 Cellobiose growth	−
	28 Raffinose growth	+ +
	65 Starch formation	+ + + +
332 Schizosaccharomyces slooffiae*	1 D–Glucose fermentation	+ +
	3 Maltose fermentation	− − − −
	11 Raffinose fermentation	−
	15 D–Xylose growth	− −
	16 L–Arabinose growth	−
	28 Raffinose growth	− − −
	32 Glycerol growth	− − −
	36 D–Mannitol growth	− − −
	44 Citrate growth	−
	46 Ethanol growth	−
	65 Starch formation	+ + + +

The identification of particular species

Species	Tests	Responses
333 Schwanniomyces occidentalis	5 Sucrose fermentation	+ +
	14 D−Ribose growth	− −
(m) 20 Candida albicans	27 Lactose growth	− −
(m) 322 Saccharomycopsis fibuligera**	29 Melezitose growth	+
(a) 312 Saccharomyces cerevisiae	31 Starch growth	+ +
	33 Erythritol growth	− −
	36 D−Mannitol growth	+
	41 5−Ketogluconate growth	− −
	42 D,L−Lactate growth	− −
	46 Ethanol growth	+
	65 Starch formation	− −

distinguished from (m) by 74− (absence of septate hyphae)
distinguished from (a) by 77+ (ascospores hat− or saturn−shaped)

Species	Tests	Responses
334 Selenozyma peltata*	1 D−Glucose fermentation	+ + +
	16 L−Arabinose growth	+ +
	26 Melibiose growth	− −
	28 Raffinose growth	−
	33 Erythritol growth	+ + + +
	35 Galactitol growth	+ + +
	38 myo−Inositol growth	− − −
	42 D,L−Lactate growth	+ + + + +
	46 Ethanol growth	− − − −
	49 Nitrate growth	− − −
	62 Growth at 37 degrees	+ +

Species	Tests	Responses
335 Sporidibolus johnsonii	1 D−Glucose fermentation	− − −
	17 D−Arabinose growth	+ +
341 Sporobolomyces holsaticus	18 L−Rhamnose growth	− − − −
(m) 304 Rhodotorula glutinis	22 Me α−D−glucoside grth	+ + + +
	28 Raffinose growth	− − −
	33 Erythritol growth	− −
	38 myo−Inositol growth	−
	49 Nitrate growth	+ + + +
	62 Growth at 37 degrees	+ +

distinguished from (m) by 74+ (presence of septate hyphae)

Species	Tests	Responses
336 Sporidibolus ruinenii	19 Sucrose growth	+
	23 Cellobiose growth	+
	29 Melezitose growth	− − − −
	33 Erythritol growth	−
	34 Ribitol growth	+
	35 Galactitol growth	+ + + +
	38 myo−Inositol growth	− −
	40 2−Ketogluconate growth	− − − −
	46 Ethanol growth	− − −
	49 Nitrate growth	+ + +
	51 Growth sans vitamins	+ + + +
	62 Growth at 37 degrees	−

Species	Tests	Responses
337 Sporobolomyces albo−rubescens	1 D−Glucose fermentation	− − −
	5 Sucrose fermentation	−
311 Rhodotorula rubra	16 L−Arabinose growth	+
	17 D−Arabinose growth	+ + + +
	20 Maltose growth	+ + + +
	26 Melibiose growth	−
	27 Lactose growth	− − −
	28 Raffinose growth	+ + + +
	31 Starch growth	− − −
	33 Erythritol growth	− − −
	35 Galactitol growth	− − − −
	49 Nitrate growth	− − − −
	60 Growth in 50% glucose	−
338 Sporobolomyces antarcticus*	12 D−Galactose growth	− −
	23 Cellobiose growth	− − −
	27 Lactose growth	+ +
	34 Ribitol growth	−
	38 myo−Inositol growth	+ + +
	49 Nitrate growth	+ + + +
	62 Growth at 37 degrees	+
339 Sporobolomyces gracilis	1 D−Glucose fermentation	− −
	12 D−Galactose growth	− − −
309 Rhodotorula pallida	15 D−Xylose growth	+ + +
	19 Sucrose growth	− −
	20 Maltose growth	−
	32 Glycerol growth	+
	33 Erythritol growth	−
	36 D−Mannitol growth	+ +
	38 myo−Inositol growth	− − −
	44 Citrate growth	+
	45 Methanol growth	− −
	46 Ethanol growth	− − −
	49 Nitrate growth	− − −
	62 Growth at 37 degrees	−
340 Sporobolomyces hispanicus	1 D−Glucose fermentation	− −
	19 Sucrose growth	− − − −
(m) 302 Rhodotorula araucariae*	21 Trehalose growth	+ +
(m) 342 Sporobolomyces odorus	23 Cellobiose growth	− − −
	37 D−Glucitol growth	+
	38 myo−Inositol growth	−
	43 Succinate growth	+ + +
	49 Nitrate growth	+ + + +
	51 Growth sans vitamins	+ + + +
	62 Growth at 37 degrees	− − − −

distinguished from (m) by 72+ (presence of filaments)
                         74+ (presence of septate hyphae)

The identification of particular species

Species	Tests	Responses
341 Sporobolomyces holsaticus	1 D−Glucose fermentation	− −
	18 L−Rhamnose growth	−
335 Sporidibolus johnsonii	20 Maltose growth	+ + +
(m) 304 Rhodotorula glutinis	26 Melibiose growth	−
(m) 345 Sporobolomyces roseus	27 Lactose growth	− −
	33 Erythritol growth	− − −
	34 Ribitol growth	+ + +
	35 Galactitol growth	− − −
	40 2−Ketogluconate growth	− − −
	43 Succinate growth	+
	49 Nitrate growth	+ + +
	51 Growth sans vitamins	+ + +

distinguished from (m) by 74+ (presence of septate hyphae)

Species	Tests	Responses
342 Sporobolomyces odorus	1 D−Glucose fermentation	−
	12 D−Galactose growth	− − − −
(m) 3 Aessosporon salmonicolor**	16 L−Arabinose growth	−
(m) 53 Candida foliarum	17 D−Arabinose growth	+ + +
(m) 302 Rhodotorula araucariae*	20 Maltose growth	− − − −
(m) 340 Sporobolomyces hispanicus	23 Cellobiose growth	− −
(m) 346 Sporobolomyces salmonicolor	32 Glycerol growth	+
	33 Erythritol growth	−
	34 Ribitol growth	+
	35 Galactitol growth	−
	45 Methanol growth	− − −
	49 Nitrate growth	+ + + +

distinguished from (m) by 66+ (pink colonies)
72+ (presence of filaments)
74− (absence of septate hyphae)

Species	Tests	Responses
343 Sporobolomyces pararoseus	1 D−Glucose fermentation	− −
	16 L−Arabinose growth	− − −
	19 Sucrose growth	+
	21 Trehalose growth	+ + +
	23 Cellobiose growth	+ +
	26 Melibiose growth	−
	27 Lactose growth	− − −
	28 Raffinose growth	+ +
	33 Erythritol growth	− − −
	43 Succinate growth	+ +
	49 Nitrate growth	− −
	51 Growth sans vitamins	+ + +

Species	Tests	Responses
344 Sporobolomyces puniceus*	12 D−Galactose growth	−
	15 D−Xylose growth	− −
	18 L−Rhamnose growth	−
	32 Glycerol growth	+
	34 Ribitol growth	+ + +
	38 myo−Inositol growth	− − −
	46 Ethanol growth	− − −
	49 Nitrate growth	+ +
	65 Starch formation	+ + + +

Species	Tests	Responses
345 Sporobolomyces roseus	1 D−Glucose fermentation	− −
	5 Sucrose fermentation	−
(m) 341 Sporobolomyces holsaticus	17 D−Arabinose growth	+ + +
	28 Raffinose growth	+ +
	29 Melezitose growth	+ + +
	31 Starch growth	+ + +
	32 Glycerol growth	+ + +
	33 Erythritol growth	− − −
	38 myo−Inositol growth	− − −
	49 Nitrate growth	+ + +
	62 Growth at 37 degrees	−

distinguished from (m) by 74− (absence of septate hyphae)

Species	Tests	Responses
346 Sporobolomyces salmonicolor	1 D−Glucose fermentation	− −
	2 Galactose fermentation	−
3 Aessosporon salmonicolor**	19 Sucrose growth	+ + + +
(m) 342 Sporobolomyces odorus	20 Maltose growth	− − − −
	21 Trehalose growth	+ + + +
	23 Cellobiose growth	−
	27 Lactose growth	−
	35 Galactitol growth	− − −
	36 D−Mannitol growth	+
	37 D−Glucitol growth	+
	40 2−Ketogluconate growth	− − − −
	49 Nitrate growth	+ + +
	51 Growth sans vitamins	+
	54 Growth sans biotin	+

distinguished from (m) by 74+ (presence of septate hyphae)

Species	Tests	Responses
347 Sporobolomyces singularis	12 D−Galactose growth	− −
	15 D−Xylose growth	− − −
	19 Sucrose growth	− −
	27 Lactose growth	+ + +
	38 myo−Inositol growth	−
	46 Ethanol growth	+
	49 Nitrate growth	−

Species	Tests	Responses
348 Stephanoascus ciferrii**	18 L−Rhamnose growth	+
	26 Melibiose growth	+ +
	27 Lactose growth	− −
	29 Melezitose growth	− −
	33 Erythritol growth	+
	35 Galactitol growth	+
	38 myo−Inositol growth	+
	65 Starch formation	− −

Species	Tests	Responses
349 Sterigmatomyces aphidis*	12 D−Galactose growth	+ +
	14 D−Ribose growth	+ +
(m) 47 Candida edax*	26 Melibiose growth	+ + + +
	33 Erythritol growth	+ +
	34 Ribitol growth	+ +
	49 Nitrate growth	+ + + +
	62 Growth at 37 degrees	+ + + +

distinguished from (m) by 74− (absence of septate hyphae)

Species	Tests	Responses
350 Sterigmatomyces elviae*	17 D−Arabinose growth	+
	19 Sucrose growth	+
(m) 421 Trichosporon cutaneum	20 Maltose growth	− − −
	27 Lactose growth	+
	28 Raffinose growth	+
	33 Erythritol growth	+ + +
	45 Methanol growth	−
	62 Growth at 37 degrees	+

distinguished from (m) by 72− (absence of filaments)

Species	Tests	Responses
351 Sterigmatomyces halophilus	1 D−Glucose fermentation	− −
	12 D−Galactose growth	+
	16 L−Arabinose growth	+
	19 Sucrose growth	− −
	28 Raffinose growth	−
	33 Erythritol growth	+ + +
	42 D,L−Lactate growth	−
	45 Methanol growth	− −
	49 Nitrate growth	+ + +
352 Sterigmatomyces indicus	1 D−Glucose fermentation	− − −
	12 D−Galactose growth	+
	16 L−Arabinose growth	+ +
	19 Sucrose growth	− −
	22 Me α−D−glucoside grth	−
	23 Cellobiose growth	− − −
	24 Salicin growth	+ + +
	27 Lactose growth	− − −
	33 Erythritol growth	+ + +
	38 myo−Inositol growth	− −
	42 D,L−Lactate growth	−
	49 Nitrate growth	− − −
353 Sterigmatomyces nectairii*	1 D−Glucose fermentation	−
	19 Sucrose growth	− − −
	21 Trehalose growth	+
	23 Cellobiose growth	− −
	24 Salicin growth	−
	27 Lactose growth	−
	32 Glycerol growth	− − −
	37 D−Glucitol growth	+
	43 Succinate growth	+
	49 Nitrate growth	+ +
	51 Growth sans vitamins	− − −
	62 Growth at 37 degrees	− − −
354 Sterigmatomyces penicillatus*	27 Lactose growth	+
	28 Raffinose growth	− −
(m) 60 Candida humicola	32 Glycerol growth	−
	33 Erythritol growth	+ +
	46 Ethanol growth	− −
	62 Growth at 37 degrees	−
	65 Starch formation	+

distinguished from (m) by 67− (cells not budding)

Species	Tests	Responses
355 Sterigmatomyces polyborus*	18 L−Rhamnose growth	+
	28 Raffinose growth	+ +
(m) 16 Bullera alba	33 Erythritol growth	+ +
(m) 130 Cryptococcus flavus	35 Galactitol growth	+
	46 Ethanol growth	− −
	62 Growth at 37 degrees	− −
	65 Starch formation	− −

distinguished from (m) by 67− (cells not budding)

Species	Tests	Responses
356 Sympodiomyces parvus*	22 Me α−D−glucoside grth	+ + +
	23 Cellobiose growth	+
	24 Salicin growth	− − −
	26 Melibiose growth	−
	28 Raffinose growth	−
	29 Melezitose growth	− − −
	33 Erythritol growth	+ + + +
	35 Galactitol growth	−
	43 Succinate growth	− −
	44 Citrate growth	−
	46 Ethanol growth	− − − −
	62 Growth at 37 degrees	−

Species	Tests	Responses
357 Torulaspora delbrueckii**	1 D−Glucose fermentation	+ + + +
	14 D−Ribose growth	− −
266 Pichia membranaefaciens	16 L−Arabinose growth	− − − −
359 Torulaspora pretoriensis**	23 Cellobiose growth	− − − −
432 Zygosaccharomyces bailii**	24 Salicin growth	− −
433 Zygosaccharomyces bisporus**	31 Starch growth	− − − −
439 Zygosaccharomyces rouxii**	33 Erythritol growth	− − − −
(m) 29 Candida bombi*	35 Galactitol growth	−
(m) 73 Candida lambica	41 5−Ketogluconate growth	− −
(m) 97 Candida rugopelliculosa*	44 Citrate growth	− − − −
(m) 99 Candida sake	49 Nitrate growth	− − − −
(m) 117 Candida valida	52 Growth sans inositol	+
(m) 217 Kluyveromyces waltii*	55 Growth sans thiamin	+ + + +
(m) 263 Pichia kudriavzevii	56 Grth sans pyridoxine	+ + +
(m) 325 Saccharomycopsis vini**	57 Growth sans niacin	+ + + +
(a) 257 Pichia fluxuum	61 Growth in 60% glucose	+ + + +
(a) 361 Torulopsis apicola	63 0.01% cyclohex. growth	−
(a) 366 Torulopsis bombicola*	65 Starch formation	− − −

distinguished from (m) by 72− (absence of filaments)
distinguished from (a) by 75+ (formation of ascospores)
    76+ (ascospores spherical, oval or reniform)

Species	Tests	Responses
358 Torulaspora globosa**	1 D−Glucose fermentation	+
	11 Raffinose fermentation	+ +
	12 D−Galactose growth	− − − −
	15 D−Xylose growth	− −
	19 Sucrose growth	+
	20 Maltose growth	−
	22 Me α−D−glucoside grth	−
	23 Cellobiose growth	− − − −
	28 Raffinose growth	+ +
	37 D−Glucitol growth	− −
	42 D,L−Lactate growth	+ + + + +
	49 Nitrate growth	−
	51 Growth sans vitamins	+
	62 Growth at 37 degrees	+
	63 0.01% cyclohex. growth	+
	64 0.1% cyclohex. growth	+ + + +

Species	Tests	Responses
359 Torulaspora pretoriensis**	5 Sucrose fermentation	+
	7 Melibiose fermentation	−
357 Torulaspora delbrueckii**	11 Raffinose fermentation	+ +
	12 D−Galactose growth	+ + +
	22 Me α−D−glucoside grth	+
	23 Cellobiose growth	− − −
	24 Salicin growth	−
	26 Melibiose growth	− −
	28 Raffinose growth	+ +
	40 2−Ketogluconate growth	+ + + +
	42 D,L−Lactate growth	+ + + +
	43 Succinate growth	− − − −
	49 Nitrate growth	−
	60 Growth in 50% glucose	+
360 Torulopsis anatomiae	1 D−Glucose fermentation	+ +
	15 D−Xylose growth	− −
	18 L−Rhamnose growth	+ + +
	19 Sucrose growth	−
	32 Glycerol growth	− −
	33 Erythritol growth	−
	37 D−Glucitol growth	− −
361 Torulopsis apicola	12 D−Galactose growth	− −
	13 L−Sorbose growth	+ + +
(m) 217 Kluyveromyces waltii*	21 Trehalose growth	− −
(a) 357 Torulaspora delbrueckii**	26 Melibiose growth	− −
	28 Raffinose growth	+ + +
	43 Succinate growth	− − −
	46 Ethanol growth	− − −
	49 Nitrate growth	− − −

distinguished from (m) by 72− (absence of filaments)
distinguished from (a) by 75− (inability to form ascospores)

Species	Tests	Responses
362 Torulopsis apis	1 D−Glucose fermentation	− − −
	15 D−Xylose growth	−
	21 Trehalose growth	− −
	22 Me α−D−glucoside grth	−
	28 Raffinose growth	+ + +
	29 Melezitose growth	−
	37 D−Glucitol growth	+
	43 Succinate growth	− − −
	46 Ethanol growth	− − −
	49 Nitrate growth	− −
363 Torulopsis auriculariae*	1 D−Glucose fermentation	− − −
	12 D−Galactose growth	− −
	15 D−Xylose growth	− − −
	19 Sucrose growth	+ +
	23 Cellobiose growth	− − −
	29 Melezitose growth	+
	32 Glycerol growth	− −
	33 Erythritol growth	−
	43 Succinate growth	+
	49 Nitrate growth	− − −

Species	Tests	Responses
364 Torulopsis austromarina*	1 D−Glucose fermentation	− −
	12 D−Galactose growth	+ +
(m) 327 Schizoblastosporion starkeyi−henricii	15 D−Xylose growth	−
	20 Maltose growth	− −
	27 Lactose growth	−
	34 Ribitol growth	−
	36 D−Mannitol growth	− −
	38 myo−Inositol growth	−
	42 D,L−Lactate growth	− −
	54 Growth sans biotin	− −
	62 Growth at 37 degrees	− −

distinguished from (m) by 69− (budding not apical)

Species	Tests	Responses
365 Torulopsis bacarum*	1 D−Glucose fermentation	− −
	12 D−Galactose growth	− − −
	16 L−Arabinose growth	+ +
	19 Sucrose growth	+
	25 Arbutin growth	−
	27 Lactose growth	−
	28 Raffinose growth	+
	33 Erythritol growth	+ + +
	38 myo−Inositol growth	− −
	45 Methanol growth	−
	49 Nitrate growth	+ +

Species	Tests	Responses
366 Torulopsis bombicola*	11 Raffinose fermentation	+ + +
	20 Maltose growth	−
(a) 312 Saccharomyces cerevisiae	21 Trehalose growth	−
(a) 357 Torulaspora delbrueckii**	23 Cellobiose growth	− −
	24 Salicin growth	−
	26 Melibiose growth	− − −
	29 Melezitose growth	− −
	32 Glycerol growth	+ + +
	36 D−Mannitol growth	+ +
	42 D,L−Lactate growth	− − −
	46 Ethanol growth	+ + +
	62 Growth at 37 degrees	+ + +

distinguished from (a) by 75− (inability to form ascospores)

Species	Tests	Responses
367 Torulopsis cantarelii	1 D−Glucose fermentation	+ + +
	12 D−Galactose growth	− − −
	13 L−Sorbose growth	+ +
	15 D−Xylose growth	− − −
	19 Sucrose growth	− −
	24 Salicin growth	−
	33 Erythritol growth	+ + + +
	44 Citrate growth	−
	45 Methanol growth	− −
	51 Growth sans vitamins	+

Species	Tests	Responses
368 Torulopsis castellii	12 D–Galactose growth	− −
	19 Sucrose growth	− −
	21 Trehalose growth	+ + +
	24 Salicin growth	−
	32 Glycerol growth	+ + + +
	36 D–Mannitol growth	−
	40 2–Ketogluconate growth	+ + + +
	43 Succinate growth	− − − −
	46 Ethanol growth	− − −
	49 Nitrate growth	−
	52 Growth sans inositol	− − − −
	62 Growth at 37 degrees	+ +
369 Torulopsis dendrica*	15 D–Xylose growth	− −
385 Torulopsis karawaiewi*	19 Sucrose growth	− − − −
	24 Salicin growth	+ + + +
	27 Lactose growth	− −
	36 D–Mannitol growth	− − − −
	42 D,L–Lactate growth	− − − −
	44 Citrate growth	+ + + +
	49 Nitrate growth	− −
	62 Growth at 37 degrees	− −
370 Torulopsis ernobii	1 D–Glucose fermentation	+ + +
	12 D–Galactose growth	− −
	16 L–Arabinose growth	−
	18 L–Rhamnose growth	+ +
	19 Sucrose growth	+
	21 Trehalose growth	− − −
	22 Me α–D–glucoside grth	+
	33 Erythritol growth	−
	34 Ribitol growth	+ +
	42 D,L–Lactate growth	− −
	45 Methanol growth	−
	49 Nitrate growth	+ +
371 Torulopsis etchellsii	2 Galactose fermentation	− −
	3 Maltose fermentation	+ + + +
	19 Sucrose growth	− −
	21 Trehalose growth	− − −
	33 Erythritol growth	−
	45 Methanol growth	−
	49 Nitrate growth	+ + + +
372 Torulopsis fragaria*	13 L–Sorbose growth	+
	16 L–Arabinose growth	+ + +
	18 L–Rhamnose growth	− −
	22 Me α–D–glucoside grth	+ + +
	27 Lactose growth	+ + +
	32 Glycerol growth	+ +
	33 Erythritol growth	− − −
	34 Ribitol growth	+
	35 Galactitol growth	−
	38 myo–Inositol growth	− − −
	49 Nitrate growth	+ + +

Species	Tests	Responses
373 Torulopsis fructus*	1 D−Glucose fermentation	+ + +
	12 D−Galactose growth	− − −
	13 L−Sorbose growth	+ + + +
	15 D−Xylose growth	+ + +
	19 Sucrose growth	− − − −
	21 Trehalose growth	+
	23 Cellobiose growth	−
	24 Salicin growth	− − −
	33 Erythritol growth	− −
	34 Ribitol growth	+
	37 D−Glucitol growth	+ +
	44 Citrate growth	+ + + +
	45 Methanol growth	−
	49 Nitrate growth	−
	56 Grth sans pyridoxine	−
374 Torulopsis fujisanensis	19 Sucrose growth	− − −
	21 Trehalose growth	− − − −
	23 Cellobiose growth	+
	27 Lactose growth	− − −
	29 Melezitose growth	−
	33 Erythritol growth	−
	35 Galactitol growth	+ + + +
	38 myo−Inositol growth	−
	43 Succinate growth	+ +
	46 Ethanol growth	+ + +
375 Torulopsis glabrata	2 Galactose fermentation	− −
	12 D−Galactose growth	− −
(a) 312 Saccharomyces cerevisiae	19 Sucrose growth	− −
	21 Trehalose growth	+ + + +
	22 Me α−D−glucoside grth	− −
	23 Cellobiose growth	− −
	24 Salicin growth	− −
	37 D−Glucitol growth	− − − −
	43 Succinate growth	− − − −
	52 Growth sans inositol	+ + + +
	57 Growth sans niacin	− − − −

distinguished from (a) by 75− (inability to form ascospores)

Species	Tests	Responses
376 Torulopsis gropengiesseri	1 D−Glucose fermentation	+ +
	5 Sucrose fermentation	+
	12 D−Galactose growth	+ +
	20 Maltose growth	−
	21 Trehalose growth	− −
	23 Cellobiose growth	+ + + +
	27 Lactose growth	−
	28 Raffinose growth	+ + +
	34 Ribitol growth	−
	36 D−Mannitol growth	+ +
	37 D−Glucitol growth	+
	42 D,L−Lactate growth	−
	43 Succinate growth	− −
	46 Ethanol growth	− − − −
	49 Nitrate growth	−

Species	Tests	Responses
377 Torulopsis haemulonii	1 D−Glucose fermentation	+ +
	5 Sucrose fermentation	+
	18 L−Rhamnose growth	+ + +
	19 Sucrose growth	+
	23 Cellobiose growth	− − −
	33 Erythritol growth	− − −
	38 myo−Inositol growth	−
	45 Methanol growth	−
378 Torulopsis halonitratophila	2 Galactose fermentation	− −
	5 Sucrose fermentation	−
	13 L−Sorbose growth	− − −
	15 D−Xylose growth	− −
	19 Sucrose growth	− −
	20 Maltose growth	− − −
	21 Trehalose growth	− − −
	24 Salicin growth	−
	34 Ribitol growth	−
	36 D−Mannitol growth	+ + +
	49 Nitrate growth	+ + +
	51 Growth sans vitamins	− − −
379 Torulopsis halophilus*	2 Galactose fermentation	+ + + +
	15 D−Xylose growth	−
	20 Maltose growth	− − − −
	24 Salicin growth	+ +
	25 Arbutin growth	+
	28 Raffinose growth	−
	33 Erythritol growth	−
	34 Ribitol growth	−
	36 D−Mannitol growth	+ + + + +
	37 D−Glucitol growth	− − − −
	38 myo−Inositol growth	−
	49 Nitrate growth	+ + + +
380 Torulopsis holmii	2 Galactose fermentation	+ + +
83 Candida milleri*	15 D−Xylose growth	−
(a) 215 Kluyveromyces polysporus	21 Trehalose growth	+
(a) 314 Saccharomyces exiguus	27 Lactose growth	−
	28 Raffinose growth	+ + + +
	36 D−Mannitol growth	− −
	37 D−Glucitol growth	− − −
	39 Gluconolactone growth	+ + + +
	43 Succinate growth	− − −
	46 Ethanol growth	− − − −
	49 Nitrate growth	− −
	60 Growth in 50% glucose	− − − −

distinguished from (a) by 75− (inability to form ascospores)

Species	Tests	Responses
381 Torulopsis humilis*	1 D–Glucose fermentation	+
	2 Galactose fermentation	+
	12 D–Galactose growth	+
	19 Sucrose growth	– – –
	21 Trehalose growth	+ +
	36 D–Mannitol growth	–
	37 D–Glucitol growth	–
	39 Gluconolactone growth	+
	42 D,L–Lactate growth	+ + +
	43 Succinate growth	– –
	53 Grth sans pantothenate	–
	63 0.01% cyclohex. growth	+ +
382 Torulopsis inconspicua	1 D–Glucose fermentation	– – –
	12 D–Galactose growth	– –
(a) 266 Pichia membranaefaciens	15 D–Xylose growth	–
	21 Trehalose growth	– –
	23 Cellobiose growth	– – –
	36 D–Mannitol growth	– –
	37 D–Glucitol growth	–
	42 D,L–Lactate growth	+ +
	56 Grth sans pyridoxine	– – –
	62 Growth at 37 degrees	+
distinguished from (a) by 75– (inability to form ascospores)		
383 Torulopsis ingeniosa	13 L–Sorbose growth	+
	16 L–Arabinose growth	– –
	17 D–Arabinose growth	– – – –
	22 Me α–D–glucoside grth	+ + +
	29 Melezitose growth	+
	31 Starch growth	+ + + +
	32 Glycerol growth	+
	33 Erythritol growth	–
	34 Ribitol growth	– –
	38 myo–Inositol growth	– –
	42 D,L–Lactate growth	+
	46 Ethanol growth	– – –
	49 Nitrate growth	+ + + +
384 Torulopsis insectalens*	12 D–Galactose growth	+ +
	15 D–Xylose growth	– – –
	19 Sucrose growth	– – –
	23 Cellobiose growth	+ +
	24 Salicin growth	+
	27 Lactose growth	–
	32 Glycerol growth	– – –
	37 D–Glucitol growth	+ +
	49 Nitrate growth	– –

Species	Tests	Responses
385 Torulopsis karawaiewi*	1 D–Glucose fermentation	+ + + +
	2 Galactose fermentation	– –
46 Candida diversa	5 Sucrose fermentation	–
369 Torulopsis dendrica*	12 D–Galactose growth	– –
(m) 39 Candida citrea*	13 L–Sorbose growth	– – – –
(m) 72 Candida krissii*	15 D–Xylose growth	– –
(m) 117 Candida valida	19 Sucrose growth	– – –
(m) 325 Saccharomycopsis vini**	21 Trehalose growth	– – – –
(a) 252 Pichia delftensis	23 Cellobiose growth	– – – –
(a) 257 Pichia fluxuum	42 D,L–Lactate growth	– – – –
(a) 266 Pichia membranaefaciens	49 Nitrate growth	– –
(a) 312 Saccharomyces cerevisiae	60 Growth in 50% glucose	– – –
	61 Growth in 60% glucose	–
	62 Growth at 37 degrees	– – – –

distinguished from (m) by 72– (absence of filaments)
distinguished from (a) by 75– (inability to form ascospores)

Species	Tests	Responses
386 Torulopsis kruisii*	1 D–Glucose fermentation	+ + +
	3 Maltose fermentation	– – –
	5 Sucrose fermentation	– – –
	12 D–Galactose growth	+ + +
	18 L–Rhamnose growth	–
	27 Lactose growth	–
	31 Starch growth	+ + +
	32 Glycerol growth	+
	33 Erythritol growth	– – –
	34 Ribitol growth	+
	38 myo–Inositol growth	– – –
	42 D,L–Lactate growth	– – –
	44 Citrate growth	+ +
	49 Nitrate growth	– –

Species	Tests	Responses
387 Torulopsis lactis–condensi	1 D–Glucose fermentation	+
	11 Raffinose fermentation	+ +
	23 Cellobiose growth	–
	24 Salicin growth	–
	32 Glycerol growth	–
	36 D–Mannitol growth	– –
	37 D–Glucitol growth	–
	46 Ethanol growth	–
	49 Nitrate growth	+ + +

Species	Tests	Responses
388 Torulopsis magnoliae	1 D–Glucose fermentation	+ +
	5 Sucrose fermentation	+ +
	15 D–Xylose growth	– –
	19 Sucrose growth	+ +
	21 Trehalose growth	– – – –
	22 Me α–D–glucoside grth	–
	23 Cellobiose growth	– –
	33 Erythritol growth	– –
	37 D–Glucitol growth	+ + + +
	49 Nitrate growth	+ + + +

Species	Tests	Responses

389 Torulopsis mannitofaciens*

1 D−Glucose fermentation	+ + + +
16 L−Arabinose growth	+ + +
19 Sucrose growth	+
20 Maltose growth	+ +
22 Me α−D−glucoside grth	−
25 Arbutin growth	− −
33 Erythritol growth	−
36 D−Mannitol growth	+
37 D−Glucitol growth	− − − −
49 Nitrate growth	+ + + +

390 Torulopsis maris

18 L−Rhamnose growth	+ + +
19 Sucrose growth	− −
21 Trehalose growth	− − −
23 Cellobiose growth	− −
24 Salicin growth	−
27 Lactose growth	− − −
33 Erythritol growth	− −
38 myo−Inositol growth	−

391 Torulopsis methanolovescens*

(a) 264 Pichia lindnerii*

12 D−Galactose growth	− −
16 L−Arabinose growth	− −
18 L−Rhamnose growth	+ +
33 Erythritol growth	− − − −
34 Ribitol growth	+ +
44 Citrate growth	+ +
45 Methanol growth	+ + + +
49 Nitrate growth	− −

distinguished from (a) by 75− (inability to form ascospores)

392 Torulopsis molischiana

(a) 177 Hansenula capsulata

16 L−Arabinose growth	+
19 Sucrose growth	−
29 Melezitose growth	+ +
34 Ribitol growth	+
45 Methanol growth	+ +

distinguished from (a) by 75− (inability to form ascospores)

393 Torulopsis multis−gemmis*

13 L−Sorbose growth	+
15 D−Xylose growth	+
16 L−Arabinose growth	+ +
19 Sucrose growth	+
22 Me α−D−glucoside grth	−
27 Lactose growth	− − −
28 Raffinose growth	− − −
36 D−Mannitol growth	− − −
37 D−Glucitol growth	+ +
38 myo−Inositol growth	− −
42 D,L−Lactate growth	− −
44 Citrate growth	+ + +
49 Nitrate growth	− − −

The identification of particular species

Species	Tests	Responses
394 Torulopsis musae*	1 D–Glucose fermentation	+ +
	5 Sucrose fermentation	− −
	12 D–Galactose growth	− − −
	13 L–Sorbose growth	+
	15 D–Xylose growth	+
	19 Sucrose growth	+
	23 Cellobiose growth	− − −
	28 Raffinose growth	−
	29 Melezitose growth	+ +
	33 Erythritol growth	−
	44 Citrate growth	+ +
	45 Methanol growth	−
	56 Grth sans pyridoxine	−
395 Torulopsis nagoyaensis*	12 D–Galactose growth	+
	13 L–Sorbose growth	+ +
	19 Sucrose growth	− −
	23 Cellobiose growth	+
	33 Erythritol growth	−
	42 D,L–Lactate growth	+ +
	44 Citrate growth	+ +
	45 Methanol growth	+ + +
396 Torulopsis navarrensis*	1 D–Glucose fermentation	+
	13 L–Sorbose growth	− − − −
	16 L–Arabinose growth	− − −
	17 D–Arabinose growth	−
	22 Me α–D–glucoside grth	+ +
	26 Melibiose growth	+ + + +
	27 Lactose growth	− − −
	29 Melezitose growth	+
	31 Starch growth	− − − −
	33 Erythritol growth	− − − −
	35 Galactitol growth	−
	36 D–Mannitol growth	+ +
	37 D–Glucitol growth	+ +
	44 Citrate growth	+ + + +
	49 Nitrate growth	−
397 Torulopsis nemodendra*	1 D–Glucose fermentation	− −
	35 Galactitol growth	+ + +
	38 myo–Inositol growth	−
	44 Citrate growth	−
	45 Methanol growth	+ + +
	62 Growth at 37 degrees	−
398 Torulopsis nitratophila	12 D–Galactose growth	+ + +
	13 L–Sorbose growth	−
	18 L–Rhamnose growth	+ +
	19 Sucrose growth	−
	33 Erythritol growth	− − −
	34 Ribitol growth	+ + +
	44 Citrate growth	− − −
	45 Methanol growth	+ + + +

Species	Tests	Responses
399 Torulopsis nodaensis*	1 D−Glucose fermentation	+ +
	3 Maltose fermentation	− − −
	13 L−Sorbose growth	+ + +
	15 D−Xylose growth	− −
	19 Sucrose growth	− − −
	21 Trehalose growth	−
	34 Ribitol growth	−
	42 D,L−Lactate growth	−
	44 Citrate growth	−
	49 Nitrate growth	+ + +
400 Torulopsis norvegica	12 D−Galactose growth	−
	13 L−Sorbose growth	− − −
(a) 188 Hansenula mrakii	15 D−Xylose growth	+ + +
	19 Sucrose growth	− − −
	21 Trehalose growth	− −
	24 Salicin growth	+
	34 Ribitol growth	− −
	37 D−Glucitol growth	+
	42 D,L−Lactate growth	+
	45 Methanol growth	−
	49 Nitrate growth	+ + +

distinguished from (a) by 75− (inability to form ascospores)

Species	Tests	Responses
401 Torulopsis pampelonensis*	1 D−Glucose fermentation	+
	13 L−Sorbose growth	+ + +
(m) 79 Candida melibiosica	14 D−Ribose growth	− −
	16 L−Arabinose growth	−
	26 Melibiose growth	+ + +
	27 Lactose growth	− −
	29 Melezitose growth	+
	31 Starch growth	− − −
	33 Erythritol growth	−
	36 D−Mannitol growth	+ +
	37 D−Glucitol growth	+
	38 myo−Inositol growth	−
	42 D,L−Lactate growth	− − −
	44 Citrate growth	+ + +
	49 Nitrate growth	− −

distinguished from (m) by 72− (absence of filaments)

Species	Tests	Responses
402 Torulopsis philyla*	1 D−Glucose fermentation	− −
	15 D−Xylose growth	− − −
	19 Sucrose growth	− − −
	21 Trehalose growth	+ + + +
	23 Cellobiose growth	− − −
	27 Lactose growth	−
	32 Glycerol growth	− − − −
	34 Ribitol growth	+ +
	44 Citrate growth	+ +
	49 Nitrate growth	− − − −
403 Torulopsis pignaliae*	1 D−Glucose fermentation	+
	12 D−Galactose growth	−
	16 L−Arabinose growth	+
	23 Cellobiose growth	− −
	33 Erythritol growth	− −
	34 Ribitol growth	+
	38 myo−Inositol growth	−
	45 Methanol growth	+ + +
	46 Ethanol growth	− −

The identification of particular species

Species	Tests	Responses
404 Torulopsis pinus	19 Sucrose growth	− −
	23 Cellobiose growth	− −
	24 Salicin growth	− −
	43 Succinate growth	− − − −
	44 Citrate growth	+ +
	45 Methanol growth	+ + + +
	49 Nitrate growth	− −
405 Torulopsis psychrophila*	1 D−Glucose fermentation	− −
	14 D−Ribose growth	− − −
	16 L−Arabinose growth	+ + +
	19 Sucrose growth	− −
	22 Me α−D−glucoside grth	− −
	23 Cellobiose growth	− − −
	27 Lactose growth	− − −
	32 Glycerol growth	+
	33 Erythritol growth	+ + + +
	38 myo−Inositol growth	−
	43 Succinate growth	− −
	44 Citrate growth	− −
406 Torulopsis pustula*	13 L−Sorbose growth	+ + +
	14 D−Ribose growth	+ + +
	17 D−Arabinose growth	+
	18 L−Rhamnose growth	−
	19 Sucrose growth	− − −
	23 Cellobiose growth	+ + +
	33 Erythritol growth	− − −
	35 Galactitol growth	−
	38 myo−Inositol growth	− −
	44 Citrate growth	+ +
	45 Methanol growth	−
	49 Nitrate growth	+ + +
407 Torulopsis schatavii*	14 D−Ribose growth	+
	19 Sucrose growth	− −
	23 Cellobiose growth	−
	27 Lactose growth	− −
	33 Erythritol growth	+ +
	44 Citrate growth	+
	45 Methanol growth	−
	51 Growth sans vitamins	+ +
	62 Growth at 37 degrees	− −
408 Torulopsis silvatica*	1 D−Glucose fermentation	− − − −
	12 D−Galactose growth	− −
	15 D−Xylose growth	− − − −
	19 Sucrose growth	− −
	21 Trehalose growth	+ + + +
	23 Cellobiose growth	− − −
	36 D−Mannitol growth	+ + + +
	42 D,L−Lactate growth	+ +
	44 Citrate growth	− −
	49 Nitrate growth	− − − −
	62 Growth at 37 degrees	+ + + +

Species	Tests	Responses
409 Torulopsis sonorensis*	12 D−Galactose growth	−
	16 L−Arabinose growth	+ +
	21 Trehalose growth	− − −
	33 Erythritol growth	− − −
	34 Ribitol growth	+ +
	45 Methanol growth	+ + + +
	46 Ethanol growth	+
	49 Nitrate growth	−
410 Torulopsis sorbophila*	1 D−Glucose fermentation	− − −
	12 D−Galactose growth	+
	13 L−Sorbose growth	+ + +
	19 Sucrose growth	− − −
	21 Trehalose growth	−
	27 Lactose growth	− −
	33 Erythritol growth	−
	34 Ribitol growth	− − −
	36 D−Mannitol growth	+ +
	37 D−Glucitol growth	+
	42 D,L−Lactate growth	− − − −
	49 Nitrate growth	−
	56 Grth sans pyridoxine	− −
	62 Growth at 37 degrees	+ + + +
411 Torulopsis spandovensis*	1 D−Glucose fermentation	+ +
	21 Trehalose growth	− − −
	23 Cellobiose growth	−
	27 Lactose growth	−
	28 Raffinose growth	+
	32 Glycerol growth	+ +
	33 Erythritol growth	− −
	35 Galactitol growth	+ + +
	43 Succinate growth	+
412 Torulopsis stellata	1 D−Glucose fermentation	+
(a) 312 Saccharomyces cerevisiae	21 Trehalose growth	− −
	22 Me α−D−glucoside grth	−
	28 Raffinose growth	+ + +
	32 Glycerol growth	−
	36 D−Mannitol growth	− −
	37 D−Glucitol growth	−
	43 Succinate growth	−
	46 Ethanol growth	− − −
	49 Nitrate growth	− − −
	52 Growth sans inositol	− −
	55 Growth sans thiamin	−
	65 Starch formation	− − −

distinguished from (a) by 75− (inability to form ascospores)

Species	Tests	Responses
413 Torulopsis tannotolerans*	2 Galactose fermentation	+ + + +
	5 Sucrose fermentation	− −
	6 Trehalose fermentation	
	14 D−Ribose growth	− − − −
	19 Sucrose growth	− −
	21 Trehalose growth	− − −
	36 D−Mannitol growth	− − −
	37 D−Glucitol growth	−
	39 Gluconolactone growth	+ + + +
	49 Nitrate growth	− − − −
	52 Growth sans inositol	+ +
	54 Growth sans biotin	+ + + +
	55 Growth sans thiamin	− − − −
	56 Grth sans pyridoxine	+ +
	57 Growth sans niacin	+ + + +
414 Torulopsis torresii	6 Trehalose fermentation	+ + +
	12 D−Galactose growth	+
	13 L−Sorbose growth	+ + +
	14 D−Ribose growth	− −
	16 L−Arabinose growth	+ + +
	19 Sucrose growth	− − −
	33 Erythritol growth	− −
	44 Citrate growth	−
	45 Methanol growth	− −
	46 Ethanol growth	+
415 Torulopsis vanderwaltii	1 D−Glucose fermentation	−
	23 Cellobiose growth	− −
	24 Salicin growth	−
	28 Raffinose growth	− −
	29 Melezitose growth	− −
	34 Ribitol growth	+ + +
	45 Methanol growth	−
	46 Ethanol growth	− − −
	49 Nitrate growth	+ + +
	51 Growth sans vitamins	− −
	56 Grth sans pyridoxine	−
416 Torulopsis versatilis	2 Galactose fermentation	+
	6 Trehalose fermentation	+
	22 Me α−D−glucoside grth	−
	34 Ribitol growth	− −
	37 D−Glucitol growth	+ +
	46 Ethanol growth	− −
	49 Nitrate growth	+ +
417 Torulopsis wickerhamii	1 D−Glucose fermentation	+ +
	12 D−Galactose growth	+ +
	18 L−Rhamnose growth	+ + + +
	19 Sucrose growth	− −
	21 Trehalose growth	− − −
	22 Me α−D−glucoside grth	− −
	27 Lactose growth	−
	34 Ribitol growth	+ +
	44 Citrate growth	− −
	45 Methanol growth	− −
	46 Ethanol growth	+ +
	49 Nitrate growth	+

| Species | Tests | Responses |

418 Torulopsis xestobii*

(m) 311 Rhodotorula rubra

Tests	Responses
12 D−Galactose growth	+ +
15 D−Xylose growth	+
17 D−Arabinose growth	− − −
22 Me α−D−glucoside grth	+ +
27 Lactose growth	− − −
32 Glycerol growth	+
34 Ribitol growth	+ +
36 D−Mannitol growth	− − −
37 D−Glucitol growth	−
38 myo−Inositol growth	−
44 Citrate growth	+ + +
49 Nitrate growth	− − −
51 Growth sans vitamins	− − −

distinguished from (m) by 66− (colonies not pink)

419 Trichosporon aquatile*

    421 Trichosporon cutaneum

Tests	Responses
1 D−Glucose fermentation	− −
14 D−Ribose growth	+
16 L−Arabinose growth	+ + +
22 Me α−D−glucoside grth	+ + +
28 Raffinose growth	− − − −
32 Glycerol growth	− −
34 Ribitol growth	− −
36 D−Mannitol growth	− − − −
38 myo−Inositol growth	−
42 D,L−Lactate growth	+ +
49 Nitrate growth	− − − −

420 Trichosporon brassicae*

Tests	Responses
12 D−Galactose growth	+
13 L−Sorbose growth	+ + +
22 Me α−D−glucoside grth	−
23 Cellobiose growth	− −
27 Lactose growth	− − − −
28 Raffinose growth	− −
33 Erythritol growth	− − −
34 Ribitol growth	− − −
35 Galactitol growth	−
38 myo−Inositol growth	+ + + +
42 D,L−Lactate growth	+
44 Citrate growth	+ + +

421 Trichosporon cutaneum

    326 Sarcinosporon inkin**
    419 Trichosporon aquatile*
(m) 2 Aessosporon dendrophilum*
(m) 19 Candida aaseri
(m) 25 Candida blankii
(m) 42 Candida curvata
(m) 52 Candida fluviotilis*
(m) 60 Candida humicola
(m) 61 Candida hydrocarbofumarica*
(m) 91 Candida podzolica*
(m) 129 Cryptococcus dimennae
(m) 131 Cryptococcus gastricus
(m) 137 Cryptococcus laurentii
(m) 138 Cryptococcus luteolus
(m) 140 Cryptococcus magnus**
(m) 149 Debaryomyces marama
(m) 227 Lipomyces lipofer
(m) 244 Pichia abadieae*
(m) 285 Pichia scolyti
(m) 350 Sterigmatomyces elviae*

Tests	Responses
1 D−Glucose fermentation	− − −
15 D−Xylose growth	+ + +
27 Lactose growth	+ + +
30 Inulin growth	− − −
45 Methanol growth	− − −
46 Ethanol growth	+ + +
49 Nitrate growth	− − −
52 Growth sans inositol	+
53 Grth sans pantothenate	+
57 Growth sans niacin	+
59 Growth sans PABA	+ + +
60 Growth in 50% glucose	− − −

distinguished from (m) by 68+ (formation of splitting cells)

Species	Tests	Responses
422 Trichosporon eriense*	15 D−Xylose growth	+ + +
	19 Sucrose growth	−
	21 Trehalose growth	− − −
	22 Me α−D−glucoside grth	−
	27 Lactose growth	− − −
	28 Raffinose growth	−
	33 Erythritol growth	+ + +
	34 Ribitol growth	− −
	36 D−Mannitol growth	+ +
	42 D,L−Lactate growth	+
	45 Methanol growth	−
423 Trichosporon fennicum*	5 Sucrose fermentation	+ + +
	26 Melibiose growth	− −
	27 Lactose growth	+ + +
	32 Glycerol growth	+ +
	33 Erythritol growth	+
	35 Galactitol growth	− − −
	42 D,L−Lactate growth	− − −
	44 Citrate growth	+
	51 Growth sans vitamins	+
	62 Growth at 37 degrees	+ + +
424 Trichosporon melibiosaceum*	3 Maltose fermentation	+ + +
	5 Sucrose fermentation	+ +
	18 L−Rhamnose growth	−
	28 Raffinose growth	+
	33 Erythritol growth	+ + +
	46 Ethanol growth	− − −
	49 Nitrate growth	−
	62 Growth at 37 degrees	− − −
425 Trichosporon pullulans	1 D−Glucose fermentation	−
(m) 125 Cryptococcus albidus	12 D−Galactose growth	+
	22 Me α−D−glucoside grth	+
	24 Salicin growth	+
	26 Melibiose growth	+
	33 Erythritol growth	+ +
	49 Nitrate growth	+ +
	62 Growth at 37 degrees	−
	65 Starch formation	+

distinguished from (m) by 68+ (formation of splitting cells)

426 Trichosporon terrestre*	1 D−Glucose fermentation	−
	26 Melibiose growth	−
	27 Lactose growth	−
	29 Melezitose growth	− −
	33 Erythritol growth	+ + +
	34 Ribitol growth	+ + +
	38 myo−Inositol growth	+ +
	42 D,L−Lactate growth	+
	49 Nitrate growth	+ + +

The identification of particular species

Species	Tests	Responses
427 Trigonopsis variabilis	1 D–Glucose fermentation	–
	13 L–Sorbose growth	+ +
	19 Sucrose growth	– –
	21 Trehalose growth	+ +
	23 Cellobiose growth	–
	27 Lactose growth	– –
	34 Ribitol growth	– –
	38 myo–Inositol growth	– –
	44 Citrate growth	+
	62 Growth at 37 degrees	+ +
428 Wickerhamia fluorescens	1 D–Glucose fermentation	+
	11 Raffinose fermentation	+
	15 D–Xylose growth	– –
	19 Sucrose growth	+
	20 Maltose growth	–
	22 Me α–D–glucoside grth	– –
	34 Ribitol growth	+ + +
	37 D–Glucitol growth	– – –
	42 D,L–Lactate growth	+ +
	44 Citrate growth	+ + +
429 Wickerhamiella domercqii**	1 D–Glucose fermentation	– –
	13 L–Sorbose growth	+ + +
	15 D–Xylose growth	– – –
	21 Trehalose growth	– – – –
	24 Salicin growth	– –
	28 Raffinose growth	–
	38 myo–Inositol growth	– –
	44 Citrate growth	+ +
	49 Nitrate growth	+ + + +
430 Wingea robertsii	2 Galactose fermentation	–
	5 Sucrose fermentation	+
	6 Trehalose fermentation	+ + + +
	11 Raffinose fermentation	+ + +
	13 L–Sorbose growth	+ + +
	18 L–Rhamnose growth	+ + + +
	26 Melibiose growth	– – –
	28 Raffinose growth	+
	33 Erythritol growth	+ + + +
	35 Galactitol growth	– –
	49 Nitrate growth	–
431 Zendera ovetensis	1 D–Glucose fermentation	– – –
	12 D–Galactose growth	+ + + +
	13 L–Sorbose growth	+ + +
	15 D–Xylose growth	– –
	19 Sucrose growth	–
	21 Trehalose growth	– – – –
	27 Lactose growth	–
	33 Erythritol growth	– –
	36 D–Mannitol growth	– –
	38 myo–Inositol growth	
	42 D,L–Lactate growth	+ + + +
	44 Citrate growth	– –
	51 Growth sans vitamins	– – – –
	60 Growth in 50% glucose	–
	62 Growth at 37 degrees	– – – –

Species | Tests | Responses

432 Zygosaccharomyces bailii**

    357 Torulaspora delbrueckii**
    433 Zygosaccharomyces bisporus**
    439 Zygosaccharomyces rouxii**
(m) 325 Saccharomycopsis vini**

Test	Response
1 D−Glucose fermentation	+ +
15 D−Xylose growth	− −
22 Me α−D−glucoside grth	−
23 Cellobiose growth	−
28 Raffinose growth	− −
33 Erythritol growth	− −
37 D−Glucitol growth	+ +
43 Succinate growth	− −
47 Ethylamine growth	+
49 Nitrate growth	− −
60 Growth in 50% glucose	+
61 Growth in 60% glucose	+

distinguished from (m) by 74− (absence of septate hyphae)

433 Zygosaccharomyces bisporus**

    266 Pichia membranaefaciens
    357 Torulaspora delbrueckii**
    432 Zygosaccharomyces bailii**
    439 Zygosaccharomyces rouxii**
(m) 325 Saccharomycopsis vini**
(a) 117 Candida valida

Test	Response
1 D−Glucose fermentation	+ + + + +
2 Galactose fermentation	− − − −
15 D−Xylose growth	− − − − −
18 L−Rhamnose growth	−
19 Sucrose growth	− − −
21 Trehalose growth	−
28 Raffinose growth	− −
33 Erythritol growth	− −
39 Gluconolactone growth	+ +
43 Succinate growth	− − − −
44 Citrate growth	− −
47 Ethylamine growth	+ +
49 Nitrate growth	− − − − −
52 Growth sans inositol	+ + +
56 Grth sans pyridoxine	+
61 Growth in 60% glucose	+ + + + +
65 Starch formation	− −

distinguished from (m) by 74− (absence of septate hyphae)
distinguished from (a) by 75+ (formation of ascospores)

434 Zygosaccharomyces cidri**

Test	Response
1 D−Glucose fermentation	+
2 Galactose fermentation	+
7 Melibiose fermentation	+
11 Raffinose fermentation	+
13 L−Sorbose growth	+
23 Cellobiose growth	− − −
25 Arbutin growth	−
26 Melibiose growth	+ + +
33 Erythritol growth	− −
42 D,L−Lactate growth	+ + + +
63 0.01% cyclohex. growth	+ +
64 0.1% cyclohex. growth	+ +

435 Zygosaccharomyces fermentati**

Test	Response
2 Galactose fermentation	+
9 Cellobiose fermentn	− −
11 Raffinose fermentation	+
12 D−Galactose growth	+
16 L−Arabinose growth	−
22 Me α−D−glucoside grth	+ +
23 Cellobiose growth	+ + +
26 Melibiose growth	− −
28 Raffinose growth	+ +
33 Erythritol growth	−
44 Citrate growth	−
46 Ethanol growth	+
52 Growth sans inositol	− − −
60 Growth in 50% glucose	+
62 Growth at 37 degrees	+

Species

436 Zygosaccharomyces florentinus

Tests	Responses
1 D–Glucose fermentation	+
7 Melibiose fermentation	+ +
13 L–Sorbose growth	+ + + + +
23 Cellobiose growth	– – –
26 Melibiose growth	+ + +
33 Erythritol growth	– –
34 Ribitol growth	– –
42 D,L–Lactate growth	– – – – –
44 Citrate growth	– –
60 Growth in 50% glucose	+ + + + +
63 0.01% cyclohex. growth	+ + +
64 0.1% cyclohex. growth	+ + + +

437 Zygosaccharomyces microellipsodes

Tests	Responses
7 Melibiose fermentation	+ + + +
20 Maltose growth	–
22 Me α–D–glucoside grth	– –
23 Cellobiose growth	– – –
26 Melibiose growth	+
33 Erythritol growth	– –
34 Ribitol growth	– –
38 myo–Inositol growth	– – –
39 Gluconolactone growth	+ + +
42 D,L–Lactate growth	+ +
47 Ethylamine growth	+ +
60 Growth in 50% glucose	+ + +
61 Growth in 60% glucose	– – – – –
62 Growth at 37 degrees	– –
63 0.01% cyclohex. growth	– –
64 0.1% cyclohex. growth	– – –

438 Zygosaccharomyces mrakii

Tests	Responses
1 D–Glucose fermentation	+ +
21 Trehalose growth	– – – –
26 Melibiose growth	+ + + +
27 Lactose growth	– –
29 Melezitose growth	– – – –
34 Ribitol growth	– –
38 myo–Inositol growth	– –
63 0.01% cyclohex. growth	+ +
64 0.1% cyclohex. growth	+ +

439 Zygosaccharomyces rouxii

266 Pichia membranaefaciens
357 Torulaspora delbrueckii**
432 Zygosaccharomyces bailii**
433 Zygosaccharomyces bisporus**
(m) 325 Saccharomycopsis vini**
(a) 117 Candida valida

Tests	Responses
1 D–Glucose fermentation	+ + + +
2 Galactose fermentation	– – – –
15 D–Xylose growth	– – – –
18 L–Rhamnose growth	–
22 Me α–D–glucoside grth	–
23 Cellobiose growth	– –
28 Raffinose growth	– – – –
33 Erythritol growth	– – –
43 Succinate growth	– – –
47 Ethylamine growth	+ +
49 Nitrate growth	– – –
52 Growth sans inositol	+ +
56 Grth sans pyridoxine	+ +
57 Growth sans niacin	+ +
60 Growth in 50% glucose	+
61 Growth in 60% glucose	+ + +
65 Starch formation	– –

distinguished from (m) by 74– (absence of septate hyphae)
distinguished from (a) by 75+ (formation of ascospores)

9
How the computing was done

The keys were constructed using the computer program Genkey (Payne, 1975, 1978). Some additional small subsidiary programs were also required. The first of these programs read the results of the tests, as abstracted from the records of the Centraalbureau voor Schimmelcultures, reorganized the information and produced the table of responses to the tests in Chapter 6. A further program converted the data into the numerical form required by Genkey.

In Chapter 6, the results of the tests and observations are given as follows.

(i)   +, positive response
(ii) −, negative response
(iii) V, variable response (i.e. either + or −)
(iv) S, slow response
(v)   D, delayed response
(vi) ?, response not known

For computing the keys, responses (iii) to (vi) were combined and recoded as equivocal responses (Barnett, 1971a,b; Barnett & Pankhurst, 1974). Such responses may be expected to differ between strains of any one species or may be particularly dependent on the precise method of testing, so that sometimes positive and sometimes negative results are obtained. These entries must appear as both positive and negative in the keys.

Irredundant test sets
It would not usually be practicable to use the keys sequentially, doing a given test in response to the result of a previous test. Generally, all tests required for the appropriate key will be done simultaneously; the key is then used solely for identifying. As the complete set of tests in the key will be done for any identification, that set should be minimal; it should contain no redundant tests, that is, those that can be omitted without making any species unidentifiable. Sets without redundant tests are termed irredundant; Genkey contains an algorithm for finding all irredundant sets of tests (reviewed by Preece, 1976; Payne & Preece, 1980). The method is illustrated below using the hypothetical results shown in Table 9.1.

Table 9.1 *Artificial data showing results of tests A−E with species 1−4*

	Test				
Species	A	B	C	D	E
1	+	+	+	−	+
2	+	−	+	−	−
3	+	−	−	+	+
4	−	+	+	+	−

First, lists are made of tests that can distinguish each pair of species as shown in Table 9.2: for example, species 1 and species 2 in Table 9.1 differ in their responses to tests B and E, so the first entry in Table 9.2 is B,E. The lists are then expressed as a sum (e.g. B + E for species 1 and 2) and multiplied together according to the rules of Boolean algebra. These rules state that A × A = A, etc.; and that if a term contains all the elements of some other term in the sum, it

Table 9.2 *List of tests that can distinguish each pair of species*

	Species			
Species	1	2	3	4
1		B,E	B,C,D	A,D,E
2			C,D,E	A,B,D
3				A,B,C,E
4				

should be deleted, for example AB + ABC = AB. Thus multiplying the lists for the pairs of species (1,2) and (1,3) gives

$$(B + E) \times (B + C + D) = BB + BC + BD + BE + CE + DE$$
$$= B + CE + DE$$

Each term in this list is a set of tests that can be used to distinguish between species 1 and species 2, and between species 1 and species 3, so that the sets are (i) test B, (ii) tests C and E, (iii) tests D and E. (This indicates the rationale behind the rules of the algebra. BB becomes B because there is no need to record a test more than once in the same set. The fact that B is a set in its own right implies that BC, BD and BE contain redundant tests C, D and E, respectively; if these redundant tests are deleted, three more instances of the set B would be obtained; this set is already in the list, so BC, BD and BE can be deleted.)

Continuing the process, irredundant sets to distinguish all the species are given by

$$(B + CE + DE) \times (A + D + E) \times (C + D + E) \times (A + B + D) \times (A + B + C + E)$$
$$= (AB + BD + BE + CE + DE) \times (C + D + E) \times (A + B + D) \times (A + B + C + E)$$
$$= (ABC + BD + BE + CE + DE) \times (A + B + D) \times (A + B + C + E)$$
$$= (ABC + BD + BE + ACE + DE) \times (A + B + C + E)$$
$$= ABC + BD + BE + ACE + DE$$

Several short cuts are available to make the method more efficient. For example, if the list of tests to distinguish between a particular pair of species contains all the tests in the list to distinguish between some other pair of species, the longer list can be deleted. (Since in order to be able to distinguish between the pair of species corresponding to the shorter list, one of the tests in the list must be in the set; this test will also distinguish between the pair of species corresponding to the longer list.) Hence, the final entry in Table 9.2, 'A,B,C,E', which contains the entry 'B,E', could have been deleted. Other short cuts are described by Willcox & Lapage (1972) and Payne & Preece (1980).

As the example shows, there can be more than one irredundant set, and the sets can be of different sizes. The numbers of sets available for the yeast keys varied from one, for the key to all the yeasts using only physiological tests, to over 1000 with the clinical yeasts, for example. When choosing which set to use for each key, an attempt was made to select sets containing the most reliable and convenient tests, so in some cases the sets selected are not of minimum size. Also, with groups of yeasts for which more than one key was constructed, sets were chosen so that the physiological tests in the key using both physiological and microscopical tests are all included in the key using physiological tests only.

How the keys were produced
Genkey allows keys to be constructed for any required group of species and allows the tests to be restricted to any specified set. Hence it was unnecessary to modify the data in order to produce the specialized keys (or to construct the appropriate irredundant test sets).

To construct a key with binary (e.g. + or −) tests, Genkey starts by selecting the test that best divides the species into two groups. The first group is composed of those species that could give negative responses to the test, that is, the species known to give negative responses plus those with equivocal responses; the second group is composed of the species with positive responses, again plus those with equivocal responses. Hence any species with responses described as equivocal will occur in more than one part of the key. Tests are then chosen to divide each group further, continuing until the groups each contain only one species or a set of species that cannot be distinguished from each other with the tests available.

If no species have equivocal responses, the best test is that which most nearly splits the group of species into two equal sub-groups (e.g. Barnett, 1957). However, with equivocal responses, the choice is no longer so straightforward. Consider the responses of six species to two tests, A and B. Suppose that with test A, four species have equivocal responses, one species gives a negative response and one species a positive response, while with test B, four species give a negative response and two species a positive response. Then test A will form equal groups, each of five species, whereas test B will form unequal groups of four and two. However, since each group formed by test B is smaller than the corresponding group formed by test A, test B gives superior discrimination. Accordingly, there is a further requirement, namely to choose a test with few equivocal responses. These (possibly conflicting) requirements are resolved by choosing the test with a minimum value of some *criterion function* – a function whose arguments, for test i, are: p_i, the proportion of species with positive responses to test i; q_i, the proportion with negative responses; and r_i, the proportion with equivocal responses. The function used for constructing the yeast keys was

$$GP_i = -\min(p_i q_i)/[\theta+(1-\theta)\{r_i + 2\min(p_i,q_i)\}]$$

devised by Gower & Payne (1975). The parameter θ, which controls the bias against equivocal responses, was set to 0.25. GP has a maximum value of 0 for tests with which either $p_i = 0$ or $q_i = 0$ (such tests are termed *indefinite* by Gower & Payne, 1975), and a minimum value of -0.5 for a test with which $p_i = q_i = 0.5$, $r = 0$, i.e. a test that divides the species into equal groups and which has no equivocal responses.

Gower & Barnett (1971) suggested that tests, whose criterion function values differ from that of the best test by less than some specified tolerance, be regarded as equivalent. The test can then be chosen from these equivalent tests according to some subsidiary criterion. The subsidiary criterion used for the yeast keys was an integer index between one and four, assigned to each test as an arbitrary assessment of its value. For example, test number 34 for ribitol utilization was assessed highly (index one) because (i) it is easy to do, as is true for most of the aerobic growth tests, (ii) very pure ribitol is readily obtained and (iii) it can be autoclaved as it is very heat-stable. On the other hand, test number 24 for utilizing salicin, a phenolic β-D-glucoside, was given a low grading because the phenolic aglycon, liberated on hydrolysis, might inhibit the growth of some yeasts. So a negative result could be obtained either because of this inhibition or because the yeast could not hydrolyse salicin. Also, the results of test 24 are highly associated (Barnett, 1976) with those of test 23, the utilization of another β-D-glucoside, cellobiose. No such objection applies to cellobiose as on hydrolysis it liberates D-glucose only. Since the two tests provide much the same information, index four was assigned to test 24. If the set of equivalent tests contained more than one test with the best reliability index, the final choice was made using the original criterion GP. As the reliability of the tests used to make each identification in the key is liable to be rather more important than the efficiency of the key, the tolerance for equivalent tests was set to the rather large value of 0.25.

Once constructed, the keys were printed in a compact form (Payne, Walton & Barnett, 1974) in which space is saved by printing both the negative and positive results of each test on the same line. The printed key thus occupies about half the number of lines of a more conventional form.

As explained above, yeasts with equivocal responses may occur more than once in a key. If there are many such yeasts, identical groups may occur in different parts of the key. Each of these groups is distinguished by the same tests so the printed key will contain an identical section each time the group occurs. To save space, Genkey stores the index of the test at the start of the first replicate section, so that when a duplicate is encountered subsequently the user is directed back to the original section and the duplicate is omitted. For example, in the first key to yeasts that do not ferment glucose (key no. 1), the section '32 Growth without thiamin...' can be reached from '31 Cellobiose growth negative' and from '39 Nitrate growth positive'. This process, termed *reticulation*, reduced the number of lines in the keys by about 10%. Payne (1977) has described the method fully.

Sets of tests to identify individual yeasts

Sets of tests to distinguish each yeast from the other yeasts are printed in Chapter 8. These can be formed using the same method as that described on p. 277. For yeast n, the sets of tests are obtained by multiplying the lists of tests to distinguish pairs of yeasts $(1,n), (2,n), \ldots, (n-1,n), (n,n+1), \ldots (n,439)$. Thus sets of tests to distinguish species 3 from the other species in Table 9.1 would be given by

$$(B + C + D) \times (C + D + E) \times (A + B + C + E)$$
$$= (BE + C + D) \times (A + B + C + E)$$
$$= BE + C + AD + BD + DE$$

As with the sets of tests for the keys, the number of sets varied from one (for *Candida humicola*) to over 1000. We have again tried to avoid undesirable tests. Furthermore, since the value of each test must depend on circumstances, additional sets have been selected to give, where feasible, an alternative to each test in the recommended set.

The test sets were formed only from physiological tests. Some yeasts can be distinguished from certain others only by tests involving microscopical examination or the production of ascospores. A subsidiary program was written to determine, for each yeast, (i) the yeasts that cannot be distinguished from that yeast by physiological tests, (ii) which of these yeasts can be distinguished by microscopical examination (and the tests required) and (iii) which can be distinguished only by tests involving the production of ascospores.

Finally, a program was written to use all this information to produce the table in Chapter 8.

10
Alternatives to keys

One of the main disadvantages of identification keys like those in this book is that they use a fixed set of tests. This is not serious for the key to all the yeasts that is based on physiological tests only; for only one irredundant set of physiological tests distinguishes between all the 439 yeasts listed in Chapter 6. Consequently, the physiological key to all the yeasts contained all the tests in this set and further physiological tests were unnecessary. However, for particular sub-groups of yeasts there were many irredundant sets of tests from which the keys might have been formed (see p. 277), and no one choice could be best for all users. Where appropriate there are additional keys to allow a choice between microscopical examination and extra physiological tests.

Polyclaves
One simple, compact way of making an identification from the results of any sufficiently large set of tests is the polyclave. This is a pack of punched cards, where each possible test result is represented by a card and a position on the cards is allocated to each species. Thus for the test results computed for this book, there would be one card for D-glucose fermentation negative, one for D-glucose fermentation positive, and so on, making 156 cards in all. Each card has holes punched in the positions corresponding to the species that can give the test result concerned. When the cards for a set of observed test results are superimposed accurately and held up to the light, the specific names that might give an appropriate identification correspond to the positions where holes appear in every card. A polyclave for the yeasts has been made for the test results listed in Chapter 6. This polyclave was constructed by R. J. Pankhurst (Department of Botany, British Museum (Natural History), Cromwell Road, London SW7 5BD) from whom copies are obtainable. A disadvantage of polyclaves is that they cannot easily be used to suggest the best tests to do next when a given set of test results is satisfied by several species.

Computer-based identification systems
Computer programs have been written that are analogous to polyclaves. The test results are fed into the computer and the program sorts through the data to see which species might give those results. The program then can assess the remaining tests, using criteria similar to those for selecting tests in keys (p. 279) and can suggest which test or tests to do next (e.g. Mullin, 1970; Morse, 1971). For identifying yeasts it would probably be more appropriate to select a set of tests, and for complete identification an irredundant set (p. 277). Various authors, including Gyllenberg (1963, 1964), Rypka, Clapper, Bowen & Babb (1967) and Rypka & Babb (1970), have discussed other methods of selecting sets of tests.

Programs can be adapted to allow for the possibility of observing uncharacteristic responses, either because the yeasts are aberrant or because there are flaws in the methods of testing. Usually only the names of species that can give results identical to those observed are regarded as possibly valid identifications. Morse (1971) enabled the user of his identification system to specify a number of tests, say n, so that the name of any species, with no more than n results differing from those observed, must also be considered a possible identification. The correct species would then be eliminated only if more than n uncharacteristic responses were observed. Payne & Preece (1977) allowed a similar specification as part of their method for detecting and correcting errors that may occur when a key is used.

Identification by matching

Campbell (1973) described a method of identifying *Saccharomyces* species, in which a fixed set of 58 tests is done and then a matching coefficient (e.g. Sneath & Sokal, 1973) is calculated for each species. This coefficient is defined as the number of tests for which the results observed with the unknown yeast agree with those of the named species, divided by the total number of tests. The yeast is identified as the species with the highest coefficient if this is greater than 85% or provisionally identified if the coefficient is between 80% and 85%. In other words, the specific name given is that for which the results differ from those observed for the fewest tests, providing these do not exceed 20% of the tests done. Campbell (1975) extended this method to species of *Candida* and *Torulopsis*.

Keys based on groups of tests

To use the keys in Chapter 6, testing can be done either non-sequentially, by performing every test in the key before the key is used, or sequentially, by performing tests only as required by the key. Sequential identification uses fewer tests but is impracticable if the tests take a long time to do.

A compromise is a key in which more than one test is applied at each stage. Genkey (Payne, 1975, 1978) selects groups of tests by considering every combination of tests, up to a specified group size, and using test selection criteria similar to those used to select single tests. However, for groups of more than about four tests, the number of combinations to be considered makes the method impracticable. An alternative method is under investigation; here the group is formed sequentially by adding at each stage the test that gives the best improvement in selection criterion.

APPENDIX:
Recently altered names of yeast species

In the table below the first column is an alphabetical list of yeast names no longer accepted, but recognized in *The Yeasts* (Lodder, 1970) or in *A New Key to The Yeasts* (Barnett & Pankhurst, 1974). The second column gives the currently accepted name of each of these yeasts, as used in this *Guide*.

Name in *The Yeasts* or *A New Key*	Name used in this *Guide*
Brettanomyces bruxellensis	*Dekkera bruxellensis*
Brettanomyces intermedius	*Dekkera intermedia*
Bullera dendrophila	*Aessosporon dendrophilum*
Candida australis	*Candida sake*
Candida ciferrii	*Stephanoascus ciferrii*
Candida claussenii	*Candida albicans*
Candida fibrae	*Hyphopichia burtonii*
Candida guilliermondii	*Pichia guilliermondii*
Candida ingens	*Pichia humboldtii*
Candida krusei	*Pichia kudriavzevii*
Candida langeronii	*Candida albicans*
Candida lipolytica	*Saccharomycopsis lipolytica*
Candida obtusa	*Candida lusitaniae*
Candida requinyii	*Pichia kudriavzevii*
Candida salmonicola	*Candida sake*
Candida shehatae	*Pichia stipitis*
Candida slooffii	*Saccharomyces telluris*
Candida sorbosa	*Pichia kudriavzevii*
Candida stellatoidea	*Candida albicans*
Cryptococcus infirmo-miniatus	*Rhodosporidium infirmo-miniatum*
Debaryomyces cantarellii	*Debaryomyces polymorphus*
Endomycopsis bispora	*Hansenula beckii*
Endomycopsis burtonii	*Hyphopichia burtonii*
Endomycopsis capsularis	*Saccharomycopsis capsularis*
Endomycopsis fibuligera	*Saccharomycopsis fibuligera*
Endomycopsis javanensis	*Arthroascus javanensis*
Endomycopsis monospora	*Ambrosiozyma monospora*
Endomycopsis ovetensis	*Zendera ovetensis*
Endomycopsis platypodis	*Hormoascus platypodis*
Endomycopsis selenospora	*Guilliermondella selenospora*
Endomycopsis vini	*Saccharomycopsis vini*
Hansenula malanga	*Saccharomycopsis malanga*
Kloeckera africana	*Hanseniaspora vineae*
Kloeckera apiculata	*Hanseniaspora uvarum*
Kloeckera corticis	*Hanseniaspora osmophila*
Kloeckera javanica	*Hanseniaspora occidentalis*
Kluyveromyces cicerisporus	*Kluyveromyces bulgaricus*
Kluyveromyces fragilis	*Kluyveromyces marxianus*
Kluyveromyces vanudenii	*Kluyveromyces lactis*
Kluyveromyces veronae	*Kluyveromyces thermotolerans*
Kluyveromyces wikenii	*Kluyveromyces bulgaricus*
Leucosporidium capsuligenum	*Filobasidium capsuligenum*
Pichia melissophila	*Debaryomyces melissophilus*
Pichia nonfermentans	*Arthroascus javanensis*

Name in *The Yeasts* or *A New Key*	Name used in this *Guide*
Pichia polymorpha	*Debaryomyces polymorphus*
Rhodosporidium sphaerocarpum	*Rhodotorula glutinis*
Rhodosporidium toruloides	*Rhodotorula glutinis*
Saccharomyces aceti	*Saccharomyces cerevisiae*
Saccharomyces amurcae	*Zygosaccharomyces cidri*
Saccharomyces bailii	*Zygosaccharomyces bailii*
Saccharomyces bayanus	*Saccharomyces cerevisiae*
Saccharomyces beticus	*Saccharomyces cerevisiae*
Saccharomyces bisporus	*Zygosaccharomyces bisporus*
Saccharomyces capensis	*Saccharomyces cerevisiae*
Saccharomyces chevalieri	*Saccharomyces cerevisiae*
Saccharomyces cidri	*Zygosaccharomyces cidri*
Saccharomyces cordubensis	*Saccharomyces cerevisiae*
Saccharomyces coreanus	*Saccharomyces cerevisiae*
Saccharomyces delbrueckii	*Torulaspora delbrueckii*
Saccharomyces diastaticus	*Saccharomyces cerevisiae*
Saccharomyces eupagycus	*Zygosaccharomyces florentinus*
Saccharomyces fermentati	*Torulaspora delbrueckii*
Saccharomyces florentinus	*Zygosaccharomyces florentinus*
Saccharomyces florenzanii	*Torulaspora delbrueckii*
Saccharomyces globosus	*Saccharomyces cerevisiae*
Saccharomyces heterogenicus	*Saccharomyces cerevisiae*
Saccharomyces hienipiensis	*Saccharomyces cerevisiae*
Saccharomyces inconspicuus	*Torulaspora delbrueckii*
Saccharomyces inusitatus	*Saccharomyces cerevisiae*
Saccharomyces italicus	*Saccharomyces cerevisiae*
Saccharomyces kloeckerianus	*Torulaspora globosa*
Saccharomyces microellipsodes	*Zygosaccharomyces microellipsodes*
Saccharomyces montanus	*Zygosaccharomyces fermentati*
Saccharomyces mrakii	*Zygosaccharomyces mrakii*
Saccharomyces norbensis	*Saccharomyces cerevisiae*
Saccharomyces oleaceus	*Saccharomyces cerevisiae*
Saccharomyces oleaginosus	*Saccharomyces cerevisiae*
Saccharomyces pretoriensis	*Torulaspora pretoriensis*
Saccharomyces prostoserdovii	*Saccharomyces cerevisiae*
Saccharomyces rosei	*Torulaspora delbrueckii*
Saccharomyces rouxii	*Zygosaccharomyces rouxii*
Saccharomyces saitoanus	*Torulaspora delbrueckii*
Saccharomyces transvaalensis	*Pachytichospora transvaalensis*
Saccharomyces uvarum	*Saccharomyces cerevisiae*
Saccharomyces vafer	*Torulaspora delbrueckii*
Saccharomycopsis guttulata	Not included in Guide
Schwanniomyces alluvius	*Schwanniomyces occidentalis*
Schwanniomyces castellii	*Schwanniomyces occidentalis*
Schwanniomyces persoonii	*Schwanniomyces occidentalis*
Selenotila intestinalis	*Metschnikowia lunata*
Selenotila peltata	*Selenozyma peltata*
Torulopsis bovina	*Saccharomyces telluris*
Torulopsis candida	*Debaryomyces hansenii*
Torulopsis colliculosa	*Torulaspora delbrueckii*
Torulopsis dattila	*Kluyveromyces thermotolerans*
Torulopsis domercqii	*Wickerhamiella domercqii*
Torulopsis pintolopesii	*Saccharomyces telluris*
Trichosporon aculeatum	*Aciculoconidium aculeatum*
Trichosporon capitatum	*Geotrichum capitatum*
Trichosporon fermentans	*Geotrichum fermentans*
Trichosporon inkin	*Sarcinosporon inkin*
Trichosporon penicillatum	*Geotrichum penicillatum*

Glossary

Below are explanations of terms as used in this *Guide*. More information is provided in the following publications: Ainsworth (1971), Cowan (1968, 1978), Snell & Dick (1957), Stearn (1973).

aglycon. The non-carbohydrate moiety liberated on hydrolysis of a glycoside, e.g.
methyl α-D-glucopyranoside $\rightarrow$ D-glucose + methanol
       (glycoside)   (sugar)   (aglycon)
anascosporogenous. Not producing ascospores.
arthroconidium. Arthrospore. One of the cells formed by fragmentation of a hypha into separate cells.
arthrospore. Arthroconidium.
ascospore. Haploid cell produced by meiotic division of diploid cell; formed within the mother-cell or ascus.
ascosporogenous. Forms ascospores.
ascus. Cell in which ascospores are formed.
assimilation test. Test of utilization of organic or nitrogenous compound, usually with aerobic growth as the criterion for utilization.
Balling and Brix scales. For hydrometric measurements; expressed as percentage by weight of a pure sucrose solution of the same density.
ballistoconidium. A cell, resembling a basidiospore (see a standard textbook of mycology), ejected from its mother-cell by a drop-excretion mechanism. Formed by yeasts of the genera *Aessosporon*, *Bullera*, *Sporidiobolus* and *Sporobolomyces*.
ballistospore. Ballistoconidium.
basidiocarp. The basidium-bearing organ of a Basidiomycete (see a standard textbook of mycology).
binary test. Test with two possible responses, + or −.
blastoconidium. Blastospore. A cell formed by budding from a hypha or pseudohypha.
blastospore. Blastoconidium.
botryose. Grouped in clusters, like bunches of grapes.
bud. An outgrowth from a cell, becoming a daughter-cell, by vegetative reproduction. Characteristic of most yeasts.
catenulate. In a chain.
chlamydospore. A round, thick-walled cell, developed from a hypha.
clamp connection. A hyphal outgrowth, produced after the formation of a septum, joining the two hyphal cells separated by that septum. By this means, a nucleus is passed from one adjacent hyphal cell to the next, so a binucleate (dikaryotic) hypha is produced.
colony. Growth on the surface of a solid medium from one cell or a small group of cells.
conidiophore. Hyphal cell on which conidia are formed.
conidium. Spore. An asexual, thin-walled cell, e.g. blastoconidium, arthroconidium, ballistoconidium.
denticle. Small, tooth-like projection from a cell.
dichotomous key. Method for identifying organisms by a consecutive series of questions and answers, each question having only two alternative answers, yes or no.
diploid. Having two sets of chromosomes (2x), as opposed to one set (haploid), three sets (triploid) or more (polyploid).
dolipore. Central pore in septum of hypha; that part of the septum around the pore is thickened.
endospore. An asexual cell formed within another cell.
equivocal responses to tests. Test answers that are not likely to be found to be certainly + or certainly − by every observer, e.g. 'variable' (some strains +, others −), 'slow', 'delayed', 'not known'.

evanescent. Having a short existence.

falcate. Sickle-shaped.

ferment. *For yeasts*, utilize sugar non-oxidatively; catabolism by the glycolytic pathway to form chiefly ethanol and carbon dioxide.

filament. Thread-like hyphal or pseudohyphal cell.

fission. Formation of two cells by division of a whole cell.

genus. That rank in hierarchical biological classification between 'family' and 'species'. It is a group of species with many characters in common. For example, in '*Saccharomyces cerevisiae*', '*Saccharomyces*' is the genus.

haploid. Having one set of chromosomes. See 'diploid'.

heterothallic. Sexual conjugation occurring only through participation of two strains, the cells of either of which are intersterile. See 'homothallic'.

holotype. The one strain used by, or designated by, the original author(s) of the name of that species as the 'type'.

homonym. The same name given to more than one kind of organism: name identical to previously existing legitimate name based on a different 'type'.

homothallic. Fertile sexual conjugation occurring between cells of a single strain. See 'heterothallic'.

hypha. An elongate cell, or chain of cells, the filament of a mycelium. See 'pseudohypha'.

irredundant test sets. Sets of tests which include none that could be omitted without making any species less distinguishable.

lectotype. A strain subsequently chosen as nomenclatural type from the original strains studied by the author(s) of a new name, the author(s) having failed to designate a holotype.

neotype. A strain chosen as nomenclatural type of a name for which none of the strains on which the original description was based has survived.

new combination. A combination of a generic and a specific name, in which the specific epithet has been moved from another genus. Abbreviated *nov. comb.* or *comb. nov.* (i.e. *combinatio nova*).

nomen conservandum. A name conserved; one whose use has been sanctioned by an International Botanical Congress, despite the name contravening the rules of nomenclature.

nomen nudum. A name which, when published, lacked an adequate description (often a Latin description) as required by the rules of nomenclature.

oblate. Flattened at the poles.

phialoconidium. Phialoconidia are conidia produced successively from a flask-shaped or cylindrical cell (phialide) by budding or by forming transverse septa.

physiological test. Used in classifying or identifying yeasts, test of some physiological character, such as (i) ability to use a substrate for aerobic growth or anaerobic fermentation, (ii) requirement for a growth factor (e.g. vitamin), (iii) ability to grow in the presence of high concentration of solute (e.g. 50% D-glucose) or at 37°C.

program. Computer program. Set of instructions for computer to carry out required computations.

promycelium. An elongate cell formed by germination of a diploid teliospore of a basidiomycetous yeast. Meiosis occurs in the promycelium, which produces haploid sporidia.

pseudohypha. Cells, usually elongate, produced by a series of buddings, each daughter-cell remaining attached to its mother-cell, so forming a chain which may be branched. Mass of pseudohyphae often called 'pseudomycelium'.

pseudomycelium. See 'pseudohypha'.

psychrophilic. A psychrophilic yeast is one that can grow at low temperatures.

pure culture. The progeny from one yeast cell; a clone.

reticulation (in keys). The user is referred back to an earlier entry in the key, instead of forward to the same entry reprinted. This device saves space.

septate. Having septa.

septum. Transverse wall dividing cell or hypha into compartments. Formed by centripetal growth of wall material.

species. Similar strains (sometimes only one strain) which have been given a binomial, that is (i) a generic name and (ii) a specific epithet.

sporidium. Basidiospore of the Ustilaginales (see a textbook of mycology).
sporophore. Conidiophore.
sterigma. A fine process from a cell from which is produced another cell.
strain. Descendants of a single colony, ideally of a single cell without intervention of sexual reproduction.
sympodula. Hyphal cell with alternate branching, first one side then the other.
syntype. Each of two or more strains listed in the original publication of a species, when the author(s) has not designated a holotype; or, each of two or more strains simultaneously designated as types.
systematics. Comparative study of living organisms; includes taxonomy.
tautonym. The name of a species in which the generic name is repeated as the specific epithet, e.g. *Candida candida*. Not permitted by the *International Code of Botanical Nomenclature*.
taxon. Any taxonomic group of organisms, e.g. species, genus, family, order.
taxonomy. That part of biological systematics concerning classification, nomenclature and identification.
teliospore. A thick-walled cell (of basidiomycetous yeasts) in which nuclear fusion occurs.
type. Nomenclatural type, type strain or type species. The strain or species chosen to represent the name of a taxon (genus, species, subspecies, variety, forma) and to which that name is permanently attached.
vegetative. Asexual.
verticil. A number of parts in a circle round an axis.
yeast. A fungus that is mainly unicellular.

References

Ahearn, D.G. (1978). Medically important yeasts. *Annual Review of Microbiology*, **32**, 59–68.

Ahearn, D.G., Roth, F.J., Fell, J.W. & Meyers, S.P. (1960). Use of shaken cultures in the assimilation test for yeast identification. *Journal of Bacteriology*, **79**, 369–371.

Ahearn, D.G., Roth, F.J. & Meyers, S.P. (1968). Ecology and characterization of yeasts from aquatic regions of south Florida. *Marine Biology*, **1**, 291–308.

Ahearn, D.G., Yarrow, D. & Meyers, S.P. (1970). *Pichia spartinae* sp. n. from Louisiana marshland habitats. *Antonie van Leeuwenhoek*, **36**, 503–508.

Ainsworth, G.C. (1971). *Ainsworth & Bisby's Dictionary of the Fungi*, sixth edition. Kew, Surrey: Commonwealth Mycological Institute.

Amano, Y., Goto, S. & Kagami, M. (1975). A strongly ethanol assimilating new yeast *Candida brassicae* nov. sp. *Journal of Fermentation Technology*, **53**, 311–314.

Asai, Y., Makiguchi, N., Shimada, M. & Kurimura, Y. (1976). New species of methanol-assimilating yeasts. *Journal of General and Applied Microbiology*, **22**, 197–202.

Babjeva, I.P. & Gorin, S.E. (1975). *Lipomyces anomalus* sp. nov. *Antonie van Leeuwenhoek*, **41**, 185–191.

Babjeva, I.P. & Reshetova, I.S. (1975). A new yeast species from the soil – *Candida podzolica* sp. n. [In Russian.] *Mikrobiologiya*, **44**, 333–338.

Ballou, C.E. (1974). Some aspects of the structure, immunochemistry, and genetic control of yeast mannans. *Advances in Enzymology and Related Areas of Molecular Biology*, **40**, 239–270.

Bandoni, R.J., Johri, B.N. & Reid, S.A. (1975). Mating among isolates of three species of *Sporobolomyces*. *Canadian Journal of Botany*, **53**, 2942–2944.

Bandoni, R.J., Lobo, K.J. & Brezden, S.A. (1971). Conjugation and chlamydospores in *Sporobolomyces odorus*. *Canadian Journal of Botany*, **49**, 683–686.

Baptist, J.N. & Kurtzman, C.P. (1976). Comparative enzyme patterns in *Cryptococcus laurentii* and its taxonomic varieties. *Mycologia*, **68**, 1195–1203.

Barker, B.T.P. (1901a). A conjugating yeast. *Proceedings of the Royal Society of London*, Series B, **68**, 345–348.

Barker, B.T.P. (1901b). A conjugating 'yeast'. *Philosophical Transactions of the Royal Society of London*, Series B, **194**, 467–486.

Barnett, J.A. (1957). Some unsolved problems of yeast taxonomy. *Antonie van Leeuwenhoek*, **23**, 1–14.

Barnett, J.A. (1971a). Identifying yeasts. *Nature (London)*, **229**, 578.

Barnett, J.A. (1971b). Selection of tests for identifying yeasts. *Nature (London)*, **232**, 221–223.

Barnett, J.A. (1976). The utilization of sugars by yeasts. *Advances in Carbohydrate Chemistry and Biochemistry*, **32**, 125–234.

Barnett, J.A. & Buhagiar, R.W.M. (1971). *Torulopsis fragaria* species nova, a yeast from fruit. *Journal of General Microbiology*, **67**, 233–238.

Barnett, J.A. & Ingram, M. (1955). Technique in the study of yeast assimilation reactions. *Journal of Applied Bacteriology*, **18**, 131–148.

Barnett, J.A. & Pankhurst, R.J. (1974). *A New Key to The Yeasts*. Amsterdam: North-Holland Publishing Company.

Batra, L.R. (1971). Two new hemiascomycetes: *Pichia crossotarsi* and *P. microspora*. *Mycologia*, **63**, 994–1001.

Beech, F.W., Carr, J.G. & Codner, R.C. (1955). A multipoint inoculator for plating bacteria or yeasts. *Journal of General Microbiology*, **13**, 408–410.

Berlese, A.N. (1895). I funghi diversi dai saccaromiceti e capaci di determinare la fermentazione alcoolica. *Giornale Viticoltura e di Enologia*, **3**, 52–55.

Blagodatskaja, V. & Kocková-Kratochvílová, A. (1973). The heterogeneity of the species *Candida lipolytica*, *Candida pseudolipolytica* n. sp. and *Candida lipolytica* var. *thermotolerans* n. var. *Biológia (Bratislava)*, **28**, 709–716.

Boedijn, K.B. (1960). On a new genus of the Endomycetaceae. *Mycopathologia et Mycologia Applicata*, **12**, 163–167.

Bonaly, R. (1974). Aperçu sur la structure des parois des levures. *Sciences Pharmaceutiques et Biologiques de Lorraine*, **2**, 25–40.

Buhagiar, R.W.M. (1975). *Torulopsis bacarum*, *Torulopsis pustula* and *Torulopsis multis-gemmis* sp. nov., three new yeasts from soft fruit. *Journal of General Microbiology*, **86**, 1–11.

Buhagiar, R.W.M. & Barnett, J.A. (1971). The yeasts of strawberries. *Journal of Applied Bacteriology*, **34**, 727–739.

Buhagiar, R.W.M. & Barnett, J.A. (1973). *Sterigmatomyces acheniorum* species nova, a yeast from strawberries. *Journal of General Microbiology*, **77**, 71–78.

Buller, A.H.R. (1958). *Researches on Fungi*, vol. 5. New York: Hafner Publishing Co.

Campbell, I. (1972). Numerical analysis of the genera *Saccharomyces* and *Kluyveromyces*. *Journal of General Microbiology*, **73**, 279–301.

Campbell, I. (1973). Computer identification of yeasts of the genus *Saccharomyces*. *Journal of General Microbiology*, **77**, 127–135.

Campbell, I. (1974). Methods of numerical taxonomy for various genera of yeasts. *Advances in Applied Microbiology*, **17**, 135–156.

Campbell, I. (1975). Numerical analysis and computerized identification of the yeast genera *Candida* and *Torulopsis*. *Journal of General Microbiology*, **90**, 125–132.

Capriotti, A. (1958). *Zygosaccharomyces mrakii* nova species. A new yeast from silage. *Archiv für Mikrobiologie*, **30**, 387–392.

Capriotti, A. (1961). *Debaryomyces cantarellii* nova species, a new yeast isolated from a Finnish soil. *Archiv für Mikrobiologie*, **39**, 123–129.

Capriotti, A. (1967). *Saccharomyces servazzii* n. sp. A new yeast from Finland soil. *Annali di Microbiologia ed Enzimologia*, **17**, 79–84.

Castelli, T. (1938). Nuovi blastomiceti isolati da mosti del chianti e zone limitrofe. *Archiv für Mikrobiologie*, **9**, 449–468.

Ciferri, R. (1925). Studi sulle Torulopsidaceae. Sui nomi generici di *Torula*, *Eutorula*, *Torulopsis*, *Cryptococcus*, e sul nome di gruppo *Torulaceae*. *Atti dell' Istituto Botanico dell' Università di Pavia*, Serie III, **2**, 129–146.

Ciferri, R. & Redaelli, P. (1929). Studies on the Torulopsidaceae. A trial general systematic classification of the asporigenous ferments. *Annales Mycologici*, **27**, 243–295.

Cowan, S.T. (1968). *A Dictionary of Microbial Taxonomic Usage*. Edinburgh: Oliver & Boyd.

Cowan, S.T. (1978). *Dictionary of Microbial Taxonomy*. Edited by L.R. Hill. Cambridge: Cambridge University Press.

Cowan, S.T. & Steel, K.J. (1966). *Manual for the Identification of Medical Bacteria*. Cambridge: Cambridge University Press.

Darwin, C. (1859). *On the origin of species by means of natural selection, or the preservation of favoured races in the struggle for life*. London: John Murray.

Diddens, H.A. & Lodder, J. (1942). *Die anaskosporogenen Hefen*, zweite Hälfte. Amsterdam: North-Holland Publishing Company.

do Carmo-Sousa, L. (1970). *Trichosporon* Behrend. In *The Yeasts. A Taxonomic Study*, pp. 1309–1352. Edited by J. Lodder. Amsterdam: North-Holland Publishing Company.

Dwidjoseputro, D. & Wolf, F.T. (1970). Microbiological studies of Indonesian fermented foodstuffs. *Mycopathologia et Mycologia Applicata*, **41**, 211–222.

Fell, J.W. (1970). Yeasts with heterobasidiomycetous life cycles. In *Recent Trends in Yeast Research*, pp. 49–66. Edited by D.G. Ahearn. Atlanta: School of Arts and Sciences, Georgia State University.

Fell, J.W. (1974). Heterobasidiomycetous yeasts *Leucosporidium* and *Rhodosporidium*, their systematics and sexual incompatibility systems. *Transactions of the Mycological Society of Japan*, **15**, 316–323.

Fell, J.W. & Hunter, I.L. (1974). *Torulopsis austromarina* sp. nov. A yeast isolated from the Antarctic Ocean. *Antonie van Leeuwenhoek*, **40**, 307–310.

Fell, J.W., Hunter, I.L. & Tallman, A.S. (1973). Marine basidiomycetous yeasts (*Rhodosporidium* spp. n.) with tetrapolar and multiple allelic bipolar mating systems. *Canadian Journal of Microbiology*, **19**, 643–657.

Fell, J.W. & Phaff, H.J. (1970). *Leucosporidium* Fell, Statzell, Hunter et Phaff. In *The Yeasts. A Taxonomic Study*, second edition, pp. 776–802. Edited by J. Lodder. Amsterdam: North-Holland Publishing Company.

Fell, J.W. & Statzell, A.C. (1971). *Sympodiomyces* gen. n. a yeast-like organism from southern marine waters. *Antonie van Leeuwenhoek*, **37**, 359–367.

Fowell, R.R. (1952). Sodium acetate agar as a sporulation medium for yeast. *Nature (London)*, **170**, 578.

Golubev, W.I. (1973). *Nadsonia commutata* nov. sp. *Mikrobiologiya*, **42**, 1058–1061.

Golubev, W.I. (1977). *Metschnikowia lunata* sp. nov. *Antonie van Leeuwenhoek*, **43**, 317–322.

Gómez, C.R. & Rico, G.S. (1956). Una nueva variedad de *Torulopsis holmii* (Jorgensen) y estudio de sus productos metabólicos. *Microbiología Española*, **9**, 147–162.

Gorin, P.A.J. & Spencer, J.F.T. (1970). Proton magnetic resonance spectroscopy – an aid in identification and chemotaxonomy of yeasts. *Advances in Applied Microbiology*, **13**, 25–89.

Goto, S. & Sugiyama, J. (1968). Studies on the Himalayan yeasts and moulds (1). A new species of *Debaryomyces* and some asporogenous yeasts. *Journal of Japanese Botany*, **43**, 102–108.

Goto, S. & Sugiyama, J. (1970). Studies on Himalayan yeasts and moulds IV. Several asporogenous yeasts including two new taxa of *Cryptococcus*. *Canadian Journal of Botany*, **48**, 2097–2101.

Goto, S., Sugiyama, J. & Iizuka, H. (1969). A taxonomic study of Antarctic yeasts. *Mycologia*, **61**, 748–774.

Goto, S., Yamasato, K. & Iizuka, H. (1974). Identification of yeasts isolated from the Pacific Ocean. *Journal of General and Applied Microbiology*, **20**, 309–316.

Gower, J.C. & Barnett, J.A. (1971). Selecting tests in diagnostic keys with unknown responses. *Nature (London)*, **232**, 491–493.

Gower, J.C. & Payne, R.W. (1975). A comparison of different criteria for selecting binary tests in diagnostic keys. *Biometrika*, **62**, 665–672.

Grinbergs, J. (1967). Zur kenntnis einer neuen Hefeart: *Candida tepae* sp. nov. *Archiv für Mikrobiologie*, **56**, 202–204.

Grinbergs, J. & Yarrow, D. (1970a). Two new *Candida* species: *Candida chilensis* sp. n. and *Candida valdiviana* sp. n. *Antonie van Leeuwenhoek*, **36**, 143–148.

Grinbergs, J. & Yarrow, D. (1970b). *Rhodotorula araucariae* sp. n. *Antonie van Leeuwenhoek*, **36**, 455–457.

Groenewege, J. (1921). Over de oorzaak van rustiness op rubber van Hevea brasiliensis. *Mededelingen van het Algemeen Proefstation voor den Landbouw*, No. 11, 5–19.

Grose, E.S. & Marinkelle, C.J. (1968). A new species of *Candida* from Colombian bats. *Mycopathologia et Mycologia Applicata*, **36**, 225–227.

Guilliermond, A. (1912). *Les Levures*. Paris: Octave Doin et Fils, Editeurs.

Gyllenberg, H. (1963). A general method for deriving determination schemes for random collections of microbial isolates. *Annales Academiae Scientiarum Fennicae* Series A IV. Biologica Part 69, 1–23.

Gyllenberg, H. G. (1964). An approach to numerical description of microbial populations. *Annales Academiae Scientiarum Fennicae* Series A IV. Biologica Part 81, 1–23.

Hedrick, L.R. (1976). *Candida fluviotilis* sp. nov. and other yeasts from aquatic environments. *Antonie van Leeuwenhoek*, **42**, 329–332.

Hedrick, L.R. & Dupont, P.D. (1968). Two new yeasts: *Trichosporon aquatile* and *Trichosporon eriense* spp. n. *Antonie van Leeuwenhoek*, **34**, 474–482.

Henninger, W. & Windisch, S. (1975a). *Pichia lindnerii* sp. n., eine neue Methanol assimilierende Hefe aus Erde. *Archives of Microbiology*, **105**, 47–48.

Henninger, W. & Windisch, S. (1975b). A new yeast of *Sterigmatomyces*, S. aphidis sp. n. *Archives of Microbiology*, **105**, 49–50.
Henninger, W. & Windisch, S. (1976a). *Torulopsis spandovensis* sp. n., a new yeast from beer. *Archives of Microbiology*, **107**, 205–206.
Henninger, W. & Windisch, S. (1976b). *Kluyveromyces blattae* sp. n., eine neue vielsporige Hefe aus *Blatta orientalis*. *Archives of Microbiology*, **109**, 153–156.
International Code of Botanical Nomenclature (1978). Adopted by the Twelfth International Botanical Congress, Leningrad, July 1975. Prepared and edited by F.A. Stafleu, V. Demoulin, W. Greuter, P. Hiepko, I.A. Linczevski, R. McVaugh, R.D. Meikle, R.C. Rollins, R. Ross, J.M. Schopf & E.G. Voss. Utrecht: Bohn, Scheltema & Holkema.
International Code of Nomenclature of Bacteria (1975). Approved by the First International Congress of Bacteriology, Jerusalem, September 1973. Edited by S.P. Lapage, P.H.A. Sneath, E.F. Lessel, V.B.D. Skerman, H.P.R. Seeliger & W.A. Clark. Washington, D.C.: American Society for Microbiology.
Jacob, F. (1969). *'Pichia' adzetii* et *'Pichia' abadieae*, nouvelles espèces de levures isolées de liqueurs tannantes végétales. *Bulletin Trimestriel de la Société Mycologique de France*, **85**, 117–127.
Jacob, F. (1970). Deux espèces nouvelles de levures asporogènes isolées de liqueurs tannantes végétales. *Annales de l'Institut Pasteur*, Paris, **118**, 207–213.
Johannsen, E. & van der Walt, J.P. (1978). Interfertility as basis for the delimitation of *Kluyveromyces marxianus*. *Archives of Microbiology*, **118**, 45–48.
Jurzitza, G. (1970). Über Isolierung, Kultur and Taxonomie einiger Anobiidensymbionten (Insecta, Coleoptera). *Archiv für Mikrobiologie*, **72**, 203–222.
Kato, K., Kurimura, Y., Makiguchi, N. & Asai, Y. (1974). Determination of methanol strongly assimilating yeasts. *Journal of General and Applied Microbiology*, **20**, 123–127.
King, D.S. & Jong, S.C. (1975). *Sarcinosporon*: a new genus to accommodate *Trichosporon inkin* and *Prototheca filamenta*. *Mycotaxon*, **3**, 89–94.
King, D.S. & Jong, S.C. (1976). *Aciculoconidium*: a new hyphomycetous genus to accommodate *Trichosporon aculeatum*. *Mycotaxon*, **3**, 401–408.
Klöcker, A. (1913). Recherches sur les organismes de fermentation I. Recherches sur quelques nouvelles espèces de *Pichia*, et remarques relatives aux descriptions spécifiques des Saccharomycètes en général. *Comptes Rendus des Travaux du Laboratoire de Carlsberg*, **10**, 207–226.
Klöcker, A. (1924). *Die Gärungsorganismen in der Theorie und Praxis der Alkoholgärungsgewerbe*, third edition. Berlin: Urban & Schwarzenberg.
Kocková-Kratochvílová, A. & Ondrušová, D. (1971). Torulopsisarten aus den Oberflächen höherer Pilze. *Torulopsis kruisii* n. sp. und *Torulopsis schatavii* n. sp. *Biológia (Bratislava)*, **26**, 447–485.
Kodama, K. (1972). Ascosporogenous yeasts from exudates of trees in Japan. *Abstracts of the Fourth International Fermentation Symposium*, 19 to 25 March 1972, Kyoto, Japan, Abstract G15-1, p. 291.
Kodama, K. (1975). New species of *Pichia* isolated from tree exudates in Japan. *Journal of Fermentation Technology*, **53**, 626–630.
Kodama, K. & Kyono, T. (1974a). Ascosporogenous yeasts isolated from tree exudates in Japan. *Journal of Fermentation Technology*, **52**, 1–9.
Kodama, K. & Kyono, T. (1974b). Ascosporogenous yeasts isolated from tree exudates in Japan (continued). *Journal of Fermentation Technology*, **52**, 605–613.
Kolfschoten, G.A. & Yarrow, D. (1970). *Brettanomyces naardenensis*, a new yeast from soft drinks. *Antonie van Leeuwenhoek*, **36**, 458–460.
Komagata, K. & Nakase, T. (1965). New species of the genus *Candida* isolated from frozen foods. *Journal of General and Applied Microbiology*, **11**, 255–267.
Komagata, K., Nakase, T. & Katsuya, N. (1964). Assimilation of hydrocarbons by yeast. II. Determination of hydrocarbon-assimilating yeasts. *Journal of General and Applied Microbiology*, **10**, 323–331.
Krassilnikov, N.A., Babjeva, I.P. & Meavahd, K. (1967). A new genus of soil yeast *Zygolipomyces* nov. gen. [In Russian.] *Mikrobiologiya*, **36**, 923–931.
Kreger-van Rij, N.J.W. (1964). *A Taxonomic Study of the Yeast Genera*

Endomycopsis, Pichia and Debaryomyces. Doctoral thesis, University of Leiden.

Kreger-van Rij, N.J.W. (1970a). *Debaryomyces* Lodder et Kreger-van Rij nom. conserv. In *The Yeasts. A Taxonomic Study*, second edition, pp. 129–156. Edited by J. Lodder. Amsterdam: North-Holland Publishing Company.

Kreger-van Rij, N.J.W. (1970b). *Endomycopsis* Dekker. In *The Yeasts. A Taxonomic Study*, second edition, pp. 166–208. Edited by J. Lodder. Amsterdam: North-Holland Publishing Company.

Kreger-van Rij, N.J.W. (1970c). *Pichia* Hansen. In *The Yeasts. A Taxonomic Study*, second edition, pp. 455–555. Edited by J. Lodder. Amsterdam: North-Holland Publishing Company.

Kreger-van Rij, N.J.W. (1973). Endomycetales, basidiomycetous yeasts, and related fungi. In *The Fungi, An Advanced Treatise*, vol. IVA, pp. 11–32. Edited by G.C. Ainsworth, F.K. Sparrow & A.S. Sussman. New York: Academic Press.

Kreger-van Rij, N.J.W. (Editor) (in preparation). *The Yeasts. A Taxonomic Study*, third edition. Amsterdam: North-Holland Publishing Company.

Kreger-van Rij, N.J.W. & Veenhuis, M. (1971). A comparative study of the cell wall structure of basidiomycetous and related yeasts. *Journal of General Microbiology*, **68**, 87–95.

Kudriavzev, V.I. (1954). *The Systematics of Yeasts*. [In Russian.] Moscow: Academy of Sciences of the USSR. [Translation: W.I. Kudrjawzew. (1960). *Die Systematik der Hefen*. Berlin: Akademie-Verlag.]

Kudrjawzew, W.I. (1960). *Die Systematik der Hefen*. Berlin: Akademie-Verlag.

Kumbhojkar, M.S. (1972). *Schizosaccharomyces slooffiae* Kumbhojkar, a new species of osmophilic yeasts from India. *Current Science*, **41**, 151–152.

Kurtzman, C.P. (1973). Formation of hyphae and chlamydospores by *Cryptococcus laurentii*. *Mycologia*, **65**, 388–395.

Kurtzman, C.P. & Kreger-van Rij, N.J.W. (1976). Ultrastructure of ascospores from *Debaryomyces melissophilus*, a new taxonomic combination. *Mycologia*, **68**, 422–425.

Kurtzman, C.P. & Smiley, M.J. (1974). A taxonomic re-evaluation of the round-spored species of *Pichia*. *Proceedings of the Fourth International Symposium on Yeasts, Vienna*, Part I, 231–232.

Kurtzman, C.P., Smiley, M.J. & Baker, F.L. (1975). Scanning electron microscopy of ascospores of *Debaryomyces* and *Saccharomyces*. *Mycopathologia et Mycologia Applicata*, **55**, 29–34.

Kurtzman, C.P., Vesonder, R.F. & Smiley, M.J. (1974). Formation of extracellular 3-D-hydroxy-palmitic acid by *Saccharomycopsis malanga*. *Mycologia*, **66**, 580–587.

Kurtzman, C.P. & Wickerham, L.J. (1972). *Pichia besseyi* sp. n. *Antonie van Leeuwenhoek*, **38**, 49–52.

Kurtzman, C.P. & Wickerham, L.J. (1973). *Saccharomycopsis crataegensis*, a new heterothallic yeast. *Antonie van Leeuwenhoek*, **39**, 81–87.

Kwon-Chung, K.J. (1977). Perfect state of *Cryptococcus uniguttulatus*. *International Journal of Systematic Bacteriology*, **27**, 293–299.

Langeron, M. & Guerra, P. (1935). Bases morphologiques et biologiques de la classification des champignons levuriformes anascospores. *Proceedings of the Sixth International Botanical Congress Amsterdam*, **2**, 165–167.

Leask, B.G.S. & Yarrow, D. (1976). *Pichia norvegensis* sp. nov. *Sabouraudia*, **14**, 61–63.

Lederberg, J. & Lederberg, E.M. (1952). Replica plating and indirect selection of bacterial mutants. *Journal of Bacteriology*, **63**, 399–406.

Lindner, [P.] (1904). Neue Erfahrungen aus dem letzten Jahre in bezug auf Hefe und Gärung. *Jahrbuch des Versuchs- und Lehransfalt für Brauerei in Berlin*, **7**, 441–464.

Lloyd, C.G. (1905). Mycological notes. No. 19. May, 1905. [Bound in *Index of the Mycological Writings of C.G. Lloyd*, Vol. II. 1905–1908. Cincinnati, Ohio.]

Lodder, J. (1932). Über einige durch das 'Centraalbureau voor Schimmelcultures' neuerworbene sporogene Hefearten. *Zentralblatt für Bakteriologie, Parasitenkunde und Infektionskrankheiten*, Zweite Abteilung, **86**, 227–253.

Lodder, J. (1934). *Die Anaskosporogenen Hefen*, erste Hälfte. Amsterdam: N.V. Noord-Hollandsche Uitgeversmaatschappij.

Lodder, J. (1938). *Torulopsis* or *Cryptococcus*? *Mycopathologia et Mycologia Applicata*, **1**, 62–67.

Lodder, J. (Editor) (1970). *The Yeasts. A Taxonomic Study*, second edition. Amsterdam: North-Holland Publishing Company.

Lodder, J. & Kreger-van Rij, N.J.W. (1952). *The Yeasts. A Taxonomic Study*, first edition. Amsterdam: North-Holland Publishing Company.

Mager, J. & Aschner, M. (1947). Biological studies on capsulated yeasts. *Journal of Bacteriology*, **53**, 283–295.

Mallette, M.F. (1969). Evaluation of growth by physical and chemical means. *Methods in Microbiology*, **1**, 521–566.

Marcilla, J., Alas, G. & Feduchy, E. (1939). Contribución al estudio de las levaduras que forman velo sobre ciertos vinos de elevado grado alcohólico. *Anales del Centro de Investigaciones vinícolas*, **1**, 1–230. [Note: this issue was put together in 1936, but not published until 1939. Consequently some authors give its date as 1936.]

Martini, A. (1973). Ibridazioni DNA/DNA tra specie di lieviti del genere *Kluyveromyces*. *Annali della Facoltà di Agraria, Università degli Studi di Perugia*, **28**, 3–15.

Mayr, E. (1942). *Systematics and the Origin of Species*. New York: Columbia University Press.

McClary, D.O., Nulty, W.L. & Miller, G.R. (1959). Effect of potassium versus sodium in the sporulation of *Saccharomyces*. *Journal of Bacteriology*, **78**, 362–368.

Meyer, S.A., Anderson, K., Brown, R.E., Smith, M.T., Yarrow, D., Mitchell, G. & Ahearn, D.G. (1975). Physiological and DNA characterization of *Candida maltosa*, a hydrocarbon-utilizing yeast. *Archives of Microbiology*, **104**, 225–231.

Meyer, S.A., Brown, R.E. & Smith, M.T. (1977). Species status of *Hanseniaspora guilliermondii* Pijper. *International Journal of Systematic Bacteriology*, **27**, 162–164.

Meyer, S.A., Smith, M.T. & Simione, F.P. (1978). Systematics of *Hanseniaspora* Zikes and *Kloeckera* Janke. *Antonie van Leeuwenhoek*, **44**, 79–96.

Meynell, G.G. & Meynell, E. (1970). *Theory and Practice in Experimental Bacteriology*, second edition. Cambridge: Cambridge University Press.

Miller, M.W., Phaff, H.J., Miranda, M., Heed, W.B. & Starmer, W.T. (1976). *Torulopsis sonorensis*, a new species of the genus *Torulopsis*. *International Journal of Systematic Bacteriology*, **26**, 88–91.

Miller, M.W., Yoneyama, M. & Soneda, M. (1976). *Phaffia*, a new yeast genus in the Deuteromycotina (Blastomycetes). *International Journal of Systematic Bacteriology*, **26**, 286–291.

Monod, J., Cohen-Bazire, G. & Cohn, M. (1951). Sur la biosynthèse de la β-galactosidase (lactase) chez *Escherichia coli*. La spécificité de l'induction. *Biochimica et Biophysica Acta*, **7**, 585–599.

Montrocher, R. (1967). Quelques nouvelles espèces et variétés du genre *Candida* (levures asporogènes). *Revue de Mycologie*, **32**, 69–92.

Moriyon, I. & Ramírez, C. (1974). New species of yeasts isolated from an acid washed brown soil. *Proceedings of the Fourth International Symposium on Yeasts, Vienna*, Part I, 233–234.

Morse, L.E. (1971). Specimen identification and key construction with time-sharing computers. *Taxon*, **20**, 269–282.

Mullin, J.K. (1970). A computer optimized question asker for bacteriological specimen identification. *Mathematical Biosciences*, **6**, 55–66.

Nadson, G. & Krassilnikov, N. (1928). Un nouveau genre d'Endomycétacées: *Guilliermondella*, nov. gen. *Comptes Rendus Hebdomadaires des Séances de l'Académie des Sciences*, **187**, 307–309.

Naganishi, H. (1928). The study of five new species of yeast. *Journal of Zymology*, **6**, 1–12 [Titles of paper and of journal are in Japanese.]

Nakase, T. (1971a). New species of yeasts resembling *Candida krusei* (Cast.) Berkhout. *Journal of General and Applied Microbiology*, **17**, 383–398.

Nakase, T. (1971b). New species of yeasts found in Japan. *Journal of General and Applied Microbiology*, **17**, 409–419.

Nakase, T. (1971c). Four new yeasts found in Japan. *Journal of General and Applied Microbiology*, **17**, 469–478.

Nakase, T. (1975). Three new asporogenous yeasts found in industrial waste water. *Antonie van Leeuwenhoek*, **41**, 201–210.

Nakase, T. & Komagata, K. (1966). New yeasts, *Endomycopsis muscicola* and *Pichia zaruensis*. *Journal of General and Applied Microbiology*, **12**, 347–352.

Nakase, T., Komagata, K. & Fukazawa, Y. (1976). *Candida pseudointermedia* sp. nov., isolated from 'Kamaboko', a traditional fish-paste product in Japan. *Journal of General and Applied Microbiology*, **22**, 177–182.

Nel, E.E. & van der Walt, J.P. (1968). *Torulopsis humilis*, sp. n. *Mycopathologia et Mycologia Applicata*, **36**, 94–96.

Nieuwdorp, P.J., Bos, P. & Slooff, W.C. (1974). Classification of *Lipomyces*. *Antonie van Leeuwenhoek*, **40**, 241–254.

Novák, E.K. & Zsolt, J. (1961). A new system proposed for yeasts. *Acta Botanica Academiae Scientiarum Hungaricae*, **7**, 93–145.

Ohara, Y., Nonomura, H. & Yamazaki, T. (1965). *Candida incommunis* sp. nov., a new yeast isolated from grape must. *Journal of General and Applied Microbiology*, **11**, 273–275.

Ohara, Y., Nonomura, H. & Yunome, H. (1960). New species and varieties of Cryptococcoideae from grape musts. *Journal of the Agricultural Chemical Society of Japan*, **34**, 709–711.

Oki, T., Kouno, K., Kitai, A. & Ozaki, A. (1972). New yeasts capable of assimilating methanol. *Journal of General and Applied Microbiology*, **18**, 295–305.

Olive, L.S. (1968). An unusual new heterobasidiomycete with *Tilletia*-like basidia. *Journal of the Elisha Mitchell Scientific Society*, **84**, 261–266.

Onishi, H. (1957). Studies on osmophilic yeasts Part III. Classification of the osmophilic soy and miso yeasts. *Bulletin of the Agricultural Chemical Society of Japan*, **21**, 151–156.

Onishi, H. (1972). *Candida tsukubaensis* sp. n. *Antonie van Leeuwenhoek*, **38**, 365–367.

Onishi, H. & Suzuki, T. (1969). *Torulopsis mannitofaciens* sp. n. isolated from soy-sauce mash. *Antonie van Leeuwenhoek*, **35**, 258–260.

Payne, R.W. (1975). Genkey: a program for constructing diagnostic keys. In *Biological Identification with Computers*, pp. 65–72. Edited by R.J. Pankhurst. London: Academic Press.

Payne, R.W. (1977). Reticulation and other methods of reducing the size of printed diagnostic keys. *Journal of General Microbiology*, **98**, 595–597.

Payne, R.W. (1978). *Genkey. A program for Constructing and Printing Identification Keys and Diagnostic Tables*. Harpenden: Rothamsted Experimental Station.

Payne, R.W. & Preece, D.A. (1977). Incorporating checks against observer error into identification keys. *New Phytologist*, **79**, 203–209.

Payne, R.W. & Preece, D.A. (1980). Identification keys and diagnostic tables: a review. *Journal of the Royal Statistical Society*. Series A (In Press.)

Payne, R.W., Walton, E. & Barnett, J.A. (1974). A new way of representing diagnostic keys. *Journal of General Microbiology*, **83**, 413–414.

Phaff, H.J. (1970a). *Hanseniaspora* Zikes. In *The Yeasts. A Taxonomic Study*, second edition, pp. 209–225. Edited by J. Lodder. Amsterdam: North-Holland Publishing Company.

Phaff, H.J. (1970b). *Sporobolomyces* Kluyver et van Niel. In *The Yeasts. A Taxonomic Study*, second edition, pp. 831–862. Edited by J. Lodder. Amsterdam: North-Holland Publishing Company.

Phaff, H.J. (1970c). *Kloeckera* Janke. In *The Yeasts. A Taxonomic Study*, second edition, pp. 1146–1160. Edited by J. Lodder. Amsterdam: North-Holland Publishing Company.

Phaff, H.J. (1971). Structure and biosynthesis of the yeast cell envelope. In *The Yeasts*, vol. 2, pp. 135–210. Edited by A.H. Rose & J.S. Harrison. London: Academic Press.

Phaff, H.J. & Fell, J.W. (1970). *Cryptococcus* Kützing emend. Phaff et Spencer. In *The Yeasts. A Taxonomic Study*, second edition, pp. 1088–1145. Edited by J. Lodder. Amsterdam: North-Holland Publishing Company.

Phaff, H.J., Miller, M.W. & Miranda, M. (1976). *Pichia scutulata*, a new species from tree exudates. *International Journal of Systematic Bacteriology*, **26**, 326–331.

Phaff, H.J., Miller, M.W., Miranda, M., Heed, W.B. & Starmer, W.T. (1974). *Cryptococcus cereanus*, a new species of the genus *Cryptococcus*. *International Journal of Systematic Bacteriology*, **24**, 486–490.

Phaff, H.J., Miller, M.W. & Shifrine, M. (1956). The taxonomy of yeasts isolated from *Drosophila* in the Yosemite region of California. *Antonie van Leeuwenhoek*, **22**, 145–161.

Phaff, H.J., Miller, M.W., Yoneyama, M. & Soneda, M. (1972). A comparative study of the yeast florae associated with trees on the Japanese islands and on the west coast of North America. *Proceedings of the Fourth International Fermentation Symposium*, 19–25 March 1972, Kyoto, Japan; Fermentation Technology Today, 759–774.

Philippov, G.S. (1932). Porcha fruktovi konservov, vizvannaya drojjevim gribkom. [Zur Frage der Verdorbenheit der Fruchtkonserven.] *Trudy Tsentral'nogo Nauchno-issledovatel'skogo Biokhimicheskogo Instituta Pishchevoi i Vkusovoi Promyshlennosti Navkomsnaba Soyuza SSR [Schriften des Zentralen Biochemischen der Nahrungs- und Genussmittelindustrie]* **2**, 26–32.

Pignal, M.C. (1970). A new species of yeast isolated from decaying insect-invaded wood. *Antonie van Leeuwenhoek*, **36**, 525–529.

Pijper, A. (1928*a*). Een nieuwe *Hanseniospora*. *Verslagen van de gewone Vergadering der Afdeeling Natuurkunde, Koninklijke Akademie van Wetenschappen*, **37**, 868–871.

Pijper, A. (1928*b*). A new *Hanseniospora*. *Proceedings, Koninklijke Akademie van Wetenschappen, Amsterdam*, **31**, 989–992.

Pitt, J.I. & Miller, M.W. (1970). Speciation in the yeast genus *Metschnikowia*. *Antonie van Leeuwenhoek*, **36**, 357–381.

Pontecorvo, G. (1949). Auxanographic techniques in biochemical genetics. *Journal of General Microbiology*, **3**, 122–126.

Powell, E.O. (1963). Photometric methods in bacteriology. *Journal of the Science of Food and Agriculture*, **14**, 1–8.

Preece, D.A. (1976). Identification keys and diagnostic tables. *Mathematical Scientist*, **1**, 43–65.

Price, C.W., Fuson, G.B. & Phaff, H.J. (1978). Genome comparison in yeast systematics: delimitation of species within the genera *Schwanniomyces*, *Saccharomyces*, *Debaryomyces*, and *Pichia*. *Microbiological Reviews*, **42**, 161–193.

Ramírez, C. (1974). A compilation of descriptions of new *Candida* species with keys to all species of the genus described up to date. *Microbiología Española*, **27**, 15–78.

Ramírez, C. & González, C. (1972). *Candida iberica* sp. n. A new species isolated from Spanish sausages. *Canadian Journal of Microbiology*, **18**, 1778–1780.

Ramírez, C. & Martinez, A.T. (1978). *Torulopsis pampelonensis* sp. nov. A new species of yeast isolated from beech forest soil. *Canadian Journal of Microbiology*, **24**, 433–435.

Ramírez, C. & Moriyon, I. (1978). *Torulopsis navarrensis* sp. nov. A new species of yeast isolated from an acid washed brown soil in the province of Navarra, Spain. *Mycopathologia*, **63**, 61–63.

Redhead, S.A. & Malloch, D.W. (1977). The Endomycetaceae: new concepts, new taxa. *Canadian Journal of Botany*, **55**, 1701–1711.

Rodrigues de Miranda, L. (1972). *Filobasidium capsuligenum* nov. comb. *Antonie van Leeuwenhoek*, **38**, 91–99.

Rodrigues de Miranda, L. (1975). Two species of the genus *Sterigmatomyces*. *Antonie van Leeuwenhoek*, **41**, 193–199.

Rodrigues de Miranda, L. & Diem, H.G. (1974). Deux nouvelles espèces de levures isolées de la phyllosphere de l'Orge. *Canadian Journal of Botany*, **52**, 279–282.

Rodrigues de Miranda, L. & Norkrans, B. (1968). *Candida suecica* sp. n. isolated from marine environment. *Antonie van Leeuwenhoek*, **34**, 115–118.

Rodrigues de Miranda, L. & Török, T. (1976). *Pichia humboldtii* sp. nov., the perfect state of *Candida ingens*. *Antonie van Leeuwenhoek*, **42**, 343–348.

Rypka, E.W. & Babb, R. (1970). Automatic construction and use of an identification table. *Medical Research Engineering*, **9**, 9–19.

Rypka, E.W., Clapper, W.E., Bowen, I.G. & Babb, R. (1967). A model for the identification of bacteria. *Journal of General Microbiology*, **46**, 407–424.

Sahm, H. (1977). Metabolism of methanol by yeasts. *Advances in Biochemical Engineering*, **6**, 77–103.

Saito, K. (1923). Beschreibung von zwei neuen Hefearten, nebst Bemerkungen über die Sporenbildung bei *Torulaspora Delbrücki*, Lindner. *Botanical Magazine (Tokyo)*, **37**, 63–66.

Santa María, J. (1970). *Saccharomyces gaditensis* y *Saccharomyces cordubensis*, dos nuevas especies de levaduras de 'flor'. *Boletín del Instituto Nacional de Investigaciones Agronómicas*, **62**, 57–66.

Santa María, J. (1971). *Candida ergatensis* nov. spec. *Anales del Instituto Nacional de Investigaciones Agrarias*, Serie: General, Núm. 1, 85–88.

Santa María, J. & García Aser, C. (1970). *Pichia castillae*, nov. spec., aislada de insectos. *Boletín del Instituto Nacional de Investigaciones Agronómicas*, **62**, 51–55.

Santa María, J. & García Aser, C. (1971). *Debaryomyces yarrowii* nov. spec. *Anales del Instituto Nacional de Investigaciones Agrarias*, Serie: General, Núm. 1, 89–92.

Schiönning, H. (1903). Nouveau genre de la famille des Saccharomycètes. *Comptes Rendus des Travaux du Laboratoire de Carlsberg*, **6**, 103–125.

Scott, D.B. & van der Walt, J.P. (1970a). *Hansenula sydowiorum* sp. n. *Antonie van Leeuwenhoek*, **36**, 45–48.

Scott, D.B. & van der Walt, J.P. (1970b). Three new yeasts from South African insect sources. *Antonie van Leeuwenhoek*, **36**, 389–396.

Scott, D.B. & van der Walt, J.P. (1971a). *Hansenula dryadoides* sp. n., a new species from South African insect sources. *Antonie van Leeuwenhoek*, **37**, 171–175.

Scott, D.B. & van der Walt, J.P. (1971b). *Pichia cicatricosa* sp. n. a new auxiliary ambrosia fungus. *Antonie van Leeuwenhoek*, **37**, 177–183.

Shehata, A.M.E.T., Mrak, E.M. & Phaff, H.J. (1955). Yeasts isolated from *Drosphila* and from their suspected feeding places in southern and central California. *Mycologia*, **47**, 799–811.

Shifrine, M., Phaff, H.J. & Demain, A.L. (1954). Determination of carbon assimilation patterns of yeasts by replica plating. *Journal of Bacteriology*, **68**, 28–35.

Smith, M.T. (1973). [In] Progress Report 1972 of the Centraalbureau voor Schimmelcultures Baarn and Delft. *Verhandelingen der Koninklijke Nederlandse Akademie van Wetenschappen*, Afdeling Natuurkunde, Tweede Reeks, **61**, 59–81.

Smith, M.T. (1974). *Hanseniaspora occidentalis* sp. nov. *Antonie van Leeuwenhoek*, **40**, 441–444.

Smith, M.T., Simione, F.P. & Meyer, S.A. (1977). *Kloeckera apis* st. nov.; the imperfect state of *Hanseniaspora guilliermondii* Pijper. *Antonie van Leeuwenhoek*, **43**, 219–223.

Smith, M.T., van der Walt, J.P. & Johannsen, E. (1976). The genus *Stephanoascus* gen. nov. (Ascoideaceae). *Antonie van Leeuwenhoek*, **42**, 119–127.

Sneath, P.H.A. & Sokal, R.R. (1973). *Numerical Taxonomy. The Principles and Practice of Numerical Classification*. San Francisco: W.H. Freeman & Co.

Snell, W.H. & Dick, E.A. (1957). *A Glossary of Mycology*. Cambridge, Massachusetts: Harvard University Press.

Sonck, C.E. (1974). *Candida savonica* sp. nov. *Antonie van Leeuwenhoek*, **40**, 543–545.

Sonck, C.E. & Yarrow, D. (1969). Two yeast species isolated in Finland. *Antonie van Leeuwenhoek*, **35**, 172–177.

Soneda, M. & Uchida, S. (1971). A survey on the yeast[s]. *Bulletin of the National Science Museum*, Ueno Park, Tokyo, **14**, 438–459. [Part 6 of Mycological reports from New Guinea and the Solomon Islands (1–11) compiled by Y. Kobayasi.]

Spencer, J.F.T., Gorin, P.A.J. & Tulloch, A.P. (1970). *Torulopsis bombicola* sp. n. *Antonie van Leeuwenhoek*, **36**, 129–133.

Stadelmann, F. (1975). A new species of the genus *Bullera* Derx. *Antonie van Leeuwenhoek*, **41**, 575–582.

Stearn, W.T. (1973). *Botanical Latin*, second edition. Newton Abbot: David & Charles.

Stelling-Dekker, N.M. (1931). Die sporogenen Hefen. *Verhandelingen der*

Koninklijke Akademie van Wetenschappen te Amsterdam, Afdeeling Natuurkunde, Tweede Sectie, **28**, 1–547.

Stokes, J.L. (1971). Influence of temperature on the growth and metabolism of yeasts. In *The Yeasts*, vol. 2, pp. 119–134. Edited by A.H. Rose & J.S. Harrison. London: Academic Press.

Sugihara, T.F., Kline, L. & Miller, M.W. (1971). Microorganisms of the San Francisco sour dough bread process. *Applied Microbiology*, **21**, 456–458.

Sugiyama, J. & Goto, S. (1969). Mycoflora in core samples from stratigraphic drillings in middle Japan IV. The yeast genera *Candida* Berkhout, *Trichosporon* Behrend, and *Rhodotorula* Harrison em. Lodder from core samples. *Journal of the Faculty of Science*, Tokyo University, Section 3, Botany, **10**, 97–116.

Turpin, [P.J.F.] (1838). Mémoire sur la cause et les effets de la fermentation alcoolique et acéteuse. *Comptes Rendus Hebdomadaires des Séances de l'Académie des Sciences*, **7**, 369–402.

van der Walt, J.P. (1967). *Wingea*, a new genus of the Saccharomycetaceae. *Antonie van Leeuwenhoek*, **33**, 97–99.

van der Walt, J.P. (1970a). The perfect and imperfect states of *Sporobolomyces salmonicolor*. *Antonie van Leeuwenhoek*, **36**, 49–55.

van der Walt, J.P. (1970b). *Kluyveromyces* van der Walt emend. van der Walt. In *The Yeasts. A Taxonomic Study*, second edition, pp. 316–378. Edited by J. Lodder. Amsterdam: North-Holland Publishing Company.

van der Walt, J.P. (1970c). *Saccharomyces* Meyen emend. Reess. In *The Yeasts. A Taxonomic Study*, second edition, pp. 555–718. Edited by J. Lodder. Amsterdam: North-Holland Publishing Company.

van der Walt, J.P. (1970d). Criteria and methods used in classification. In *The Yeasts. A Taxonomic Study*, second edition, pp. 34–113. Edited by J. Lodder. Amsterdam: North-Holland Publishing Company.

van der Walt, J.P. (1972). The yeast genus *Ambrosiozyma* gen. nov. (Ascomycetes). *Mycopathologia et Mycologia Applicata*, **46**, 305–315.

van der Walt, J.P. (1973). *Aessosporon dendrophilum* sp. nov., the perfect state of *Bullera dendrophila*. *Antonie van Leeuwenhoek*, **39**, 455–460.

van der Walt, J.P. (1978). The genus *Pachytichospora* gen. nov. (Saccharomycetaceae). *Bothalia*, **12**, 563–564.

van der Walt, J.P. & Johannsen, E. (1975a). *Hansenula lynferdii* sp. nov. *Antonie van Leeuwenhoek*, **41**, 13–16.

van der Walt, J.P. & Johannsen, E. (1975b). *Pichia philogaea* sp. nov. *Antonie van Leeuwenhoek*, **41**, 173–177.

van der Walt, J.P. & Johannsen, E. (1975c). *Trichosporon terrestre* sp. nov. *Antonie van Leeuwenhoek*, **41**, 361–365.

van der Walt, J.P. & Johannsen, E. (1975d). The genus *Torulaspora* Lindner. *Council of Scientific and Industrial Research*, Research Report No. 325, pp. 1–23, Microbiology Research Group Bulletin 2. Pretoria, South Africa.

van der Walt, J.P., Johannsen, E. & Nakase, T. (1973). *Candida naeodendra*, a new species of the *Candida diddensii* group. *Antonie van Leeuwenhoek*, **39**, 491–495.

van der Walt, J.P. & Liebenberg, N.V.D.W. (1973). The yeast genus *Wickerhamiella* gen. nov. (Ascomycetes). *Antonie van Leeuwenhoek*, **39**, 121–128.

van der Walt, J.P. & Nakase, T. (1973). *Candida homilentoma*, a new yeast from South African insect sources. *Antonie van Leeuwenhoek*, **39**, 449–453.

van der Walt, J.P. & Nel, E.E. (1968). *Candida edax* sp. n. *Antonie van Leeuwenhoek*, **34**, 106–108.

van der Walt, J.P. & Scott, D.B. (1970). *Bullera dendrophila* sp. n. *Antonie van Leeuwenhoek*, **36**, 383–387.

van der Walt, J.P. & Scott, D.B. (1971a). *Pichia ambrosiae* sp. n. a new auxiliary ambrosia fungus. *Antonie van Leeuwenhoek*, **37**, 15–20.

van der Walt, J.P. & Scott, D.B. (1971b). The yeast genus *Saccharomycopsis* Schiönning. *Mycopathologia et Mycologia Applicata*, **43**, 279–288.

van der Walt, J.P. & Scott, D.B. (1971c). *Saccharomycopsis synnaedendra*, a new yeast from South African insect sources. *Mycopathologia et Mycologia Applicata*, **44**, 101–106.

van der Walt, J.P., Scott, D.B. & van der Klift, W.C. (1971a). Four new, related *Candida* species from South African insect sources. *Antonie van Leeuwenhoek*, **37**, 449–460.

van der Walt, J.P., Scott, D.B. & van der Klift, W.C. (1971b). Five new *Torulopsis* species from South African insect sources. *Antonie van Leeuwenhoek*, **37**, 461–471.

van der Walt, J.P., Scott, D.B. & van der Klift, W.C. (1972). Six new *Candida* species from South African insect sources. *Mycopathologia et Mycologia Applicata*, **47**, 221–236.

van der Walt, J.P. & Tscheuschner, I.T. (1957). *Hanseniaspora vineae* sp. nov. Transactions of the British Mycological Society, **40**, 211–212.

van der Walt, J.P. & van der Klift, W.C. (1972). *Pichia melissophila* sp. n., a new osmototolerant yeast from apiarian sources. *Antonie van Leeuwenhoek*, **38**, 361–364.

van Uden, N. & Buckley, H. (1970). *Candida* Berkhout. In *The Yeasts. A Taxonomic Study*, second edition, pp. 893–1087. Edited by J. Lodder. Amsterdam: North-Holland Publishing Company.

van Uden, N. & Vidal-Leiria, M. (1970). *Torulopsis* Berlese. In *The Yeasts. A Taxonomic Study*, second edition, pp. 1235–1308. Edited by J. Lodder. Amsterdam: North-Holland Publishing Company.

von Arx, J.A. (1972). On *Endomyces*, *Endomycopsis* and related yeast-like fungi. *Antonie van Leeuwenhoek*, **38**, 289–309.

von Arx, J.A. (1977). Notes on *Dipodascus*, *Endomyces* and *Geotrichum* with the description of two new species. *Antonie van Leeuwenhoek*, **43**, 333–340.

von Arx, J.A., Rodrigues de Miranda, L., Smith, M.T. & Yarrow, D. (1977). *Studies in Mycology No. 14. The Genera of Yeasts and the Yeast-like Fungi*. Baarn: Centraalbureau voor Schimmelcultures, Institute of the Royal Netherlands Academy of Arts and Sciences.

von Arx, J.A. & van der Walt, J.P. (1976). The ascigerous state of *Candida chodatii*. *Antonie van Leeuwenhoek*, **42**, 309–314.

Wentink, P. & La Rivière, J.W.M. (1962). An automatic turbidimeter for studies of microbial growth in aerated cultures. *Antonie van Leeuwenhoek*, **28**, 85–90.

Wickerham, L.J. (1951). *Taxonomy of yeasts*. United States Department of Agriculture Technical Bulletin, No. 1029, 56 p.

Wickerham, L.J. (1970). *Hansenula* H. et P. Sydow. In *The Yeasts. A Taxonomic Study*, second edition, pp. 226–315. Edited by J. Lodder. Amsterdam: North-Holland Publishing Company.

Wickerham, L.J. & Burton, K.A. (1948). Carbon assimilation tests for the classification of yeasts. *Journal of Bacteriology*, **56**, 363–371.

Wickerham, L.J. & Burton, K.A. (1956). Hybridization studies involving *Saccharomyces lactis* and *Zygosaccharomyces ashbyi*. *Journal of Bacteriology*, **71**, 290–295.

Wickerham, L.J. & Kurtzman, C.P. (1971). Two new Saturn-spored species of *Pichia*. *Mycologia*, **63**, 1013–1018.

Wickerham, L.J., Kurtzman, C.P. & Herman, A.I. (1970). Sexuality in *Candida lipolytica*. In *Recent Trends in Yeast Research*, pp. 81–92. Edited by D.G. Ahearn. Atlanta: School of Arts and Sciences, Georgia State University.

Willcox, W.R. & Lapage, S.P. (1972). Automatic construction of diagnostic tables. *The Computer Journal*, **15**, 263–267.

Yamada, K., Furukawa, T. & Nakahara, T. (1970). Studies on the utilization of hydrocarbons by microorganisms Part XVIII. Fermentative production of fumaric acid by *Candida hydrocarbofumarica* n. sp. *Agricultural and Biological Chemistry*, **34**, 670–675.

Yamada, Y., Nojiri, M., Matsuyama, M. & Kondo, K. (1976). Coenzyme Q system in the classification of the ascosporogenous yeast genera *Debaryomyces*, *Saccharomyces*, *Kluyveromyces* and *Endomycopsis*. *Journal of General and Applied Microbiology*, **22**, 325–337.

Yarrow, D. (1968). *Torulopsis peltata* sp. n. *Antonie van Leeuwenhoek*, **34**, 81–84.

Yarrow, D. (1969a). *Candida steatolytica* sp. n. *Antonie van Leeuwenhoek*, **35**, 24–28.

Yarrow, D. (1969b). *Selenotila peltata* comb. n. *Antonie van Leeuwenhoek*, **35**,

418–420.

Yarrow, D. (1972). Four new combinations in yeasts. *Antonie van Leeuwenhoek*, **38**, 357–360.

Yarrow, D. (1975). [In] Centraalbureau voor Schimmelcultures Baarn and Delft, Progress Report 1974. *Verhandelingen der Koninklijke Nederlandse Akademie van Wetenschappen*, Afdeling Natuurkunde, Tweede Reeks, **66**, 56–78.

Yarrow, D. (1978). *Candida milleri* spec. nov. *International Journal of Systematic Bacteriology*, **28**, 608–610.

Yarrow, D. & Ahearn, D.G. (1971). *Brettanomyces abstinens* sp. n. *Antonie van Leeuwenhoek*, **37**, 296–298.

Yarrow, D. & Meyer, S.A. (1978). A proposal for amendment of the diagnosis of the genus *Candida* Berkhout nom. cons. *International Journal of Systematic Bacteriology*, **28**, 611–615.

Addendum
List of new taxa, added in proof April 1979

These taxa have come to notice since compiling the original lists.

CANDIDA AURIGIENSIS

Santa María, J. (1978). Biotaxonomic studies on yeast. *Comunicaciones del Instituto Nacional de Investigaciones Agronómicas*, Serie: General, 3, 53–57.

CANDIDA CITRICA nom. nud.

Furukawa, T., Matsuyoshi, T., Minoda, Y. & Yamada, K. (1977). Fermentative production of citric acid from *n*-paraffins by yeast. *Journal of Fermentation Technology*, 55, 356–363.

CANDIDA FERMENTICARENS[1]

van der Walt, J.P. & Baker, P.B. (1978). *Candida fermenticarens* – a new yeast from arboricole lichen. *Bothalia*, 12, 561–562.

CANDIDA FUSIFORMATA

Buhagiar, R.W.M. (1979). *Candida fusiformata* sp. nov., a new yeast from cabbages and cauliflowers. *Journal of General Microbiology*, 110, 91–97.

CANDIDA METHANOPHILUM[1] nom. nud.

Mimura, A., Wada, M., Nakano, T., Hayakawa, S. & Iguchi, T. (1978). Isolation and characterization of a methanol-assimilating yeast, *Candida methanophilum* sp. nov. *Journal of Fermentation Technology*, 56, 443–450.

CANDIDA PARARUGOSA

Nakase, T., Komagata, K. & Fukazawa, Y. (1978). *Candida pararugosa*, a new species of asporogenous yeast. *Journal of General and Applied Microbiology* 24, 17–25.

CANDIDA QUERETANA

Herrera, T. & Ulloa, M. (1978). Descripcion de una especie nueva de levadura, *Candida queretana*, aislada de tepache de queretara, Mexico. *Boletin de la Sociedad Mexicana de Micologia*, 12, 13–18. [The type strain, CBS 6990, has now been identified as *Candida boidinii*.]

CANDIDA SONCKII

Hopsu-Havu, V.K., Tunnela, E. & Yarrow, D. (1978). *Candida sonckii* sp. nov. *Antonie van Leeuwenhoek*, 44, 435–438.

CRYPTOCOCCUS BACILLISPORUS Kwon-Chung et Bennett

Kwon-Chung, K.J., Bennett, J.E. & Theodore, T.S. (1978). *Cryptococcus bacillisporus* sp. nov.: serotype B-C of *Cryptococcus neoformans*. *International Journal of Systematic Bacteriology*, 28, 616–620.

DEBARYOZYMA

van der Walt, J.P. & Johannsen, E. (1978). The genus *Debaryozyma* van der Walt & Johannsen, nom. nov. *Persoonia*, 10, 146–148. [(i) *Debaryomyces* Klöcker is synonymous with *Torulaspora* Lindner and (ii) *Debaryomyces* Lodder et Kreger-van Rij is illegitimate. So the name *Debaryozyma* has been introduced and substituted for *Debaryomyces*, giving the following specific names: *Debaryozyma hansenii* (Zopf) van der Walt et Johannsen, *Debaryozyma castellii* (Capriotti) van der Walt et Johannsen, *Debaryozyma coudertii* (Saëz) van der Walt et Johannsen, *Debaryozyma polymorpha* (Klöcker) van der Walt et Johannsen, *Debaryozyma tamarii* (Ohara et Nonomura) van der Walt et Johannsen, *Debaryozyma vanriji* van der Walt et Tscheuschner) van der Walt et Johannsen. *Pichia pseudopolymorpha* becomes *Debaryozyma pseudopolymorpha* (Ramírez et Boidin) van der Walt et Johannsen.]

FILOBASIDIELLA

Kwon-Chung, K.J. (1975). A new genus, *Filobasidiella*, the perfect state of *Cryptococcus neoformans*. *Mycologia*, **67**, 1197–1200. [This genus has hyphae with clamp connections on which are produced non-septate basidia which have swollen apices and bear basipetal chains of basidiospores. This sexual state has been observed in the two species so far described only after mating strains of opposite sex.]

FILOBASIDIELLA BACILLISPORA

Kwon-Chung, K.J. (1976). A new species of *Filobasidiella*, the sexual state of *Cryptococcus neoformans* B and C serotypes. *Mycologia*, **68**, 942–946.

FILOBASIDIELLA NEOFORMANS

Kwon-Chung, K.J. (1976). A new species of *Filobasidiella*, the sexual state of *Cryptococcus neoformans* B and C serotypes. *Mycologia*, **68**, 942–946. [Both species may be treated as *Cryptococcus neoformans*.]

FILOBASIDIUM UNIGUTTULATUM

Kwon-Chung, K.J. (1977). Perfect state of *Cryptococcus uniguttulatus*. *International Journal of Systematic Bacteriology*, **27**, 293–299. [This sexual state of *Cryptococcus uniguttulatus* is produced after mating strains of opposite sex.]

PICHIA AMETHIONINA

Starmer, W.T., Phaff, H.J., Miranda, M. & Miller, M.W. (1978). *Pichia amethionina*, a new heterothallic yeast associated with the decaying stems of cereoid cacti. *International Journal of Systematic Bacteriology*, **28**, 433–441.

PICHIA CACTOPHILA

Starmer, W.T., Phaff, H.J., Miranda, M. & Miller, M.W. (1978). *Pichia cactophila*, a new species of yeast found in decaying tissue of cacti. *International Journal of Systematic Bacteriology*, **28**, 318–325.

PICHIA HEEDII

Phaff, H.J., Starmer, W.T., Miranda, M. & Miller, M.W. (1978). *Pichia heedii*, a new species of yeast indigenous to necrotic cacti in the North American Sonoran Desert. *International Journal of Systematic Bacteriology*, **28**, 326–331.

PICHIA METHANOTHERMO[1] Minami et Yamamura

Minami, K., Yamamura, M., Shimizu, S., Ogawa, K. & Sekine, N. (1978). A new methanol-assimilating, high productive, thermophilic yeast. *Journal of Fermentation Technology*, **56**, 1–7.

PICHIA SEGOBIENSIS

Santa María, J. & García Aser, C. (1977). *Pichia segobiensis*, sp. nov. *Anales del Instituto Nacional de Investigaciones Agrarias*, Serie General, **5**, 45–50.

SACCHAROMYCES ABULIENSIS

Santa María, J. (1978). Biotaxonomic studies on yeast. *Comunicaciones del Instituto Nacional de Investigaciones Agronómicas*, Serie: General, **3**, 15–50. [The type strain has now been identified as *Saccharomyces cerevisiae*.]

SACCHAROMYCES ALBASITENSIS

Santa María, J. (1978). Biotaxonomic studies on yeast. *Comunicaciones del*

Instituto Nacional de Investigaciones Agronómicas, Serie: General, **3**, 15–50. [The type strain has now been identified as *Zygosaccharomyces cidri*.]

SACCHAROMYCES ASTIGIENSIS

Santa María, J. (1978). Biotaxonomic studies on yeast. *Comunicaciones del Instituto Nacional de Investigaciones Agronómicas*, Serie: General, **3**, 15–50. [The type strain has now been identified as *Zygosaccharomyces cidri*.]

SACCHAROMYCES HISPALENSIS

Santa María, J. (1978). Biotaxonomic studies on yeast. *Comunicaciones del Instituto Nacional de Investigaciones Agronómicas*, Serie: General, **3**, 15–50. [The type strain has now been identified as *Saccharomyces cerevisiae*.]

SPOROPACHYDERMIA

Rodrigues de Miranda, L. (1978). A new genus: *Sporopachydermia*. *Antonie van Leeuwenhoek*, **44**, 439–450. [This genus was described to accommodate the sexual states of *Cryptococcus lactativorus* and *Cryptococcus cereanus* on account of the particularly thick wall of their ascospores. These yeasts bud multilaterally, do not form filaments, do not ferment, and do not grow with nitrate as nitrogen source. The ascospores are round, oval or elliptical with a smooth, thick wall. Two species were described, *Sporopachydermia lactativora* and *Sporopachydermia cereana*.]

TORULOPSIS ARMENTI

Kocková-Kratochvílová, A., Sláviková, E. & Beránek, J. (1977). *Torulopsis armenti* n. sp. *Zeitschrift für Allgemeine Mikrobiologie*, **17**, 429–431. [Examination of a culture of the type strain sent to the Centraalbureau voor Schimmelcultures revealed the presence of many rough-walled ascospores and the strain has been identified as *Debaryomyces hansenii*.]

TORULOPSIS AZYMA

van der Walt, J.P., Johannsen, E. & Yarrow, D. (1978). *Torulopsis geochares* and *Torulopsis azyma*, two new, haploid species of ascomycetous affinity. *Antonie van Leeuwnhoek*, **44**, 97–104.

TORULOPSIS GEOCHARES

van der Walt, J.P., Johannsen, E. & Yarrow, D. (1978). *Torulopsis geochares* and *Torulopsis azyma*, two new, haploid species of ascomycetous affinity. *Antonie van Leeuwenhoek*, **44**, 97–104.

TORULOPSIS METHANODOMERCQII[1] Abe et Yokote

Yokote, Y., Sugimoto, M. & Abe, S. (1974). Yeasts utilizing methanol as a sole carbon source. *Journal of Fermentation Technology*, **52**, 201–209.

TORULOPSIS METHANOSORBOSA Abe et Yokote

Yokote, Y., Sugimoto, M. & Abe, S. (1974). Yeasts utilizing methanol as a sole carbon source. *Journal of Fermentation Technology*, **52**, 201–209. [The type strain of this species has now been identified as *Torulopsis nagoyaensis*. However, *Torulopsis methanosorbosa* was described two years before *Torulopsis nagoyaensis* and therefore has priority; so the correct name of this species is *Torulopsis methanosorbosa*.]

---

[1] Strains of these species were not available for examination at the Centraalbureau voor Schimmelcultures.

The following names are listed by the American Type Culture Collection (Catalogue of strains 1, thirteenth edition, 1978), but the present authors do not know whether descriptions have been validly published.

CANDIDA PERIPHELOSUM Harada *et al.*
[Strain ATCC 20314 is indistinguishable from *Debaryomyces hansenii*.]

CANDIDA PETROPHILUM Takeda *et al.*
[Strain ATCC 20226 mates with strain CBS 6124—1 and has been identified as *Saccharomycopsis lipolytica*.]

TORULOPSIS METHANOPHILES Urakami
[Strain ATCC 20433 has been identified as *Torulopsis molischiana*.]

TORULOPSIS METHANOTHERMO Urakami
[Strain ATCC 20434 forms many hat-shaped ascospores on maize meal agar in slide cultures and has been identified as *Hansenula polymorpha*.]

TORULOPSIS PETROPHILUM Takeda *et al.*
[Strain ATCC 20225 appears to be a non-filamentous strain of *Saccharomycopsis lipolytica*; all other characteristics conform with those of this species.]

Author index

Abe, S., 302
Ahearn, D.G., 17, 20, 29, 33, 43, 288, 293, 299
Ainsworth, G.C., 285, 288
Alas, G., 6, 293
Amano, Y., 17, 288
Anderson, K., 20, 293
Asai, Y., 25, 28, 36, 288, 291
Aschner, M., 22, 293

Babb, R., 281, 295, 296
Babjeva, I.P., 20, 26, 288, 291
Baker, F.L., 23, 292
Baker, P.B., 300
Ballou, C.E., 5, 288
Bandoni, R.J., 13, 288
Baptist, J.N., 22, 288
Barker, B.T.P., 14, 38, 288
Barnett, J.A., 1, 27, 30, 35, 39, 42, 43, 46, 72, 277, 279, 283, 288, 289, 290, 294
Batra, L.R., 17, 288
Beech, F.W., 43, 288
Bennett, J.E., 300
Beránek, J., 302
Berlese, A.N., 7, 288
Blagodatskaja, V.M., 7, 289
Boedijn, K.B., 31, 289
Bonaly, R., 5, 289
Bos, P., 11, 26, 294
Bowen, I.G., 281, 296
Brezden, S.A., 13, 288
Brown, R.E., 20, 24, 293
Buckley, H., 7, 20, 28, 29, 31, 33, 298
Buhagiar, R.W.M., 30, 34, 35, 36, 39, 43, 288, 289, 300
Buller, A.H.R., 5, 289
Burton, K.A., 3, 43, 46, 298

Campbell, I., 7, 8, 282, 289
Capriotti, A., 23, 31, 38, 289
Carr, J.G., 43, 288
Castelli, T., 4, 38, 289
Ciferri, R., 7, 289
Clapper, W.E., 281, 296
Codner, R.C., 43, 288
Cohen-Bazire, G., 42, 293
Cohn, M., 42, 293
Cowan, S.T., v, 3, 44, 285, 289

Darwin, C., 3, 289
Demain, A.L., 43, 296
Dick, E.A., 285, 296
Diddens, H.A., 6, 22, 289
Diem, H.G., 19, 295
do Carmo-Sousa, L., 23, 24, 32, 289
Dupont, P.D., 37, 290
Dwidjoseputro, D., 31, 32, 289

Eliot, T.S., v

Feduchy, E., 6, 293

Fell, J.W., 12, 13, 22, 23, 30, 33, 34, 39, 43, 288, 289, 290, 294
Fowell, R.R., 29, 290
Fukazawa, Y., 21, 294, 300
Furukawa, T., 19, 298, 300
Fuson, G.B., 3, 8, 295

García Aser, C., 23, 28, 296, 301
Golubev, W.I., 27, 73, 290
Gómez, C.R., 20, 290
González, C., 19, 295
Gorin, P.A.J., 5, 34, 290, 296
Gorin, S.E., 26, 288
Goto, S., 17, 20, 22, 23, 32, 36, 288, 290, 297
Gower, J.C., 279, 290
Grinbergs, J., 18, 21, 22, 30, 290
Groenewege, J., 22, 290
Grose, E.S., 18, 290
Guerra, P., 6, 292
Guilliermond, A., 31, 290
Gyllenberg, H.G., 281, 290

Hayakawa, S., 300
Hedrick, L.R., 19, 37, 290
Heed, W.B., 22, 37, 293, 295
Henninger, W., 26, 28, 33, 37, 290, 291
Herman, A.I., 31, 298
Herrera, T., 300
Hopsu-Havu, V.K., 300
Hunter, I.L., 30, 34, 290

Iguchi, T., 300
Iizuka, H., 20, 32, 36, 290
Ingram, M., 42, 43, 46, 288

Jacob, F., 27, 36, 37, 291
Johannsen, E., 2, 5, 7, 10, 13, 20, 25, 29, 33, 34, 37, 38, 291, 296, 297, 300, 302
Johri, B.N., 13, 288
Jong, S.C., 9, 12, 15, 32, 291
Jurzitza, G., 35, 37, 291

Kagami, M., 17, 288
Kato, K., 28, 291
Katsuya, N., 20, 291
King, D.S., 9, 12, 15, 32, 291
Kitai, A., 35, 294
Kline, L., 20, 297
Klöcker, A., 23, 31, 291
Kocková-Kratochvílová, A., 7, 35, 36, 289, 291
Kodama, K., 26, 28, 29, 291
Kolfschoten, G.A., 17, 291
Komagata, K., 20, 21, 25, 32, 33, 291, 294
Kondo, K., 5, 298
Kouno, K., 35, 294
Krassilnikov, N.A., 24, 26, 291, 293
Kreger-van Rij, N.J.W., 1, 2, 4, 5, 7, 14, 16, 23, 24, 25, 26, 31, 32, 34, 38, 39, 45, 291, 292, 293
Kudriavzev, V.I., 2, 3, 4, 38, 292

Kudrjawzew, W.I., see Kudriavzev, V.I.
Kumbhojkar, M.S., 32, 292
Kurimura, Y., 25, 28, 36, 288, 291
Kurtzman, C.P., 9, 22, 23, 28, 29, 31, 32, 288, 292, 298
Kwon-Chung, K.J., 10, 292, 300, 301
Kyono, T., 26, 29, 291

Langeron, M., 6, 292
Lapage, S.P., 278, 298
la Rivière, J.W.M., 42, 298
Leask, B.G.S., 29, 292
Lederberg, E.M., 43, 292
Lederberg, J., 43, 292
Liebenberg, N.V.D.W., 13, 38, 297
Lindner, P., 34, 292
Lloyd, C.G., 2, 292
Lobo, K.J., 13, 288
Lodder, J., 1, 2, 4, 6, 7, 9, 14, 15, 22, 24, 26, 34, 38, 39, 43, 45, 73, 178, 283, 289, 292, 293

McClary, D.O., 45, 293
Mager, J., 22, 293
Makiguchi, N., 25, 28, 36, 288, 291
Mallette, M.F., 42, 293
Malloch, D.W., 13, 38, 295
Marcilla, J., 6, 293
Marinkelle, C.J., 18, 290
Martinez, A.T., 36, 295
Martini, A., 10, 293
Matsuyama, M., 5, 298
Matsuyoshi, T., 300
Mayr, E., 3, 293
Meavahd, K., 26, 291
Meyer, S.A., 6, 7, 10, 13, 18, 20, 24, 35, 36, 293, 296, 299
Meyers, S.P., 29, 33, 43, 288
Meynell, E., 42, 293
Meynell, G.G., 42, 293
Miller, G.R., 45, 293
Miller, M.W., 2, 3, 11, 20, 22, 24, 27, 29, 37, 293, 294, 295, 297, 301
Mimura, A., 300
Minami, K., 300
Minoda, Y., 300
Miranda, M., 22, 29, 37, 293, 294, 295, 301
Mitchell, G., 20, 293
Monod, J., 42, 293
Montrocher, R., 17, 20, 293
Moriyon, I., 36, 293, 295
Morse, L.E., 281, 293
Mrak, E.M., 24, 296
Mullin, J.K., 281, 293

Nadson, G., 24, 293
Naganishi, H., 2, 4, 293
Nakahara, T., 19, 298
Nakano, T., 300
Nakase, T., 17, 18, 19, 20, 21, 25, 32, 33, 34, 35, 37, 291, 293, 294, 297, 300

Nel, E.E., 18, 35, 294, 297
Nieuwdorp, P.J., 11, 26, 294
Nojiri, M., 5, 298
Nonomura, H., 19, 22, 294
Norkrans, B., 21, 295
Novák, E.K., 23, 31, 38, 294
Nulty, W.L., 45, 293

Ogawa, K., 301
Ohara, Y., 19, 22, 294
Oki, T., 35, 294
Olive, L.S., 10, 23, 294
Ondrušová, D., 35, 36, 291
Onishi, H., 22, 35, 36, 294
Ozaki, A., 35, 294

Pankhurst, R.J., 1, 27, 277, 281, 283, 288
Payne, R.W., 72, 277, 278, 279, 280, 281, 282, 290, 294
Persoon, C.H., 7
Phaff, H.J., 2, 3, 5, 8, 15, 22, 23, 24, 27, 29, 30, 37, 39, 43, 290, 293, 294, 295, 296, 301
Philippov, G.S., 26, 295
Pignal, M.C., 28, 295
Pijper, A., 24, 295
Pitt, J.I., 3, 295
Pontecorvo, G., 43, 295
Powell, E.O., 42, 295
Preece, D.A., 277, 278, 281, 294, 295
Price, C.W., 3, 8, 295

Ramírez, C., 19, 36, 293, 295
Redaelli, P., 7, 289
Redhead, S.A., 13, 38, 295
Reid, S.A., 13, 288
Reshetova, I.S., 20, 288
Rico, G.S., 20, 290
Rodrigues de Miranda, L., 2, 6, 10, 12, 14, 19, 21, 23, 24, 28, 30, 32, 33, 38, 295, 298, 302
Roth, F.J., 33, 43, 288
Rypka, E.W., 281, 295, 296

Sahm, H., 154, 296

Saito, K., 4, 296
Santa María, J., 6, 19, 23, 28, 296, 300, 301, 302
Schiönning, H., 31, 296
Scott, D.B., 10, 12, 15, 16, 17, 18, 19, 20, 21, 25, 28, 31, 32, 33, 35, 36, 37, 296, 297
Sekine, N., 301
Shehata, A.M.E.T., 24, 296
Shifrine, M., 2, 24, 43, 295, 296
Shimada, M., 25, 36, 288
Shimizu, S., 301
Simione, F.P., 7, 10, 24, 293, 296
Slávikova, E., 302
Slooff, W.C., 11, 26, 294
Smiley, M.J., 23, 32, 292
Smith, M.T., 2, 7, 10, 12, 13, 14, 20, 22, 23, 24, 32, 33, 38, 293, 296, 298
Sneath, P.H.A., 7, 282, 296
Snell, W.H., 285, 296
Sokal, R.R., 7, 282, 296
Sonck, C.E., 21, 33, 37, 296
Soneda, M., 11, 18, 27, 29, 293, 295, 296
Spencer, J.F.T., 5, 34, 290, 296
Stadelmann, F., 9, 17, 296
Starmer, W.T., 22, 37, 293, 295, 301
Statzell, A.C., 13, 33, 34, 290
Stearn, W.T., 285, 296
Steel, K.J., 44, 289
Stelling-Dekker, N.M., 31, 34, 38, 296
Stokes, J.L., 39, 297
Sugihara, T.F., 20, 297
Sugimoto, M., 302
Sugiyama, J., 20, 22, 23, 32, 36, 290, 297
Suzuki, T., 35, 294

Tallman, A.S., 30, 290
Theodore, T.S., 300
Török, T., 28, 295
Tscheuschner, I.T., 24, 298
Tulloch, A.P., 34, 296
Tunnela, E., 300
Turpin, P.J.F., 7, 297

Uchida, S., 18, 29, 296
Ulloa, M., 300

van der Klift, W.C., 17, 18, 19, 20, 21, 22, 23, 35, 36, 298
van der Walt, J.P., 1, 3, 5, 7, 9, 10, 12, 13, 15, 16, 17, 18, 19, 20, 21, 22, 23, 24, 25, 26, 27, 28, 29, 31, 32, 33, 34, 35, 36, 37, 38, 39, 41, 45, 46, 291, 294, 296, 297, 298, 300, 302
van Uden, N., 7, 20, 28, 29, 31, 33, 35, 36, 38, 298
Veenhuis, M., 5, 292
Vesonder, R.F., 32, 292
Vidal-Leiria, M., 7, 35, 36, 38, 298
von Arx, J.A., 2, 9, 10, 12, 14, 15, 16, 23, 24, 25, 26, 32, 38, 298

Wada, M., 300
Walton, E., 72, 279, 294
Wentink, P., 42, 298
Wickerham, L.J., 3, 24, 28, 29, 31, 43, 45, 46, 292, 298
Willcox, W.R., 278, 298
Windisch, S., 26, 28, 33, 37, 290, 291
Wolf, F.T., 31, 32, 289

Yamada, K., 19, 298, 300
Yamada, Y., 5, 298
Yamamura, M., 301
Yamasato, K., 20, 290
Yamazaki, T., 19, 294
Yarrow, D., 2, 7, 8, 12, 13, 14, 17, 18, 20, 21, 22, 23, 24, 25, 26, 29, 30, 31, 32, 33, 35, 36, 37, 38, 288, 290, 291, 292, 293, 296, 298, 299, 300, 302
Yokote, Y., 302
Yoneyama, M., 11, 27, 293, 295
Yunome, H., 22, 294

Zsolt, J., 23, 31, 38, 294

Index of taxa

*Numbers in bold type
refer to main entries in Table 8.1.*

Aciculoconidium, 9, 15
Aciculoconidium aculeatum, 15, 48–49, 284
   identity check, **179**
   in keys no. 1, 2, 3, 4
Aessosporon, 9, 15, 40
Aessosporon dendrophilum, 15, 48–49, 283
   identity check, **179**, 272
   in keys no. 1, 2
Aessosporon salmonicolor, 15, 48–49
   identity check, **179**, 255, 256
   in keys no. 1, 2, 9
Ambrosiozyma, 5, 9, 15–16
Ambrosiozyma ambrosiae, 27
Ambrosiozyma cicatricosa, 16, 48–49
   identity check, **179**
   in keys no. 3, 4, 6
Ambrosiozyma monospora, 16, 48–49, 283
   identity check, **179**
   in keys no. 1, 2, 3, 4, 6
Ambrosiozyma philentoma, 16, 48–49
   identity check, **180**
   in keys no. 1, 2, 3, 4, 6
Ambrosiozyma platypodis, see *Hormoascus platypodis*
Arthroascus, 9, 16
Arthroascus javanensis, 16, 48–49, 283
   identity check, **180**, 204, 231, 237
   in keys no. 1, 2, 6

Botryoascus, 9, 16
Botryoascus synnaedendrus, 16–17, 48–49
   identity check, **180**
   in keys no. 1, 2, 6
Brettanomyces, 9
Brettanomyces abstinens, 4, 17, 48–49
   identity check, **180**
   in keys no. 3, 4
Brettanomyces anomalus, 48–49
   identity check, **180**
   in keys no. 3, 4, 17, 18
Brettanomyces bruxellensis, 283
Brettanomyces claussenii, 48–49
   identity check, **180**
   in keys no. 3, 4, 16, 17, 18
Brettanomyces custersianus, 48–49
   identity check, **181**, 244
   in keys no. 1, 2, 3, 4, 17, 18
Brettanomyces custersii, 48–49
   identity check, **181**, 213
   in keys no. 3, 4, 16, 17, 18
Brettanomyces intermedius, 283
Brettanomyces lambicus, 4, 48–49
   identity check, **181**, 213
   in keys no. 3, 4, 16, 17, 18
Brettanomyces naardenensis, 17, 48–49
   identity check, **181**
   in keys no. 1, 2, 3, 4
Bullera, 4, 9, 17, 40
   sexual state, see *Aessosporon*

Bullera alba, 48–49
   identity check, **182**, 191, 207, 208, 209, 258
   in keys no. 1, 2
Bullera dendrophila, 15, 283
Bullera piricola, 17, 48–49
   identity check, **182**
   in keys no. 1, 2
Bullera tsugae, 48–49
   identity check, **182**
   in keys no. 1, 2

Candida, 4, 7, 9
Candida aaseri, 48–49
   identity check, **182**, 188, 272
   in keys no. 1, 2, 10, 11, 14
Candida albicans, 4, 48–49, 283
   identity check, **182**, 196, 203, 206, 253
   in keys no. 3, 4, 10, 11, 13, 14, 16
Candida amylolenta, 17, 48–49
   identity check, **183**, 191
   in keys no. 1, 2
Candida aquatica, 48–49
   identity check, **183**
   in keys no. 1, 2, 3, 4
Candida aurigiensis, 300
Candida australis, 283
Candida beechii, 48–49
   identity check, **183**, 200
   in keys no. 3, 4, 10, 11, 17, 18
Candida berthetii, 48–49
   identity check, **183**
   in keys no. 1, 2, 3, 4
Candida blankii, 48–49
   identity check, **183**, 191, 272
   in keys no. 1, 2, 3, 4, 10, 11
Candida bogoriensis, 48–49
   identity check, **184**
   in keys no. 1, 2, 16
Candida boidinii, 48–49
   identity check, **184**
   in keys no. 3, 4, 12, 15, 16
Candida boleticola, 17, 48–49
   identity check, **184**
   in keys no. 3, 4, 15
Candida bombi, 17, 48–49
   identity check, **184**, 258
   in keys no. 3, 4
Candida brassicae, 17, 48–49
   identity check, **184**
   in keys no. 3, 4
Candida brumptii, 6, 48–49
   identity check, **185**
   in keys no. 1, 2, 3, 4, 13, 14, 16
Candida buffonii, 48–49
   identity check, **185**
   in keys no. 1, 2
Candida buinensis, 18, 48–49
   identity check, **185**
   in keys no. 3, 4
Candida butyri, 18, 48–49
   identity check, **185**, 188, 239
   in keys no. 3, 4, 15

Candida cacaoi, 48–49
   identity check, **185**
   in keys no. 3, 4, 10, 11, 15
Candida candida, illegitimacy of, 6
Candida catenulata, 6, 48–49
   identity check, **186**, 199, 200
   in keys no. 1, 2, 3, 4, 10, 11, 13, 14, 15, 16
Candida chilensis, 18, 48–49
   identity check, **186**
   in keys no. 3, 4
Candida chiropterorum, 18, 48–49
   identity check, **186**, 210
   in keys no. 1, 2, 10, 11
Candida ciferrii, 33, 283
Candida citrea, 18, 50–51
   identity check, **186**, 204, 237, 265
   in keys no. 3, 4, 15
Candida citrica, 300
Candida claussenii, 283
Candida conglobata, 50–51
   identity check, **187**
   in keys no. 3, 4, 10, 11
Candida curiosa, 50–51
   identity check, **187**
   in keys no. 3, 4, 15
Candida curvata, 50–51
   identity check, **187**, 272
   in keys no. 1, 2, 10, 11, 14, 15
Candida dendronema, 18, 50–51
   identity check, **187**
   in keys no. 3, 4, 10, 11
Candida diddensii, 50–51
   identity check, 182, 185, **188**, 192, 203, 205
   in keys no. 1, 2, 3, 4, 10, 11, 15
Candida diffluens, 50–51
   identity check, **188**
   in keys no. 1, 2
Candida diversa, 50–51
   identity check, **188**, 265
   in keys no. 3, 4, 16
Candida edax, 18, 50–51
   identity check, **188**, 256
   in keys no. 1, 2, 10, 11
Candida entomaea, 18, 50–51
   identity check, **188**
   in keys no. 3, 4
Candida entomophila, 18, 50–51
   identity check, **189**
   in keys no. 3, 4, 10, 11
Candida ergatensis, 18, 50–51
   identity check, **189**
   in keys no. 3, 4
Candida famata, 6
Candida fermenticarens, 300
Candida fibrae, 283
Candida flavificans, 19, 50–51
   identity check, **189**
   in keys no. 1, 2, 3, 4
Candida fluviotilis, 19, 50–51
   identity check, **189**, 272
   in keys no. 1, 2, 3, 4

Candida foliarum, 50–51
  identity check, **189**, 255
  in keys no. 1, 2, 10, 11
Candida fragicola, 19, 50–51
  identity check, **190**
  in keys no. 3, 4, 15
Candida freyschussii, 50–51
  identity check, **190**
  in keys no. 3, 4
Candida friedrichii, 50–51
  identity check, **190**
  in keys no. 3, 4, 10, 11, 15
Candida fusiformata, 300
Candida glaebosa, 50–51
  identity check, **190**
  in keys no. 1, 2, 10, 11, 15
Candida graminis, 19, 50–51
  identity check, **190**
  in keys no. 1, 2
Candida guilliermondii, 283
Candida heveanensis, 22
Candida homilentoma, 19, 50–51
  identity check, **191**, 203
  in keys no. 3, 4
Candida humicola, 50–51
  identity check, 182, 183, **191**, 209, 257, 272
  in keys no. 1, 2, 10, 11, 14, 15, 16
Candida hydrocarbofumarica, 19, 50–51
  identity check, 183, **191**, 272
  in keys no. 1, 2, 11
Candida hylophila, 19, 50–51
  identity check, **191**
  in keys no. 1, 2
Candida iberica, 19, 50–51
  identity check, **192**, 206, 244
  in keys no. 1, 2, 3, 4, 15
Candida incommunis, 19, 50–51
  identity check, **192**
  in keys no. 1, 2, 3, 4, 10, 11, 16
Candida ingens, 28, 283
Candida inositophila, 19, 50–51
  identity check, **192**
  in keys no. 3, 4, 10, 11
Candida insectamans, 20, 50–51
  identity check, **192**
  in keys no. 1, 2, 3, 4
Candida insectorum, 20, 50–51
  identity check, 188, **192**, 197, 201, 203, 205, 238
  in keys no. 3, 4
Candida intermedia, 50–51
  identity check, **193**
  in keys no. 3, 4, 10, 11, 14, 15, 16, 17, 18
Candida ishiwadae, 20, 50–51
  identity check, **193**
  in keys no. 3, 4
Candida javanica, 50–51
  identity check, **193**
  in keys no. 1, 2, 10, 11
Candida kefyr, 50–51
  identity check, **193**
  in keys no. 3, 4, 15
Candida krissii, 20, 50–51
  identity check, **193**, 265
  in keys no. 1, 2, 3, 4
Candida krusei, 283
Candida lambica, 50–51
  identity check, **194**, 204, 235, 237, 258
  in keys no. 3, 4, 15, 17, 18
Candida langeronii, 283
Candida lipolytica, 8, 31, 283

Candida lusitaniae, 50–51, 283
  identity check, **194**
  in keys no. 3, 4, 10, 11, 14, 15
Candida macedoniensis, 50–51
  identity check, **194**
  in keys no. 3, 4, 14, 15, 17, 18
Candida maltosa, 20, 50–51
  identity check, **194**, 224
  in keys no. 3, 4, 10, 11
Candida marina, 50–51
  identity check, **195**
  in keys no. 1, 2
Candida maritima, 50–51
  identity check, **195**
  in keys no. 3, 4, 15
Candida melibiosica, 50–51
  identity check, **195**, 268
  in keys no. 3, 4, 10, 11, 14
Candida melinii, 50–51
  identity check, **195**, 216, 217, 222, 246
  in keys no. 1, 2, 16
Candida membranaefaciens, 50–51
  identity check, **196**
  in keys no. 3, 4, 10, 11, 14
Candida mesenterica, 50–51
  identity check, **196**
  in keys no. 1, 2, 10, 11, 15
Candida methanophilum, 300
Candida milleri, 20, 52–53
  identity check, **196**, 224, 225, 248, 263
  in keys no. 3, 4
Candida mogii, 52–53
  identity check, 182, **196**
  in keys no. 3, 4, 10, 11
Candida muscorum, 52–53
  identity check, **196**
  in keys no. 1, 2
Candida naeodendra, 20, 52–53
  identity check, 192, **197**, 203, 238
  in keys no. 1, 2, 3, 4
Candida norvegensis, 52–53
  identity check, **197**, 238
  in keys no. 1, 2, 3, 4, 15
Candida obtusa, 283
Candida oleophila, 20, 52–53
  identity check, **197**, 200, 230
  in keys no. 3, 4, 17, 18
Candida oregonensis, 52–53
  identity check, **197**
  in keys no. 3, 4, 10, 11
Candida parapsilosis, 52–53
  identity check, **198**
  in keys no. 3, 4, 10, 11, 13, 14, 15, 16, 17, 18
Candida pararugosa, 300
Candida periphelosum, 303
Candida petrophilum, 303
Candida podzolica, 20, 52–53
  identity check, **198**, 209, 272
  in keys no. 1, 2
Candida pseudointermedia, 21, 52–53
  identity check, **198**
  in keys no. 3, 4
Candida pseudolipolytica, 7–8
Candida pseudotropicalis, 52–53
  identity check, **198**, 225
  in keys no. 3, 4, 14, 15, 17, 18
Candida punicea, 32
Candida quercuum, 21, 52–53
  identity check, **199**
  in keys no. 1, 2, 3, 4

Candida queretana, 300
Candida ravautii, 6, 52–53
  identity check, 186, **199**
  in keys no. 1, 2, 3, 4, 10, 11, 13, 14
Candida requinyii, 283
Candida rhagii, 52–53
  identity check, **199**
  in keys no. 3, 4, 10, 11
Candida rugopelliculosa, 21, 52–53
  identity check, **199**, 237, 258
  in keys no. 3, 4
Candida rugosa, 52–53
  identity check, 186, **200**
  in keys no. 1, 2, 10, 11, 14, 15, 16
Candida sake, 20, 52–53, 283
  identity check, 197, **200**, 229, 230, 258
  in keys no. 3, 4, 10, 11, 15, 16, 17, 18
Candida salmanticensis, 52–53
  identity check, **200**
  in keys no. 3, 4
Candida salmonicola, 283
Candida santamariae, 52–53
  identity check, 183, **200**
  in keys no. 3, 4, 10, 11, 15
Candida savonica, 21, 52–53
  identity check, **201**, 206
  in keys no. 1, 2, 3, 4
Candida shehatae, 283
Candida silvae, 52–53
  identity check, **201**, 205, 251
  in keys no. 1, 2, 14, 15
Candida silvanorum, 21, 52–53
  identity check, 192, **201**
  in keys no. 3, 4, 10, 11
Candida silvicultrix, 21, 52–53
  identity check, **202**
  in keys no. 3, 4
Candida slooffii, 283
Candida solani, 52–53
  identity check, **202**
  in keys no. 3, 4, 15, 16, 17, 18
Candida sonckii, 300
Candida sorbosa, 283
Candida sorboxylosa, 21, 52–53
  identity check, **202**, 237, 248
  in keys no. 3, 4, 15
Candida steatolytica, 21, 52–53
  identity check, **202**
  in keys no. 3, 4, 14, 16
Candida stellatoidea, 283
Candida suecica, 21, 52–53
  identity check, **202**
  in keys no. 1, 2, 3, 4
Candida tenuis, 52–53
  identity check, 188, 191, 192, 197, **203**, 205, 238
  in keys no. 3, 4, 10, 11, 14, 15, 16, 17, 18
Candida tepae, 21, 52–53
  identity check, **203**
  in keys no. 1, 2
Candida terebra, 21–22, 52–53
  identity check, 188, 192, **203**
  in keys no. 1, 2, 3, 4
Candida tropicalis, 52–53
  identity check, 182, **203**, 206
  in keys no. 3, 4, 10, 11, 13, 14, 15, 16, 17, 18
Candida tsukubaensis, 22, 52–53
  identity check, **204**
  in keys no. 1, 2

Candida utilis, 7, 39, 52–53
  identity check, **204**
  in keys no. 3, 4, 14, 15, 16, 17, 18
Candida valdiviana, 22, 52–53
  identity check, **204**
  in keys no. 1, 2, 3, 4, 10, 11
Candida valida, 52–53
  identity check, 180, 186, 194, **204**, 214, 236, 237, 242, 248, 258, 265, 275, 276
  in keys no. 1, 2, 3, 4, 15, 16, 17, 18
Candida vartiovaarai, 52–53
  identity check, **204**
  in keys no. 3, 4, 17, 18
Candida veronae, 52–53
  identity check, 188, 192, 203, **205**
  in keys no. 3, 4, 10, 11, 16
Candida vinaria, 22, 52–53
  identity check, **205**
  in keys no. 1, 2
Candida vini, 52–53
  identity check, 201, **205**, 234, 235, 251
  in keys no. 1, 2, 15, 16
Candida viswanathii, 52–53
  identity check, 182, 203, **206**
  in keys no. 3, 4, 14
Candida zeylanoides, 52–53
  identity check, 192, 201, **206**
  in keys no. 1, 2, 3, 4, 10, 11, 14, 15, 16
Citeromyces, 9
Citeromyces matritensis, 52–53
  identity check, **206**
  in keys no. 3, 4, 5, 15, 16
Cryptococcus, 5, 9, 44
Cryptococcus albidus, 40, 52–53
  identity check, **206**, 208, 214, 245, 273
  in keys no. 1, 2, 13, 14, 15, 16
Cryptococcus ater, 54–55
  identity check, 182, **207**
  in keys no. 1, 2, 14
Cryptococcus bacillisporus, 300
Cryptococcus bhutanensis, 22, 54–55
  identity check, **207**
  in keys no. 1, 2
Cryptococcus cereanus, 22, 54–55, 302
  identity check, **207**
  in keys no. 1, 2
Cryptococcus dimennae, 54–55
  identity check, **207**, 272
  in keys no. 1, 2
Cryptococcus flavus, 54–55
  identity check, 182, **207**, 258
  in keys no. 1, 2, 15
Cryptococcus gastricus, 54–55
  identity check, **208**, 272
  in keys no. 1, 2, 14
Cryptococcus heveanensis, 22, 54–55
  identity check, 206, **208**
  in keys no. 1, 2
Cryptococcus himalayensis, 22, 54–55
  identity check, **208**
  in keys no. 1, 2
Cryptococcus hungaricus, 54–55
  identity check, 182, **208**, 209
  in keys no. 1, 2, 9
Cryptococcus infirmo-miniatus, 30, 283
Cryptococcus kuetzingii, 54–55
  identity check, **208**
  in keys no. 1, 2, 14, 15
Cryptococcus lactativorus, 54–55, 302
  identity check, **209**
  in keys no. 1, 2

Cryptococcus laurentii, 9, 22, 54–55
  identity check, 182, 191, **209**, 272
  in keys no. 1, 2, 9, 13, 14, 15, 16
Cryptococcus luteolus, 54–55
  identity check, 182, 191, 208, **209**, 272
  in keys no. 1, 2, 16
Cryptococcus macerans, 54–55
  identity check, **209**
  in keys no. 1, 2, 9
Cryptococcus magnus, 22, 54–55
  identity check, 182, 198, **209**, 272
  in keys no. 1, 2
Cryptococcus melibiosum, 54–55
  identity check, **210**
  in keys no. 1, 2
Cryptococcus neoformans, 54–55, 300, 301
  identity check, 186, **210**
  in keys no. 1, 2, 13, 14, 15, 16
Cryptococcus skinneri, 54–55
  identity check, **210**
  in keys no. 1, 2
Cryptococcus terreus, 54–55
  identity check, **210**
  in keys no. 1, 2, 15
Cryptococcus uniguttulatus, 54–55, 301
  identity check, **210**
  in keys no. 1, 2, 13, 14, 15
Cyniclomyces, 10
  omission from keys, 1, 72

Debaryomyces, 5, 10, 300
Debaryomyces cantarellii, 23, 283
Debaryomyces castellii, 54–55
  identity check, **211**
  in keys no. 3, 4, 5, 10, 11
Debaryomyces coudertii, 54–55
  identity check, **211**
  in keys no. 1, 2, 5
Debaryomyces hansenii, 54–55, 284, 302, 303
  identity check, **211**, 212, 213, 240
  in keys no. 1, 2, 3, 4, 5, 10, 11, 13, 14, 15, 16, 17, 18
Debaryomyces marama, 54–55
  identity check, **211**, 212, 228, 272
  in keys no. 1, 2, 3, 4, 5, 10, 11, 14, 17, 18
Debaryomyces melissophilus, 22–23, 54–55, 283
  identity check, **212**
  in keys no. 1, 2, 5
Debaryomyces nepalensis, 23, 54–55
  identity check, 211, **212**
  in keys no. 1, 2, 3, 4, 5
Debaryomyces phaffii, 54–55
  identity check, **212**, 213
  in keys no. 1, 2, 3, 4, 5, 10, 11
Debaryomyces polymorphus, 23, 54–55, 283, 284
  identity check, **212**
  in keys no. 3, 4, 5, 15, 16
Debaryomyces tamarii, 54–55
  identity check, **212**
  in keys no. 1, 2, 3, 4, 5, 15
Debaryomyces vanriji, 54–55
  identity check, 211, 212, **213**, 228
  in keys no. 1, 2, 3, 4, 5, 10, 11
Debaryomyces yarrowii, 23, 54–55
  identity check, **213**
  in keys no. 1, 2, 5
Debaryozyma, 300
Debaryozyma castellii, 300
Debaryozyma coudertii, 300

Debaryozyma hansenii, 300
Debaryozyma polymorpha, 301
Debaryozyma pseudopolymorpha, 301
Debaryozyma tamarii, 301
Debaryozyma vanriji, 301
Dekkera, 5, 10
Dekkera bruxellensis, 54–55, 283
  identity check, 181, **213**
  in keys no. 3, 4, 6, 16, 17, 18
Dekkera intermedia, 54–55, 283
  identity check, 181, **213**
  in keys no. 3, 4, 6, 16, 17, 18
Dipodascus, 10

Endomycetales, 4
Endomycopsella, 31
Endomycopsis, 31
Endomycopsis bispora, 24, 283
Endomycopsis burtonii, 10, 26, 283
Endomycopsis capsularis, 31, 283
Endomycopsis fibuligera, 31, 283
Endomycopsis javanensis, 9, 16, 283
Endomycopsis lipolytica, 31
Endomycopsis monospora, 16, 283
Endomycopsis muscicola, 25
Endomycopsis ovetensis, 13, 38, 283
Endomycopsis platypodis, 10, 25, 283
Endomycopsis selenospora, 10, 24, 283
Endomycopsis vini, 32, 283

Filobasidiella, 301
Filobasidiella bacillispora, 301
Filobasidiella neoformans, 301
Filobasidium, 10, 23
Filobasidium capsuligenum, 23, 54–55, 283
  identity check, **214**
  in keys no. 1, 2, 3, 4, 8, 16, 17, 18
Filobasidium floriforme, 23, 54–55
  identity check, 206, **214**
  in keys no. 1, 2
Filobasidium uniguttulatum, 301

Geotrichum, 10
Geotrichum capitatum, 23, 54–55, 284
  identity check, 204, **214**, 237
  in keys no. 1, 2, 13, 14, 15
Geotrichum fermentans, 23, 54–55, 284
  identity check, **214**
  in keys no. 1, 2, 3, 4, 15, 16
Geotrichum penicillatum, 24, 54–55, 284
  identity check, **214**
  in keys no. 1, 2, 3, 4
Guilliermondella, 10, 24
Guilliermondella selenospora, 24, 54–55, 283
  identity check, **215**
  in keys no. 1, 2, 3, 4, 5

Hanseniaspora, 4, 10
Hanseniaspora guilliermondii, 24, 56–57
  identity check, **215**
  in keys no. 3, 4, 6
Hanseniaspora melligeri, 24
Hanseniaspora occidentalis, 24, 56–57, 283
  identity check, **215**
  in keys no. 3, 4, 6, 16
Hanseniaspora osmophila, 24, 56–57, 283
  identity check, **215**, 216
  in keys no. 3, 4, 5, 15, 16
Hanseniaspora uvarum, 24, 56–57, 283
  identity check, **215**
  in keys no. 3, 4, 5, 6, 15, 16, 17, 18

Index of taxa

Hanseniaspora valbyensis, 24, 56–57
  identity check, **216**
  in keys no. 3, 4, 6, 14, 15, 16, 17, 18
Hanseniaspora vineae, 24, 56–57, 283
  identity check, 215, **216**
  in keys no. 3, 4, 5, 15, 16, 17, 18
Hansenula, 5, 10
Hansenula anomala, 56–57
  identity check, **216**, 217
  in keys no. 3, 4, 6, 15, 16, 17, 18
Hansenula beckii, 24, 56–57, 283
  identity check, 195, **216**, 217, 222, 246
  in keys no. 1, 2, 3, 4, 6
Hansenula beijerinckii, 56–57
  identity check, **216**
  in keys no. 3, 4, 6
Hansenula bimundalis, 56–57
  identity check, 195, 216, **217**, 222, 246
  in keys no. 1, 2, 3, 4, 6
Hansenula californica, 56–57
  identity check, **217**
  in keys no. 3, 4, 6, 15, 16
Hansenula canadensis, 56–57
  identity check, 195, 216, **217**, 222, 246
  in keys no. 1, 2, 6
Hansenula capsulata, 56–57
  identity check, **217**, 266
  in keys no. 3, 4, 6, 12
Hansenula ciferrii, 56–57
  identity check, 216, **217**
  in keys no. 3, 4, 6
Hansenula dimennae, 56–57
  identity check, **218**
  in keys no. 3, 4, 6
Hansenula dryadoides, 25, 56–57
  identity check, **218**
  in keys no. 1, 2, 6
Hansenula fabianii, 56–57
  identity check, **218**
  in keys no. 3, 4, 6, 15, 17, 18
Hansenula glucozyma, 56–57
  identity check, **218**
  in keys no. 1, 2, 3, 4, 6, 12
Hansenula henricii, 56–57
  identity check, **218**
  in keys no. 1, 2, 3, 4, 6, 12
Hansenula holstii, 56–57
  identity check, **219**
  in keys no. 3, 4, 6, 15
Hansenula jadinii, 56–57
  identity check, **219**
  in keys no. 1, 2, 3, 4, 6, 14
Hansenula lynferdii, 25, 56–57
  identity check, **219**
  in keys no. 3, 4, 6
Hansenula malanga, 31, 283
Hansenula minuta, 56–57
  identity check, **219**
  in keys no. 1, 2, 3, 4, 6, 12, 15
Hansenula mrakii, 56–57
  identity check, **219**, 268
  in keys no. 3, 4, 6
Hansenula muscicola, 25, 56–57
  identity check, **220**
  in keys no. 3, 4, 6
Hansenula nonfermentans, 56–57
  identity check, **220**
  in keys no. 1, 2, 6, 12
Hansenula ofunaensis, 25, 56–57
  identity check, **220**
  in keys no. 1, 2, 6, 12

Hansenula petersonii, 56–57
  identity check, **220**
  in keys no. 3, 4, 6, 14
Hansenula philodendra, 25, 56–57
  identity check, **220**
  in keys no. 1, 2, 3, 4, 6, 12
Hansenula polymorpha, 56–57, 303
  identity check, **220**
  in keys no. 3, 4, 6, 12
Hansenula saturnus, 56–57
  identity check, **221**
  in keys no. 3, 4, 6, 16
Hansenula silvicola, 56–57
  identity check, **221**
  in keys no. 3, 4, 6, 16
Hansenula subpelliculosa, 56–57
  identity check, **221**
  in keys no. 3, 4, 6, 15, 16, 17, 18
Hansenula sydowiorum, 25, 56–57
  identity check, **221**
  in keys no. 3, 4, 6
Hansenula wickerhamii, 56–57
  identity check, **221**
  in keys no. 1, 2, 6, 12
Hansenula wingei, 56–57
  identity check, 195, 216, 217, **222**, 246
  in keys no. 1, 2, 6
Hormoascus, 5, 10, 25
Hormoascus platypodis, 25–26, 56–57, 283
  identity check, **222**
  in keys no. 1, 2, 3, 4, 6
Hyphopichia, 10, 26
Hyphopichia burtonii, 26, 56–57, 283
  identity check, **222**
  in keys no. 1, 2, 3, 4, 6, 10, 11, 14, 15, 16

Kloeckera, see *Hanseniaspora* species, 24, 283
Kloeckera africana, 283
Kloeckera apiculata, 283
Kloeckera apis, 24
Kloeckera corticis, 283
Kloeckera javanica, 24, 283
Kloeckera occidentalis, 24
Kluyveromyces, 10
Kluyveromyces aestuarii, 56–57
  identity check, **222**
  in keys no. 3, 4, 5
Kluyveromyces africanus, 56–57
  identity check, **223**, 248
  in keys no. 3, 4, 5
Kluyveromyces blattae, 26, 56–57
  identity check, **223**, 248
  in keys no. 3, 4, 5
Kluyveromyces bulgaricus, 56–57, 283
  identity check, **223**, 224, 225
  in keys no. 3, 4, 5, 15, 17, 18
Kluyveromyces cicerisporus, 283
Kluyveromyces delphensis, 58–59
  identity check, **223**, 237
  in keys no. 3, 4, 5, 15
Kluyveromyces dobzhanskii, 58–59
  identity check, **224**
  in keys no. 3, 4, 5
Kluyveromyces drosophilarum, 58–59
  identity check, **224**
  in keys no. 3, 4, 5
Kluyveromyces fragilis, 283
Kluyveromyces lactis, 58–59, 283
  identity check, 194, 223, **224**, 225, 226
  in keys no. 3, 4, 5, 14, 15, 16

Kluyveromyces lodderi, 58–59
  identity check, 196, **224**, 225
  in keys no. 3, 4, 5
Kluyveromyces marxianus, 58–59, 283
  identity check, 198, 223, 224, **225**
  in keys no. 3, 4, 5, 13, 14, 15, 16, 17, 18
Kluyveromyces phaffii, 58–59
  identity check, **225**
  in keys no. 3, 4, 5
Kluyveromyces phaseolosporus, 58–59
  identity check, 223, 224, **225**
  in keys no. 3, 4, 5
Kluyveromyces polysporus, 58–59
  identity check, 196, 224, **225**, 263
  in keys no. 3, 4, 5
Kluyveromyces thermotolerans, 26, 58–59, 283, 284
  identity check, **226**
  in keys no. 3, 4, 5, 15, 16
Kluyveromyces vanudenii, 283
Kluyveromyces veronae, 26, 283
Kluyveromyces waltii, 26, 58–59
  identity check, **226**, 258, 259
  in keys no. 3, 4, 5
Kluyveromyces wickerhamii, 58–59
  identity check, 224, **226**
  in keys no. 3, 4, 5
Kluyveromyces wikenii, 283

Leucosporidium, 11, 39
Leucosporidium antarcticum, 42, 58–59
  identity check, **226**
  in keys no. 1, 2, 8
Leucosporidium capsuligenum, 23, 283
Leucosporidium frigidum, 58–59
  identity check, **226**
  in keys no. 1, 2, 3, 4, 8
Leucosporidium gelidum, 58–59
  identity check, **227**, 244
  in keys no. 1, 2, 3, 4, 8
Leucosporidium nivalis, 58–59
  identity check, **227**
  in keys no. 1, 2, 3, 4, 8
Leucosporidium scottii, 58–59
  identity check, **227**
  in keys no. 1, 2, 8, 10, 11, 15, 16
Leucosporidium stokesii, 58–59
  identity check, **227**, 245
  in keys no. 1, 2, 3, 4, 8
Lipomyces, 11
Lipomyces anomalus, 26, 58–59
  identity check, **227**
  in keys no. 1, 2, 5
Lipomyces kononenkoae, 58–59
  identity check, **228**
  in keys no. 1, 2, 5
Lipomyces lipofer, 58–59
  identity check, **228**, 272
  in keys no. 1, 2, 5
Lipomyces starkeyi, 58–59
  identity check, **228**
  in keys no. 1, 2, 5, 16
Lipomyces tetrasporus, 26, 58–59
  identity check, 211, 213, **228**
  in keys no. 1, 2, 5
Lodderomyces, 5, 11
Lodderomyces elongisporus, 58–59
  identity check, **228**
  in keys no. 1, 2, 3, 4, 5, 10, 11, 16

Malassezia, 4

Metschnikowia, 3, 11
Metschnikowia biscuspidata, 58–59
  identity check, 200, **229**, 230
  in keys no. 3, 4, 7
Metschnikowia krissii, 58–59
  identity check, **229**
  in keys no. 1, 2, 7
Metschnikowia lunata, 27, 58–59, 284
  identity check, **229**, 230
  in keys no. 1, 2, 3, 4, 7, 10, 11
Metschnikowia pulcherrima, 7, 58–59
  identity check, 197, 200, 229, **230**
  in keys no. 3, 4, 7, 10, 11, 15, 16
Metschnikowia reukaufii, 58–59
  identity check, 200, 229, **230**
  in keys no. 3, 4, 7, 10, 11, 16
Metschnikowia zobellii, 58–59
  identity check, 229, **230**
  in keys no. 1, 2, 3, 4, 7, 15

Nadsonia, 11
Nadsonia commutata, 27, 58–59
  identity check, 180, **231**, 244, 251
  in keys no. 1, 2, 5, 12
Nadsonia elongata, 58–59
  identity check, **231**, 237
  in keys no. 3, 4, 5, 16
Nadsonia fulvescens, 58–59
  identity check, **231**
  in keys no. 3, 4, 5
Nematospora, 11
Nematospora coryli, 58–59
  identity check, **231**, 248
  in keys no. 1, 2, 3, 4, 7

Oosporidium, 11
  omission from keys, 1, 72

Pachysolen, 11
Pachysolen tannophilus, 58–59
  identity check, **232**
  in keys no. 3, 4, 6
Pachytichospora, 5, 11, 27
Pachytichospora transvaalensis, 27, 58–59, 284
  identity check, **232**, 248
  in keys no. 3, 4, 5, 16
Phaffia, 11, 27
Phaffia rhodozyma, 27, 60–61
  identity check, **232**
  in keys no. 1, 2, 3, 4, 9
Pichia, 4, 11
Pichia abadieae, 27, 60–61
  identity check, **232**, 272
  in keys no. 1, 2, 3, 4, 5, 6, 7
Pichia acaciae, 60–61
  identity check, **232**
  in keys no. 3, 4, 6
Pichia ambrosiae, 27–28, 60–61
  identity check, **233**
  in keys no. 1, 2, 3, 4, 6
Pichia amethionina, 301
Pichia angophorae, 60–61
  identity check, **233**
  in keys no. 3, 4, 6
Pichia besseyi, 28, 60–61
  identity check, **233**
  in keys no. 3, 4, 6
Pichia bovis, 60–61
  identity check, **233**
  in keys no. 3, 4, 6

Pichia burtonii, see *Hyphopichia burtonii*
Pichia cactophila, 301
Pichia castillae, 28, 60–61
  identity check, **233**
  in keys no. 1, 2, 6, 10, 11
Pichia chambardii, 60–61
  identity check, **234**
  in keys no. 1, 2, 6
Pichia cicatricosa, 15
Pichia delftensis, 60–61
  identity check, 205, **234**, 265
  in keys no. 1, 2, 3, 4, 6
Pichia dispora, 60–61
  identity check, **234**
  in keys no. 3, 4, 6
Pichia etchellsii, 60–61
  identity check, **234**
  in keys no. 1, 2, 3, 4, 5, 10, 11, 14, 15, 16
Pichia farinosa, 60–61
  identity check, **235**
  in keys no. 3, 4, 5, 10, 11, 13, 14, 15, 16, 17, 18
Pichia fermentans, 60–61
  identity check, 194, **235**, 237
  in keys no. 3, 4, 6, 15, 16, 17, 18
Pichia fluxuum, 60–61
  identity check, 205, **235**, 248, 258, 265
  in keys no. 1, 2, 3, 4, 6
Pichia guilliermondii, 60–61, 283
  identity check, **235**
  in keys no. 3, 4, 6, 10, 11, 13, 14, 15, 16, 17, 18
Pichia haplophila, 60–61
  identity check, **236**
  in keys no. 1, 2, 3, 4, 6
Pichia heedii, 301
Pichia heimii, 28, 60–61
  identity check, **236**
  in keys no. 3, 4, 6
Pichia humboldtii, 28, 60–61, 283
  identity check, **236**
  in keys no. 1, 2, 5, 16
Pichia kluyveri, 60–61
  identity check, **236**, 237
  in keys no. 3, 4, 6, 15
Pichia kudriavzevii, 60–61, 283
  identity check, 204, **236**, 237, 242, 248, 258, 283
  in keys no. 3, 4, 5, 13, 14, 15, 16, 17, 18
Pichia lindnerii, 28, 60–61
  identity check, **237**, 266
  in keys no. 1, 2, 3, 4, 6, 12
Pichia media, 60–61
  identity check, **237**
  in keys no. 1, 2, 6
Pichia melissophila, 22, 283
Pichia membranaefaciens, 60–61
  identity check, 180, 186, 194, 199, 202, 204, 214, 223, 231, 235, 236, **237**, 242, 243, 248, 249, 258, 264, 265, 275, 276
  in keys no. 1, 2, 3, 4, 5, 6, 14, 15, 16, 17, 18
Pichia methanolica, 28, 60–61
  identity check, **237**
  in keys no. 3, 4, 6, 12
Pichia methanothermo, 301
Pichia microspora, 17
Pichia mucosa, 28, 60–61
  identity check, **238**
  in keys no. 3, 4, 6
Pichia naganishii, 28–29, 60–61
  identity check, **238**
  in keys no. 3, 4, 6, 12

Pichia nakazawae, 29, 60–61
  identity check, 192, 197, 203, **238**
  in keys no. 3, 4, 6
Pichia nonfermentans, 283
Pichia norvegensis, 29, 60–61
  identity check, 197, **238**
  in keys no. 1, 2, 3, 4, 6, 13, 14
Pichia ohmeri, 60–61
  identity check, **239**
  in keys no. 3, 4, 6, 15, 17, 18
Pichia onychis, 60–61
  identity check, **239**
  in keys no. 3, 4, 6, 15, 17, 18
Pichia pastoris, 60–61
  identity check, **239**
  in keys no. 3, 4, 6, 12
Pichia philogaea, 29, 60–61
  identity check, 185, **239**
  in keys no. 3, 4, 6
Pichia pijperi, 60–61
  identity check, **240**
  in keys no. 3, 4, 6, 15
Pichia pinus, 60–61
  identity check, **240**
  in keys no. 1, 2, 3, 4, 6, 12
Pichia polymorpha, 23, 284
Pichia pseudopolymorpha, 60–61, 301
  identity check, 211, **240**
  in keys no. 1, 2, 3, 4, 5
Pichia quercuum, 60–61
  identity check, **240**
  in keys no. 1, 2, 3, 4, 6
Pichia rabaulensis, 29, 60–61
  identity check, **240**
  in keys no. 3, 4, 6
Pichia rhodanensis, 60–61
  identity check, **241**, 243
  in keys no. 3, 4, 6
Pichia robertsii, 5
Pichia saitoi, 60–61
  identity check, **241**, 248
  in keys no. 3, 4, 5, 6
Pichia salictaria, 60–61
  identity check, **241**
  in keys no. 1, 2, 6
Pichia sargentensis, 29, 60–61
  identity check, **241**
  in keys no. 3, 4, 6
Pichia scolyti, 62–63
  identity check, **241**, 272
  in keys no. 1, 2, 3, 4, 6
Pichia scutulata, 29, 62–63
  identity check, 204, 236, 237, **242**
  in keys no. 1, 2, 3, 4, 5
Pichia segobiensis, 301
Pichia spartinae, 29, 62–63
  identity check, **242**
  in keys no. 3, 4, 6
Pichia stipitis, 62–63, 283
  identity check, **242**
  in keys no. 3, 4, 6
Pichia strasburgensis, 62–63
  identity check, **242**
  in keys no. 3, 4, 6
Pichia terricola, 62–63
  identity check, 237, **243**
  in keys no. 1, 2, 3, 4, 5, 15
Pichia toletana, 62–63
  identity check, **243**
  in keys no. 1, 2, 3, 4, 6

Index of taxa

Pichia trehalophila, 62–63
  identity check, **243**
  in keys no. 3, 4, 6, 12
Pichia veronae, 29, 62–63
  identity check, 241, **243**
  in keys no. 3, 4, 6
Pichia vini, 62–63
  identity check, **243**
  in keys no. 1, 2, 5, 14, 16
Pichia wickerhamii, 62–63
  identity check, **244**
  in keys no. 3, 4, 6
Pityrosporum, 4, 11
  omission from keys, 1, 72
Prosaccharomyces, 31

Rhodosporidium, 12
Rhodosporidium bisporidiis, 30, 62–63
  identity check, 227, **244**
  in keys no. 1, 2, 8, 9
Rhodosporidium capitatum, 30, 62–63
  identity check, **244**, 245
  in keys no. 1, 2, 8, 9
Rhodosporidium dacryoidum, 30, 62–63
  identity check, 181, 192, 231, **244**, 247, 251
  in keys no. 1, 2, 8, 9
Rhodosporidium diobovatum, see *Rhodotorula glutinis*
Rhodosporidium infirmo-miniatum, 30, 62–63, 283
  identity check, 206, 227, 244, **245**
  in keys no. 1, 2, 8, 9, 15
Rhodosporidium malvinellum, 30, 62–63
  identity check, **245**
  in keys no. 1, 2, 8, 9
Rhodosporidium sphaerocarpum, 284
Rhodosporidium toruloides, 284
Rhodotorula, 12
Rhodotorula acheniorum, 6, 30, 62–63
  identity check, **245**
  in keys no. 1, 2, 9
Rhodotorula araucariae, 30–31, 62–63
  identity check, **245**, 254, 255
  in keys no. 1, 2, 9, 10, 11
Rhodotorula aurantiaca, 62–63
  identity check, **245**, 246
  in keys no. 1, 2, 9, 14, 15, 16
Rhodotorula glutinis, 62–63, 284
  identity check, 195, 216, 217, 222, 245, **246**, 253, 255
  in keys no. 1, 2, 8, 9, 10, 11, 13, 14, 15, 16
Rhodotorula graminis, 62–63
  identity check, **246**
  in keys no. 1, 2, 9, 10, 11
Rhodotorula lactosa, 62–63
  identity check, **246**
  in keys no. 1, 2, 9
Rhodotorula marina, 62–63
  identity check, **246**
  in keys no. 1, 2, 9, 15
Rhodotorula minuta, 62–63
  identity check, **246**
  in keys no. 1, 2, 9, 14, 15, 16
Rhodotorula pallida, 62–63
  identity check, 244, **247**, 250, 254
  in keys no. 1, 2, 9, 15, 16
Rhodotorula pilimanae, 62–63
  identity check, **247**
  in keys no. 1, 2, 9, 10, 11
Rhodotorula rubra, 62–63
  identity check, **247**, 254, 272
  in keys no. 1, 2, 9, 13, 14, 15, 16

Rhodozyma montanae, 27

Saccharomyces, 12
  asci, 5
  family of, 4
  identifying by computer matching, 282
  interfertility of species, 3
  numerical analysis, 8
Saccharomyces abuliensis, 301
Saccharomyces aceti, 284
Saccharomyces albasitensis, 301
Saccharomyces amurcae, 284
Saccharomyces astigiensis, 302
Saccharomyces bailii, 38, 284
Saccharomyces bayanus, 284
Saccharomyces beticus, 6, 284
Saccharomyces bisporus, 38, 284
Saccharomyces capensis, 284
Saccharomyces carlsbergensis, 2
Saccharomyces cerevisiae, 2, 4, 39, 40, 62–63, 284, 301, 302
  identity check, 196, 202, 204, 223, 231, 232, 235, 236, 237, 241, **248**, 249, 251, 253, 260, 262, 265, 270
  in keys no. 3, 4, 5, 13, 14, 15, 16, 17, 18
Saccharomyces chevalieri, 284
Saccharomyces cidri, 38, 284
Saccharomyces cordubensis, 284
Saccharomyces coreanus, 284
Saccharomyces dairensis, 62–63
  identity check, 223, **248**
  in keys no. 3, 4, 5, 15
Saccharomyces delbrueckii, 8, 34, 284
Saccharomyces diastaticus, 284
Saccharomyces eupagycus, 284
Saccharomyces exiguus, 62–63
  identity check, 196, **248**, 263
  in keys no. 3, 4, 5, 15, 16, 17, 18
Saccharomyces fermentati, 4, 8, 38, 284
Saccharomyces florentinus, 38, 284
Saccharomyces florenzanii, 284
Saccharomyces globosus, 284
Saccharomyces heterogenicus, 284
Saccharomyces hienipiensis, 284
Saccharomyces hispalensis, 302
Saccharomyces inconspicuus, 8, 284
Saccharomyces inusitatus, 284
Saccharomyces italicus, 284
Saccharomyces kloeckerianus, 8, 34, 284
Saccharomyces kluyveri, 62–63
  identity check, **249**
  in keys no. 3, 4, 5, 16
Saccharomyces microellipsodes, 38, 284
Saccharomyces montanus, 2, 38, 284
Saccharomyces mrakii, 38, 284
Saccharomyces norbensis, 284
Saccharomyces oleaceus, 284
Saccharomyces oleaginosus, 284
Saccharomyces polymorpha, 23
Saccharomyces pretoriensis, 8, 34, 284
Saccharomyces prostoserdovii, 284
Saccharomyces rosei, 8, 284
Saccharomyces rouxii, 38, 284
Saccharomyces saitoanus, 284
Saccharomyces servazzii, 31, 62–63
  identity check, **249**
  in keys no. 3, 4, 5
Saccharomyces telluris, 62–63, 283, 284
  identity check, 237, 248, **249**
  in keys no. 3, 4, 5, 13, 14
Saccharomyces transvaalensis, 5, 11, 27, 284

Saccharomyces unisporus, 62–63
  identity check, **249**
  in keys no. 3, 4, 5, 15, 16, 17, 18
Saccharomyces uvarum, 2, 284
Saccharomyces vafer, 8, 284
Saccharomyces veronae, 26
Saccharomycetaceae, 4
Saccharomycodes, 4, 12
Saccharomycodes ludwigii, 62–63
  identity check, **250**
  in keys no. 3, 4, 5, 16
Saccharomycopsis, 12, 31
Saccharomycopsis capsularis, 31, 62–63, 283
  identity check, **250**
  in keys no. 1, 2, 3, 4, 6
Saccharomycopsis crataegensis, 31, 62–63
  identity check, 247, **250**
  in keys no. 1, 2, 3, 4, 6
Saccharomycopsis fibuligera, 31, 62–63, 283
  identity check, **250**, 253
  in keys no. 1, 2, 3, 4, 6, 15, 16
Saccharomycopsis guttulata, 10, 284
Saccharomycopsis lipolytica, 8, 31, 62–63, 283, 303
  identity check, **250**
  in keys no. 1, 2, 6, 14, 15, 16
Saccharomycopsis malanga, 31–32, 62–63, 283
  identity check, **251**
  in keys no. 1, 2, 3, 4, 6, 15
Saccharomycopsis synnaedendra, 16
Saccharomycopsis vini, 32, 64–65, 283
  identity check, 201, 205, 231, 248, **251**, 258, 265, 275, 276
  in keys no. 1, 2, 3, 4, 6, 15
Sarcinosporon, 12, 32
Sarcinosporon inkin, 32, 64–65, 284
  identity check, **251**, 272
  in keys no. 1, 2, 14
Schizoblastosporion, 12
Schizoblastosporion starkeyi-henricii, 64–65
  identity check, 244, **251**, 260
  in keys no. 1, 2, 14
Schizosaccharomyces, 4, 12
Schizosaccharomyces japonicus, 64–65
  identity check, **252**
  in keys no. 3, 4, 5, 16
Schizosaccharomyces malidevorans, 64–65
  identity check, **252**
  in keys no. 3, 4, 5, 15, 16
Schizosaccharomyces octosporus, 64–65
  identity check, **252**
  in keys no. 3, 4, 5, 15, 16
Schizosaccharomyces pombe, 64–65
  identity check, **252**
  in keys no. 3, 4, 5, 15, 16, 17, 18
Schizosaccharomyces slooffii, 32, 64–65
  identity check, **252**
  in keys no. 3, 4, 5
Schwanniomyces, 12
Schwanniomyces alluvius, 284
Schwanniomyces castellii, 284
Schwanniomyces occidentalis, 64–65, 284
  identity check, 182, 248, 250, **253**
  in keys no. 3, 4, 6
Schwanniomyces persoonii, 284
Selenotila intestinalis, 284
Selenotila peltata, 32, 284
Selenozyma, 4, 12, 32
Selenozyma peltata, 32, 64–65, 284
  identity check, **253**
  in keys no. 3, 4, 10, 11

Sporidiobolus, 12, 40
Sporidiobolus johnsonii, 64–65
  identity check, 246, **253**, 255
  in keys no. 1, 2, 8, 9, 10, 11
Sporidiobolus ruinenii, 64–65
  identity check, **253**
  in keys no. 1, 2, 8, 9, 10, 11
Sporobolomyces, 4, 13, 40
  sexual state, see *Aessosporon*
Sporobolomyces albo-rubescens, 64–65
  identity check, 247, **254**
  in keys no. 1, 2, 9
Sporobolomyces antarcticus, 32, 64–65
  identity check, **254**
  in keys no. 1, 2, 10, 11, 12
Sporobolomyces gracilis, 64–65
  identity check, 247, **254**
  in keys no. 1, 2, 9
Sporobolomyces hispanicus, 13, 64–65
  identity check, 245, **254**, 255
  in keys no. 1, 2, 9, 10, 11
Sporobolomyces holsaticus, 64–65
  identity check, 246, 253, **255**, 256
  in keys no. 1, 2, 9, 10, 11, 15
Sporobolomyces odorus, 13, 64–65
  identity check, 179, 189, 245, 254, **255**, 256
  in keys no. 1, 2, 9, 10, 11
Sporobolomyces pararoseus, 13, 64–65
  identity check, **255**
  in keys no. 1, 2, 9, 10, 11, 15, 16
Sporobolomyces puniceus, 32–33, 64–65
  identity check, **255**
  in keys no. 1, 2, 9, 15
Sporobolomyces roseus, 64–65
  identity check, 255, **256**
  in keys no. 1, 2, 9, 14, 16
Sporobolomyces salmonicolor, 13, 15, 64–65
  identity check, 179, 255, **256**
  in keys no. 1, 2, 9, 10, 11, 14, 15, 16
Sporobolomyces singularis, 64–65
  identity check, **256**
  in keys no. 1, 2
Sporobolomycetaceae, 40
Sporopachydermia, 302
Sporopachydermia cereana, 302
Sporopachydermia lactativora, 302
Stephanoascus, 13, 33
Stephanoascus ciferrii, 33, 64–65, 283
  identity check, **256**
  in keys no. 1, 2, 6, 10, 11
Sterigmatomyces, 4, 13
Sterigmatomyces acheniorum, 6, 30
Sterigmatomyces aphidis, 33, 64–65
  identity check, 188, **256**
  in keys no. 1, 2
Sterigmatomyces elviae, 33, 64–65
  identity check, **257**, 272
  in keys no. 1, 2, 14
Sterigmatomyces halophilus, 64–65
  identity check, **257**
  in keys no. 1, 2, 14
Sterigmatomyces indicus, 64–65
  identity check, **257**
  in keys no. 1, 2
Sterigmatomyces nectairii, 33, 64–65
  identity check, **257**
  in keys no. 1, 2
Sterigmatomyces penicillatus, 33, 64–65
  identity check, 191, **257**
  in keys no. 1, 2

Sterigmatomyces polyborus, 33, 64–65
  identity check, 182, 207, **258**
  in keys no. 1, 2
Sympodiomyces, 13, 33–34
Sympodiomyces parvus, 34, 64–65
  identity check, **258**
  in keys no. 1, 2

Torula, 7
Torula heveanensis, 22
Torulaspora, 5, 13, 34, 300
Torulaspora delbrueckii, 34, 38, 66–67, 284
  identity check, 184, 199, 200, 204, 226, 235, 236, 237, 251, **258**, 259, 260, 275, 276
  in keys no. 3, 4, 5, 15, 16, 17, 18
Torulaspora globosa, 34, 66–67, 284
  identity check, **258**
  in keys no. 3, 4, 5, 16
Torulaspora manchurica, 2
Torulaspora pretoriensis, 34, 66–67, 284
  identity check, 258, **259**
  in keys no. 3, 4, 5
Torulopsis, 7, 13, 282
Torulopsis anatomiae, 66–67
  identity check, **259**
  in keys no. 3, 4
Torulopsis apicola, 66–67
  identity check, 226, 258, **259**
  in keys no. 3, 4, 15, 16
Torulopsis apis, 66–67
  identity check, **259**
  in keys no. 1, 2, 10, 11
Torulopsis armenti, 302
Torulopsis auriculariae, 34, 66–67
  identity check, **259**
  in keys no. 1, 2
Torulopsis austromarina, 34, 66–67
  identity check, 251, **260**
  in keys no. 1, 2, 12
Torulopsis azyma, 302
Torulopsis bacarum, 34, 66–67
  identity check, **260**
  in keys no. 1, 2
Torulopsis bombicola, 34, 66–67
  identity check, 248, 258, **260**
  in keys no. 3, 4, 10, 11
Torulopsis bovina, 284
Torulopsis candida, 6, 284
Torulopsis cantarellii, 66–67
  identity check, **260**
  in keys no. 3, 4, 16
Torulopsis castellii, 66–67
  identity check, **261**
  in keys no. 3, 4
Torulopsis colliculosa, 7, 284
Torulopsis dattila, 284
Torulopsis dendrica, 35, 66–67
  identity check, **261**, 265
  in keys no. 1, 2, 3, 4
Torulopsis domercqii, 13, 284
Torulopsis ernobii, 66–67
  identity check, **261**
  in keys no. 3, 4, 17, 18
Torulopsis etchellsii, 66–67
  identity check, **261**
  in keys no. 3, 4, 14, 15
Torulopsis fragaria, 35, 66–67
  identity check, **261**
  in keys no. 1, 2, 15
Torulopsis fructus, 35, 66–67
  identity check, **262**
  in keys no. 3, 4, 15

Torulopsis fujisanensis, 66–67
  identity check, **262**
  in keys no. 1, 2, 9
Torulopsis geochares, 302
Torulopsis glabrata, 66–67
  identity check, 248, **262**
  in keys no. 3, 4, 13, 14, 15, 16
Torulopsis gropengiesseri, 66–67
  identity check, **262**
  in keys no. 3, 4, 10, 11, 15
Torulopsis haemulonii, 66–67
  identity check, **263**
  in keys no. 3, 4, 15
Torulopsis halonitratophila, 66–67
  identity check, **263**
  in keys no. 1, 2, 3, 4
Torulopsis halophilus, 35, 66–67
  identity check, **263**
  in keys no. 3, 4
Torulopsis heveanensis, 22
Torulopsis holmii, 66–67
  identity check, 196, 225, 248, **263**
  in keys no. 3, 4, 14, 15
Torulopsis humilis, 35, 66–67
  identity check, **264**
  in keys no. 3, 4, 17, 18
Torulopsis inconspicua, 66–67
  identity check, 237, **264**
  in keys no. 1, 2, 14, 15, 16
Torulopsis ingeniosa, 66–67
  identity check, **264**
  in keys no. 1, 2
Torulopsis insectalens, 35, 66–67
  identity check, **264**
  in keys no. 1, 2
Torulopsis karawaiewi, 35, 66–67
  identity check, 186, 188, 193, 204, 234, 235, 237, 248, 251, 261, **265**
  in keys no. 3, 4
Torulopsis kruisii, 35, 66–67
  identity check, **265**
  in keys no. 3, 4
Torulopsis lactis-condensi, 66–67
  identity check, **265**
  in keys no. 3, 4, 15
Torulopsis magnoliae, 66–67
  identity check, **265**
  in keys no. 3, 4, 10, 11, 14, 15
Torulopsis mannitofaciens, 33, 66–67
  identity check, **266**
  in keys no. 3, 4
Torulopsis maris, 66–67
  identity check, **266**
  in keys no. 1, 2, 10, 11
Torulopsis methanodomercqii, 302
Torulopsis methanolovescens, 35, 66–67
  identity check, 237, **266**
  in keys no. 1, 2, 3, 4, 12
Torulopsis methanophiles, 303
Torulopsis methanosorbosa, 302
Torulopsis methanothermo, 303
Torulopsis mogii, see *Zygosaccharomyces rouxii*
Torulopsis molischiana, 66–67, 303
  identity check, 217, **266**
  in keys no. 3, 4, 12, 15
Torulopsis multis-gemmis, 35, 66–67
  identity check, **266**
  in keys no. 1, 2, 3, 4
Torulopsis musae, 35, 66–67
  identity check, **267**
  in keys no. 3, 4, 15

Torulopsis nagoyaensis, 35–36, 66–67
    identity check, **267**
    in keys no. 3, 4, 12
Torulopsis navarrensis, 36, 66–67
    identity check, **267**
    in keys no. 3, 4
Torulopsis nemodendra, 36, 66–67
    identity check, **267**
    in keys no. 1, 2, 12
Torulopsis nitratophila, 66–67
    identity check, **267**
    in keys no. 1, 2, 3, 4, 10, 11, 12
Torulopsis nodaensis, 36, 66–67
    identity check, **268**
    in keys no. 3, 4, 12
Torulopsis norvegica, 66–67
    identity check, 219, **268**
    in keys no. 1, 2, 3, 4, 10, 11, 14, 16, 17,18
Torulopsis pampelonensis, 36, 68–69
    identity check, 195, **268**
    in keys no. 3, 4
Torulopsis peltata, 32
Torulopsis petrophilum, 303
Torulopsis philyla, 36, 68–69
    identity check, **268**
    in keys no. 1, 2
Torulopsis pignaliae, 36, 68–69
    identity check, **268**
    in keys no. 3, 4, 12
Torulopsis pintolopesii, 284
Torulopsis pinus, 68–69
    identity check, **269**
    in keys no. 1, 2, 10, 11, 12
Torulopsis psychrophila, 36, 68–69
    identity check, **269**
    in keys no. 1, 2
Torulopsis pulcherrima, 7
Torulopsis pustula, 36, 68–69
    identity check, **269**
    in keys no. 1, 2
Torulopsis rosea, 7
Torulopsis schatavii, 36, 68–69
    identity check, **269**
    in keys no. 1, 2, 3, 4
Torulopsis silvatica, 36, 68–69
    identity check, **269**
    in keys no. 1, 2
Torulopsis sonorensis, 36, 68–69
    identity check, **270**
    in keys no. 3, 4, 12
Torulopsis sorbophila, 37, 68–69
    identity check, **270**
    in keys no. 1, 2
Torulopsis spandovensis, 37, 68–69
    identity check, **270**
    in keys no. 3, 4

Torulopsis stellata, 68–69
    identity check, 248, **270**
    in keys no. 3, 4, 15, 16, 17, 18
Torulopsis tannotolerans, 37, 68–69
    identity check, **271**
    in keys no. 3, 4
Torulopsis torresii, 68–69
    identity checks, **271**
    in keys no. 3, 4, 10, 11
Torulopsis vanderwaltii, 68–69
    identity check, **271**
    in keys no. 1, 2, 10, 11, 16
Torulopsis versatilis, 68–69
    identity check, **271**
    in keys no. 3, 4, 15, 16, 17, 18
Torulopsis wickerhamii, 68–69
    identity check, **271**
    in keys no. 3, 4
Torulopsis xestobii, 37, 68–69
    identity check, 247, **272**
    in keys no. 1, 2, 3, 4, 10, 11
Trichosporon, 4, 13
Trichosporon aculeatum, 9, 15, 284
Trichosporon aquatile, 37, 68–69
    identity check, **272**
    in keys no. 1, 2, 10, 11, 12
Trichosporon brassicae, 37, 68–69
    identity check, **272**
    in keys no. 1, 2, 15
Trichosporon capitatum, 23, 284
Trichosporon cutaneum, 68–69
    identity check, 179, 182, 183, 187, 189, 191, 198, 207, 208, 209, 211, 228, 232, 241, 251, 257, **272**
    in keys no. 1, 2, 13, 14, 15, 16
Trichosporon eriense, 37, 68–69
    identity check, **273**
    in keys no. 1, 2
Trichosporon fennicum, 37, 68–69
    identity check, **273**
    in keys no. 3, 4, 10, 11
Trichosporon fermentans, 23, 284
Trichosporon inkin, 12, 32, 284
Trichosporon melibiosaceum, 37, 68–69
    identity check, **273**
    in keys no. 3, 4, 10, 11
Trichosporon penicillatum, 24, 284
Trichosporon pullulans, 68–69
    identity check, 206, **273**
    in keys no. 1, 2, 15, 16
Trichosporon terrestre, 37, 68–69
    identity check, **273**
    in keys no. 1, 2, 12
Trichosporon variabile, see *Hyphopichia burtonii*
Trigonopsis, 13

Trigonopsis variabilis, 4, 68–69
    identity check, **274**
    in keys no. 1, 2

Wickerhamia, 13
Wickerhamia fluorescens, 68–69
    identity check, **274**
    in keys no. 3, 4, 6, 7
Wickerhamiella, 13
Wickerhamiella domercqii, 38, 68–69, 284
    identity check, **274**
    in keys no. 1, 2, 5, 16
Wingea, 5, 13
Wingea robertsii, 68–69
    identity check, **274**
    in keys no. 3, 4, 5

Zendera, 13, 38
Zendera ovetensis, 38, 68–69, 283
    identity check, **274**
    in keys no. 1, 2, 5, 12
Zygolipomyces lactosus, 26
Zygolipomyces tetrasporus, 26
Zygosaccharomyces, 14, 38
Zygosaccharomyces bailii, 38, 68–69, 284
    identity check, 251, 258, **275**, 276
    in keys no. 3, 4, 5, 14, 15, 16, 17, 18
Zygosaccharomyces bisporus, 38, 68–69, 284
    identity check, 204, 237, 251, 258, **275**, 276
    in keys no. 3, 4, 5, 15, 16
Zygosaccharomyces cidri, 38, 68–69, 284, 302
    identity check, **275**
    in keys no. 3, 4, 5
Zygosaccharomyces fermentati, 2, 4, 38, 68–69, 284
    identity check, **275**
    in keys no. 3, 4, 5
Zygosaccharomyces florentinus, 4, 38, 68–69, 284
    identity check, **276**
    in keys no. 3, 4, 5, 16
Zygosaccharomyces microellipsodes, 38, 68–69, 284
    identity check, **276**
    in keys no. 3, 4, 5, 16
Zygosaccharomyces mrakii, 38, 70–71, 284
    identity check, **276**
    in keys no. 3, 4, 5
Zygosaccharomyces rouxii, 38, 70–71, 284
    identity check, 204, 237, 251, 258, 275, **276**
    in keys no. 3, 4, 5, 15, 16, 17, 18
Zygosaccharomyces thermotolerans, 26

Subject index

acetate agar, 41, 45
acid production test, 5
acids, organic, 41
actidione, 5
adaptation, 44
aeration for growth tests, 41
aerobic growth tests, 5, 41–43
affinity for substrate, 41
agar
  washed, 42
  see also media
alditols, 41
altered names of species, 283–284
alternatives to keys, 281–282
American Type Culture Collection, 303
amino acids in defined media, 46
p-aminobenzoic acid, 44, 46
ammonia, 43
ammonium sulphate, 46
anaerobic fermentation, 43–44
arthroconidia, 4, 265
arthrospores, see arthroconidia
ascospores, 4, 5, 40–41, 285
ascosporogenous yeasts
  cell wall structure, 5
  with hat- or Saturn-shaped ascospores, 142
  with round, oval or reniform ascospores, 137
  with ascospores of other shapes, 145
L-asparagine, 46
assimilation tests, 5, 41, 285
auxanograms, 41, 42, 43

bacteria, International Code of Nomenclature, 3, 291
ballistoconidia, 4, 17, 40, 285
ballistospores, see ballistoconidia
basidiomycetous yeasts, 5, 146
beer, see brewing, yeasts associated with
binary tests, 278, 285
binomials, 4
biochemical characteristics, 5
biotin, 44, 46
bipolar budding, 4
blastoconidia, 4, 5, 285
blastospores, see blastoconidia
borate, 46
botanical nomenclature, International Code, 4, 291
brewing, yeasts associated with, 172
budding, 4

calcium chloride, 46
carbon base medium, 43, 46
carbon dioxide, 44
carbon sources, 41, 42, 43
cell wall, 4, 5
cellobiose, 279
Centraalbureau voor Schimmelcultures, 1, 15, 72, 277
changes in classification, 1, 2
changes in nomenclature, 6, 7

characteristics for classifying yeasts, 4, 5, 9–14
clamp connections, 5, 285
classes, 4
classification, 3
classifying yeasts, 4, 5, 9–14
clinical yeasts, 155, 156
coenzyme Q, 5
combinatio nova, 6, 286
computer identification, 281, 282
computing methods, 1, 277–280
concentration of test substrate, 41, 42, 43, 44
conserved taxa, 7
  see also nomen conservandum, 286
copper sulphate, 46
corn meal agar, 45
criterion function for selecting tests, 279
cycloheximide, 5

derepression, 44
divisions, 4
DNA
  hybridization, 3, 8
  sequence complementarity, 7
Durham tubes, 43

electron transport and ubiquinone, 5
equivalent names of yeasts, 283–284
equivocal responses to tests, 277, 279, 285
ethylamine, 43

false results from tests, 41
families, 4
fat splitting, 5
fermentation of sugars, 5, 43–44, 286
fermenting yeasts, 100–102
ferric chloride, 46
fertility and species, 3
filamentous growth, 4, 40, 286
fission, 4, 286
folic acid, 44, 46
food, yeasts associated with, 159–160

genera
  characteristics, 9–14
  nouns, 4
  omitted, 1
  type of, 6
Genkey computer program, 277, 278, 282
genotypic and phenotypic characteristics, 3, 7
genus, see genera
D-glucose
  in defined media, 46
  fermenting yeasts, 100–102
  non-fermenting yeasts, 75–76
  50% and 60%, growth tests, 44
ß-D-glucosides, 279
Gorodkowa agar, 41, 45
groups of tests for keys, 282
growth
  factors, 44
  filamentous, 40
  temperature, 39
  tests, 5, 39, 41–43

L-histidine, 46
holotype, 6, 286
homonym, 6, 7, 286
hybridization, 3
hydrocarbon-using yeasts, 147–148
hyphae, 4, 40, 286
  septal pores, 4–5

identification, 3
  checking, 177–276, 280
  computer, 281
  methods, 39–46
illegitimate names, 6, 7
incubation temperature, 39, 40, 41, 44
induction, 44
inhibition by nitrogen compounds, 43
inoculation, 41, 42, 44
inositol, 44, 46
interbreeding, 7
International Code of Botanical Nomenclature, 4, 6, 291
International Code of Nomenclature of Bacteria, 3, 291
iodide, 46
iodine test, 44
irredundant tests, 277–278, 286
isoprene units of ubiquinone, 5

key
  alternatives to, 281–282
  compact form, 1, 72, 280
  construction, 278
  groups of tests, 282
  reticulation, 1, 72, 280
K_m for substrate, 41

laboratory methods, 39–46
Latin description, 6
lectotype, 6, 286
Lugol's iodine, 44
L-lysine, 43

magnesium sulphate, 46
maintaining yeasts, 39
maize meal agar, 40, 45
malt extract, 39, 40, 41, 45
malt–yeast–glucose–peptone medium, 39, 40, 41, 42, 45
manganese sulphate, 46
mannan of cell wall, 5
matching by computer, 282
media, 45–46
  acetate agar, 41, 45
  ascosporulation, 41
  carbon base, 43, 46
  chemically defined, 46
  corn meal agar, see maize meal agar
  filamentous growth, 40
  Gorodkowa agar, 41, 45
  growth test, 41, 42, 46
  maintaining yeasts, 39
  malt extract, 39, 40, 41, 45

Subject index

malt–yeast–glucose–peptone, 39, 40, 41, 42, 45
morphology agar, 46
nitrogen base, 41, 42, 43, 46
potato–glucose agar, 40, 45
V-8 agar, 41, 45
vitamin-free, 46
yeast extract, 44
yeast–glucose–peptone, 39, 40, 45
medical yeasts, *see* clinical yeasts
methanol-using yeasts, 154
DL-methionine, 46
methods for identifying yeasts, 39–46
microscopical appearance, 4, 39
microscopical examination, for ascospores, ballistoconidia, filaments, 40
mitochondria and ubiquinone, 5
molybdate, 46
Monod tubes, 42
multilateral budding, 4
mycelium, *see* hyphae
myo-inositol, 44, 46

names of yeasts
alteration, 3, 283–284
new, 15–38
naming yeasts, 6
neotype, 6, 286
nephelometer, 42
new combination, 2, 6, 286
new taxa, 1, 6, 15–38
nicotinic acid, 44, 46
nitrate, 43
nitrite, 43
nitrogen base medium, 41, 42, 43, 46
nitrogen compounds for aerobic growth, 43
nitrous acid, 43
nomen conservandum, 7, 286
nomen dubium, 7
nomen nudum, 6, 286
nomenclature, 3
nomenclatural changes, 6, 7
non-fermenting yeasts, 75–76
nova combinatio, 6, 286
nucleic acid hybridization, 3, 8
numerical taxonomy, 7

objective classification, 7, 8
omissions from keys, 1
orders, 4

pantothenate, 44, 46
pH and nitrite utilization, 43
phenotypic and genotypic characteristics, 3, 7
phosphate in defined media, 46
photoelectric assessment of growth, 42
physiological characteristics, 5, 6, 286
pink yeasts, 146
piridoxine, 44, 46
polyclave, 281
potassium iodide, 44, 46
potato–glucose agar, 40, 45
pseudohyphae, 4, 40, 286
pseudomycelium, *see* pseudohyphae
psychrophilic yeasts, 42, 286

red pigmented yeasts, 7
see also pink yeasts
redundant tests, 277
reliability of tests, 279
renaming yeasts, 6–7
replica plating, 41, 42–43
respiratory chain and ubiquinone, 5
reticulation of keys, 1, 72, 280, 286
ribitol, 279
riboflavin, 44, 46

salicin, 279
salts in defined media, 46
selection of tests, criterion function, 279
septal pores, 4–5
sexual reproduction, 3, 5
slide cultures, 40, 41
sodium chloride, 46
species
altered names, 3, 283–284
authority for, 4
meaning, 3, 286
mention in publication, 4
names, 4
new, 15–38
numbers of strains, 4
omitted, 1
type of, 6
spores, *see* arthroconidia; ascospores; ballistoconidia; blastoconidia; teliospores
sporophores, 5, 287
sporulation, *see* ascospores
starch-like compounds, 5, 44
sterigmata, 4, 287
strains, 3, 4, 287

substrates for growth tests, 41, 42
suffixes for taxa, 4
sugars
fermentation of, 5, 43–44
used as carbon source, 41
syntype, 6, 287
systematics, 3, 287

tautonym, 6, 287
taxa, 3, 287
new, 15–38
taxonomy, 3, 287
teliospores, 4, 5, 287
temperature, incubation, 39, 40, 41, 44
tests
criterion function for selecting, 279
equivocal responses, 277, 279, 285
fermentation, 5, 43–44
groups for keys, 282
growth, 39, 41–43, 44
reliability, 279
uncharacteristic responses, 281
thiamin, 44, 46
trace elements, 46
DL-tryptophan, 46
T-tubes, 41–42
type, nomenclatural, 6, 287

ubiquinone, 5
uncharacteristic responses to tests, 281
unipolar budding, 4
urea hydrolysis, 5

V-8 agar, 41, 45
vegetative reproduction, 4, 287
vitamin-free medium, 46
vitamin requirements, 5, 44

wine, yeasts associated with, 166–167

yeast extract medium, 44
yeast–glucose–peptone medium, 39, 40, 45
yeast morphology agar, 46
YM media, 45

zinc sulphate, 46